Topics in
Current Physics

40

Topics in Current Physics   Founded by Helmut K. V. Lotsch

Microscopic Methods in Metals

Edited by U. Gonser

With 253 Figures

Springer-Verlag Berlin Heidelberg New York
London Paris Tokyo

Professor Dr. Ulrich Gonser

Angewandte Physik, Universität des Saarlandes
D-6600 Saarbrücken, Fed. Rep. of Germany

ISBN-13: 978-3-642-46573-4   e-ISBN-13:978-3-642-46571-0
DOI: 10.1007/978-3-642-46571-0

Library of Congress Cataloging-in-Publication Data. Microscopic methods in metals. (Topics in current physics ; 40) 1. Metallography–Methodology. 2. Microscope and microscopy–Methodology. I. Gonser, U. II. Series. TN690.M498 1986 669'.95'0282 86-3844

Softcover reprint of the hardcover 1st edition 1986

2153/3150-543210

Preface

Methods of scientific investigation can be divided into two categories: they are either macroscopic or microscopic in nature. The former are generally older, classical methods where the sample as a whole is studied and various local properties are deduced by differentiation. The microscopic methods, on the other hand, have been discovered and developed more recently, and they operate for the most part on an atomistic scale.

Glancing through the shelves of books on the various scientific fields, and, in particular, on the field of physical metallurgy, we are surprised at how little consideration has been given to the microscopic methods. How these tools provide new insight and information is a question which so far has not attracted much attention. Similar observations can be made at scientific conferences, where the presentation of papers involving microscopic methods is often pushed into a far corner. This has led users of such methods to organize their own special conferences.

The aim of this book is to bridge the present gap and encourage more interaction between the various fields of study and selected microscopic methods, with special emphasis on their suitability for investigating metals. In each case the principles of the method are reviewed, the advantages and successes pointed out, but also the shortcomings and limitations indicated.

Our book is addressed to physicists, chemists, engineers – in short, to all interested in a basic understanding of the application of microscopic methods to the study of metals. Of course, teachers on related subjects should also find the book useful as a guideline. Emphasis has therefore been placed on well-selected examples. Since lack of space did not allow complete coverage of recent sophistications, the book is not intended for peers or experts in the various areas. However, with his inherent scientific curiosity, the specialist in one method might like to take a look over the fence into the fundamentals of other "rival" methods.

I am indebted to the Japan Society for the Promotion of Science and to my friend Professor Francisco Eiichi Fujita for sponsoring my sabbatical at the University of Osaka. I should also like to thank Mrs. Carolyn Schneider for her skilful and patient help in the book's preparation.

Saarbrücken, May 1986 *Ulrich Gonser*

Table of Contents

List of Contributors

Chappert, Jacques
DRF/Service de Physique, CEA, Centre d'Etudes Nucléaires, 85 X,
F-38041 Grenoble, France

Ehrlich, Gert
Coordinated Science Laboratory, 1101 W. Springfield Avenue,
University of Illinois at Urbana-Champaign,
Urbana, IL 61801, USA

Ernst, Norbert
Fritz-Haber-Institut der Max-Planck-Gesellschaft, Faradayweg 4–6,
D-1000 Berlin 33, FRG

Fujita, Francisco E.
Faculty of Engineering Science, Osaka University, Toyonaka, Osaka 560, Japan

Gonser, Ulrich
Angewandte Physik, Universität des Saarlandes, D-6600 Saarbrücken, FRG

Hirabayashi, Makoto
Research Institute for Iron, Steel and Other Metals,
Tohoku University, Sendai 980, Japan

Hoppe, Martin
Ernst Leitz Wetzlar GmbH, Postfach 2020, D-6330 Wetzlar 1, FRG

Panissod, Pierre
Laboratoire de Magnétisme et de Structure Electronique des Solides,
Université Louis Pasteur, 4, rue Blaise-Pascal,
F-67070 Strasbourg, France

Recknagel, Ekkehard
Fakultät für Physik, Universität Konstanz, Bücklestraße 13,
D-7750 Konstanz, FRG

Seah, Martin P.
Division of Materials Applications, National Physical Laboratory, Middlesex,
Teddington, TW11 0LW, United Kingdom

Stearns, Daniel G.
Lawrence Livermore National Laboratory, P.O. Box 5508, L-473,
Livermore, CA 94550, USA

Stearns, Mary B.
Department of Physics, Arizona State University, Tempe, AZ 95287, USA

Thaer, Andreas
Ernst Leitz Wetzlar GmbH, Postfach 2020, D-6330 Wetzlar 1, FRG

Triftshäuser, Werner
Institut für Nukleare Festkörperphysik, Universität der Bundeswehr München,
D-8014 Neubiberg, FRG

Wagner, Christian N.J.
School of Engineering and Applied Science, University of California,
6531 Boelter Hall, Los Angeles, CA 90024, USA

Wertheim, Gunther K.
AT&T Bell Laboratories, Murray Hill, NJ 07974, USA

Wichert, Thomas
Fakultät für Physik, Universität Konstanz, Postfach 5560,
D-7750 Konstanz, FRG

Yaouanc, Alain
Service de Physique, Groupe Magnétisme et Diffusion par Interactions Hyper-
fines, Centre d'Etudes Nucléaires, 85 X, F-38041 Grenoble, France

1. Concerning Methods

U. Gonser

Man is a tool-using animal,
without tools he is nothing

Thomas Carlyle

Although we have a far greater understanding of nature than our forebears, they could claim to have a much more universal view. Our rapidly growing knowledge in natural science has led to a narrowing of the individual view: as we dig deeper we lose breadth. The trend is towards sophistication and specialization. Nowadays it is impossible for any one person to keep up with various fields *and* with various methods. For an experimentalist this means that he must concentrate either on the one or the other, and he thus becomes a field or a method specialist. Another factor promoting this trend towards specialization is the increasing cost of modern equipment. Parallel to this development research itself has changed from an individual type of creative engagement to a collective type of team work.

In general, the subjects of books and conferences tend to focus on the various fields of study, such as magnetism, semiconductivity, defects, or superconductivity. It is relatively seldom that the emphasis is on techniques, as in the case of the present book. In recent years the number of methods has greatly increased, and writing on methods must therefore be selective. Thus, in consideration of this factor, we have restricted ourselves to those methods which are microscopic by nature and in particular to those which are most applicable to metals. Furthermore, the methods chosen should be considered as typically representative of the wide range of modern methods available.

The tools of a scientist are his methods, which enable him to probe the properties of his surroundings. In general, new methods are the result of the advancement, refinement and combination of certain aspects of theoretical reflection, experimental sophistication and engineering ingenuity.

In most instances a number of people are involved in the creation, development and improvement of a method. Often it is rather difficult to trace all the individual contributions which have led to the ensuing success of a method. Sometimes problems arise which are connected with priorities, national prestige or with the question of whether the original idea – the *Gedankenexperiment* – or the successful experimental verification was the decisive step in the creation of a new method.

To take a closer look at the various methods, we might divide them formally into the following three groups: descriptive methods, abbreviated methods and name-tag methods.

1

1.1 Descriptive Methods

In the first category, we find mostly the old and classical methods which allow the investigation of electrical, magnetic, mechanical and thermodynamic properties. Typical examples are electrical and heat conductivity, calorimetry, specific heat, magnetization, strength, and fracture.

1.2 Abbreviated Methods

Abbreviations have become popular everywhere. In science, the English abbreviations are used to designate units, properties and, in particular, the rapidly increasing number of methods. Nowadays we need a guide to find our way through the jungle of abbreviations and this jungle is nothing but a mass of stumps that have been left over after the trees were cut down. It seems necessary to standardize the numerous new abbreviations on an international basis.

Table 1.1 lists some of the abbreviated methods with their characteristic probe particles. Of course, this table is far from being complete and merely exemplifies the situation in general. Special applications of a method might have their own abbreviations such as PAC for perturbed angular correlation and TDPAC for time differential perturbed angular correlation. In some cases one abbreviation is used for different methods as, for instance, SAM for scanning Auger electron microscopy and also for scanning acoustic microscopy. In other cases, different abbreviations are used for one and the same method or slightly different methods.

Table 1.1. Techniques and their characteristics

Acronym	Technique	Probe particle
ADPD	Angular-dependent photoelectron diffraction	Photon
AES	Auger electron spectroscopy	Electron
APS	Appearance potential spectroscopy	Electron
ARPES	Angle-resolved photo-emission spectroscopy	Photon
ARUPS	Angle-resolved ultraviolet photo-emission spectroscopy	Photon
BIS	Bremsstrahlung isochromate spectroscopy	Electron
CEMS	Conversion electron Mössbauer spectroscopy	Photon
CT	Computer tomography	Photon, phonon
DCEMS	Depth-selective conversion electron Mössbauer spectroscopy	Photon
DLTS	Deep level transient spectroscopy	Electron
DNMR	Double nuclear magnetic resonance	Photon
EAPFS	Extended appearance potential fine structure	Electron
EDPD	Energy-dependent photoelectron diffraction	Photon
EDXD	Energy dispersive x-ray diffraction	Photon
EELS	Electron energy loss spectroscopy	Electron
ELDOR	Electron double resonance	Photon
ENDOR	Electron nuclear double resonance	Photon
EPMA	Electron probe micro-analyzer	Electron
EPR	Electron paramagnetic resonance	Photon
ESCA	Electron spectroscopy for chemical analysis	Photon

ESD	Electron-stimulated desorption	Electron
ESDIAD	Electron-stimulated desorption ion angular distribution	Electron
ESR	Electron spin resonance	Photon
EXAFS	Extended x-ray absorption fine structure	Photon
EXELFS	Extended x-ray-edge energy-loss fine structure	Electron
FEM	Field electron microscopy	Electron
FES	Field emission spectroscopy	Electron
FIM	Field ion microscopy	Atom
FMR	Ferro-magnetic resonance	Photon
FQHE	Fractional quantum Hall effect	Electron
GDOS	Glow discharge optical spectroscopy	Atom
HEAD	Helium-atom diffraction	Atom
HEED	High-energy electron diffraction	Electron
HREELS	High-resolution electron-energy-loss spectroscopy	Electron
HREM	High-resolution electron microscopy	Electron
HVEM	High-voltage electron microscopy	Electron
IAES	Ion-induced Auger spectroscopy	Atom
IAS	Inelastic atom scattering	Atom
ICISS	Impact collision ion-scattering spectroscopy	Atom
IIEE	Ion-induced electron emission	Atom
ILEED	Inelastic low-energy electron diffraction	Electron
IMMA	Ion-microprobe mass analysis	Atom
IMPACT	Implantation perturbed angular correlation technique	Atom
INS	Ion neutralization spectroscopy	Atom
IPES	Inverse photo-emission spectroscopy	Electron
IQHE	Integral quantum Hall effect	Electron
ISS	Ion scattering spectroscopy	Atom
LEED	Low-energy electron diffraction	Electron
LEELS	Low-energy energy loss spectroscopy	Electron
LEERM	Low-energy electron reflection microscopy	Electron
LEIS	Low-energy ion scattering	Atom
LEPD	Low-energy positron diffraction	Positron
LOES	Laser optical emission spectroscopy	Photon
MODOR	Microwave optical double resonance spectroscopy	Photon
μSR	Muon spin rotation	Muon
NEXAFS	Near-edge x-ray absorption fine structure	Photon
NIS	Neutron inelastic scattering	Neutron
NMR	Nuclear magnetic resonance	Photon
NQR	Nuclear quadrupole resonance	Photon
ODMR	Optical double magnetic resonance	Photon
PAC	Perturbed angular correlation	Photon
PES	Photo-electron spectroscopy	Photon
PIXE	Proton-induced x-ray emission	Proton
PMDR	Phosphorescence microwave double resonance spectroscopy	Photon
PSD	Photon-stimulated desorption	Photon
PSID	Photon-stimulated ion desorption	Photon
RAIRS	Reflection-absorption infrared spectroscopy	Photon
RHEED	Reflection high-energy electron diffraction	Electron
RSMR	Rayleigh scattering Mössbauer radiation	Photon
SAM	Scanning Auger electron microscopy	Electron
SAM	Scanning acoustic microscopy	Phonon
SANS	Small-angle neutron scattering	Neutron
SAXS	Small-angle x-ray scattering	Photon
SEM	Scanning electron microscopy	Electron
SEXAFS	Surface extended x-ray absorption fine structure	Photon
SHEED	Scanning high-energy electron diffraction	Electron
SIMS	Secondary ion mass spectroscopy	Atom
SLAM	Scanning laser acoustic microscopy	Photon
SNMS	Secondary neutral mass spectroscopy	Atom
SPAM	Scanning photo-acoustic microscopy	Photon

Table 1.1.

SPLEED	Spin polarization low-energy electron diffraction	Electron
STEM	Scanning transmission electron microscopy	Electron
STM	Scanning tunneling microscopy	Electron
SXAPS	Soft x-ray appearance potential spectroscopy	Electron
SXS	Soft x-ray emission spectroscopy	Electron
TDPAC	Time differential perturbed angular correlation	Photon
TDS	Thermal desorption spectroscopy	Phonon
TELS	Transmission energy loss spectroscopy	Electron
TEM	Transmission electron microscopy	Electron
TOFAP	Time-of-flight atom-probe	Atom
UPS	Ultraviolet photo-electron spectroscopy	Photon
UT	Ultrasonic testing	Phonon
WAXS	Wide-angle x-ray scattering	Photon
XANES	x-ray absorption near-edge structure	Photon
XAS	x-ray absorption spectroscopy	Photon
XES	x-ray emission spectroscopy	Electron
XFS	x-ray fluorescence spectroscopy	Photon
XPS	x-ray photo-electron spectroscopy	Photon
XRF	x-ray Fluorescence analysis	Photon

1.3 Name-Tag Methods

These methods are generally called after their discoverers and this gives them a flair of "nobility". They are noble in two senses of the word: in one sense they are per se truly aristocratic methods with a name, and in the other, it is for this reason that so many of them have earned the coveted "Nobel" distinction. Outstanding examples of name-tag methods are: The Mössbauer effect, Rutherford backscattering, the Laue and Bragg methods, and the v. Klitzing effect. In some cases "hybrids" have developed where the name-tag is incorporated in an abbreviation, for instance, CEMS for conversion electron Mössbauer spectroscopy or RSMR for Rayleigh scattering of Mössbauer radiation.

Another distinction that can be made is to divide the methods into macroscopic and microscopic methods. The older descriptive methods are mainly macroscopic by nature, i.e. the sample as a whole is investigated. The more recent abbreviated methods are mostly microscopic, i.e. in the observation atomic dimensions are approached. Because in many cases the probe particles are electrons or photons with strong interacton and small penetration depth in solids, these microscopic methods scan the surface and are useful in obtaining information on surfaces. The distinction between *micro* and *macro* is rather artificial. Here, we might adopt a dividing line which restricts the microscopic methods to those with resolutions well below one micrometer (10^{-6}m).

In addition to the spatial resolution, *micro* and *macro* could also be related to time. Methods normally have a characteristic time, that is the time of interaction of a probe particle or the time the information is collected and transmitted to a measuring instrument. Here the microsecond (10^{-6}s) might serve as the dividing line.

The chapters of this book have been arranged in the order of increasing resolution, although, of course, it must be realized that the term *resolution* is in many cases rather fuzzy. At the beginning stands ultrasound microscopy, a method now rapidly striding forward. Images obtained by ultrasound and optics are very similar, in contrast to the underlying physical principles, which are completely different. This comment might also serve as a bridge into the past where optical microscopy played the dominant rôle in metallography.

The microscopic methods involve incident radiations: electrons, photons (x-rays, γ-rays and other electromagnetic radiations), neutrons, atoms (ions), phonons, muons, positrons, etc. In most cases the principle of the method concerned can be visualized and treated as a scattering problem wherein an incident probe "particle" (photon, phonon, electron, ion etc.) interacts with the specimen and the emitting "particle" carries with it a certain piece of information. The incoming and outgoing particles might be the same. However, considering the methods listed in the table, we often have the situation of different incoming and outgoing particles. In particular, the probe particle is quite frequently an electron and the scattered particle a photon or vice versa – as in the photo effect – where a photon is the incident probe and an electron is emitted.

Each chapter stands on its own as a complete unit in which the principles of the method are explained, and use is made of the typical nomenclature and symbols. Of course, within the limited space available, many points have had to be generalized, but ample references are given where more details can be found. Also described are typical applications which have made it possible to deepen our understanding of the metallic state and its characteristic properties; in particular, information on surface and lattice defects was obtained. Many examples involve amorphous metals, a field now attracting considerable interest from a scientific as well as from a technological point of view. Indeed, many of these microscopic methods have been used in order to elucidate the nature and topological structure of amorphous metals.

2. Scanning Acoustic Microscopy

M. Hoppe and A. Thaer

With 24 Figures

No doubt, light microscopy still plays the leading role in the qualitative and quantitative microscopical examination of metals. The introduction of quantitative techniques like manual and automated image analysis, microscope photometric reflectance measurements, analysis of elliptically polarized light by polarizing microscopy, interference microscopy and microhardness testing, has significantly contributed to the support of this role. In contrast to earlier expectations, scanning electron microscopy [2.1] has turned out to be much more complementary than competitive. Its higher lateral resolution and the various analytical information is restricted to object structures at the object surface or at very small distances below that surface.

It may be expected, therefore, that a new microscopical principle with the unique potential of delivering information *below* the surface will more or less immediately gain significant importance among the microscopical principles available for the microanalysis of metals. The scanning acoustic microscopy becoming available for many fields of applied microscopy now, has this unique potential of revealing information in depth. Moreover, the kind of interaction between acoustic radiation and object structures is dominantly influenced by their elastic (i.e., mechanical) properties, thus contributing to the characterization of metal samples in a way very complementary to the information obtained by light and scanning electron microscopy. The utilization of this complementary information will above all depend on the possibility of meaningfully combining these different microscopic principles. The fact that scanning acoustic microscopy has reached - and under certain special circumstances already exceeded - light microscopical resolution now, may be considered as an important prerequisite of this combination.

2.1 Principle of Scanning Acoustic Microscopy (SAM)

Acoustic Imaging is widely used at *macroscopic* dimensions for medical diagnosis and material testing purposes as well. This imaging is performed with acoustic frequencies up to about 10 MHz. At this frequency, the acoustic wavelength as resolution determining factor does not exceed 150 μm in water and in living biological tissue and - because of the higher acoustic propagation velocity - 200 to 600 μm, in metals.

In comparison, light microscopy is performed at wavelengths of $> 0.7\,\mu m$, and even much smaller wavelengths are utilized in electron microscopy.

Therefore, a decisive breakthrough towards acoustic *microscopy* could not be achieved before much higher acoustic frequencies above 500–1000 MHz became available for imaging object structures at the micro scale.

An intermediate step towards real microscopical resolution is the development of the *Scanning Laser Acoustic Microscope* (SLAM [2.2]). Figure 2.1 outlines the principle.

The fact that in SLAM the acoustic radiation has to pass the sample leads to some inherent limitations, mainly due to acoustic attenuation during passage which increases quadratically at higher frequencies; i.e., light-frequency operation is restricted to very thin samples. The final breakthrough with regard to high lateral resolution was reached by the *Scanning Acoustic Microscope* (SAM) introduced by *Lemons* and *Quate* in 1974 [2.3].

Figure 2.2 illustrates the basic principle. Acoustic frequencies up to 2 GHz and higher can be excited in the transducer situated on the rear side of the sapphire disc. The plane acoustic waves emerging from the transducer pass the sapphire disc towards the very tiny spherical cavity on the front side and are focussed by the water-filled spherical cavity onto the object. The focussing

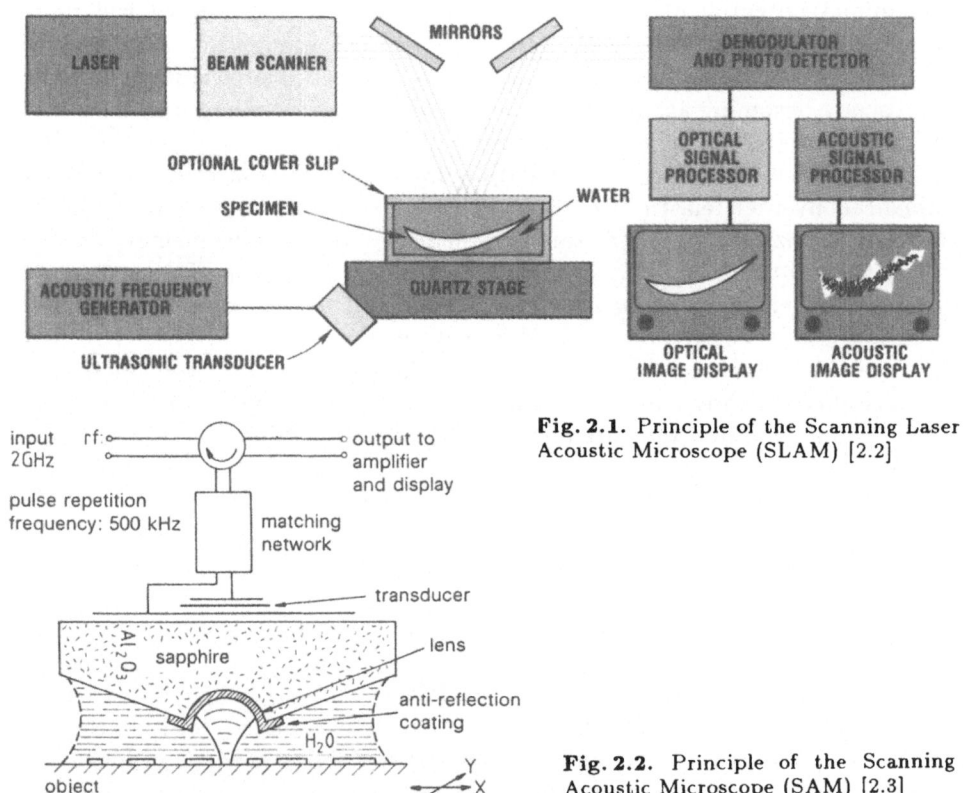

Fig. 2.1. Principle of the Scanning Laser Acoustic Microscope (SLAM) [2.2]

Fig. 2.2. Principle of the Scanning Acoustic Microscope (SAM) [2.3]

effect corresponds to a hypothetical optical lens made of glass with a refractive index of about 7.0. Thus, spherical aberrations can practically be neglected. At 2 GHz, the lateral diameter of the acoustic focussing spot reaches $0.6\,\mu$m, thus equalizing light microscopical resolution with medium-power lenses.

Part of the acoustic energy concentrated on the object is remitted towards the acoustic lens carrying information on the properties of the object structures contributing to the remission of acoustic energy. The remitted acoustic radiation picked up by the lens is guided towards the transducer, which now acts as a microphone and transforms the incoming acoustic signals into electrical ones of the same frequency. This signal representing one image point is then fed into the receiver part of the microwave chain for amplification and image formation.

In case of continuous-wave (CW) excitation of the transducer the object signal cannot satisfactorily be separated from the electrically exciting high-frequency signal, all the more because the average intensity level of the object signal is below that of the primary exciting energy by the factor 10^{-8} to 10^{-9}. Therefore, the excitation energy is released to the transducer in the form of very short pulses with a half width of 10 ns and a repetition cycle of 500 kHz. The interval between two exciting pulses provides for the acoustic object-signal sufficient time to reach the transducer for re-transformation without interfering with the primary exciting pulses (Fig. 2.3).

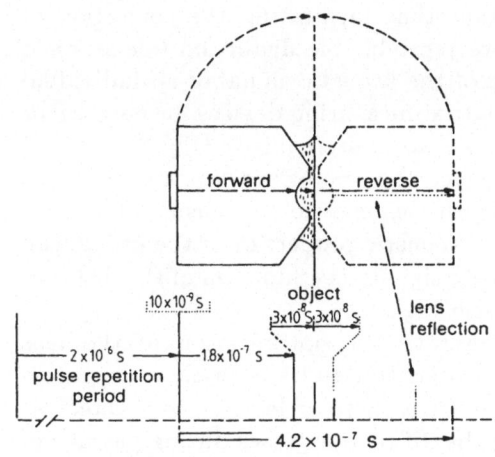

Fig. 2.3. Pulse repetition cycle in SAM: Pulse repetition period: 2×10^{-6}s, pulse half width: 10×10^{-9}s, transmission through sapphire: 1.8×10^{-7}s transmission through H_2O : 3×10^{-8}s one way

For acoustically imaging an object area, the acoustic focus must be scanned laterally across this area in the x- and y-directions, correlating the acoustic signal of each object point to the corresponding image point in the displayed image (Fig. 2.4). In using a scanning frequency of 50 Hz = 50 lines/s, due to the pulse repetition frequency of 500 kHz, one line contains 10000 pulses (pixels). Recording an image with 1024 lines would then take 20 seconds.

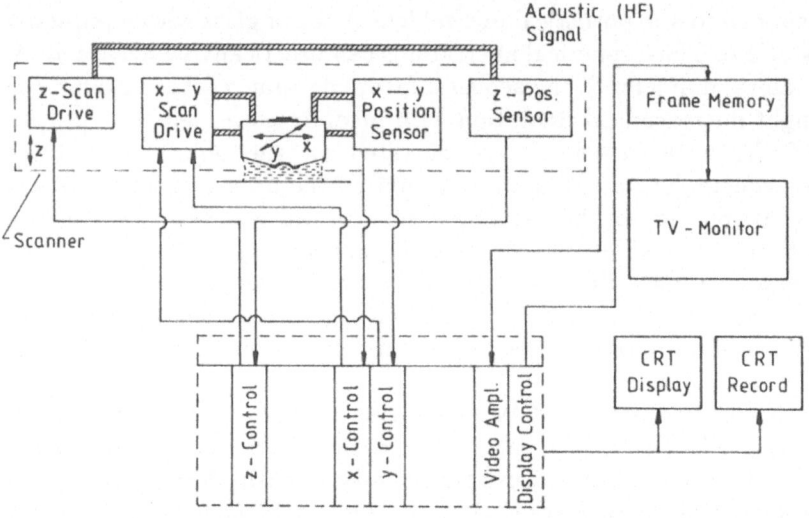

Fig. 2.4. Control of image display by the x-, y- and z-scanning positions of the acoustic focus relative to the scanned object area

In principle, the mechanical scanning movement can be made either by the lens or the object or by both. In practice, scanning with the lens is preferrred, because the mass of the lens is constant thus facilitating the correction of scanning parameters in order to achieve reproducible distortion-free scanned images. Nevertheless, for exact probing of the acoustic signal of an individual image point within the image display matrix, measuring devices for permanent control of the respective scanning positions are a prerequisite (Fig. 2.4).

In contrast to the light microscope, the motorized control of focus movement along the z axis, i.e., in depth, belongs to the basic equipment, too, because key information on specific acoustic properties of the object can be gained by recording subsurface acoustic signals as a function of the distance between object surface and focus (see also Fig. 2.14).

Particular attention has to be paid to the optimization of the lens parameters including the acoustic transducer. Due to its acoustic propagation parameters (velocity, attenuation), sapphire is more or less the best choice as lens material. However, for minimizing the diffraction losses during passage of acoustic pulses (back and forth), the acoustic waves have to travel between transducer and spherical cavity parallel to the optical axis of the trigonal sapphire crystal. Therefore, the sapphire discs to be prepared as the first step in producing the acoustic lens have to be cut exactly perpendicular to that axis.

By far the most significant losses of acoustic energy are caused during its passage through the water as the coupling medium between lens and object surface. In view of the desired acoustic properties

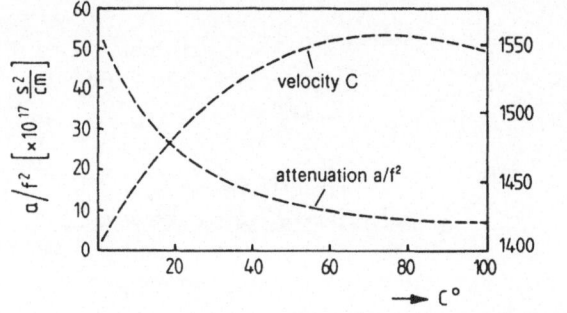

Fig. 2.5. Acoustic attenuation and propagation speed of water as a function of temperature

- propagation speed (as small as possible to keep the acoustic wavelength as small as possible)
- attenuation (as small as possible)
- wetting properties

water turned out to be the best compromise.[1] Its attenuation and acoustic propagation speed as a function of temperature are shown in Fig. 2.5. Therefore, heating the water to 70°C leads to a considerable reduction of attenuation without changing the propagation speed significantly.

In addition, however, the radius of the spherical cavity has to be kept as small as feasible for reducing the acoustic energy consuming pathway of the acoustic waves in the water-filled cavity (back and forth) as much as possible. This is particularly true at higher acoustic frequencies. At 2 GHz the lens radius should not be extended above 40 μm, at 1 GHz 80 – 100 μm is a good choice. Figure 2.6 illustrates the resulting free working distance between lens and object surfaces. The unavoidably small working distances require considerably more attention during the focussing operation compared to high-power lenses for light microscopy, all the more because *focussing and thus imaging structures below the object surface* are a unique domain of acoustic microscopy.

A major part of the acoustic energy is lost at the sapphire/water interface in both directions because of the high impedance differences between these

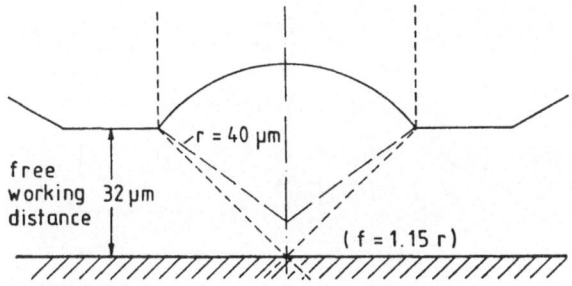

Fig. 2.6. Free working distances between lens front side and object surface at a lens radius of 40 μm

---

[1] Other coupling media like liquid gallium and mercury have been proposed and used but they do not seem to be applicable to practical acoustic microscopy because of their bad wetting properties and high toxicity [2.4].

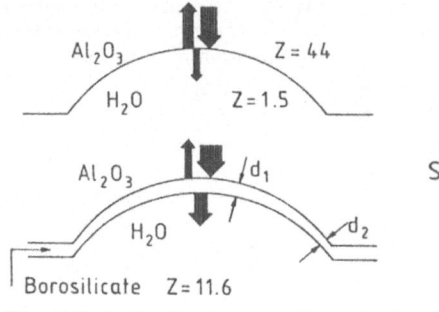

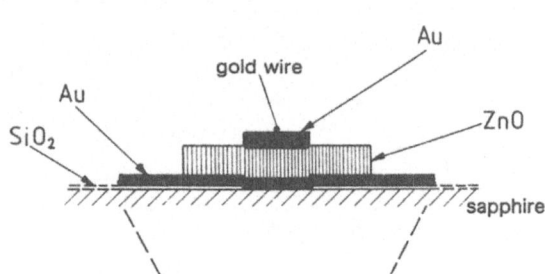

Fig. 2.7. Antireflex layer on the sapphire-lens surface for reducing acoustic energy loss at the interface sapphire/water (Z: acoustic impedance). *Top*: without antireflex layer: two-way loss of 18 dB. *Bottom*: with antireflex layer: two-way loss of 3 dB

Fig. 2.8. Arrangement (schematic) of the transducer at the rear side of the sapphire disc

two media. In complete analogy to light optics antireflex layers evaporated onto the sapphire lens surface can reduce this loss down to an acceptable level (Fig. 2.7). Certainly, the antireflex phenomenon depends on the observation of the phase shift $\lambda/4$ (one way). In principle, this condition permits antireflex-optimization within a restricted frequency band-width only, until broad-band antireflex coatings – as in optics – will become available.

Figure 2.8 schematically illustrates the arrangement of the transducer placed at the rear side of the sapphire disc exactly opposite the lens cavity. Again, for a certain frequency band-width the piezo-electric crystalline ZnO

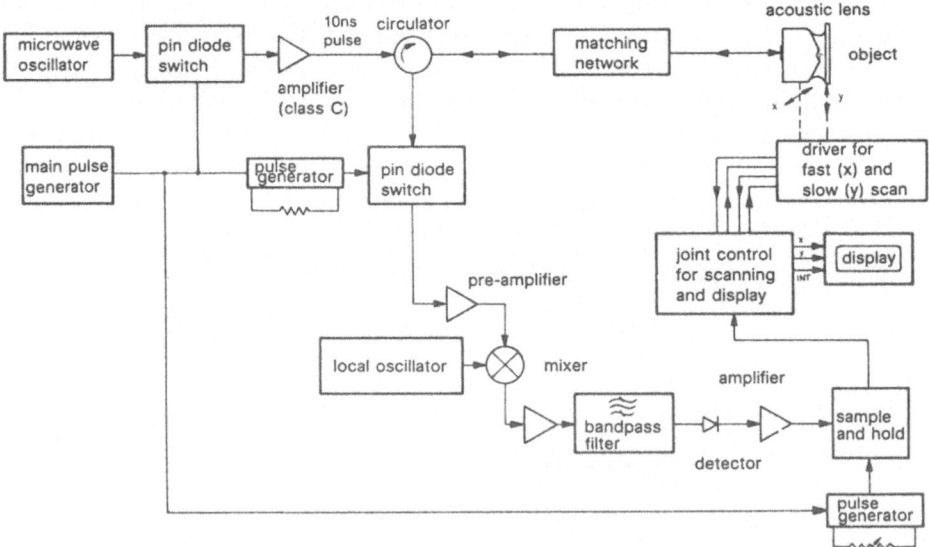

Fig. 2.9. Typical block diagram of the electronic parts of SAM, in particular of the emitter and receiver part of the RF chain

layer between the Au electrodes has to be carefully controlled with respect to its thickness for achieving optimum energy transformation in both directions. In addition, special care has to be taken to control the required orientation of the crystallites in the ZnO layer and in the Au electrode layers as well to obtain a very compact arrangement in order to achieve an optimum output and a maximum load of the transducer by the radio frequency (RF) pulses.

In view of the fact that the intensity differences between the electrical input and output signals at the transducer sum up to at least 10^{-8}, a relatively high effort has to be undertaken for an efficient separation of the pulses carrying information on the object from the primary RF pulses, but also for amplification of the image forming signals at a low noise level. Therefore, not only careful discrimination by exact time gating and frequency sensitive amplification of these signals are performed but also noise reduction by applying the heterodyne principle (mixing the preamplified signals with a different fixed frequency and further amplifying the signals at the intermediate frequency (Fig. 2.9). To display an image resulting from these final signals and the corresponding x and y coordinates delivered by the scanner use is made either of a conventional cathode ray tubes (CRT) as in Scanning Electron Microscopes (SEM) [2.1] or image storage devices in connection with TV monitors.

The equipment shown in Fig. 2.10 may be taken as an example for a commercially available SAM. Special attention has been paid to easy light-microscopical visualization and documentation of exactly the same object structures to be investigated by SAM.

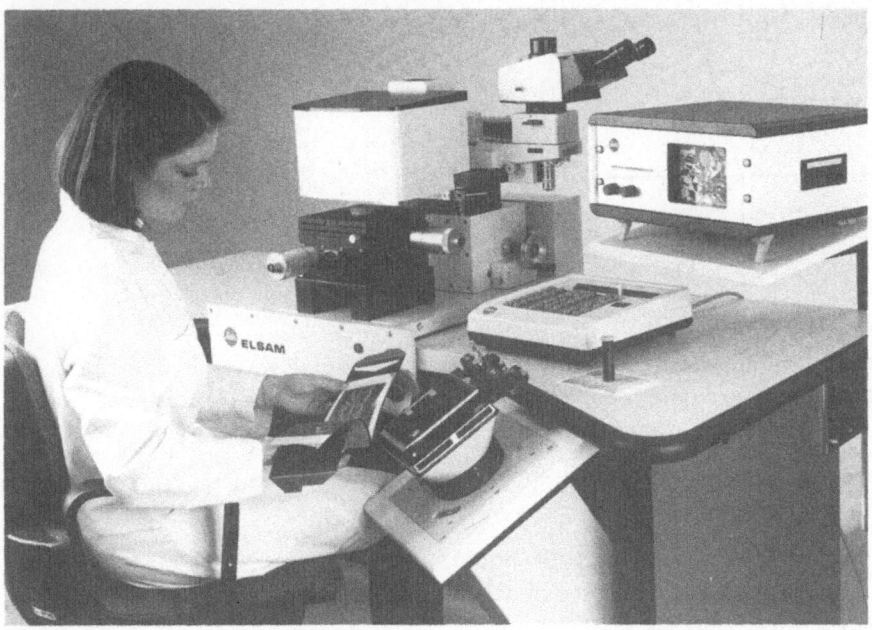

Fig. 2.10. Equipment for scanning acoustic microscopy ("ELSAM" of Ernst Leitz Wetzlar GmbH)

A microprocessor-controlled handling system facilitates ease of operation and permits quick selection of the desired imaging parameters like frequency (between 800–2000 MHz, extendable below 800 MHz down to 100 MHz), ranges of x and y scan determining the lateral magnification, and of the z scan for recording the (V)z curve (Fig. 2.14), as well as amplification, offset level, fast scan and reduced number of lines for observation, and slow scan at 1024 lines for photographic documentation or read-in of the image into the image storage device. Apart from electronic contrast control typical image enhancement operations are possible, too.

An autoniated image analyzer can be connected via a special interface.

2.2 The Image Contrast of Solids in the Reflection Scanning Acoustic Microscope; V(z)-Curves

In the reflection scanning acoustic microscope, image brightness and contrast are determined by the amplitude and the phase of the acoustic waves reflected by the specimen. For illustrating the mechanisms determining amplitude and phase, Fig. 2.11 shows an acoustic wave incident upon a liquid-solid interface.

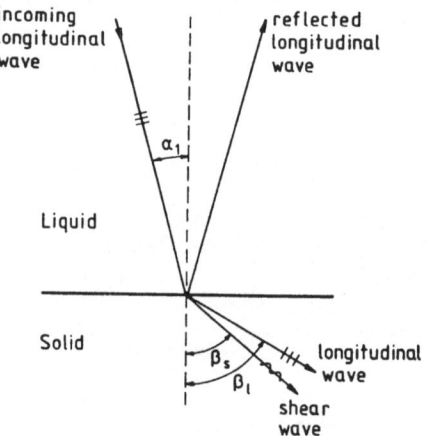

Fig. 2.11. Ray diagram of an acoustic wave crossing a liquid-solid interface

In the liquid, the wave propagates as a longitudinal wave. Inside the solid, generally there are two propagation modes: longitudinal and shear waves, travelling at different velocities. They are refracted following Snell's law:

$$\sin\alpha_1/V_0 = \sin\beta_L/V_L = \sin\beta_s/V_s \tag{2.1}$$

with
V_0 : longitudinal wave velocity in liquid,
V_L : longitudinal wave velocity in solid, and
V_s : shear wave velocity in the solid.
The wave velocities are linked to the elastic constants(c) and the density of the propagation medium (ϱ) as follows (solid: elastically isotropic):

14

$$V_0 = \sqrt{\frac{c_0}{\varrho_0}}, \qquad V_L = \sqrt{\frac{c_{11}}{\varrho}}, \qquad V_s = \sqrt{\frac{c_{44}}{\varrho}}. \tag{2.2}$$

In Fig. 2.11, there is also a reflected ray. The ratio of reflected and transmitted amplitudes is determined by the ratio of the acoustic impedances Z.

For $\alpha_1 = 0$, the acoustic impedances are defined by

$$Z_0 = \varrho_0 \cdot V_0, \qquad Z_L = \varrho \cdot V_L, \qquad Z_s = \varrho \cdot V_s \tag{2.3}$$

and the reflected amplitude R_0 is:

$$R_0 = \frac{Z_L - Z_0}{Z_L + Z_0} \qquad \text{(normal incidence)}. \tag{2.4}$$

From (2.2‑4) one can deduce: The higher the wave velocities, the elastic constants and the density of a particular material. the higher the acoustic reflectivity and the image brightness.

As will be shown later, this applies to *surface images* only (focus on the surface).

Except for materials with largely different impedances the differences in acoustic reflectivity are quite small in a high-resolution scanning acoustic microscope with a wide lens opening-angle.

This is due to the fact that only a small portion of the rays around normal incidence ($\alpha_1 = O$) contributes to material specific variations in acoustic reflectivity. Figure 2.12 shows a typical plot of amplitude and phase of the reflection coefficient vs. angle of incidence.

Only rays between 0 and 30 degrees have reflection amplitudes < 1, which are material specific. Rays with angles of incidence > 30 degrees are totally reflected; their reflected amplitude is 1.

At $\alpha_1 = 30°$, shear waves undergo total reflection (critical angle). A peak at $\alpha_1 = 13°$ marks the critical angle for longitudinal waves.

More important than the amplitude is the *phase* of the reflected wave. The transducer is a phase-sensitive detector, yielding a maximum output signal for

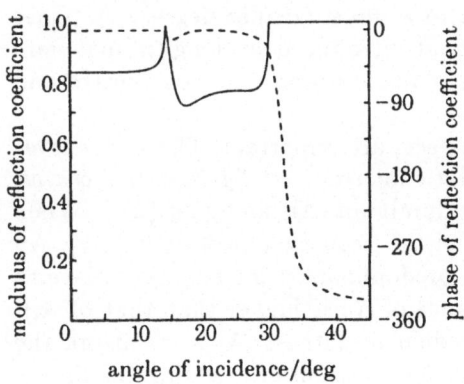

Fig. 2.12. Amplitude (modulus) and phase of the reflection coefficient vs. angle of incidence for aluminium (courtesy of *Illet* et al. [2.5]

15

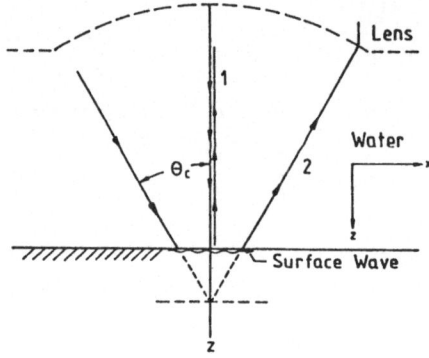

Fig. 2.13. Leaky-wave model: (1) central ray, (2) "leaking" surface wave entering the lens, θ_C critical angle for SAW excitation

plane waves with wave fronts parallel to the transducer plane. The more the incoming wave is distorted, the more the output signal will be decreased.

Figure 2.12 shows a typical plot of the phase of the reflection coefficient vs. angle of incidence. A considerable phase shift occurs at $\alpha_1 = 32°$. This is due to the excitation of a surface acoustic waves (SAWs). There is a combination of shear and longitudinal waves, with predominance of the shear wave component.

Surface acoustic waves are excited at a specific critical angle θ_c (Fig. 2.13). They propagate along the surface of the specimen, sampling a depth of approximately 1 wavelength. After a finite lateral displacement, they radiate back into the coupling medium at the critical angle. The re-radiating SAWs are recollected by the acoustic lens when the geometrical focus is shifted below the surface. Figure 2.13 shows the geometrical ray diagram (leaky-wave model, after *Parmon* and *Bertoni* [2.6].

Under conditions depicted here, only the central ray, reflected at the liquid-solid interface (or at a reflector inside the sample), and the recollected SAW component yield maximum transducer excitation (wave fronts parallel to the transducer plane).

When the position of the focus is moved vertically, phase and amplitude of both interfering rays and thus the output signal of the transducer vary. By plotting the output signal V vs. vertical movement z of the lens, the V(z) curve pattern is obtained. Typical V(z) curves are shown in Fig. 2.14. Caused by the SAW interference mentioned above and, to a much smaller degree, by phase and amplitude shifts in the reflected signal introduced by longitudinal- and shear-wave excitations, the V(z) curve is made up of a series of peaks and dips.

i) For practical work, the following aspects are important: The V(z) curves explain the contrast behaviour of the acoustic microscope. Figure 2.14 shows an example of superimposed V(z) curves of aluminum (Al) and gold (Au). When the two materials are in the focal plane of the acoustic objective (z = 0), Au will appear brighter than Al (at z = 0, predominantly impedance variations are displayed, see above, i.e., the impedance of Au is higher than that of Al). At $z = -1.5\,\mu m$ both materials exhibit equal brightness. At $z = -2\,\mu m$, the contrast is reversed compared to z = 0. The z position yielding maximum

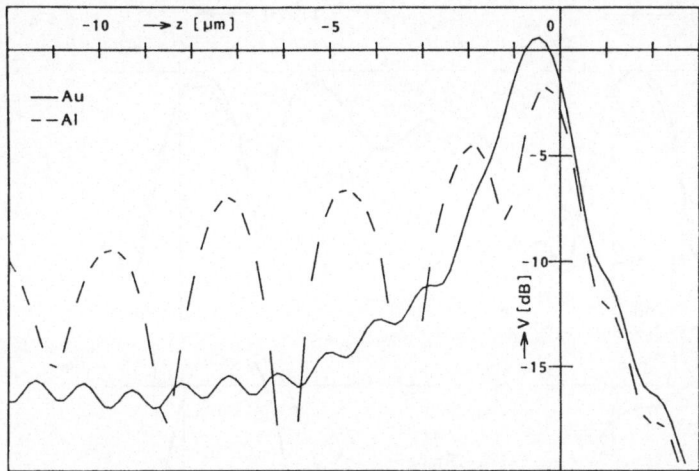

Fig. 2.14. V(z) curves of gold (Au) and aluminum (Al) (frequency: 2 GHz)

contrast between the two materials is given by the point of maximum V difference between the curves ($z = -7\,\mu$m in Fig. 2.14).

Similar considerations apply to polycrystalline single-phase materials. If the crystal grains are elastically anisotropic, the V(z) curve is orientation dependent and grain orientation contrast may easily be obtained in unetched samples. Unlike material contrast, orientation contrast is hardly obtainable at $z = 0$ since impedance variations due to grain anisotropy are generally quite small. A certain amount of defocussing into the material is needed to get the leaky-wave condition of Fig. 2.13, which for most materials is very sensitive to grain orientation.

ii) A thin film on a substrate causes perturbation of the SAW and a change of impedance, both resulting in a change of the V(z) curve compared to the pure substrate (e.g., in Fig. 2.15). The V(z) variation is strongly dependent on the film thickness; thus nondestructive film-thickness monitoring even of opaque films is possible by comparison of the experimental V(z) curves with a set of calculated or experimental reference curves for different thicknesses.

Further uses of the V(z) curve have been reported:

a) If the critical angle for SAW excitation is within the opening angle of the lens, the SAW velocity can be calculated from the spacing of the minima in the V(z) curve [2.7].
b) SAW attenuation may be deduced from V(z) curves [2.8].
c) Using a line-focus beam instead of a point source, or means for directed SAW excitation, the anisotropy of SAW velocity can be measured [2.9].
d) Structural changes in solid materials, such as plastic deformation, can be detected [2.10].

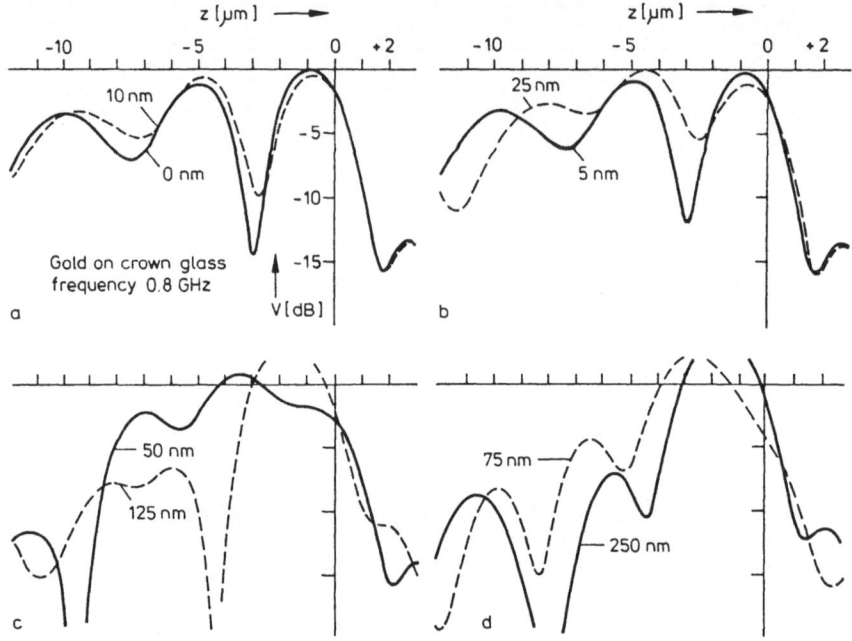

Fig. 2.15. V(z) curves of gold layers on crown glass. Gold layer thicknesses are indicated in the picture. (Frequency: 0.8 GHz)

2.3 Examples of Practical Applications of Reflection Scanning Acoustic Microscopy

Two basic aspects determine the range of applicability of the scanning acoustic microscope in metallurgy:

1. Mechanical properties and density of the specimen determine the acoustic image contrast.
2. Penetration into opaque materials reveals subsurface details and defects nondestructively.

These abilities are useful for a range of applications:

2.3.1 Grain Structure

In almost any sample the scanning acoustic microscope can reveal grain structures of monophase, unetched polycrystalline materials which may be optically isotropic or anisotropic. For most samples, both grain orientation and boundary contrast are obtained and superimposed. Twin boundaries are imaged with high contrast. Typical examples are given in Figs. 2.16–18. This ability is expecially useful when etching is difficult, too time-consuming or impossible to reproduce. Grain orientation and boundary contrast are obtained only when focussed slightly into the material to get the leaky-wave condition described in Sect. 2.2.

Fig. 2.16a,b. Alloy Inconel 600, polished unetched section. (a) optical, differential interference contrst, (b) acoustic, frequency 2.0 GHz; picture width: 580 μm

In addition, the scanning acoustic microscope can also provide information on grain contact. Again, this sensitivity is based on excitation of leaky surface acoustic waves. These waves interact sensitively with inhomogeneities approximately perpendicular to the surface of the sample. Grain boundaries can be such inhomogeneities (Fig. 2.18). It is obvious that an open boundary will cause a disruption of the surface waves, resulting in high boundary contrast. A closed boundary will exhibit good surface-wave transmission, yielding only minimum boundary contrast. In most cases, however, a basic boundary contrast exists due to the abrupt change in lattice orientation of adjacent grains, which affects the propagation of surface waves, too.

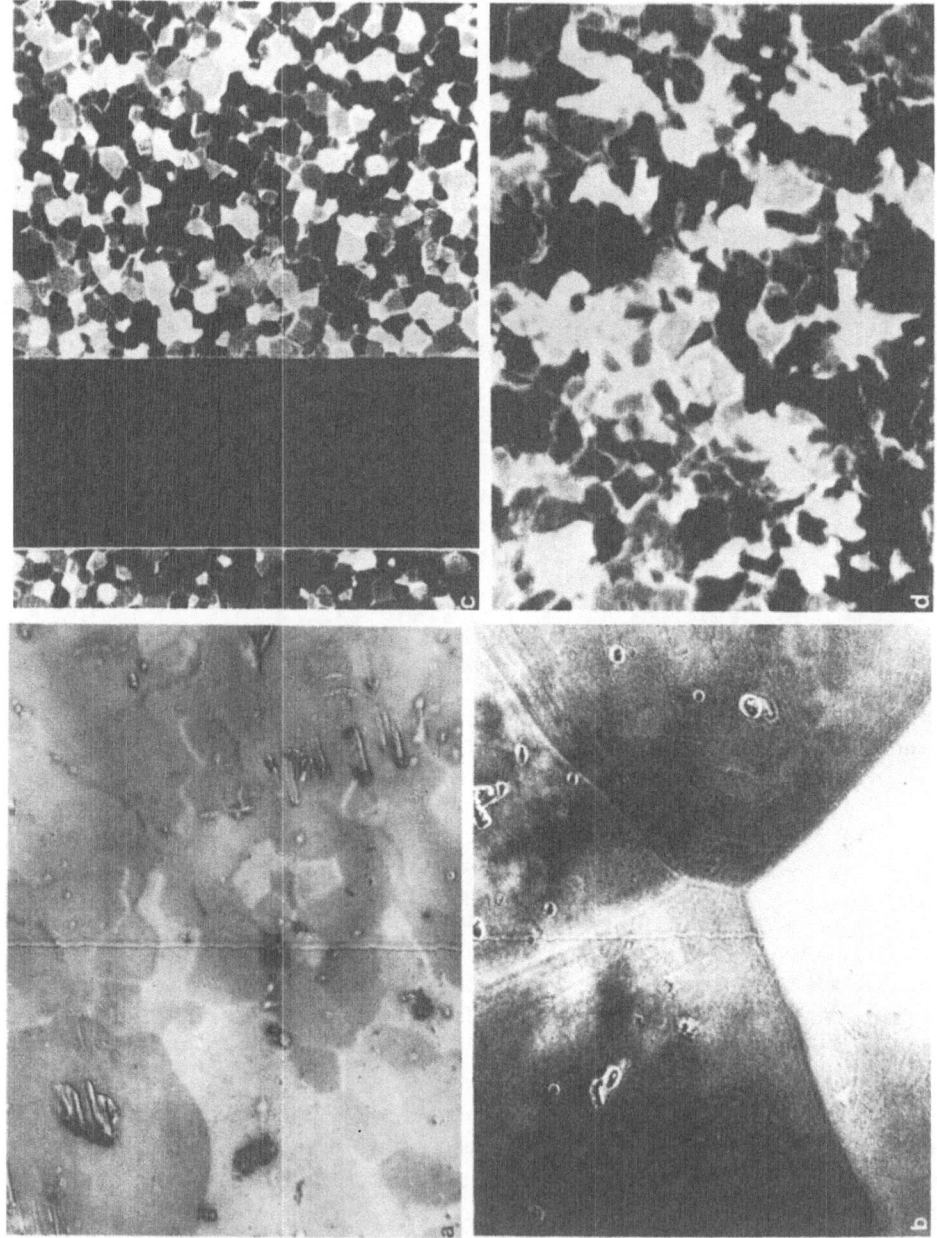

Fig. 2.17a–d. Acoustic micrographs of grain structures in polished unetched sections.
(a) Aluminium, frequency: 1 GHz, width: 580 μm;
(b) brass, frequency: 0.8 GHz, width: 400 μm;
(c) Pb-La-zirconate-titanate (PLZT) ceramic. Black vertical bar is a gold electrode. Frequency: 2.0 GHz, width: 400 μm;
(d) Steel, frequency: 0.9 GHz, width: 400 μm

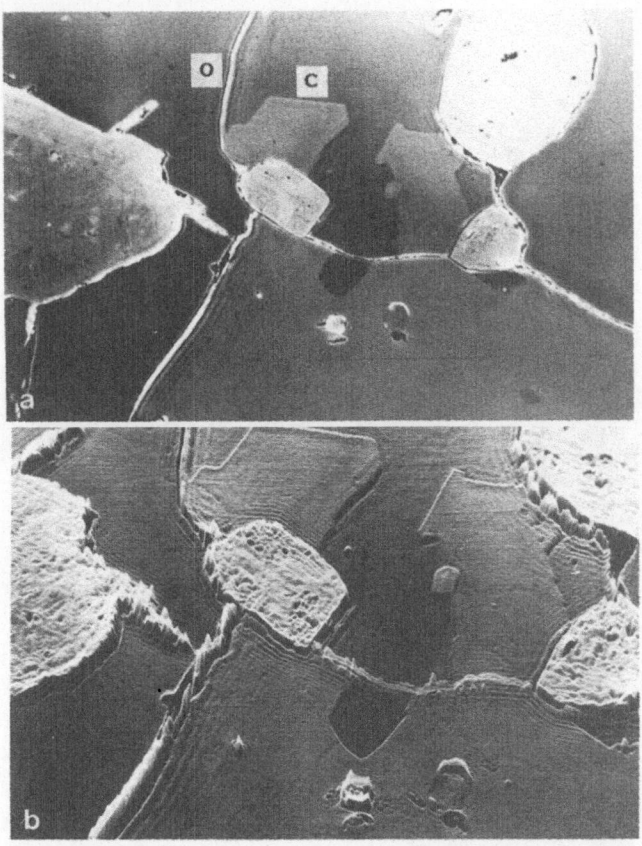

Fig. 2.18a,b. Quartzite with open and closed grain boundaries; acoustic micrographs, frequency: 1.6 GHz. (a) normal image; one open boundary is marked "o", one closed boundary is marked "c"; width: 312 μm. (b) deflection modulation image (intensity profiles are superimposed to normal image); width: 200 μm

2.3.2 Diffusion Zones

The scanning acoustic microscope can be utilized to outline diffusion zones without special sample preparation. A typical example is depicted in Fig. 2.19, showing acoustic micrographs of a Triballoy powder metal with diffusion zones around Laves phase particles (black regions around grains marked "L" in Fig. 2.19b).

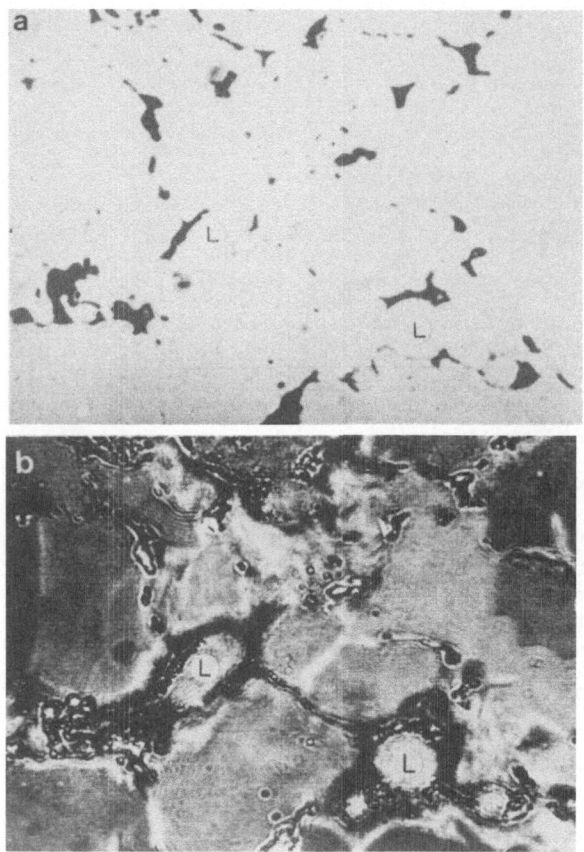

Fig. 2.19a,b. Triballoy powder metal (TPM) with diffusion zones around the Laves phase ("L"); (a) acoustic micrograph, frequency: 1 GHz, width: 230 μm, focus on surface; (b) acoustic micrograph, frequency: 1 GHz, width: 230 μm, focus 1 μm below surface: grain structure and diffusion zones (black areas around particles marked "L") are visible

2.3.3 Materials Defects

The microscope, operated at high frequency (≈ 2 GHz) can often detect surface defects even when other methods of imaging have failed. Reducing the frequency reduces the surface detection sensitivity but enhances penetration (Sect. 2.3.3b), thus making subsurface defect detection possible.

2.3.3.1. Microcracks, Microporosity, Lattice Defects

The outstanding detection sensitivity for these defects is mainly caused by scattering and reflection of leaky surface acoustic waves. Reflected surface acoustic waves often result in standing-wave patterns tracing the defects (Figs. 2.20b and 21c). The detection sensitivity is far better than the resolution

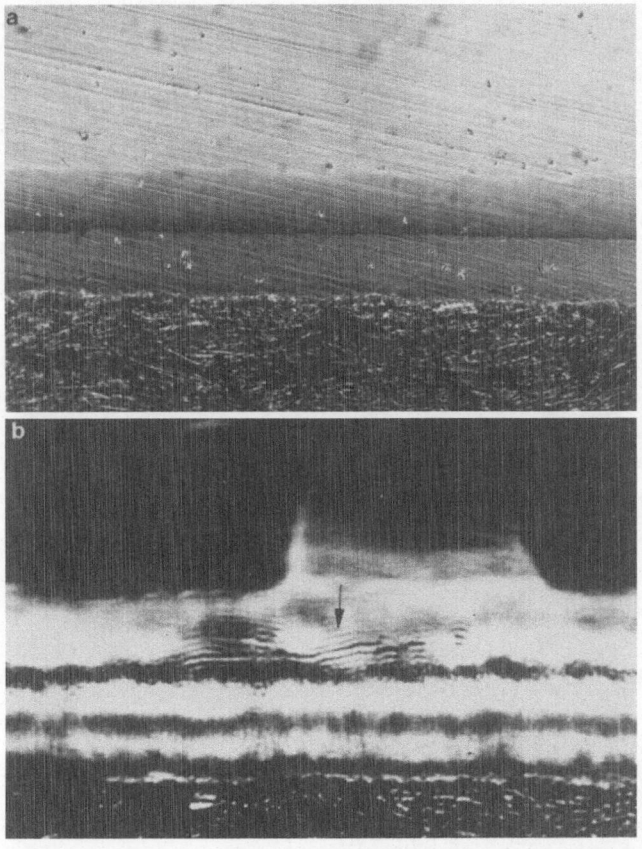

Fig. 2.20a,b. Steel with two nitride layers, cross section, (a) optical, different interference contrast. No defect is visible, (b) acoustic, frequency: 1 GHz, width: 120 μm. Arrow points to microcrack in the nitride layer

limit (resolution limit at 2 GHz: 0.6 μm, detection sensitivity for microcracks: approximately 2 nm). In addition, artefacts like shallow surface scratches can be distinguished from defects easily, since shallow scratches do not have much influence on surface wave propagation, leaving them with low contrast in acoustic images.

Two examples of lattice defects in single crystals are given in Figs. 2.22 and 23. In Fig. 2.22b, precipitates along cleavage planes are clearly revealed. Figure 2.23 depicts dislocations in silicon, which may be imaged without special sample preparation, although with very weak contrast.

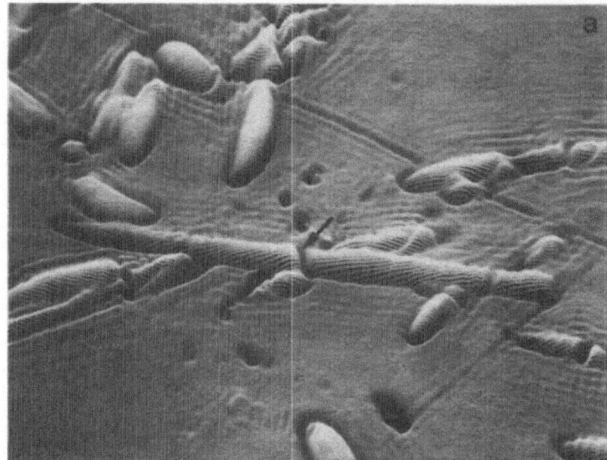

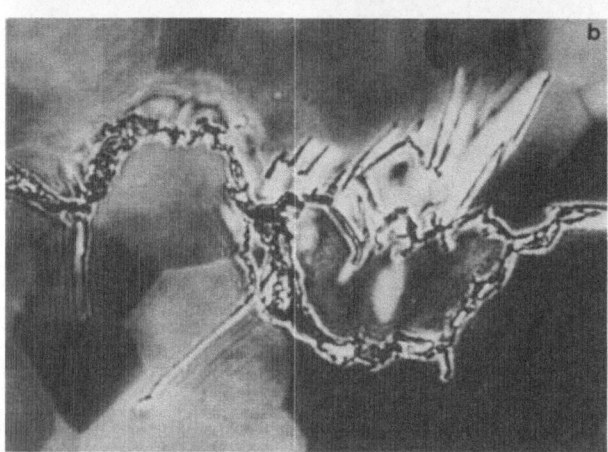

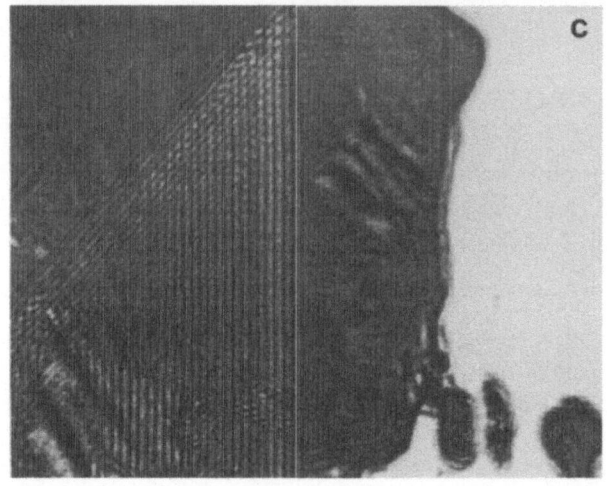

Fig. 2.21a–c. Acoustic images of microcracks in different materials:
(a) in Al_2O_3 – fibre reinforced aluminium (arrow) frequency: 1 GHz, width: 120 μm.
(b) In steel, frequency: 1 GHz, width: 230 μm.
(c) In PbS (galena), traced by standing surface acoustic waves, frequency: 2GHz, width: 230 μm

Fig. 2.22a,b. AgGaSe$_2$ single crystal with precipitates along cleavage planes, (a) acoustic, frequency: 1 GHz, width: 230 μm, focus on surface, and (b) acoustic, frequency: 1 GHz, width: 230 μm, focus 0.5 μm beneath surface: defects are visible

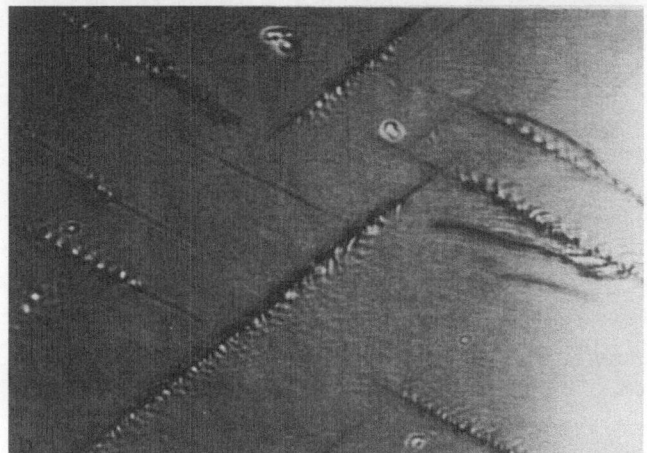

Fig. 2.23. Dislocations in silicon (unetched sample) acoustic, frequency: 1 GHz, width: 55 μm. Three defects are marked by arrows

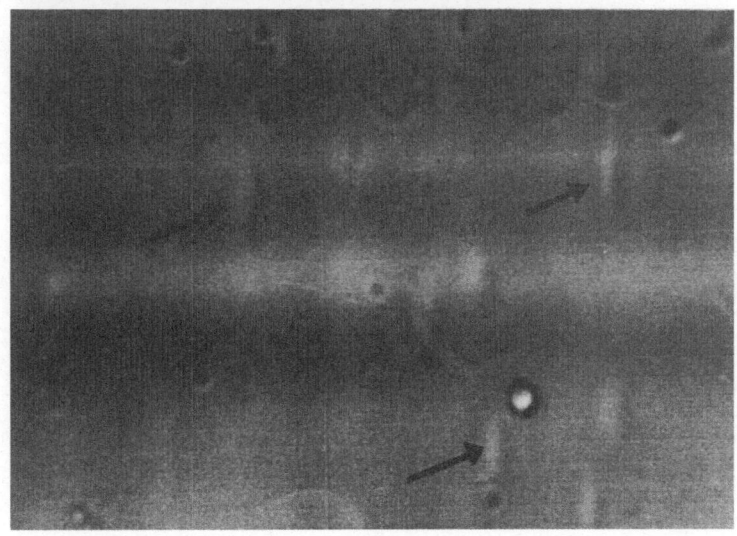

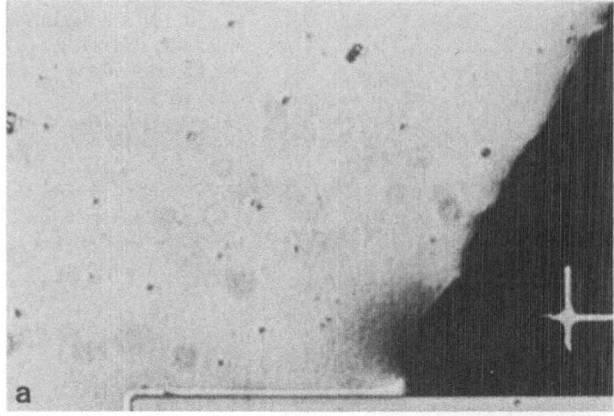

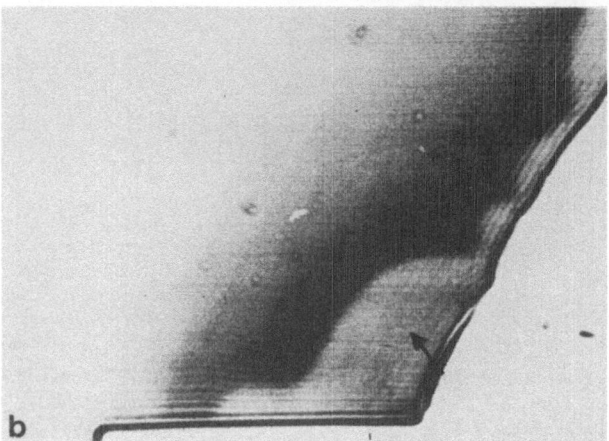

Fig. 2.24a,b. Delaminated permalloy layer on glass substrate (a) optical, and (b) acoustic, frequency: 0.85 GHz, width: 110 μm. Focus is 1.5 μm below surface and debonded area is marked by an arrow

2.3.3.2. Bonding and Adhesion of Surface Layers

Checking contact properties of substrate-layer interfaces is already a classical application of the scanning acoustic microscope; both the penetration ability and the sensitivity to mechanical properties are needed. A typical example is given in Fig. 2.24. For this application, the frequency- and material-dependent maximum penetration depth must exceed the thickness of the layer under investigation. Typical maximum penetration depths of a high-resolution scanning acoustic microscope are in the order of $1 \ldots 3\,\mu$m at 2 GHz and $50 \ldots 100\,\mu$m at 0.2 GHz for metals. Corresponding values of surface resolution are: $0.6\,\mu$m at 2 GHz and $6\,\mu$m at 0.2 GHz. As a result, a tradeoff between penetration and resolution has to be made in order to select a frequency which is optimally matched to the sample.

2.4 Outlook

The scanning acoutic microscope – in particular that of the reflection mode – has gained rapidly increasing interest in the field of material analysis. Resolving power of the light microscope has been reached. This fact demonstrates that, in addition to its "subsurface potential", sufficient lateral resolution of acoustic microscopy is needed last but not least for correlating acoustically imaged and analyzed structures with the well-known corresponding light microscopical image. At present, the whole frequency range between 50 MHz and 2 GHz corresponding to 25 μm and about 0.60 μm lateral resolution in water as coupling medium is covered by the SAM systems hitherto available commercially. Because of the decrease of penetration depth with increasing frequency a compromise between lateral resolution and penetration depth according to the analytical problem to be solved is required.

Whereas towards lower frequencies there is a more or less continuous transition to the frequency range which is used for acoustic material testing and medical diagnostics, a meaningful extension of SAM towards frequencies above 2 GHz is severely limited to, say, 4–5 GHz when using water as coupling medium. Because of the rapidly increasing acoustic attenuation in water at higher frequencies the radius of the spherical lens cavities must be reduced to values which become increasingly difficult to produce and to handle.

By replacing water as coupling medium by liquid nitrogen and still better by liquid helium at temperatures of 0.4 K and below [2.11], the door to much higher resolving power of SAM has now been opened widely. In addition to the very low propagation speed (238 m/s) and the extremely low acoustic attenuation of liquid helium in its superfluid phase, acoustic frequencies up to 80–90 GHz become meaningful [2.12]. Certainly, the introduction of cryoacousto microscopy into microscopical practice will be hampered by the provisions required to operate the lens, part of the scanner and the object, under these conditions.

References

2.1   L. Reimer: *Scanning Electron Microscopy*, Springer Ser. Opt. Sci., Vol. 45 (Springer, Berlin, Heidelberg 1985)

2.1   L.W. Kessler, D.E. Yuhas: Principles and Analytical Capabilities of Scanning Laser Acoustic Microscope (SLAM), Scanning Electron Microscopy **1**, 555 (1978)

2.3   R.A. Lemons, C.F. Quate: Acoustic Microscope–Scanning Version, Appl. Phys. Lett. **24**, 163 (1974)

2.4   J. Attal: Acoustic Microscopy: Imaging Microelectronic Circuits with Liquid Metals, In *Scanned Image Microscopy*, ed. by E.A. Ash (Academic, London 1980) p. 97

2.5   C. Ilett, M.G. Somekh, G.A.D. Briggs: Acoustic Microscopy of Elastic Discontinuities, Proc. R. Soc., London, A 393 (1984) p. 171

2.6   W. Parmon, H.L. Bertoni: Ray Interpretation of the Material Signature in the Acoustic Microscope, Electr. Lett. **15**, 684 (1979)

2.7   A. Atalar: An Angular-Spectrum Approach to Contrast in Reflection Acoustic Microscopy, J. Appl. Phys. **49**, 5130 (1978)

2.8   K. Yamanaka: Analysis of SAW Attenuation Measurement using Acoustic Microscopy, Electr. Lett. **18**, 587 (1982)

2.9   J.A. Hildebrand, L.K. Lam: Directional Acoustic Microscopy for Observation of Elestic Anisotrophy, Appl. Phys. Lett. **42**, 413 (1983)

2.10   R.D. Weglein: Meteorology and Imaging in the Acoustic Microscope, In *Scanned Image Microscopy*, ed. by E.A. Ash (Academic, London 1980) p. 127

2.11   D. Rugar, J. Foster, J. Heiserman: Acoustic Microscopy at Temperature less than 0.2 K, In *Acoustical Imaging*, ed. by E.A. Ash, C.R. Hill (Plenum, New York 1982) p.13

2.12   J.S. Foster, D. Rugar: Low-Temperature Acoustic Microscopy IEEE Trans. SU–32, 139 (1985)

3. High-Resolution Electron Microscopy

F.E. Fujita and M. Hirabayashi

With 30 Figures

In this chapter, recent progress in the techniques and the applications of high-resolution electron microscopy in the study of metals and alloys is introduced. Since the construction of the first electron microscope in 1938 in Germany, many remarkable improvements have been made successfully regarding the resolving power, the theory of image formation, combination with other kinds of analyzers and the production of accessible commercial microscopes. One of the most brilliant developments among them is that of high-resolution electron microscopy, which enables us today to directly observe the atom positions in the crystal structure and lattice defect structures. It must also be noted that the high-resolution technique is strongly coupled with the theory of microscope image formation of materials based on the theory of electron scattering and diffraction and that, as in other fields of science, computer image simulations are commonly used and compared with extremely enlarged photographic images.

In the sections following the introduction, basic theories of atomic image formation are explained and then applications of high-resolution microscopy, such as to the study of lattice defects, amorphous structure, ordered structure, surface and grain boundaries in metals, are mentioned. The readers will immediately find that high-resolution electron microscopy is not a simple technique to identify the atomic structure of solids by the powerful magnifying lenses but is an exact science supported by various basic theories.

Other techniques of microscopy giving highly magnified images of solid structures, such as field ion microscopy, scanning analytical microscopy and scanning tunnelling microscopy, are not mentioned in this chapter, although they are more or less related to high-resolution transmission microscopy and the relations will become closer in the future. The techniques and applications of high-resolution microscopy are still in rapid progress.

3.1 Background

3.1.1 Historical Development

The application of high-resolution electron microscopy to the study of metals seems to be the historical consequence of a long succession of discoveries and inventions. Metallography using the optical microscope and photography was initiated around 1850, shortly after the invention of the silver plate method by L. Daguerre (1839) or the daguerreotype. The first microscopic observation of heat-treated steel by H.C. Sorby (1864) was made with a magnification of less than ten. The development of optical microscopy, supported by L. Seidel's theory of lens abberations (1850), E. Abbe's oil immersion method (around 1880) and many other improvements and inventions, finally made metallographic observations possible with a maximum magnification of the order of $2,000 \times$, which was unavoidably limited by the wavelength of the light, about $0.5\,\mu m$ $(= 0.5 \times 10^{-6}\,m)$, as is well known. After the highest magnification and resolution in optical microscopy were reached at the end of the last century, new searches and attempts were soon started to obtain still higher resolution and magnification to overcome the limitation due to the light wavelength.

The theoretical prediction of the material wave by L. de Broglie (1924) followed by the discovery of the electron-diffraction phenomenon by C.J. Davisson and L.H. Germer (1927) (and G.P. Thomson, 1928; S. Kikuchi, 1928) and the design of the electric-field and magnetic-field lens for an electron beam by H. Busch (1926) immediately initiated the conception of the electron microscope, since the de Broglie wavelength $(\lambda = h/p)$ of electrons accelerated by 50 kV was suggested to be about 0.0053 nm. Actually, a possibility for the construction of an electron microscope was already discussed by M. Knoll in Berlin in 1931 and the first paper using the word, "Elektronenmikroskop", appeared in 1932 [3.1,2]. In 1938 a transmission-type electron microscope (TEM), quite similar in basic features as today's TEM, was built by *Ruska* and *Borries*[3.3]. This success in the construction of an electron microscope in Germany was immediately followed by the work of many investigators in many countries, and within some ten years resolution and magnification reached, say 3 nm $(= 3 \times 10^{-9}\,m = 30\,\text{Å})$ and $20,000 \times$, respectively, even though the development was interrupted by World War II.

In the early stages of the application of electron microscopy to the study of metals, until around 1950, the surface-replica method combined with the metallographical etching technique was used. A notable example of metallurgical application of electron microscopy was the observation of extremely small precipitates in an age-hardened aluminium alloy by *Castaing* [3.4]. In the meantime, the improved electro-thinning technique allowed transmission microscopy of metal specimens and finally, in 1956, *Bollmann* [3.5] and *Hirsch* et al. [3.6a] had success in the observation of dislocations in metal crystals. The image contrast of the crystal dislocation was well explained theoretically by *Hirsch* et al. [3.66]. Since the phase shift caused by the lattice strain developing around the dislocation (and the stacking fault associated with it) was the origin of the image contrast, the image of a dislocation unavoidably spreads as wide as 10 nm. A later improvement, the weak-beam method [3.7], narrowed it to 2 nm or in favourable cases even less. The simple kinematical theory of diffraction-image contrast [3.8a] was also improved to the multiple-beam dynamical theory [3.8b] and better understanding of the image contrasts of lattice defects in metals and other crystalline materials was achieved. It should be noted that the dynamical diffraction theory itself was already presented by H. Bethe in 1928, but its application to the problems of microscope image contrast was made much later, in 1961 by *Howie* and *Whelan* [3.9].

For higher resolution in electron microscopy, sufficient resolving power to analyze directly the positions of the atoms in a crystal was naturally required. This was already realized by the field ion microscope (FIM) which was invented by E.W. Müller in 1951 and showed a resolving power of the order of 0.1 nm (Chap. 4). In the case of the transmission electron microscopy, the first step was the discovery of the lattice-plane-image contrast before resolution reached 0.2 nm. *Menter* [3.10] first observed the well resolved lattice planes of copper-phthalocyanine, which has the planar distance of d = 1.19 nm. The lattice-fringe images were also found to contain dislocations. The lattice-plane fringes can be obtained by interference between the primary (transmitted) beam and a Bragg-reflected beam as the simplest case, this being called the two-beam case [3.11]. When many nonsystematic index reflections fall into the objective aperture, a cross grating image of lattice planes forms. and in this way the crystal structure can be visualized. This is quite analogous to the diffraction technique in which the crystal structure is determined by the inverse Fourier transform of the diffraction pattern or the Bragg reflection intensities. The differences are that in the microscope the inverse Fourier transform is automatically performed, without loss of phase information, to form the structure image, and that in the process of magnification by electron lenses image deformation by spherical and chromatic aberrations and defocussing unavoidably take place. Today, the front runner of transmission electron microscopy is the crystal-structure imaging technique which is usually called "high-resolution electron microscopy". And, it is noteworthy that the theories of spherical and chromatic aberrations, astigmatism and the focussing condition, originating in optical microscopy, again play an important role in modern electron microscopy to achieve the highest resolution. The *Scherzer* defocus condition [3.12] will be frequently mentioned later in this regard. Today's best commercial electron microscopes have reached point-to-point (or aperiodic) resolution of less than 0.2 nm and a lattice-fringe image resolution of less than 0.1 nm for an acceleration voltage of 200 kV. From the electron micrographs of crystal structure images depicted in the present chapter, the reader can estimate the level of the highest resolutions obtainable today (1985).

Further improvement of the resolution in electron microscopy is still being attempted for the direct observation of individual atoms, vacancies and impurity atoms in metal lattices. In 1970, *Crewe* et al. [3.13] showed the image of small clusters of atoms by a scanning transmission electron microscope with an electron beam as fine as 0.5 nm in diameter, and in the next year by means of ordinary transmission microscopes, *Hashimoto* et al. [3.14], *Henkelman* and *Ottersmayer* [3.15], and *Formanech* et al. [3.16] reported independently on the bright- and dark-field electron images of heavy metal ions in molecules and in crystals. *Hashimoto* further improved the technique and the theory, proposing the aberration-free focus (AFF) method [3.17,18], and showed a possibility of identifying a foreign atom in a metal crystal from a characteristic image contrast, as will be mentioned later.

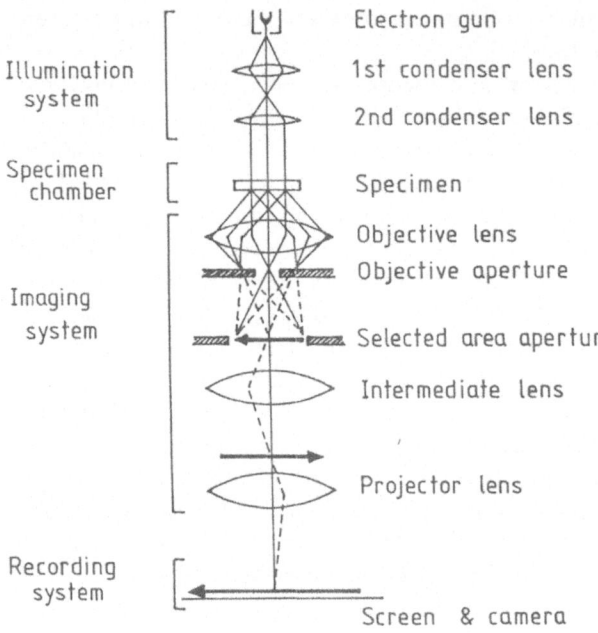

Illumination system

Electron gun

1st condenser lens

2nd condenser lens

Specimen chamber

Specimen

Objective lens
Objective aperture

Imaging system

Selected area aperture

Intermediate lens

Projector lens

Recording system

Screen & camera

Fig. 3.1. A sketch of the basic elements of a transmission electron microscope

3.1.2 Conventional vs. High-Resolution Electron Microscopy

A sketch of the basic elements of a transmission electron microscope is shown in Fig. 3.1 to help the reader's understanding of the techniques of conventional and high-resolution electron microscopy. In modern microscopes, the electron gun with a pointed filament, lanthanum hexaboride filament or a field emission source is used instead of the ordinary hairpin-type filament in order to increase the brightness and the coherency of the illuminating electron beam. A higher and higher degree of vacuum is required in the specimen chamber to avoid the contamination of thin specimens for better resolution. Likewise, no vibration and no drift of the specimen are desired in the specimen-holder design. In the imaging system for high resolution, the most important feature is the objective lens, which leads to very large spherical aberration, chromatic aberration and astigmatism because of the very large aperture angle, as compared to other lenses. Correction of astigmatism is not difficult, but the aberrations are difficult to eliminate and must be carefully taken into consideration in the high-resolution technique, as will be shown later. The image recorder system has been improved by employing a high-sensitivity recording film, an image intensifier and/or TV camera observation and recording.

In principle, the above precautions must be taken for both conventional and high-resolution electron microscopy. Nevertheless, the correction or compensation of the spherical aberration is more directly connected with the theory of image formation in the latter than in the former, as the following descriptions will show.

In conventional techniques, the image contrasts of lattice defects, small precipitates, domain boundaries, etc. in metal crystals arise from the lattice distortions or atomic displacements, which can be understood by the simple kinematical column theory as follows. Consider that the incident electron beam with the wave amplitude ψ_0 and the wave vector $\boldsymbol{k}_0$ is scattered in the direction $\boldsymbol{k}$ by the lattice atoms in a thin column along the $\boldsymbol{k}$ direction. The total scattered amplitude at the end of the column, that is the bottom of the thin specimen foil, is given by

$$\psi = \sum_j f_j \exp\left[2\pi i(\boldsymbol{k} - \boldsymbol{k}_0)\cdot\boldsymbol{r}_j\right] = \sum_j f_j \exp\left[2\pi i(\boldsymbol{g} + \boldsymbol{s})\cdot\boldsymbol{r}_j\right], \qquad (3.1)$$

where f_j is the atomic scattering amplitude, $\boldsymbol{r}_j$ the jth atom's position, $\boldsymbol{g}$ the reciprocal lattice vector, and $\boldsymbol{s}$ a small deviation of the scattering vector, $\boldsymbol{k} - \boldsymbol{k}_0$, from $\boldsymbol{g}$ [3.6b]. In the case of electron microscopy, the diffraction angle is so small that the reflecting lattice planes are almost vertical. The geometry of scattering in the real and the reciprocal space is schematically shown in Fig. 3.2. Because of the lattice periodicity we may use the crystal structure amplitude F, instead of f_j, and thereby $\boldsymbol{r}_j$ is taken as the unit-cell position. When $\boldsymbol{s} = 0$, $\boldsymbol{g}\cdot\boldsymbol{r}_j$ in (3.1) becomes an integer, that is the Bragg condition, ψ gives the amplitude of a Bragg reflection, and, correspondingly, we have the equal inclination fringes or the Bragg contours in the electron-microscope image of a thin film. When $\boldsymbol{s} \neq 0$, $\boldsymbol{s}\cdot\boldsymbol{r}_j$ gives a varying part of ψ, $(\sin\pi st)/\pi s$, which changes periodically with the film thickness t, and therefore corresponding to the intensity $\psi\psi^*$, we observe the equal thickness fringes at the periphery of the thin foil specimen.

In the case of the distorted lattice associated with lattice imperfection, a displacement $\boldsymbol{R}$ must be added to the position vector of each unit cell of the regular lattice, and the total scattered amplitude at the bottom of the specimen foil will be

$$\psi = F \sum_j \exp\left[2\pi i(\boldsymbol{g} + \boldsymbol{s})\cdot(\boldsymbol{r}_j + \boldsymbol{R})\right]$$

$$= F \sum_j \exp\left(2\pi i\boldsymbol{s}\cdot\boldsymbol{r}_j\right) \exp\left(2\pi i\boldsymbol{g}\cdot\boldsymbol{R}\right), \qquad (3.2)$$

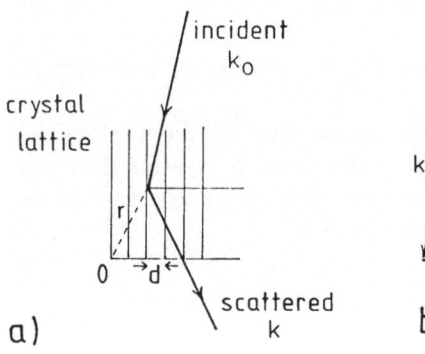

a)

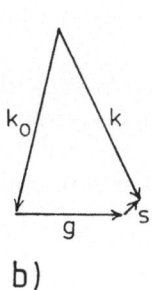

b)

Fig. 3.2. Geometry of electron scattering in a specimen in the real (a) and reciprocal space (b)

in which $\boldsymbol{g \cdot r}_j$ and $\boldsymbol{s \cdot R}$ do not appear because the former makes the exponential term unity, as mentioned above, and the latter is negligibly small. We take the integration, instead of the summation, from the top to the bottom of the film thickness, the z components of $\boldsymbol{s}$ and $\boldsymbol{r}_j$, s and z, respectively, and the phase angle α representing $2\pi \boldsymbol{g \cdot R}$, so that (3.2) becomes

$$\psi = \mathrm{F} \int_0^t \exp{(\mathrm{i}\alpha)} \, \exp{(2\pi \mathrm{i} sz)} \mathrm{d}z. \tag{3.3}$$

When we fix the film thickness t, the periodic change arising from $\exp{(2\pi \mathrm{i} sz)}$ does not appear but, instead, the non-periodic term $\exp{(\mathrm{i}\alpha)}$ plays the role in producing the spacewise changing phase shift. Accordingly, relatively broad strain or displacement contrasts of the respective lattice imperfections show up [3.66]. For instance, the lattice strain and therefore the displacement $\boldsymbol{R}$ at a lattice point 100 nm away from a dislocation in a metal crystal could produce a phase angle α of the order of 10^{-2} radians, which is large enough to give an appreciable phase contrast. This means that the dislocation contrast in conventional electron microscopy could spread as wide as 100 nm, the core position and structure being unidentifiable. One should note that the above description concerns the scattered beam which ultimately gives rise to the dark-field image contrast. The bright-field image is obtained by subtracting the scattered intensity from the transmitted beam. It must also be noted that the effect of extinction, i.e., the successive energy transfer from the incident beam to the diffracted beam, is not taken into consideration in this description. In the case of the stacking fault, a definite amount of phase shift arising from a uniform shear displacement takes place across the stacking-fault plane, and, together with the periodic term, it produces a striped constrast over the faulted area, which is called the stacking-fault fringes. Likewise, the grain or phase boundary obliquely crossing the specimen foil produces the boundary-fringe contrast. To have the finer image contrast of dislocations, it is useful to make the illumination of the far off-Bragg condition, very weakly exciting a Bragg reflection, and take the dark-field image. This is called the weak-beam method [3.8], in which only the near-core part of dislocations contributes to the contrast formation, resulting in the dislocation images as narrow as 2 nm. Better resolving power of the electron microscope is undoubtedly desirable to obtain the finer image contrast. In any case in conventional electron microscopy, however, the elements reducing the resolution, such as aberration, the beam divergence, the amount of defocus, etc., do not always decisively control the quality of the electron images.

In high-resolution electron microscopy, the situation is different. Consider the technique for obtaining the lattice fringe image from a crystal. Here the incident beam and a low-order Bragg reflected one fall in the aperture to be synthesized at the image plane. The total amplitude is

$$\psi = \exp{(2\pi \mathrm{i} \boldsymbol{k}_0 \cdot \boldsymbol{r})} + \phi_\mathrm{g} \exp{(2\pi \mathrm{i} \boldsymbol{k} \cdot \boldsymbol{r})}$$
$$= \exp{(2\pi \mathrm{i} \boldsymbol{k}_0 \cdot \boldsymbol{r})}[1 + \phi_\mathrm{g} \exp{(2\pi \mathrm{i} \boldsymbol{g} \cdot \boldsymbol{r})}], \tag{3.4}$$

where $k - k_0 = g$ and all changes which the reflected beam undergo, such as structure amplitude, phase shift, effect of thickness (and of extinction), etc., are carried by ϕ_g. Rewriting $\phi_g = R \exp(i\varepsilon)$, R containing the before-mentioned $(\sin \pi st)/\pi s$ and ε being the phase angle depending on s and t, the resultant beam intensity is given by

$$I = |\psi|^2 = 1 + R^2 + 2R \cos(2\pi g \cdot r - \varepsilon). \tag{3.5}$$

In the last part of this expression, the reciprocal lattice vector g has the dimension $1/d$, and if r is taken as the distance along g or normal to the reflecting planes, we have a sinusoidal intensity change in the electron image due to the angular term $2\pi x/d$. This is no other than the lattice-fringe image with the lattice periodicity d ([3.11] and [Ref. 3.19, p. 328]). By allowing more reflections in the same zone, that is of parallel g's, to get through the aperture and reach the image, more detailed structure of the lattice planes can be obtained from the fringe image, and, by taking more nonsystematic reflections, the cross-grid image or the lattice-structure image can be obtained.

In the above technique the resolution of the microscope plays an essential part in the image formation, and controls the quality of the image obtained since the lattice periodicity is already comparable to the resolving power. The effect of resolution can be readily taken into the theory of lattice-image formation. For instance, assuming perfect coherency of the illumination electron waves, *Scherzer* [3.12] first introduced the phase-transfer function in the image formation by a lens, which has the form

$$\exp(-i\gamma) = \exp[-\pi i/2\lambda \cdot (C_s \alpha_g^4 - 2\Delta f \alpha_g^2)], \tag{3.6}$$

where λ, C_s, Δf and α_g are the electron wavelength, the spherical aberration coefficient, the amount of defocus, and the angle of scattering or diffraction to the optical axis, respectively. The transfer function carries the phase shift into the total amplitude in (3.4), as ϕ_g does into the scattered amplitude, and reduces the quality of the image contrast. The above expression suggests that the spherical aberration could be counterbalanced by adjusting the amount of defocus to reduce the phase shift to zero for the diffraction angle α_g. Actually, in the case of a dynamic-scattering object, corresponding to thick specimens, the balancing amount of defocus to yield $\cos\gamma = \pm 1$ and therefore to make the contrast transfer function unity is given by

$$\Delta f = (4n)^{1/2} C_s^{1/2} \lambda^{1/2}, \tag{3.7}$$

where n is an integer. In the case of a kinematical-scattering object or a weak-phase object, a phase shift $\pi/2$ produced by scattering must be added to $-\gamma$ and the desirable amount of defocus to yield $\sin\gamma = \pm 1$ is

$$\Delta f = C_s^{1/2} \lambda^{1/2}, \tag{3.8}$$

35

where n is taken as unity. This is the *Scherzer* defocus condition [3.12] which will be discussed in Sects. 3.2.2 and 3.3.1,2. The chromatic aberration envelopes, reduces and terminates the oscillating transfer function at high scattering angles, and the beam divergence arising from the deficiency of the condenser system broadens, blunts and terminates it as well. The chromatic aberration constant and the beam divergence of today's good microscopes are about 1.4 nm and 4×10^{-4} radians, respectively. Remarkable improvements have been made in the theory and practice of high-resolution electron microscopy, obtaining an image resolution good enough to see atoms and atom rows in metal and compound crystals, as will be shown in the following sections.

3.2 Basic Principles of High-Resolution Electron Microscopy

As previously mentioned, the term "high-resolution TEM" has been applied to many microscopy techniques such as weak-beam dark-field imaging, lattice-fringe imaging and scanning transmission imaging. In this section, we confine ourselves to giving an outline of the fundamental aspects of high-resolution TEM which provides images of the crystal-lattice planes or the projection of atom rows in crystals. The former is the lattice fringe image, and the latter a many-beam or multi-beam lattice image. The term "crystal-structure image" is sometimes used in a particular sense for the latter.

3.2.1 Formation of Lattice Fringe Images

Lattice fringe images are formed by optical interferences of two or more electron beams with either tilted or untilted illumination. Figure 3.3 illustrates the lattice fringe imaging with use of the tilted direct beam and one of the diffracted beams.

An example of lattice fringe image is shown in Fig. 3.4, which represents an alignment of dislocations at a low-angle tilt boundary. The lattice fringes in both grains are obtained by the simultaneous excitation of paired diffraction beams ($\pm g$) with the untilted illumination.

The image intensity given in (3.5) is rewritten as

$$I(x) = 1 + |\phi_{\mathbf{g}}(x)|^2 + 2|\phi_{\mathbf{g}}(x)| \cos(2\pi \mathbf{g} \cdot \mathbf{r} - \varepsilon), \tag{3.9}$$

when the diffracted beam is expressed taking account of the scattering phase angle ε as

$$\phi_{\mathbf{g}} \equiv \phi_{\mathbf{g}}(x) \equiv |\phi_{\mathbf{g}}(x)| \exp(i\varepsilon), \tag{3.10}$$

where x is a coordinate in the object parallel to the vector g, and ε is $-\pi/2$ for the kinematical case.

For the tilted illumination, both direct and diffracted beams are symmetrical with respect to the optic axis of the microscope (Fig. 3.3). In this case (symmetrical Laue case), we can ignore the phase shift arising from

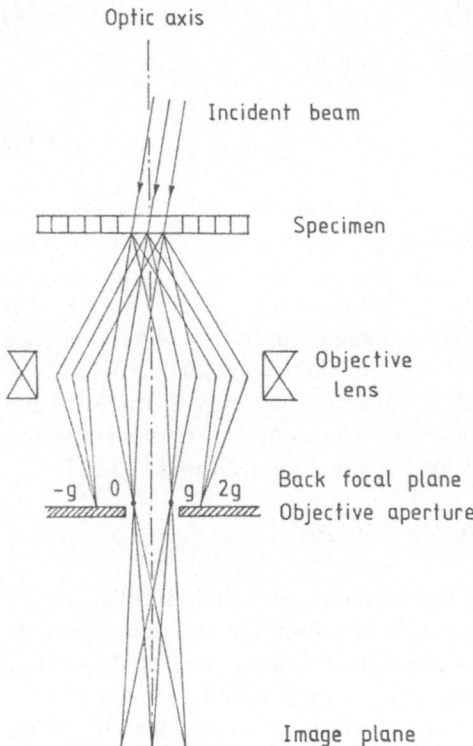

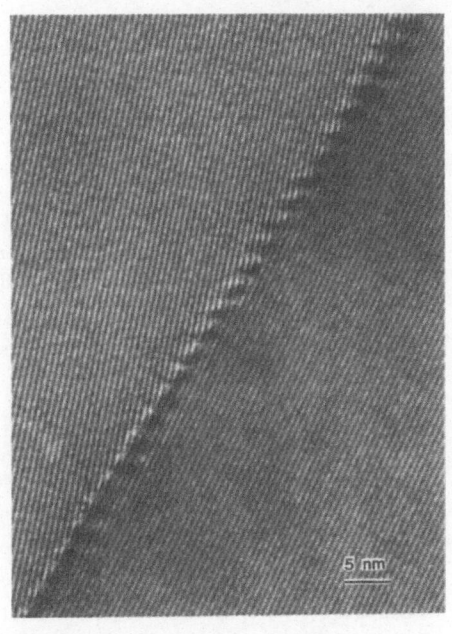

Fig. 3.3. Two-beam lattice fringe imaging with tilted illumination

Fig. 3.4. Lattice fringe image of SiC, showing a low angle tilt boundary (Courtesy of K. Hiraga)

spherical aberration and the focussing condition of the objective lens, because it is cancelled out as this phase shift is a radially symmetric function of the scattering angle α_g as described in (3.6).

For the untilted illumination or the normal incidence, however, the intensity of the two-beam lattice fringe is modified by the phase shift $\chi(\alpha)$ as

$$I(x) = 1 + |\phi_g(x)|^2 + 2|\phi_g(x)|\cos[(2\pi x/d_{hkl}) - \varepsilon - \chi(\alpha)]. \quad (3.11)$$

This expression indicates that any change in $\chi(\alpha)$ produces a sideways movement of the lattice fringe; a parallel displacement of the fringes occurs as the focus control (Δf) changes. Consequently, we note again that the two-beam lattice fringe is by no means the projection of atomic positions in a lattice plane but is nothing more than the interference image reproducing the periodicity of the diffracting lattice planes.

The above fact is revealed from more extensive calculations by the two-beam dynamical theory of electron diffraction. According to the calculation by *Hashimoto* et al. [3.20], the image intensity of a lattice fringe in the exact Bragg condition is expressed as

$$I(x) = 1 - \sin(2\pi t / \xi_g) \sin(2\pi x / d_{hkl}), \tag{3.12}$$

where ξ_g is the so-called extinction distance given by

$$\xi_g = \frac{\pi v_c}{\lambda F_g} \tag{3.13}$$

in a non-relativistic approximation. Here v_c is the volume of the unit cell, and F_g the crystal structure factor for reflection *g*. The lattice fringe disappears for a thickness t equal to multiples of $\xi_g/2$, and the highest contrast can be observed for $\xi_g/4, 3\xi_g/4, \ldots$. It is also shown that the fringe spacing agrees with the crystal lattice spacing only at the exact Bragg conditions. Therefore, we must be cautious of quantitative measurements of the fringe spacing from two-beam lattice images in which the diffraction condition varies with the localized lattice distortion as in the vicinity of dislocation lines and of Guinier-Preston zones. More details have been discussed by *Cockayne* and *Gronsky* [3.21].

3.2.2 Formation of Many-Beam Lattice Images

As noted above, the two-beam lattice image provides very little implication of atomic structures in crystals. If a large number of reflections can participate in the imaging, we may obtain more information concerning the projected position of an atom or group of atoms in a crystal. This fact is readily known from a comparison with a conventional method of x-ray structure analysis, in which the crystal structure is determined by applying an inverse Fourier transform to the integrated intensities of Bragg reflections. The atomic positions can be determined with an accuracy of the order of 0.0001 nm by accumulating as many diffraction intensity data as possible, even though the phase information of diffracted beams is lost in the intensity measurement. The many-beam lattice image, which is the inverse Fourier transform from many diffracted beams at the back focal plane of the objective lens, is formed directly inside the electron microscope without any loss of phase. This is an advantage of many-beam lattice images for investigating the crystal structure, although it is not straightforwardly possible to obtain an exact interpretation of observed images in terms of the atomic configurations in a crystal. The accuracy of atomic positions obtainable by electron microscopy is far less than by x-ray analysis because of the limited number of electron beams participating for the imaging, as compared with the x-ray case.

Many-beam lattice images of metal crystals have first been reported by *Komoda* [3.22] using an off-axial illumination for gold. It should be noted that the intensity maxima in lattice images coincide with the projected atom positions only when the untilted incident beam is parallel to the atom row [3.23]. This is important for obtaining interpretable many-beam lattice images. We outline here the basic principles of many-beam image formation. More complete expositions of the dynamical diffraction theory based on electron-wave optics have been given by *Cowley* [3.24] and *Spence* [3.25].

Many-beam imaging of a crystal is a special case of the phase-contrast image formation. When an incident electron wave passes through a foil of thickness t, a phase shift relative to a wave in the vacuum occurs as

$$\delta = \frac{2\pi}{\lambda}(n - 1)t. \tag{3.14}$$

Here n is an electron-optical refractive index which is slightly larger than unity,

$$n = \frac{\lambda}{\lambda'} \simeq 1 + \frac{\phi(r)}{2E}, \tag{3.15}$$

where λ' is the electron wavelength inside the specimen, E the accelerating energy of the electrons, and $\phi(r)$ the electrostatic potential due to the atomic arrangement within the specimen. As the potential $\phi(r)$ is positive, but much smaller than E, the phase of the electron wave is advanced by an amount depending on the actual potential as

$$\delta = \frac{2\pi}{\lambda}\int_0^t (n - 1)dz \quad = \frac{\pi\phi(x, y)t}{\lambda E} \equiv \sigma\phi(x, y)t, \tag{3.16}$$

where σ is called the interaction constant given by

$$\sigma = \frac{\pi}{\lambda E} = \frac{2\pi me\lambda}{h^2}, \tag{3.17}$$

being 0.00924 and 0.00537 for 100 and 1000 keV electrons, respectively. For a perfect crystal, the projected potential $\phi(x, y) = \int_{-t/2}^{t/2} \phi(x, y, z)dz$ is a periodic function representing the atomic alignment in the specimen, as schematically illustrated in Fig. 3.5. When the crystal is thin enough, it is a phase object in which the amplitude of the electron wave is unaffected but only the phase is altered in passing through the crystal. The modification of the incident electron

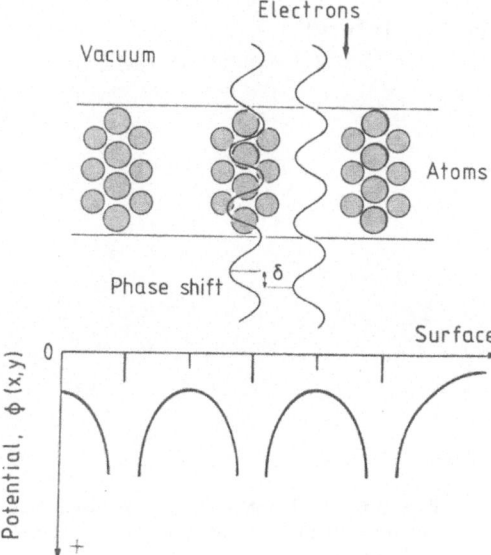

Fig. 3.5. Schematic illustrations of phase shift of electrons passing through atom-row-potentials in a crystal

wave is given by the transmission function

$$q(x, y) = \exp[i\sigma\phi(x, y)t]. \tag{3.18}$$

If $\sigma\phi(x, y)t \ll 1$,

$$q(x, y) \simeq 1 + i\sigma\phi(x, y)t. \tag{3.19}$$

This is called the weak-phase approximation. The projected potential $\phi(x, y)$ is directly recorded on a photographic film at the image plane, if the imaging system of the electron microscope is ideally perfect. However, the aberration of the objective lens in current electron microscopes prevents the observed image from being a true reproduction of $\phi(x, y)$.

At the back-focal plane of an objective lens, an electron-diffraction pattern (Fraunhofer diffraction) is formed. The amplitude of the diffraction pattern $Q(u, v)$ is the Fourier transform of $q(x, y)$ (Fig. 3.6)

$$\begin{aligned} Q(u, v) &= \mathcal{F}\{q(x, y)\} \\ &= \delta(u, v) + i\sigma\phi(u, v), \end{aligned} \tag{3.20}$$

where $\delta(u, v)$ indicates the direct beam and the second term the diffracted beams. In passing through the objective lens, the diffracted beams are modified by the phase shift

$$\chi(\alpha) \equiv \chi(u, v) = \frac{\pi}{\lambda}\left[\Delta f(u^2 + v^2)\lambda^2 - \frac{1}{2}C_s(u^2 + v^2)^2\lambda^4\right], \tag{3.21}$$

where a positive value of Δf implies an under-focussed lens. This expression is sometimes called the wave aberration [3.19]. The diffraction pattern is accordingly expressed as

$$\begin{aligned} R(u, v) &= Q(u, v) \exp[i\chi(u, v)] \\ &= \{\delta(u, v) + i\sigma\phi(u, v)\} \exp[i\chi(u, v,)] \\ &= \delta(u, v) + \sigma\phi(u, v) \exp[(i\pi/2) + i\chi(u, v)]. \end{aligned} \tag{3.22}$$

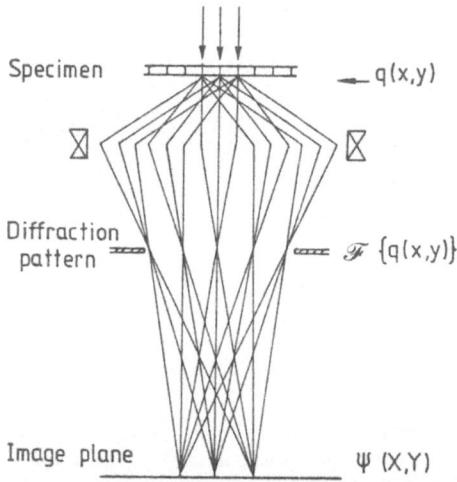

Specimen ⟶ q(x,y)

Diffraction pattern ⟶ $\mathcal{F}\{q(x,y)\}$

Image plane ⟶ Ψ (X,Y)

Fig. 3.6. A sketch showing the relations between diffraction pattern, electron image and Fourier transforms

Then the amplitude distribution in the image plane is given by the Fourier transform

$$\Psi(x, y) = \mathcal{F}\{R(u, v) \cdot A(u, v)\}$$
$$= 1 + \sigma \mathcal{F}\{\phi(u, v) \exp[i(\pi/2) + \chi(u, v) \cdot A(u, v)]\}. \tag{3.23}$$

Here $A(u, v)$ is the aperture function, i.e.

$$A(u, v) = \begin{cases} 1 & \text{for } (u^2 + v^2)^{1/2} \leq r_0 \\ 0 & \text{for } (u^2 + v^2)^{1/2} > r_0 \end{cases}, \tag{3.24}$$

where r_0 is the aperture radius in reciprocal space.

The intensity distribution of the image is given as

$$I(x, y) = \Psi(x, y)\Psi^*(x, y) = |\Psi(x, y)|^2$$
$$= |1 + \sigma \mathcal{F}[\phi(u, v)T(u, v)A(u, v)]|^2, \quad \text{where} \tag{3.25}$$

$$T(u, v) = \exp[i(\pi/2) + i\chi(u, v)]. \tag{3.26}$$

$T(u, v)$ is the phase contrast transfer function (PCTF). If

$$\chi(u, v) = \pi/2 \cdot \text{ and } \quad A(u, v) = 1,$$

the intensity is simplified as

$$I(x, y) = |1 - \sigma \mathcal{F}\{\phi(u, v)\}|^2 = |1 - \sigma \phi(-x, -y)|^2$$
$$\simeq 1 - 2\sigma \phi(-x, -y). \tag{3.27}$$

Consequently, the image intensity is proportional to the projected crystal potential; the dark region in the observed image corresponds to the high potential in the crystal, and vice versa.

As $\chi(u, v)$ is not generally $\pi/2$, however, we must consider the phase modification of diffracted beams due to $T(u, v)$. Examples of the real and imaginary parts of $T(u, v)$ for 200 kV and 1000 kV electron microscopes are shown in Fig. 3.7. For the optimum condition, that is of the Scherzer focus given by (3.8), the phase shift $\chi(u, v)$ adds up to $\pi/2$ over an angular range of scattering as large as possible so that the different beams can interfere with the direct beam to give the intuitive phase contrast. Figure 3.7a shows that phase shift $\chi(u, v)$ approximately equal to $\pi/2$ is extended over the range of $2 - 5 \, \text{nm}^{-1}$. This range is called a pass band or a phase-contrast transfer interval. The pass band of the curve in Fig. 3.7c is less expanded, as compared with Fig. 3.7a. Consequently high-voltage electron microscopy is greatly preferred for obtaining interpretable images with high spatial resolution in the atomic scale.

Here, the effects of divergence of the incident beam and of chromatic aberration should be mentioned. The latter arises mainly from the instabilities of the accelerating voltage and the energy spread of the incident electrons. In a first approximation, these effects on the phase contrast are represented by en-

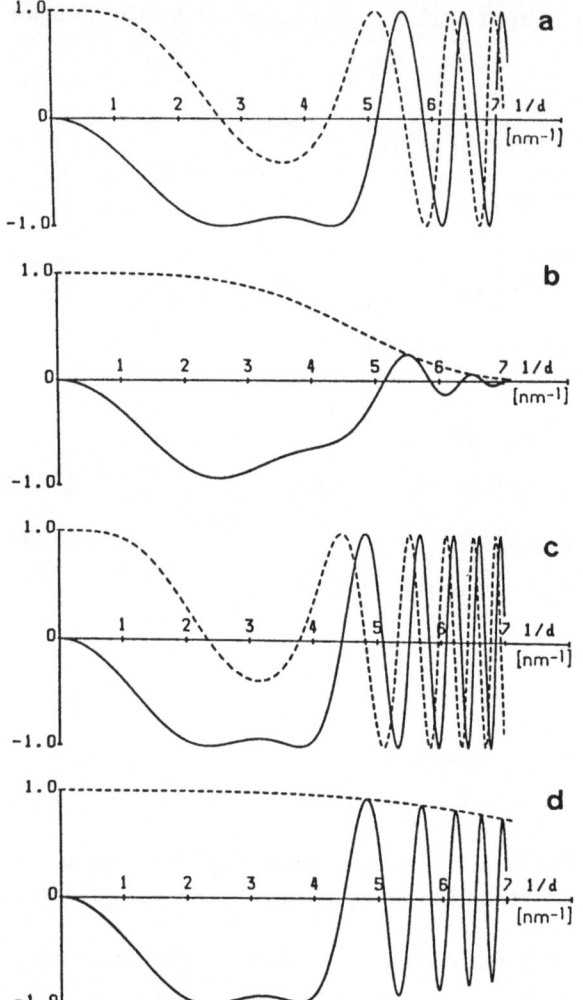

Fig. 3.7a–d. Phase contrast transfer functions. (a) and (b) at 1000 kV, C_s = 11 mm, Δf = 110 nm. (c) and (d) at 200 kV, C_s = 0.8 mm, Δf = 50 nm. Full and dotted curves correspond to $\cos(\chi + \pi/2)$ and $\sin(\chi + \pi/2)$, respectively. (b) and (d) indicate an envelope function $G(k)$ = $\exp(-2^{-1}\pi^2\lambda^2\Delta^2 k^4)$, where the chromatic aberration coefficient Δ is taken as 20 nm and 2 nm, respectively, and $G(k) \cdot \cos(\chi + \pi/2)$

velope functions [3.26,27] which make the pass band narrower and lowers the experimental resolution limit, as shown in Figs. 3.7b and d.

With the Scherzer focus condition, a many-beam image with axial illumination can be resolved down to spacing given approximately by

$$\Delta \sim 0.7 C_s^{1/4} \lambda^{3/4}, \tag{3.28}$$

called the Scherzer resolution limit. The effect of the reduced wavelength at high voltage is more significant even if the spherical aberration coefficient is usually greater for high-voltage instruments.

3.2.3 Image Simulation by the Multislice Method

In general, it is hardly possible to prepare specimens of metals and alloys for high-resolution TEM thin enough (below 1 nm thick) to adopt the weak-phase object approximation. As for rather thick crystals, the many-beam lattice images can not reflect the projected potentials, but their contrast is affected inevitably by the dynamical diffraction effect of electrons. For an interpretation of the experimental images of such specimens, comparisons with simulations computed from dynamical electron diffraction theory are necessary. The multislice method based on the physical-optics theory developed by *Cowley* and *Moodie* [3.28] is commonly utilized for the image simulation (3.24,25). In this method, the specimen is considered to consist of many very thin slices of the thickness ΔZ perpendicular to the incident beam (Fig. 3.8), and the electrons are assumed to be scattered at the top surface layer having the crystal potential $\phi(x, y)$ and propagated by Fresnel diffraction passing through the vacuum of the thickness ΔZ. The transmission function (3.18) of the jth slice is given as

$$q_j(x, y) = \exp\left[i\sigma\phi(x, y)\Delta Z\right]. \tag{3.29}$$

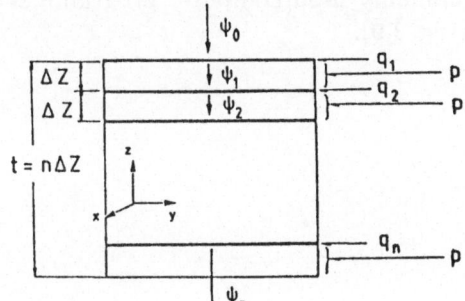

Fig. 3.8. Multislices of crystal

The propagation function of the wave from the jth slice to the $(j + 1)^{th}$ slice is expressed by

$$p(x, y) = \frac{i}{\lambda\Delta Z} \exp\left(i\pi \frac{(x^2 + y^2)}{\lambda\Delta Z}\right). \tag{3.30}$$

Thus the wave function at the potential layer of the 2nd slice is given by the convolution of $p(x, y)$ and $q_1(x, y)$ as

$$\psi_1(x, y) = q_1(x, y) * p(x, y). \tag{3.31}$$

The wave function at the exit surface of the 2nd slice is

$$\begin{aligned}
\psi_2(x, y) &= q_2(x, y)\psi_1(x, y) \\
&= q_2(x, y)[q_1(x, y) * p(x, y)].
\end{aligned} \tag{3.32}$$

In a successive way, the wave function at the exit surface of the crystal is derived as

$$\psi_n(x, y) = [\, q_n(x, y) \underset{n-1}{[}\; q_{n-1}(x, y) \underset{n-2}{[}\; \cdots$$
$$q_2(x, y)[q_1(x, y) * p(x, y)\,] \underset{1}{]} \cdots \underset{n-1}{]} * p(x, y)\,\underset{n}{]}. \tag{3.33}$$

In reciprocal space, the expression is

$$\psi_n(u, v) = \mathcal{F}\{\phi(x, y)\} = \underset{n-1}{[}\; Q_n(u, v) * \cdots$$
$$[Q_2(u, v) * Q_1(u, v)P(u, v)] \underset{1}{]} P(u, v) \cdots \underset{n-1}{]}\; P(u, v),$$

where

$$Q_n(u, v) = \mathcal{F}\{q_n(x, y)\}$$
$$P(u, v) = \mathcal{F}\{p(x, y)\}. \tag{3.34}$$

Performing the comprehensive multislice computation and the associated image simulation is time-consuming. Commonly used computer programmes can be set up in the following sequence (Fig. 3.9).

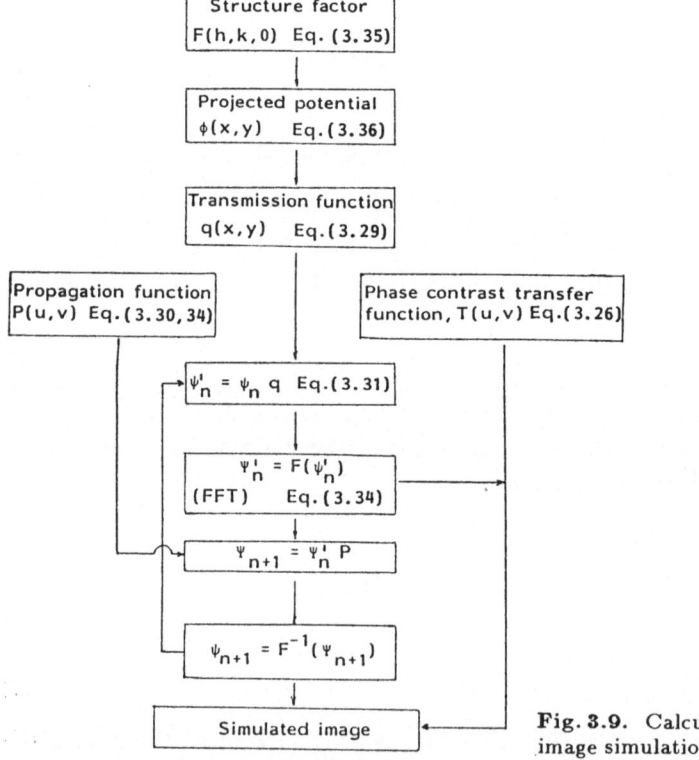

Fig. 3.9. Calculation sequences of image simulation

1. Calculation of the structure factor $\mathcal{F}(hk0)$: When the incident beam is taken parallel to the c axis of an orthorhombic lattice, the crystal structure factor is expressed as

$$\mathcal{F}(hk0) = \sum_j f_j \exp[-2\pi i(hx_j + ky_j)], \qquad (3.35)$$

where the atomic scattering factor f_j is taken from [3.29].

2. Calculation of the projected potential $\phi(x, y)$ by the Fourier transform

$$\phi(x, y) = \frac{1}{ab} \sum_{hk} \mathcal{F}(hk0) \exp 2\pi i(hx + ky). \qquad (3.36)$$

3. Calculation of the transmission function $q_j(x, y)$ using $\phi(x, y)$.
4. Calculation of the propagation function $p(x, y)$ and its Fourier transform $P(u, v)$.
5. Multislice calculation by iteration. The fast Fourier transform (FFT) algorithm [3.30] developed by *Ishizuka* and *Uyeda* [3.31] is a useful iteration method.
6. Using the phase-contrast transfer function $T(u, v)$, which is claculated in advance, the outputs of many-beam images are obtained as a function of crystal thickness and of defocus value.

Examples of the simulated images are shown in the following sections.

3.3 Applications

3.3.1 Defect and Defect Analysis

One of the most exciting problems of high-resolution electron microscopy is to observe directly the local disarrangements of atoms in the otherwise regular crystalline lattice. The disarrangements or the lattice defect structures appear in different dimensions: The one-dimensional defect is the dislocation, but the lattice strains around the dislocation line spread **three-dimensionally** and could be analyzed by directly determining the atomic displacements by the high-resolution electron microscope, which is in contrast to the strain-contrast method by the ordinary electron microscope mentioned in Sect. 3.1. Typical two-dimensional defects are the stacking fault, the monolayer platelet precipitate like the Guinier-Preston zone, the grain boundary, and the crystalline surface. The last two will be mentioned in Sects. 3.3.5 and 6, respectively. Three-dimensional defects are, for instance, the small spherical precipitate, the spherical or polyhedral void, the stacking fault tetrahedron, and the small disordered region spiked by particle irradiation. All of the above mentioned defect structures are interesting and more or less resistant objectives for high-resolution electron microscopy. The most tenable target among others is the smallest defect unit, such as the atomic vacancy, substitutional impurity atom, and the self or impurity interstitial atom.

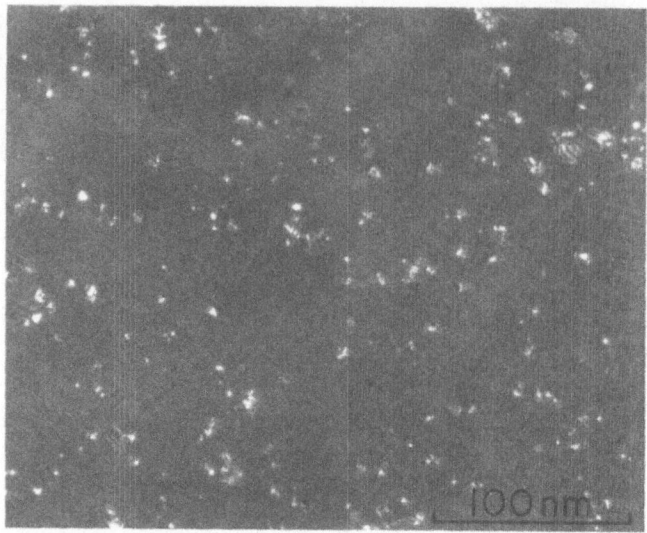

Fig. 3.10. A weak beam dark field image of small stacking fault tetrahedra (and other kinds of defects) in a Au film irradiated by D-T fusion neutrons of $1.0 \times 10^{21}\,\mathrm{m}^{-2}$ at 300 K [3.32]. (Courtesy of M. Kiritani)

a) Stacking Fault Tetrahedron

In this subsection, let us begin with the smallest lattice defect which could be observed by a good ordinary microscope without using the lattice image technique. In Fig. 3.10 is shown a weak-beam image of pure Au irradiated by D-T neutrons at around 300 K, in which the smallest stacking fault tetrahedra appear as bright dots. *Kiritani* [3.32] measured their sizes and found that their values were discrete for the smallest ones, for instance, 0.5, 0.75, 1.0 nm, and so on in diameter. This means that the observed tetrahedra could be associated with three, six, ten vacancies and so on, respectively. The atoms in the interior part of the tetrahedron are displaced toward the platelet of the missed atoms or the vacancies, producing the stacking fault across every tetrahedral plane, so that the electron waves passing through the tetrahedron will suffer a phase shift of $2\pi/3$ and, according to the term, $\exp{(2\pi i \boldsymbol{g} \cdot \boldsymbol{R})}$ in (3.2) or $\exp{(i\alpha)}$ in (3.3), a distinct phase contrast will be expected. This is probably the reason why the extremely small stacking fault tetrahedra can be observed and associated vacancies measured by conventional electron microscopy. It may be worthwhile to compare the conventional dark field image of Fig. 3.10 with the high-resolution images of stacking fault tetrahedra in Fig. 3.11 taken by *Hashimoto* and *Takai* [3.33]. These images exhibit a strong dark and bright contrast between the tetrahedra and the matrix lattice. The smallest tetrahedron they could identify was of three vacancies. The dislocations accompanied by the stacking fault tetrahedron are the multipole integrant forming the edges of the tetrahedron, so that their contribution to the strain contrast image formation would not reach very far.

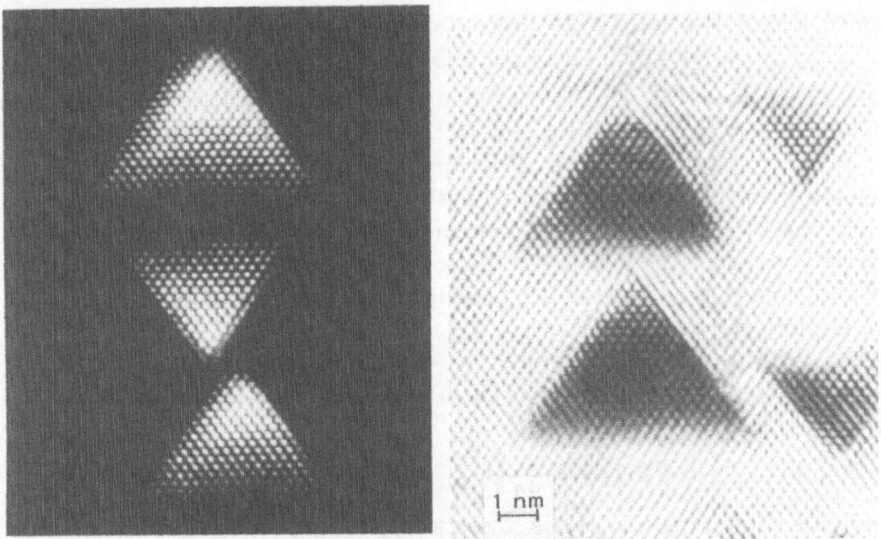

Fig. 3.11. The high-resolution images of small stacking fault tetrahedra [3.33]

b) Aberration-Free Focussing Condition and Defect Observation

As already mentioned in Sect. 3.1, the phase shifts of scattered waves brought by the spherical aberration, C_s and the amount of defocus, Δf, into the total amplitude at the back-focal plane control the quality of the image contrast, as shown by (3.6). In the case of dynamical scattering in a crystal, the highest contrast condition is given by (3.7), and by selecting the integer in the equation and thereby the amount of defocus one can make the value of the contrast transfer function, $\cos\gamma$, unity at the angles of Bragg reflections to be used for the high-resolution image formation. Since the Bragg angle, $\theta = \alpha/2$, is small, the Bragg condition is rewritten as

$$2d \ \sin\theta = \lambda \simeq d\alpha. \tag{3.37}$$

By substituting it and the high-contrast condition, $2m\pi = \gamma$, into (3.6), we obtain the focussing condition,

$$\Delta f = \frac{C_s\lambda^2}{2d^2} + m\left(\frac{2d^2}{\lambda}\right). \tag{3.38}$$

The two equations (3.7 and 3.38) are compatible, and one can fit all the Bragg reflections participating in the image formation to the best contrast condition, $\cos\gamma = 1$, at the same time. This is called the aberration free focus (AFF) condition by *Hashimoto* et al. [3.17] and is recommended by them for the lattice defect observation [3.18], because all the strong high-angle scattered or reflected waves caused by the irregular atomic arrangement around the defect can be set into the best contrast condition in this way and the high-resolution images of the atoms could be exactly represented by the dotty contrasts.

c) Dislocations Resolved in High-Resolution Electron Microscopy

An example of a high-resolution image of dislocation in an Au film is shown in Fig. 3.12 which was taken by *Hashimoto* and *Takai* [3.33]. The film plane is probably nearly (110), the Burgers vector $a/2 \cdot [1\bar{1}0]$, and the line direction perpendicular to the (110) plane. According to them, the atom (row) positions represented by bright dots agreed well with those calculated by the continuum elasticity theory except in the region near the core. They also found by successively taken pictures that the dislocation moved on the (001) slip plane slowly and smoothly at a speed of about 0.3 nm/s and concluded that it was a mobile Lomer-type dislocation.

Another interesting example is an extended normal dislocation accompanied by a stacking fault shown in Fig. 3.13. Splitting into two partials is represented by

$$a/2 \cdot [10\bar{1}] \rightarrow a/6 \cdot [21\bar{1}] + a/6 \cdot [1\bar{1}\bar{2}]. \tag{3.39}$$

As the figure shows, the two partial dislocations separated by about ten atomic distances have darker contrast on the average, presumably because of their own strain fields. Also, the $(1\bar{1}1)$ planes stacking has the fault ABCACABC... between the partials, as denoted in the picture. Further analysis exhibited that

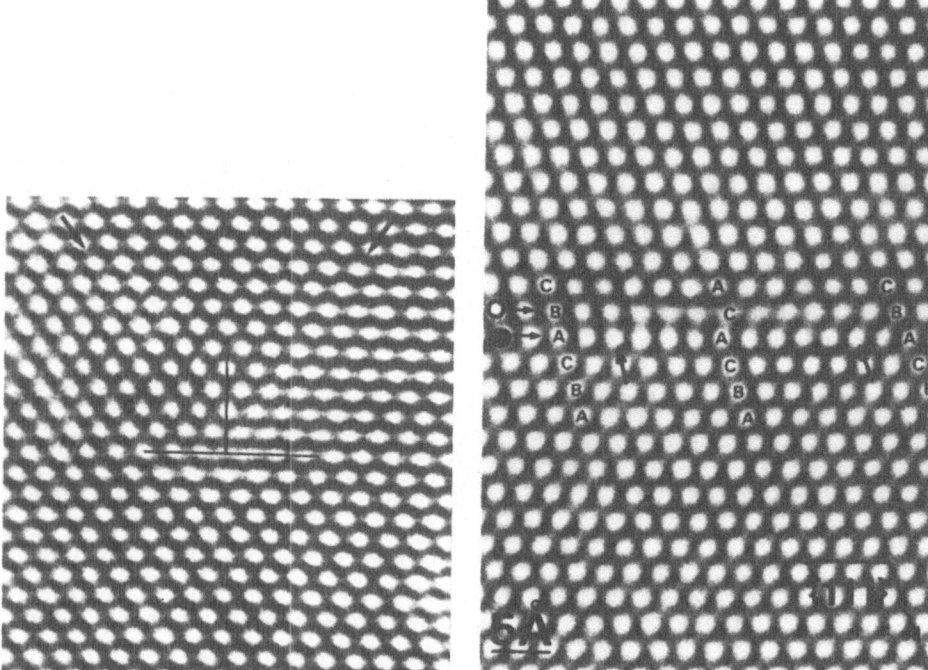

Fig. 3.12. High-resolution image of an edge dislocation in Au film [3.33]. The film plane is nearly in (110), the Burgers vector $a/2 \cdot [1\bar{1}0]$, and the line direction [110]

Fig. 3.13. High-resolution image of an extended dislocation associated with a stacking fault plane. Note that a stacking disorder ABCACABC... takes place only across the fault plane [3.33]

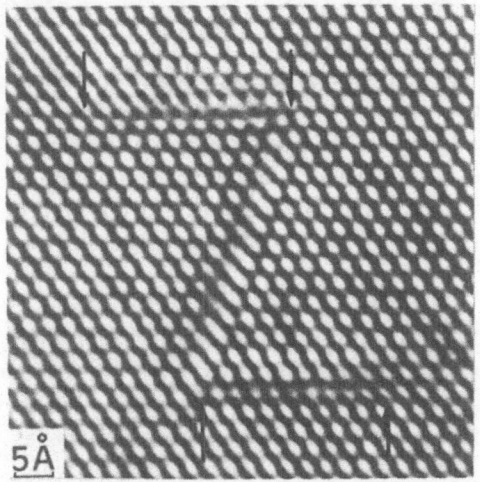

Fig. 3.14. High-resolution image of a Z-shape faulted dislocation dipole containing two $a/6 \cdot [0\bar{1}1]$ type stair rod dislocations [3.33]

this dislocation is a 60° mixed-type mobile dislocation. Screw dislocations and their extension and motion were also frequently observed.

Dislocations of other types observed by them were $a/6 \cdot [0\bar{1}1]$ type stair rod dislocations produced by the association of $a/6 \cdot [\bar{1}12]$ and $a/6 \cdot [1\bar{2}\bar{1}]$ type partials, the Z-shape faulted dislocation dipole containing two $a/6 \cdot [0\bar{1}1]$ type stair rod dislocations, the $a/6 \cdot [130]$ type stair rod dislocation, etc. The Z-shape dislocation dipole is shown in Fig. 3.14. It should be noted that the so-called immobile dislocations, such as stair rod dislocations, were mostly mobile under the condition of the high-resolution electron microscopic observation. By the same technique, they observed very thin twin platelets and their growth and shrinkage by the atomic rearrangements at the side and the tip boundaries. The motion of the side boundary was associated with the generation and gliding of partial dislocations, but that of the tip was not clear because its structure was incoherent and not always resolvable by electron microscopy.

d) Observation of Point Defects

Except the attempt to see the internal structure of an atom or the distribution of core electrons around a nucleus, the final goal of high-resolution electron microscopy is probably to envisage and identify by electron images the individual atoms and their arrangements in crystals or any aggregate structures. This is the most direct and finest chemical and structural analysis. Nevertheless, before the goal is reached, one must try to observe directly in a regular arrangement of atoms a missing or excess atom which is no other than the point defect in the structure.

In the structure image by atom-resolving high-resolution microscopy, exact reproduction of the atom positions is not always assured by their spotty images unless the conditions of image formation, including the defocus condition, the effects of chromatic aberration and beam divergence on the phase shifts of the incident and diffracted waves in the crystal which produce the shifts of atom image positions, are carefully examined. In addition, the effects of the crystal thickness, the depth of the defect position, and the interference of surrounding atoms must be taken into consideration in the image analysis. These difficulties mostly arise from the dynamical diffraction effect, which plays an essential part in the high-resolution image formation. At present, the result of the observation of a single vacancy in metal lattices is not decisive, although the computer simulation of the electron image of a vacancy has been done and compared with the experiment by *Hashimoto* et al. [3.34–36]. To eliminate the shift of an atom image and obtain the highest resolution, they employed the tilt-beam illumination aberration-free focus (TAFF) condition [3.35] in the observation of an Au thin film irradiated by 2 MeV electrons and found some atom images with less bright contrast as are indicated by small arrows in Fig. 3.15. The computer-calculated image contrast of a vacancy is shown in Fig. 3.16a–c, in each of which a vacancy is set up in the centre. The depth positions of the vacancy were set at 8, 6 and 4 nm in three calculations shown in the Fig. 3.16a–c, respectively, the model crystal being 11 nm thick. The displacements of surrounding atoms

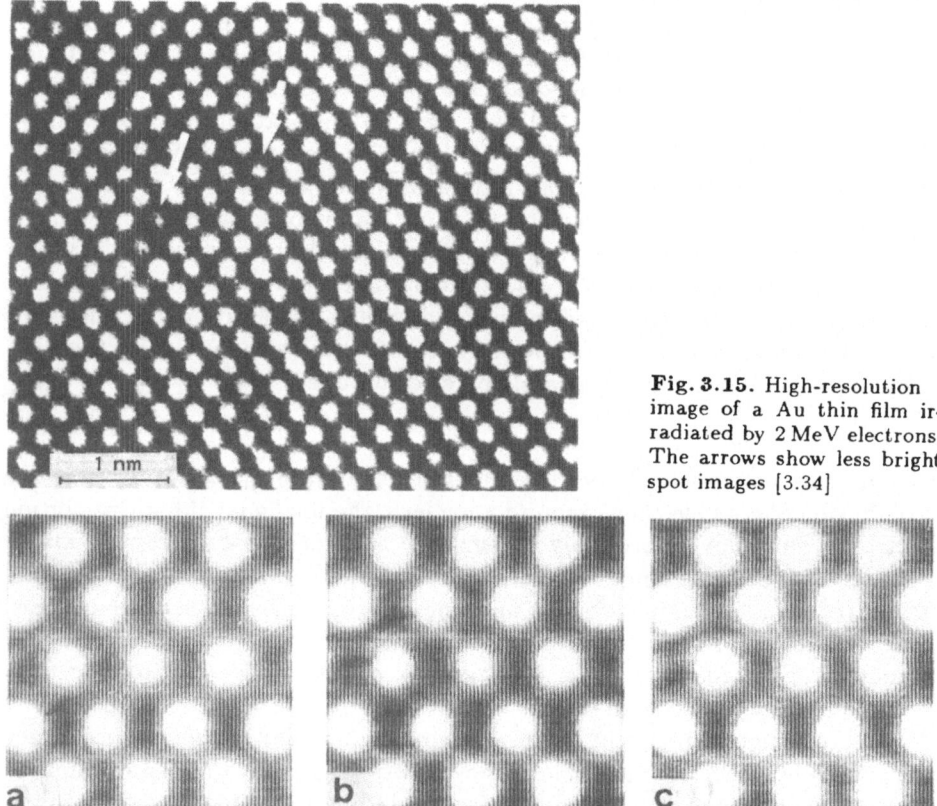

Fig. 3.15. High-resolution image of a Au thin film irradiated by 2 MeV electrons. The arrows show less bright spot images [3.34]

Fig. 3.16. The calculated image contrast of an 11 nm thick metal lattice containing a vacancy in the centre at the depth of 8 nm (a), 6 nm (b) and 4 nm (c), respectively [3.34]

were taken into account, and 16,384 excited beams and 000 and two 111 Bragg reflections were used for the image formation. The calculation shows that the observation of a single vacancy is possible and the vacancy contrast becomes clear as the depth position becomes close to the bottom of the crystal. From a comparison between experiment and calculation, they concluded that the image shown in Fig. 3.15 was that of a single vacancy located close to the bottom surface of the specimen. Further attempts to see the image of the substitutional impurity atom and that of the surface atom have been made. *Takai* et al. [3.36] carried out an electron microscope observation of thin Si crystals containing Sb atoms of 10^{18-22} cm^{-3}. Some bright spots were seen, and under the beam irradiation their image contrast changed, which could be due to the migration of Sb atoms in the direction normal to the crystal surface. The change in contrast was recorded in VTR. No attempt to observe the interstitial atom has been made. More careful examination of the image contrasts of point defects seems to be quite necessary as well as further accumulation of experimental data for the final conclusion on the point defect observation.

e) Less-Common Defects

In the following we introduce less-common defects mostly in less-common crystal structures which can hardly be investigated by techniques other than high-resolution electron microscopy.

The first of these are the stacking disorder and partial dislocations in the Laves phase alloys, which have the close-packing layer structure consisting of two kinds of atoms with different radii. Stacking faults or stacking modifications occur frequently in this structure on the c-plane of the hexagonal cell and have been identified not only by the x-ray diffraction technique but also by high-resolution electron microscopy, as Fig. 3.17 shows [3.37]. *Kitano* et al. [3.38]

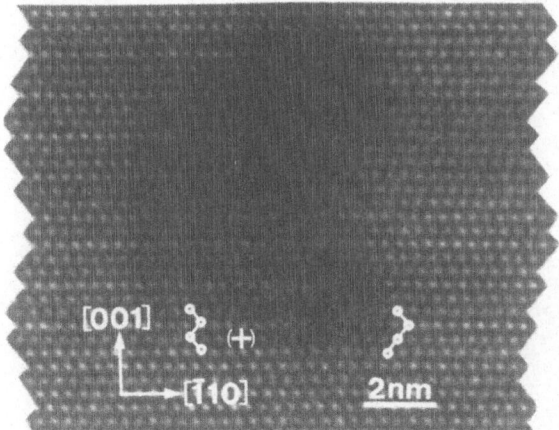

Fig. 3.17. Structure image showing an imperfect edge-type dislocation accompanying a stacking fault in a Mg(Cu-Zn)$_2$ Laves phase alloy. Note that the stacking sequences on the right and left hand side are different as the linked open circles and the zigzag cuttings on the both side edges show [3.37]

analyzed the structure images of Mg-based Laves phase alloys and found good correspondence between the c-plane stacking sequences and the patterns of arrangement of white dots, which is shown in the figure by the linked open circles and also by the zigzag cutting of the right and left side edges of the picture. It must be noted that the sequences on both sides are different from each other and the dotty image in the centre is dark. This means that a dislocation exists between the two stacking modifications, producing a strain field around it. They used a modified Burgers circuit to determine the types and the Burgers vectors of dislocations and found various kinds of stacking fault dislocations in the Laves phase, including screw-type dislocations. The stacking faults and associated partial dislocations in various less-common alloy structures can be directly analyzed in this way by high-resolution electron microscopy. Actually, studies of microstructures of various compounds using this technique were reported recently by *Komura* [3.39].

The second subject is that of the Guinier-Preston (G.P.) zones in the Al-Cu age-hardenable alloy. These zones are the first form of precipitation from the supersaturated state [3.40]. Since a G.P.(I) zone consists of a thin disc-shaped plate of Cu atoms with a collapse of the lattice on both sides of the zone, like a small vacancy type dislocation loop, because the solute Cu atoms are smaller than the solvent Al atoms, such a zone appears in the conventional bright-field micrograph as a line-shaped contrast due to the accompanying strain field and in the weak-beam image as a pair of sharp bright lines, when it is viewed edge-on [3.41,42]. The structure imaging technique has been applied by *Yoshida* et al. [3.43] to clarify the atomic arrangement in the G.P.(I) zone, for which different structures were proposed by the x-ray diffuse scattering [3.44] and the field-ion microscope study [3.45]. An example of the high-resolution images obtained is shown in Fig. 3.18. From the image simulation analysis, it was concluded that

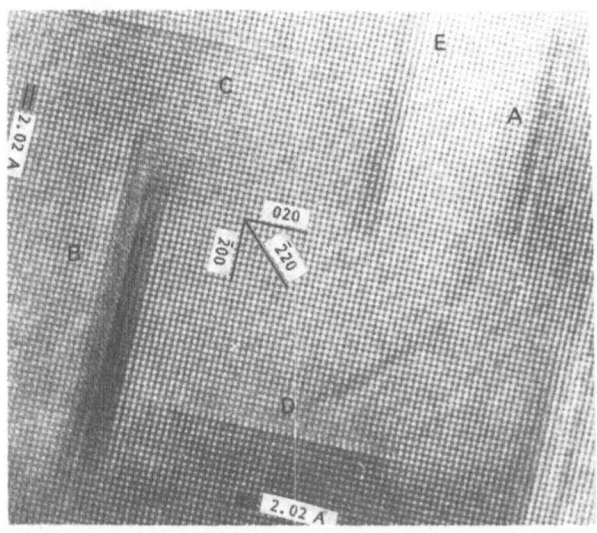

Fig. 3.18. High-resolution image of G.P.(I) zones in an age hardened Al-3.97wt.% Cu alloy [3.43]

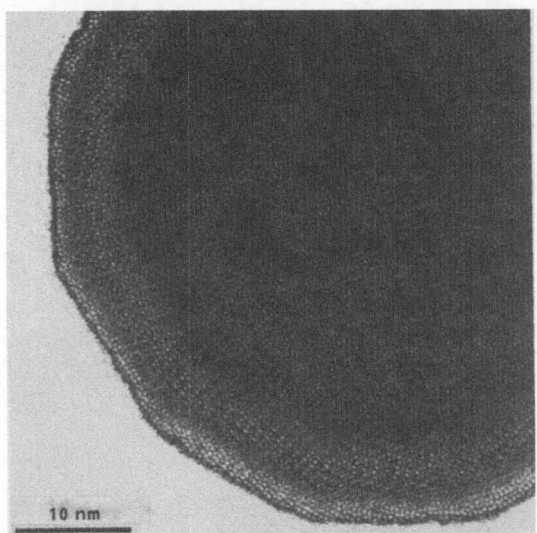

Fig. 3.19. Structure image of a Ni-Cr fine particle, which shows 12-fold symmetry in diffraction. The basic structure is that of the σ phase [3.46]

the structure of G.P.(I) is a monolayer of Cu atoms lying in a $\{100\}$ plane associated with the displaced matrix Al atoms on both sides, which will be dealt with more precisely in Sect. 3.3.4a.

Another interesting subject in high-resolution electron microscopy closely related with less-common defects is a strange compatibility of the long-range orientational order and the lack of translational symmetry, which has been reported on two different alloy systems. Small particles of a Ni-Cr alloy showed 12-fold symmetry in diffraction patterns [3.46a], and quenched specimens of an alloy with Al–14 at.% Mn composition showed sharp 10-fold diffraction symmetry [3.46b]. These unusual symmetries are inconsistent with the translational symmetry of a crystal. *Ishimasa* et al. [3.46b] studied the Ni-Cr alloy by taking the many-beam structure images of small particles containing multiply microtwinned σ-phase domains, an example of which is shown in Fig. 3.19. It was found from the dotty-image analysis that the multiple twin domains had a common c-axis and rotations of $30°$ around it. A model of the structure, which could have a far-reaching orientational order and no translational symmetry to exhibit a sharp 12-fold symmetry diffraction pattern, was deduced by tesselating the square and triangle structural units so as to keep exactly twelve directions of the edges of the two kinds of the tilting units. Domains of the crystalline σ-phase with as few as four unit cells frequently appeared with highly disordered local structures in their model. They concluded that the structure is in an intermediate state between the crystalline and the amorphous, and they called it "crystalloid" after the definition by *Mackay* [3.48]. The other odd structure of an Al-Mn alloy with 10-fold diffraction symmetry was successfully analyzed by *Hiraga* et al. [4.39] by the high-resolution many-beam image technique. Figure 3.20 shows a structure image micrograph of a melt-quenched and electrolytically thinned Al–14.3 at.% Mn alloy specimen together with an electron-diffraction pattern (on the left) and an optical diffractogram (on the right). In the structure image, bright dots lie

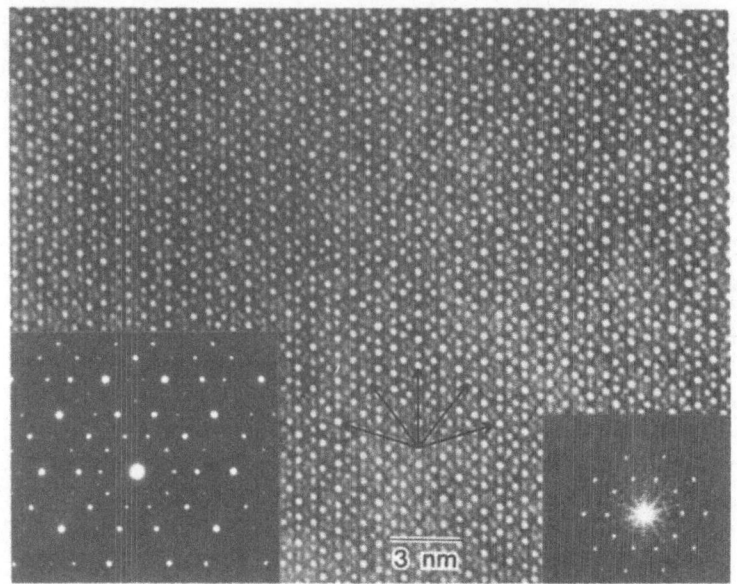

Fig. 3.20. Structure image of rapidly melt-quenched Al-14.3at.% Mn alloy, which shows 10-fold diffraction symmetry in the diffraction pattern (*left-hand side inlet*) and in the optical diffractogram (*right-hand side inlet*) [3.50]

along 5-fold directions, as indicated by arrows, but no translational symmetry exists. A multiply-twinned structure found by *Ishimasa* et al. [3.46] in the Ni-Cr σ-phase particles is not seen but 10-fold double rings and pentagons of bright dots interconnected with each other frequently appear in the picture. They considered this two-dimensional distribution of dots as a projection of a three-dimensional aggregation of icosahedrons which are the structural units of the Al-Mn quasicrystal and proposed an icosahedral quasicrystal structural model in which fundamental icosahedrons consisting of one Mn and twelve Al atoms aggregate three-dimensionally with edge-sharing. They could almost fill up the three-dimensional space by connecting the 13-atoms icosahedrons and the larger icosahedrons, each consisting of twelve of the former. An interesting problem would be how to fill further the gaps between the structural units or leave them as structural defects. It also seems to be valuable to compare the defective structures introduced or suggested by the above studies with the more defective structure, almost random packing aggregation of atoms, which will be introduced in the next section[1].

3.3.2 Amorphous Metals and Alloys

a) Study of Amorphous Structure by Diffraction Techniques

By means of diffraction techniques, atomic arrangements in random condensed matters have been extensively studied, especially for liquids. Since the amor-

---

[1] Rapid progress is proceeding in this field. See the additional references with titles at the end of the book.

phous metallic films were first made by vacuum evaporation on the cold substrate by R. Hilsch and W. Buckel (1950) and by rapid quenching from the melt by P. Duwez (1960), the diffraction study of amorphous metals and alloys has also become quite popular [3.50].

The diffraction patterns from the amorphous substances are represented by blurred concentric rings, called the halo pattern. The principal part of the theory of diffraction to be used to analyze the halo pattern and to find the atomic arrangement in the amorphous structure is as follows. For the total scattered amplitude of any radiation with the wave vector k from an amorphous specimen, we may use the same expression as (3.1), where the scattering vector $k - k_0 = q$ does not correspond to any clear Bragg reflection because there is no clear lattice periodicity. The diffracted intensity from the N-single-element atoms aggregate is, therefore,

$$I(N) = \psi\psi^* = \sum_i^N f \, \exp\left(-2\pi i q \cdot r_i\right) \times \sum_i^N f \, \exp\left(2\pi i q \cdot r_j\right)$$

$$= Nf^2 + f^2 \sum_{i \neq j} \sum \exp\left[-2\pi i q \cdot (r_i - r_j)\right]. \tag{3.40}$$

In order to find the radial distribution of atoms in the aggregate, we employ the probability function $\varrho_0^2 g(r_{ij}) dv_i dv_i$ to find two atoms, respectively, in the small volumes dv_i and dv_j separated by the distance $r_{ij} = r_i - r_j$, where ϱ_0 is the average atomic density. By introducing it into (3.40) and using integrations instead of summations, the normalized intensity distribution $S = I/N|f|^2$, or the interference function is obtained as

$$S(q) = 1 + 4\pi\varrho_0 \int_0^\infty \{g(r) - 1\} r^2 \frac{\sin 2\pi qr}{2\pi qr} dr. \tag{3.41}$$

When a random-structure model is constructed, one can measure all interatomic distances to compose the two-body correlation function $g(r)$, and obtain the interference function $S(q)$, by the above Fourier transformation, to compare them with experiment. And, conversely, from the intensity distribution of the experimental halo diffraction pattern, one can obtain the two-body correlation function by the Fourier inverse transformation. This is the ordinary technique used in diffraction experiments and in calculations to find $g(r)$ from $S(q)$, and vice versa (Chap. 5).

b) Electron Image of Disordered and Ordered Structure in Amorphous Materials

The structure of amorphous materials has been studied over a decade by high-resolution electron microscopy, since it was expected to yield information on the local atomic arrangement, especially on local order, projected to two dimensions. Note that the diffraction data, $S(q)$ and $g(r)$, obtained from the halo diffraction patterns are basically one-dimensional. In an early stage of high-resolution electron microscopy, the tilted-beam setting was used and only the large-scale spotty or fringy images were observed and interpreted in terms of the microcrystalline model [3.51]. A more careful work [3.52] showed, however, that the number and the orientations of the fringes depended on the angle of the tilting, suggesting that they were mainly due to an instrumental effect. On the other hand, under the condition of bright-field axial illumination, allowing a part of the halo diffraction beam to pass symmetrically through the aperture, some investigators have observed local lattice-like fringes among the atomic-size dotty contrast even in the as-quenched amorphous alloy films. For instance, *Ishida* et al. [3.53] examined the electrolytically thinned films of the iron, cobalt, and palladium based amorphous alloys by the high-resolution technique with the bright-field axial illumination and observed lattice-like images which varied with the alloy compositions, the quenching method and the quenching rate, even though the electron-diffraction patterns were typical of amorphous structure. They concluded that the lattice-like images arose from the microcrystalline or medium-range order regions of a few nanometers in diameter. An example of local-fringy or lattice-like image contrasts found in the as-quenched amorphous $Fe_{86}B_{14}$ alloy thin film by *Ichinose* and *Ishida* [3.54] is shown in Fig. 3.21. *Hirotsu* and *Akada* [3.55] made similar observations on as-quenched amorphous $Fe_{84}B_{16}$ and $Pd_{77.5}Cu_6Si_{16.5}$ alloys by carefully selecting the values of the defocus distance, which are shown in Fig. 3.22a and b, respectively. The lattice and fringy images of an average diameter of 1.2 nm, average spacing of 0.2 nm, and four- or six-fold symmetry are seen in Fig. 3.22a and those of 1.2 nm in diameter, 0.19 nm spacing and four-fold symmetry in Fig. 3.22b. They explained these images in terms of microcrystals with the beam incidence along $\langle 001 \rangle$ or $\langle 111 \rangle$ of the bcc structure in (a) and with that along

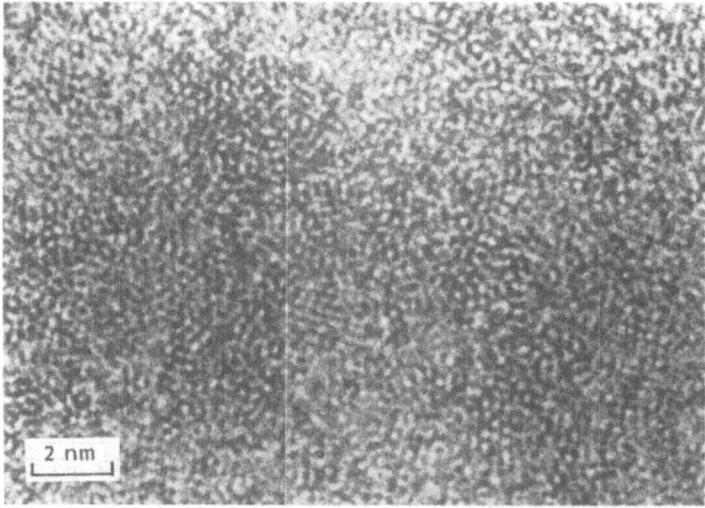

Fig. 3.21. High-resolution image of an as-quenched amorphous $Fe_{86}B_{14}$ alloy thin film showing local ordered regions [3.54]

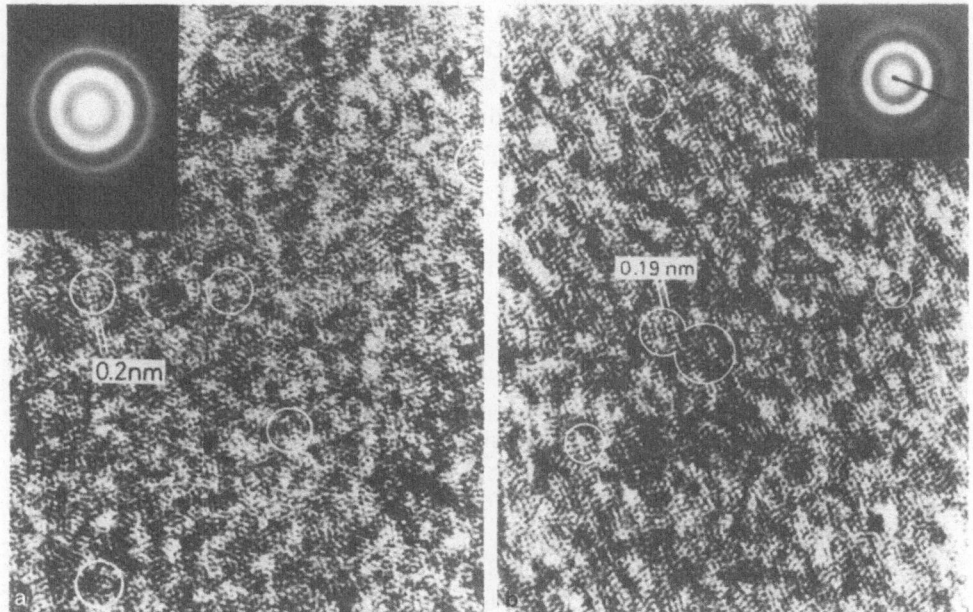

Fig. 3.22. High-resolution images of as-quenched amorphous $Fe_{84}B_{16}$ alloy (a) and $Pd_{77.5}Cu_6Si_{16.5}$ alloy (b) [3.55]

$\langle 100 \rangle$ of the fcc structure in (b). Lattice-like images with five-fold symmetry were not found at all. This means that the icosahedral cluster of thirteen atoms, which has been supposed by some investigators to be a fundamental unit of the amorphous structure, does not appear frequently in the real amorphous alloys or that the icosahedral atomic arrangement is not suitable to produce the lattice-like image.

When the quenching rate from the melt is not fast enough, the lattice-like or fringy regular patterns are larger and more frequent, sometimes almost covering the whole area observed under the electron microscope. This is nothing but the state of supersaturated crystalline aggregates, and naturally the diffraction pattern from the limited area contains spotty reflections. When a very rapidly quenched specimen is annealed at a fairly low temperature, say 100 K below the crystallization temperature, corresponding to the range of structural relaxation, the lattice-like or fringy images spread wider and appear more frequently. This strongly suggests that the structural relaxation is not always a small potential relaxation but also associated with the atomic rearrangements to increase the degree of crystalline order in the random structure [3.56].

c) Calculation of High-Resolution Electron Images from Amorphous Structure

A model calculation is necessary for the interpretation of the high-resolution electron micrographs of amorphous materials as well as for the analysis of crystalline materials. The kinematical approximation is valid for the calculation of electron images from very thin amorphous films, as described by *Howie* and coworkers [3.57,58]. The multislice method introduced in Sect. 3.2.3 is not appropriate for image simulation, since the atom positions, especially their vertical coordinates (z_j) are random, spoiling the basis of selection of the thickness of slices, Δz, and, therefore, the dynamical diffraction effect is difficult to take into account.

The total scattered wave amplitude at a point $\boldsymbol{r}$ is, in single scattering, given by

$$\psi_s(\boldsymbol{r}) = \frac{i\lambda A_q}{4\pi^2} \sum_j \sum_q f_j(q) \, \exp\left[2\pi i\boldsymbol{q}\cdot(\boldsymbol{r}-\boldsymbol{r}_j) - i\chi(\alpha)\right], \qquad (3.42)$$

which is a modification of (3.1), adding the phase transfer function of (3.6) or the exponential of the Scherzer phase shift of (3.21), χ, the summation over q on a two-dimensional grid of points within the objective aperture, and the unit area of this grid, A_q. For the bright field images, the scattered waves interfere with the transmitted waves and give rise to the image-intensity distribution $I = |1 + \psi_s|^2$, as mentioned in Sects. 3.1.2 and 3.3.1. In the case of amorphous materials, instead of sharp Bragg reflections, the blurred scattering which makes halo rings could be regarded as a large number of reflected beams and still be treated by the many-beam structure imaging technique. In view of this technique it is desirable to illuminate axially and to take the halo diffraction concentrically in the aperture with the incident, as actually employed in the before-mentioned experiments.

For the image simulation for the $Fe_{86}B_{14}$ amorphous alloy, two model assemblies are constructed using a Morse-type atomic potential; one is a dense random packing (d.r.p.) sphere of 2 nm in diameter containing 334 atoms and the other contains a medium-range order (m.r.o.) cluster of bcc structure in its centre but is otherwise the same as the former [3.59]. The S(q) and g(r) functions calculated from the two models, especially the latter, agree well with the experimentally obtained diffraction functions [3.60]. For the calculation of the phase transfer function, $C_s = 1.0$ mm and $\lambda = 0.00251$ nm have been used. The $\sin\chi$ functions calculated with the defocus values of −50 nm and −150 nm are shown by the broken and the solid lines in Fig. 3.23, respectively. The former corresponds to the Scherzer defocus in (3.8), but does not match the q values of the first halo ring represented by a line segment in the lowest part of the figure, while the latter has values of almost −1 at the halo position, showing the best contrast condition. The effect of the defocus condition on the quality of the structure image is shown in Figs. 3.24 and 25: Fifteen core atoms of a bcc (m.r.o.) cluster are displayed by the projected nine broken circles, and the calculated image contrast distribution is represented by the solid (above the level of $I = 1$) and dotted (below $I = 1$) contour lines. Figure 3.24 shows that, when $\Delta f = -50$ nm is used, noticeable fringes appear in the outer part, revealing that the lattice fringe image is displaced outside due to the mismatched spherical aberration and defocus distance. When $\Delta f = -150$ nm, a high-contrast regular

58

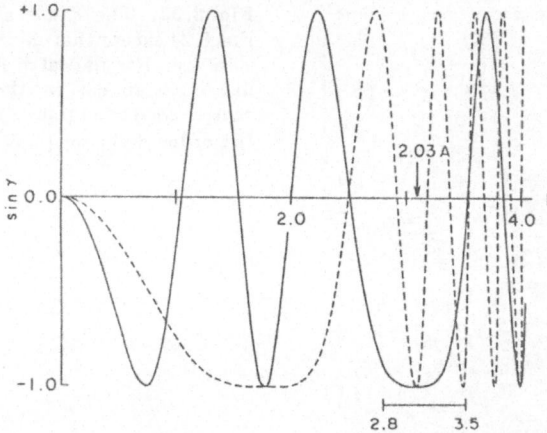

Fig. 3.23. Sinχ functions used for the image simulation of Fe-B amorphous structure containing a medium range order. $C_S = 1.0$ mm, $\lambda = 0.00251$ nm, and Δf is -50 nm (broken line) and -150 nm (full line). $K = 2\pi q$

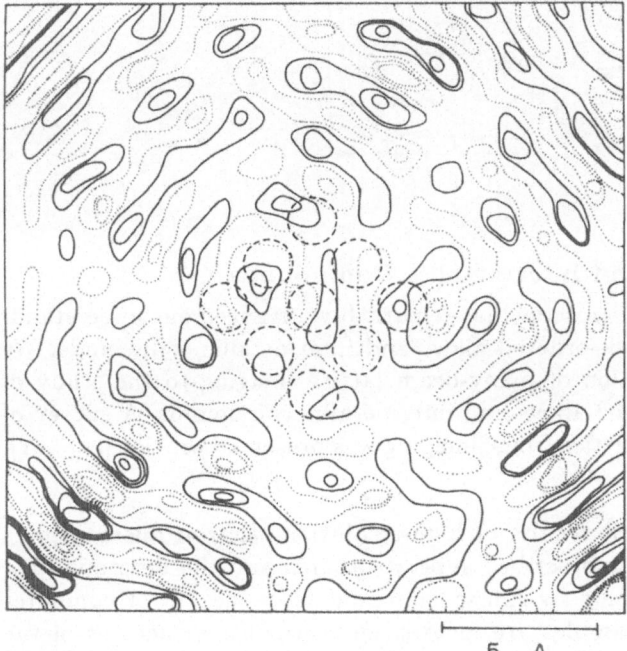

Fig. 3.24. A simulated high-resolution contrast image map for Fe-B amorphous structure containing a bcc type medium-range order cluster in the centre. Sinχ function with $\Delta f = -50$ nm in Fig. 3.23 is used. The cluster is shown by the projected 9 broken circles, and the contour lines represent the contrasts above (full lines) and below (dotted lines) the level, I $=$ 1, respectively. Note that lattice-like fringes appear in the outer part of the figure

arrangement of dots rises in the central part, corresponding to the structure of the m.r.o. cluster, as Fig. 3.25 shows.

The above result of the high-resolution image calculation satisfactorily interprets the before-mentioned dotty and fringy image contrast in the high-resolution electron micrographs of Figs. 3.21 and 22. It is, therefore, concluded that the medium-range order or the quasicrystalline clusters of some tens of atoms really exist in the as-quenched amorphous metals, as a thermodynamical theory predicts [3.56].

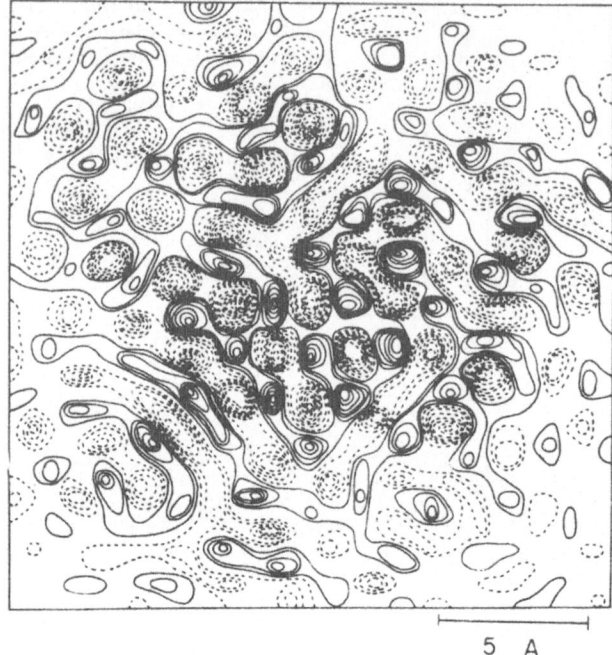

Fig. 3.25. The same as Fig. 3.24 except that $\Delta f = -150$ nm. Regular lattice-like dots appear in the centre, corresponding to the ordered cluster [3.59]

5 A

3.3.3 Ordered Alloys and Intermetallic Compounds

High-resolution electron microscopy has remarkably advanced our understanding of the real structures of ordered alloys and intermetallic compounds. We describe here the application of many-beam lattice imaging to the study of atomic structures in ordered alloys and intermetallic compounds. A survey of the early studies with the lattice fringe imaging was presented previously [3.61].

a) Ordered Alloys

One of the most significant results of high-resolution electron microscopy is the detection of hitherto undescribed superstructures and the determination of their atomic arrangements by image matching with the aid of computer simulations. Particular examples are long-period antiphase structures of the Au-based alloys, as described below.

In the Au-Mn system near 25 at.%Mn, the ordered structures are very sensitive to composition and heat-treatment. Varieties of incommensurate superstructures have been proposed from x-ray and electron diffraction studies, but the atomic configurations in the structures at 20–28 at.%Mn have been left unresolved so far. High-resolution studies with the use of a 1 MV electron microscope revealed two new superstructures $Au_{31}Mn_9$ and $Au_{22}Mn_6$ having the basic fcc structure. An experimental many-beam image of $Au_{31}Mn_9$ is presented in Fig. 3.26 [3.62]. If bright dots in the image correspond to Mn-atom rows projected down parallel to the incident beam, we can readily derive the $Au_{31}Mn_9$ superstructure shown in Fig. 3.27a. This structure is described as

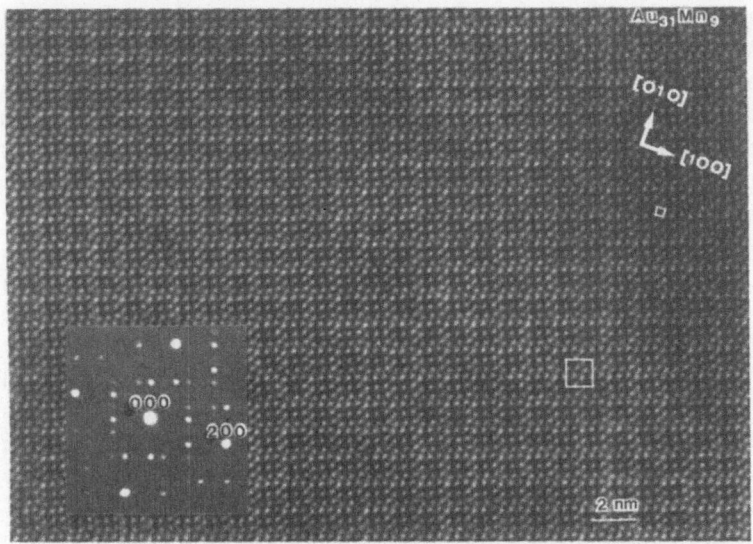

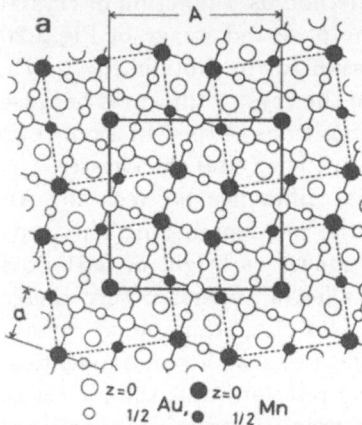

Fig. 3.26. Many-beam image of 2d-APS of $Au_{31}Mn_9$, recorded at 1 MV [3.62]. Large and small squares indicate unit cells of the superstructure and the basic fcc structure, respectively. A selected-area electron diffraction pattern is inserted

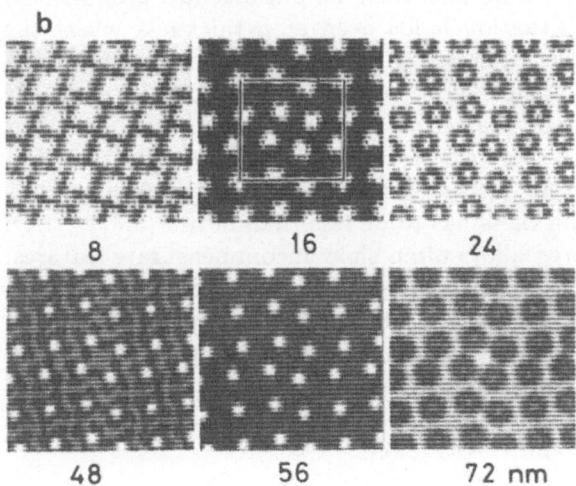

Fig. 3.27. (a) Projected model of 2d-APS of $Au_{31}Mn_9$. $A = \sqrt{10}\,a = 1.26\,nm$, where $a = 0.40\,nm$, the lattice constant of the basic fcc cell. (b) Simulated images for specimen thicknesses from 8 to 72 nm at the Scherzer defocus $\Delta f = 110\,nm$ at 1 MV

a two-dimensional anti-phase structure (2d-APS) composed of square-shaped islands of the Au_4Mn structure of the $Ni_4Mo(D1a)$ type which contains 3×3 rows of Mn atoms. The $Au_{22}Mn_6$ superstructure which is a 1d-APS based on the Au_4Mn structure was proposed by a similar method [3.63]. High-resolution electron microscopy of the long period APSs of the Au-Mn alloys was described in detail by *Watanabe* and *Terasaki* [3.64], and *Amelinckx* et al. [3.65].

The determination of APSs which incorporate features derived from basic superstructures has also been demonstrated for the Pt-rich Pt-Ti alloy [3.65]. Two intermediate superstructures $Pt_{13}Ti_2$ and $Pt_{21}Ti_4$ were found in between the known structures Pt_8Ti and Pt_3Ti in a quite analogous situation to the case of the Au-Mn system mentioned above.

The structure determination of ordered alloys by high-resolution electron microscopy is based on the one-to-one correspondence between the bright dots in the observed image and the projected positions of minority atom-rows in the crystal (minority atom-row imaging). The correspondence is confirmed by fitting the observation with computer simulation on the basis of the dynamical-scattering theory. The images shown in Fig. 3.27b are simulations for the $Au_{31}Mn_9$ structure obtained by the multislice method as a function of crystal thickness at the Scherzer focus. We see that the observed image of Fig. 3.26 agrees well with a computed image for a thickness of 16 nm. Note in Fig. 3.27b, however, that the image contrasts are sensitive to the crystal thickness because of the dynamical-scattering effect; the projected positions of the Mn atom-rows appear as either bright or dark dots depending on the crystal thicknesses.

The image of Fig. 3.26 reflects the potential difference between the Au and Mn atom-rows projected down parallel to the incident beam. This type of image, in which the minority-atom positions appear as bright or dark dots depending on the experimental conditions, is called a superstructure image [3.66]. The superstructure image of an ordered alloy is obtainable at certain specimen thicknesses when some superlattice reflections with almost identical spatial frequencies (d_{hkl}^{-1}) have nearly kinematical relationships; the phases of these reflections are almost the same and their amplitudes are proportional to the structure factors. The situation holds even at a thickness where the transmitted beam is extinct very much due to the dynamical-scattering effect. This condition is retained at rather thick foils when the ordered alloy has a smaller difference in the atomic scattering factors of the constituents. For 1 MV electron microscopy, the superstructure image of Au_3Cd with the DO_{23} type structure is obtained even for a specimen as thick as about 60 nm, while for the Au_3Mg specimen the thickness has to be less than 30 nm [3.67].

Long period APSs of ordered alloys often show incommensurate features; the antiphase domain sizes (M) which are usually estimated from the split superlattice reflections in x-ray or electron diffraction patterns are generally non-integral multiples of the fundamental cell. The fact that the non-integral or incommensurate value of M in a 1d-APS arises from a uniform mixture of domains with different commensurate sizes has been revealed by the lattice fringe images on Ag_3Mg [3.68] and Cu-Al [3.69].

For an incommensurate 2d-APS, however, it is generally impossible to construct a model of the real structure from the lattice fringe image. The inherent limitation is overcome with the use of many-beam imaging. Actually the domain configurations in the incommensurate 2d-APS of Au-based alloys are revealed by high-voltage, high-resolution electron microscopy [3.70,71]. In an Au-30 at.%Cd alloy, for example, a number of domains with the commensurate hexagonal close packed superstructure are arranged in such a characteristic way that the incommensurate periods are produced in three principal directions in the close packed plane.

For an ideally ordered alloy, the superstructure image exhibits a regular alignment of minority atom-rows projected down parallel to the incident beam, as seen in Fig. 3.26. When an A_3B alloy is not perfectly ordered, certain fractions of the lattice sites for minority atoms B are replaced by major atoms A. In such cases, the distribution of bright dots is not perfectly periodic, and the brightness of dots differs from site to site. An example is presented in Fig. 3.28. A multislice calculation for the Au_3Mn alloy indicates that the brightness of dots is roughly proportional to the replacement fraction of Mn atoms by Au atoms, if the necessary kinematical conditions are satisfied [3.72].

Initial stages of ordering processes in alloys are of interest for high-resolution electron microscopy [3.73]. In the short-range ordered stage of Ni_4Mo, diffuse scattering intensities exhibit peaks at $1\,^1/_2\,0$-type positions which do not coincide with the superlattice reflections. Two basic interpretations, the microdomain model [3.74] and the static concentration wave-packet model [3.75], have been proposed for the controversial problem. Future work using high-resolution electron microscopy is promising to resolve the unsettled issues of the short-range ordered structures.

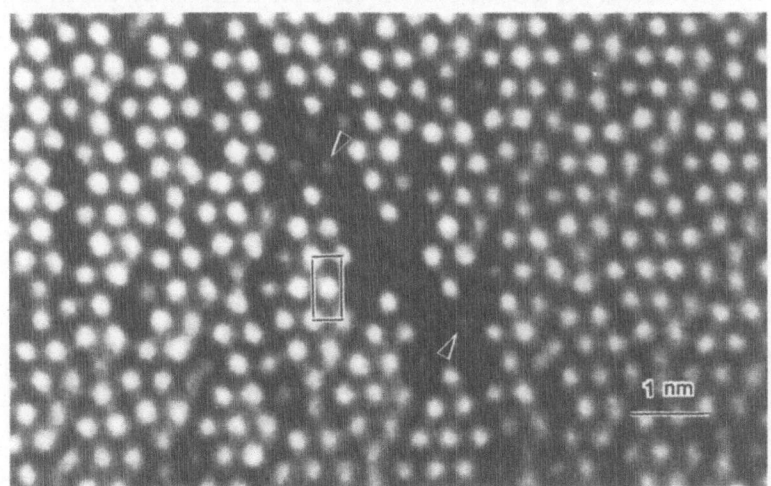

Fig. 3.28. Many-beam image of Au-22.6% Mn alloy, showing partially disordered atomic arrangement, recorded at 1 MV [3.72]. Arrows indicate dim dots. A rectangle corresponds to a unit cell of the DO_{22} superstructure

b) Intermetallic Compounds

The many-beam image investigations of intermetallic compounds have commenced with the σ, μ, M and P phases in Fe-Mo [3.76,77], Fe-Cr [3.78] and Mo-Co [3.79] systems. These phases have the so-called tetrahedrally close-packed (TCP) structures, which consist of polyhedra with the coordination number 14, 15 and 16. These TCP phases are of practical importance, because their occurrence in alloys causes severe embrittlement of the materials. By resolving a group of atoms forming a characteristic coordination polyhedron, microtwins, stacking faults and intergrowth interfaces are revealed in the sub-cell scale.

Structure defects and modulations in TCP compounds were investigated on the σ-phase formed in a duplex high-chromium stainless steel by *Southwick* and *Stobbs* [3.80], on the Laves phase $Mg(Cu_{0.1}Zn_{0.9})_2$ by *Takeda* et al. [3.81], and on precipitates in a Ni-based superalloy by *Kuo* and his colleagues [3.82].

Intermetallic compounds consisting of rare-earth and transition metals are also of practical importance as ferromagnetic materials. *Komura* and his colleagues [3.83] investigated intergrowth and stacking faults in Sm_2Ni_7 and Sm_5Ni_{19}, and found the occurrence of new polytypic structures 5T, 7T, 15R and 18R of Sm_5Ni_{19}. A Sm-Co based permanent magnet was studied by *Hiraga* et al. [3.84]. Interface habits between Sm_2Co_{17} and $SmCo_5$ in the magnet with a high coercivity are coherent, but accompanied by lattice distortion, as seen in Fig. 3.29. This image suggests strongly that the distorted interfaces act as pinning sites for magnetic domain walls.

Using a 1 MV electron microscope, *Kuwano* et al. [3.85] studied the crystal structure of $CePd_5$, and found two polytypes and a long-range modulation of the stacking structure.

Superconducting intermetallic compounds Nb_3Sn and V_3Si of the A15 type structure are current objects of high-resolution electron microscopy [3.86–89].

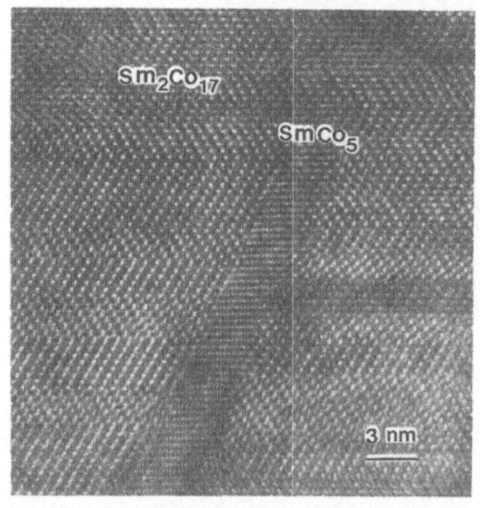

Fig. 3.29. Many-beam image of Sm-Co based permanent magnet, showing end-on view of coherent interface habit between Sm_2Co_{17} and $SmCo_5$, and frequent occurrence of stacking faults in Sm_2Co_{17} [3.84]

Linear chains of transition-metal atoms lying along the cubic axes are observed as a square mesh in a many-beam image with the incidence parallel to the third principal axis.

Recently *Zhang* et al. [3.90], and *Ye* and *Amenlinckx* [3.91] have investigated the arrangements of Ti atoms in the intermetallic compounds Cu_4Ti and Cu_8Ti_5 by taking superstructure images.

Apart from the category of "intermetallic" interstitial compounds such as sub-nitrides, oxides and carbides like Ti_3O [3.92], V_8N [3.93], Cu_4O [3.94], Cu_8O [3.95] and Fe_3C [3.96] are also targets for many-beam imaging. With the aid of image simulations by the multislice method, the atomic structures of these compounds were elucidated. In image matching between observations and simulations, one should be careful as the image contrast is sensitive to local relaxation of the host-metal atoms surrounding the interstitial atoms [3.93].

3.3.4 Phase Transformation

a) Guinier-Preston Zone Formation and Precipitation

The sequences in the precipitation process through the Guinier-Preston (G.P.) zone formation from a saturated solid solution of Al-Cu alloy have been thoroughly studied by many investigators with conventional bright-field images, weak-beam images and two-beam lattice images. As mentioned before, atom resolution images of G.P.(I) and G.P.(II) zones in an Al-4 at.%Cu alloy were observed, as in Fig. 3.18, under the condition that the defocus of the objective lens was adjusted so as to pass principal diffracted beams as 020, 200 and 220 through the pass band of $\chi = \pi/2$ in the oscillating PCTF curve [3.43,97,98]. Under this condition, which is the before-mentioned aberration-free focus (AFF) condition [3.17], the observed images of G.P.(I) and G.P.(II) zones are interpreted in comparison with simulations by the multislice method based on proposed structure models; G.P.(I) zones consisting of a single Cu layer and G.P.(II) zones having two Cu-rich layers separated by three, sometimes two, Al layers. A few zones are found to have atypical structures, i.e. multi-layered or shear-displaced [3.97]. The interfaces between the θ' (Al_2Cu) phase and the Al-rich fcc matrix were investigated with the edge-on incidence by *Wu* [3.99].

The subject of precipitation processes of carbide from martensitic Fe-C steel has been investigated by metallurgists for more than half a century. *Nagakura* and his coworkers [3.100] carried out a series of high-resolution TEM observations on the tempering stages of Fe-C martensite, and found microsyntactic intergrowth of θ-Fe_3C and χ-Fe_5C_2 in the precipitate formed at 400°C (the third stage of tempering). At a very early stage of tempering below 100°C, they proposed a structural modulation; carbon atom clusters are formed randomly on planes nearly parallel to (102), and the carbon-rich planar regions are spaced regularly at an interval of about 1 to 2 nm. In addition, static displacements of iron atoms were proposed to be associated with the concentration fluctuation of the carbon atoms.

b) Martensitic Transformation

Direct observations of internal twins and interfaces between austenite and martensite phases are particularly interesting in connection with the shape-memory behaviour associated with martensitic transformations of alloys. Lattice fringe imaging with tilted-beam illumination was utilized for the study of NiTi [3.101] and Cu-Zn-Si [3.102] alloys.

Knowles [3.103] investigated the structure of twin interfaces in the martensite phase of NiTi by taking two-dimensional lattice images with axial illumination in the edge-on direction, and presented evidence for the step and ledge structure of twin interface. The ledges are the nearest low-index rational plane to the irrational twin plane. Further, *Knowles* et al. [3.104] observed the step structure in an Au-Cd alloy.

The high-resolution imaging conditions for the stacking sequence in the β' martensite (18R) of Cu-Al alloys were studied by *Lovey* et al. [3.105] by comparison with computer simulations. The 18R martensite in a Cu-Zn-Al alloy was investigated by *Cook* et al. [3.106] using a high-voltage, high-resolution microscope.

Premartensitic phenomena are known to be characterized with the so-called "tweed" or "mottled" pattern in the CTEM images; fine striations, typical of a few nm periodicity, lie parallel to {110} traces of the parent cubic phases. *Tanner* et al. [3.107] has listed martensitic systems exhibiting such patterns, and interpreted them as due to the occurrence of incommensurate shear strains of $\{110\}\langle 1\bar{1}0\rangle$ type related to the concurrent temperature dependent softening of the parent lattice; these cubic phases in many cases experience an anomalous decrease of the elastic constants $(c_{11} - c_{12})/2$ with decreasing temperature approaching the martensite starting temperature M_s. Recently the tweed patterns of Fe-Pd alloys have been attributed to be microscopic embryos in the precursor stage of fcc-fct martensitic transition [3.108]. Similar tweed patterns are also observed in an A 15 type superconducting compound V_3Si near 20 K [3.86]. Atom resolution images of the tweed patterns would be valuable for understanding the basic nature of premartensitic phenomena.

3.3.5 Surface, Grain Boundary and Interface

a) Surface

Electron microscopy of surface structures is a recent revival, using ultra-high vacuum (UHV) transmission and reflection imaging. In the 1970's, direct observations of surfaces at a monatom-high level were elaborated by CTEM of metal films deposited on a thin substrate in vacuum [3.109]. In-situ clean surface preparation in the UHV specimen chamber of an electron microscope has allowed the observation of monatom-high terraces on close-packed plane surfaces of metal films. Since a thin (111) metal surface has a thickness of either 3N or 3N±1 atom layers of close-packed planes (N is an integer), two contrast levels are observed by a dark-field TEM corresponding to the two kinds of surface-terrace image of a (111) metal surface [3.110]. Even if the spatial resolution in dark-field images is not so high, the boundary of different contrasts in the observed images indicates monatom-high steps of the terraces.

Surface images of monatom-high steps of noble metals were obtained also by dark-field reflection electron microscopy using one of the diffraction spots in reflection high-energy electron diffraction (RHEED) [3.111]. Edges of the surface terraces are seen as wavy lines in the images, of which the high-resolution possibilities refer only to the directions perpendicular to the incident beam. Surface-reflection electron microscopy and diffraction studies were reviewed by *Cowley* [3.112].

A marked progress towards high-resolution reflection imaging of clean surfaces has been made by *Yagi* and his colleagues using an UHV in-situ specimen chamber [3.113]. Clean well-defined (001) and (111) surfaces of Ag crystallites deposited on thin graphite films in the UHV electron microscope were observed at normal incidence. In an image of an Ag (111) platelet, hexagonally arranged bright dots with a spacing of 0.28 nm are seen in one part, while they are obscure in the other part. The two parts are, respectively, the regions of the thicknesses of 3N and 3N±1 atom layers mentioned above, and the bright dots correspond to atomic images of one surface layer.

The above results are obtained in the transmission mode with the incident beam perpendicular to thin films. An alternative method for studying the atomic-scale morphology of metal surfaces is the transmission observation of ultrafine particles which are tilted to bring their atom rows in the surfaces parallel to the incident beam direction, and hence the surface planes are imaged as their edge-on projections. Using a 600 kV high-resolution electron microscope, *Marks* and *Smith* [3.114] observed the surface morphology of small particles of Ag and Au with the [110] incidence. Well-developed facetting, rafts and ledges are revealed at the atomic level. Furthermore, the relaxation of atom alignment in the surface layer is observed in the absence of carbon contamination. By comparison with image simulations, it is estimated that the outermost gold atoms are relaxed outwards by about 20%.

Surface-profile imaging of fine particles is a highly promising technique for observing surface morphology on the atomic scale. In fact, dynamic changes of structure defects on the (100) surface of Au particles were recorded by *Smith* and *Marks* [3.115]; a series of successive images indicated the surface diffusion by the atom movement. Current literature on high-resolution surface microscopy has been presented in their paper.

Recently *Iijima* and *Ichihashi* [3.116] observed random walk of surface atoms on an ultrafine Au particle with a diameter of less than 10 nm using a VTR-TV monitor system at a resolution of about 0.23 nm. They could ascertain from the real-time VTR recording that the Au particle containing about 940 atoms has the shape of a cuboctahedron with well-defined crystal habits of {100} and {111} planes, and the motion of surface atomic steps was observed under electron beam illumination.

b) Grain Boundary and Interface

Commonly, CTEM images have been utilized to examine interfacial dislocations which form a network in grain boundaries and interphase interfaces [3.117]. The usefulness of the lattice-imaging technique in the analysis of the atomic configuration in grain boundaries and interfaces has been demonstrated by many investigators [3.118].

Tilted grain boundaries in evaporated Au films were studied by many-beam imaging with edge-on incidence by *Ishida* and his colleagues [3.109]. The atomic structures at the three types of boundaries, i.e., a coherent twin, an incoherent twin and a $\Sigma 11$ coincidence boundary, were analyzed by systematic comparison with the multislice computations based on reliable interatomic model potentials. They concluded that the configurational identification of grain boundaries is achieved if the alternate atomic structures differ significantly and if a set of micrographs at specified defocus values and specimen orientations is available experimentally.

Edge-on observations of twin boundaries were carried out with the axial illumination of multiply-twinned particles (MTPs) which are normally formed in the early stages of the particle growth of fcc metals [3.120]. The MTPs usually have the decahedral and icosahedral forms consisting of five and twenty tetrahedra, respectively, with twinning on their (111) planes. In MTPs of Au and Ag, strain-relieving partial dislocations are observed at the twin boundaries [3.121]; the tetrahedra with a perfect fcc structure cannot be arranged to be completely space-filling.

Teuho et al. [3.122] found that the twin-like boundaries in an ordered Au_3Zn alloy which appear parallel to {110} planes in CTEM images consist of multiple steps lying alternately along (100) and (010) planes. The atomic configuration at the steps of the unit-cell size was proposed directly from the edge-on images. The usefulness of the edge-on observation of interfacial boundaries was also demonstrated by *Hiraga* et al. [3.123] on a sintered Fe-Nd-B permanent magnet. This attracts the current interest of researchers because of its very high energy product $(BH)_{max}$. It was revealed that the grains of the ferromagnetic compound $Fe_{14}Nd_2B$ are covered with thin Fe-rich layers, and the atomic structure of interphase interfaces between these layers plays a dominant role in obtaining the high magnetic coercivity.

Because of its technological importance, high-resolution electron microscopy was utilized to observe the atomic structure of the transition region between the metal substrate and its surface oxide film. As for the Al_2O_3 surface film on an Al-based industrial alloy, *Timsit* et al. [3.124] found that the interface with the (110) and (111) Al surface planes is slightly rough on the atomic scale but is otherwise stoichiometrically abrupt. The localized roughness was attributed to the presence of atomic ledges on the metal surface.

Finally we mention a new application of high-resolution many-beam imaging in the field of high-performance semiconductor devices and artificially prepared periodic superlattice structures, which are usually prepared by molecular beam epitaxy (MBE) deposition, metalorganic chemical vapor deposition

(MOCVD) and others. Many-beam imaging is potentially the most powerful and direct method for examining whether or not the interfaces of alternate ultra-thin layers are atomically abrupt and free of islands in the monatomic layer. It is anticipated that cross-sectional structure images should be readily applicable to a wide-range study of interfaces in "superlattices" like GaAs/AlAs [3.125–127]. As for metal multilayers [3.127], the applicability of high-resolution images has been discussed by *Baxter* and *Stobbs* [3.128].

3.4 Outlook

By reading the text and observing the pictures of this chapter, the readers may have the impression that fifty years' accumulation of the science and technology of electron microscopy has built up a new field of useful and refined technique, that of high-resolution electron microscopy, which enables us to observe directly in favourable cases the two-dimensional projection of atomic arrangement in crystal structures. Many more examples of crystal-structure images of semi-conductors, ceramics, minerals and other chemical compounds could be shown herewith, if the subject was not confined to the problems of metals and alloys. Although further improvements in the resolution of an electron microscope and in the theory of image formation, in addition to careful comparison between the observed electron image and the computer simulated image, are necessary, one could already predict the next stage of development in high-resolution electron microscopy for at least three new problems.

The first problem introduced in the following was already suggested by *Hashimoto* et al. [3.129] as visualized in the electron image of Au atoms in Fig. 3.30. The enlarged image of each atom (row) is not a spot with uniform

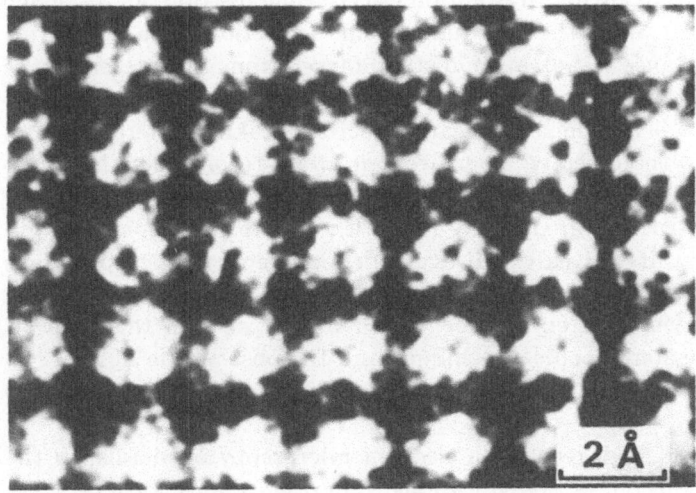

Fig. 3.30. Enlarged structure image of Au thin film [3.49]

contrast but has a hollow in the centre, suggesting that the scattering of electrons by outer-core electrons of each atom could produce the ring image reflecting the electronic structure of the atom. Such an image would be called the atomic-structure image or the electronic-structure image of an atom. In order to examine this problem further, it is necessary to take many aberration-free atom images at different acceleration voltages, including the high voltage which can partially produce the inner core and nuclear scattering, and to analyze them by the theory of image formation, which takes into account the electronic configuration of the atom and accordingly the atomic scattering amplitude. The electron images of isolated heavy atoms were already successfully obtained by some investigators [3.130–133], but, for the study of the atomic-structure image, the use of a thin metallic crystal seems to be more advantageous because of better utilization of the diffraction theory, as is readily seen in the preceding sections.

The second problem, the chemical analysis of impurity atoms in the metallic lattice, could be solved when the above-mentioned analysis of the atomic-structure image is achieved. However, in situ identification of foreign atoms in a crystal would be more easily realized by the parallel use of high-resolution electron microscopy and very local atomic excitation to measure the characteristic absorption or emission of the atom or atoms in the problem than with atomic-structure image analysis. If the identification of the positions of foreign atoms in a crystal and their chemical analysis, for instance, of Sn or Bi atoms segregated at the grain boundary or the dislocation in the Fe lattice, were obtained, it would be an extremely useful technique in various fields of metallurgy related to a very small amount of impurities. High-resolution scanning electron microscopy [3.134], which is not treated in the present chapter, seems to be applicable to this problem.

The third problem, we wish to point out, is the three-dimensional structure analysis of materials by high-resolution electron microscopy. As discussed in the last part of Sect. 3.3.3, the two-dimensional projection of an atomic arrangement in a substance is not a far distant target from the present stage of development. Therefore, if the aberration-free structure images of a crystal were taken with different incident-beam directions, taking different sets of Bragg reflections into the objective aperture, one could reconstruct the three-dimensional atomic arrangement in the crystal. In principle, this is quite the same as the crystal structure analysis by the diffraction technique, since high-resolution microscopy first gives the Bragg reflections and then reconstructs the structure image from them by an inverse Fourier transform automatically accomplished in the lense system, as mentioned in Sect. 3.1. In this respect, the application of the stereoscopical technique and the observation of noncrystalline materials mentioned before seem to be especially interesting as the first step in this problem.

More extensions of high-resolution electron microscopy of metals, to the analyses of atomic diffusion, lattice vibrations, spin orientations, surface irregularities and other phenomena seem to be possible, but to realize these, further improvement of the resolving power of the electron microscope, of the theory and technique of image formation and, in addition, the employment of different methods, such as the electron-energy selection, would be necessary.

To see both matter and phenomenon directly is always an indispensably important element in physics, and in the world of atoms high-resolution electron microscopy is realizing this on a finer and finer scale, even nearly reaching that region where the uncertainty principle emerges.

References

3.1   M. Knoll, E. Ruska: Z. Physik **78**, 318 (1932)
3.2   E. Ruska: Z. Physik **87**, 580 (1934)
3.3   B. von Borries, E. Ruska: Siemens-Werk **17**, 99 (1938)
3.4   R. Castaing: Comptes Rendus Acad. Sci. **22**, 1341 (1949); Nature **165**, 390 (1950); R. Castaing, P. Laborie: Comptes Rendus **237**, 1330 (1953)
3.5   W. Bollmann: Phys. Rev. **103**, 1588 (1956)
3.6   P.B. Hirsch, R.W. Horne, M.J. Whelan: Phil. Mag. **1**, 677 (1956)
3.7   P.B. Hirsch, A. Howie, M.J. Whelan: Phil. Trans. Roy. Soc. (London) A **252**, 499 (1960)
       D.J.H. Cockayne, I.L.F. Ray, M.J. Whelan: Phil. Mag. **20**, 1265 (1969)
3.8   Z.G. Pinsker: *Dynamical Scattering of X-Rays in Crystals*, Springer Ser. Solid-State Sci., Vol. 3 (Springer, Berlin, Heidelberg 1978)
       Sh.-L. Chang: *Multiple Diffraction of X-Rays in Crystals*, Springer Ser. Solid-State Sci., Vol. 50 (Springer, Berlin, Heidelberg 1984)
3.9   A. Howie, M.J. Whelan: Proc. Roy. Soc. A **267**, 206 (1962)
3.10   J.W. Menter: Proc. Roy. Soc. **236**, 119 (1956)
3.11   T. Komoda: Optik **21**, 93 (1964)
3.12   O. Scherzer: J. Appl. Phys.**20**, 20 (1949)
3.13   A.V. Crewe, J. Wall, J. Langmore: Microscopie Electronique **1**, 485 (1970)
3.14   H. Hashimoto, A.K. Kumao, K. Hino, H. Yotsumoto, A. Ono: Jpn. J. Appl. Phys. **10**, 1115 (1971)
3.15   R.M. Henkelman, F.P. Ottensmayer: Proc. Nat. Acad. Sci. (USA) **68**, 3000 (1971)
3.16   H. Formanech, M. Muller, M.H. Hann, T. Koller: Naturwiss. **58**, 339 (1971)
3.17   H. Hashimoto, H. Endo, Y. Takai, H. Tomioka, Y. Yokota: Chemica Scripta **14**, 23 (1978/79)
3.18   H. Hashimoto, Y. Yokota, H. Ende, A. Kumao: Chemica Scripta **14**, 125 (1978/79)
3.19   L. Reimer: *Transmission Electron Microscopy*, Springer Ser. Opt. Sci. Vol. 36 (Springer, Berlin, Heidelberg 1984)
3.20   H. Hashimoto, M. Mannami, T. Naiki: Phil. Trans. R. Soc. A **253**, 459 (1961)
3.21   D.J.H. Cockayne, R. Gronsky: Phil. Mag. **44**, 159 (1981)
3.22   T. Komoda: Jpn. J. Appl. Phys. **5**, 603 (1966)
3.23   S. Miyake, K. Fujiwara, M. Tokonami, F. Fujimoto: Jpn. J. Appl. Phys. **3**, 276 (1964)
3.24   J.M. Cowley: *Diffraction Physics*, 2nd ed. (North-Holland, Amsterdam 1984)
3.25   J.C.H. Spence: *Experimental High-Resolution Electron Microscopy of Materials* (Clarendon, Oxford 1981)
3.26   J .Frank: Optik **38**, 519 (1973)
3.27   P.J. Frejes: Acta Cryst. A **33**, 109 (1977)
3.28   J.M. Cowley, A.F. Moodie: Acta Cryst. **10**, 609 (1957)
3.29   J.A. Ibers, W.C. Hamilton (eds.): *International Tables for X-Ray Crystallography*, Vol. IV (Kynoch Press, Birmingham 1974)
3.30   H.J. Nussbaumer: *Fast Fourier Transform and Convolation Algorithms*, 2nd ed., Springer Ser. Inform. Sci., Vol. 2 (Springer, Berlin, Heidelberg 1982)
3.31   K. Ishizuka, N. Ueda: Acta Cryst. A **33**, 740 (1977)

3.32   M. Kiritani: Reports on Radiation Damage of Materials with D-T Fusion Neutrons from RTNS-II (Japan-US Fusion Cooperation Program, 1984); J. Nucl. Mat.
133/134, 395 (1985)

3.33   H. Hashimoto, Y. Takai: Bull. Jpn. Inst. Metals **22**, 595 (1983);
Y. Takai, H. Hashimoto, H. Endo: Acta Cryst. **A39**, 516 (1983)

3.34   H. Hashimoto, Y. Takai, N. Ajika, Y. Yokota, H. Endo: In *Point Defects Interaction
in Metals*, ed. by J. Takamura, M. Doyama and M. Kiritani (Univ. Tokyo Press,
Tokyo 1982) p. 698

3.35   H. Hashimoto, Y. Takai, Y. Yokota, H. Endo: Jpn J. Appl. Phys. **19**, L1 (1980)

3.36   Y. Takai, N.D. Zakharov, H. Hashimoto: Proc. 10th Intern. Congr. Elect. Micros.,
Hamburg (1982) p. 375
H. Endo, H. Hashimoto, Y. Takai: Trans. JIM **24**, 307 (1983)

3.37   Y. Kitano, M. Takata, Y. Komura: *Proc. Yamada Conf. on Dislocations in Solids*,
Tokyo (1984), (Univ. Tokyo Press, Tokyo 1985) p.521

3.38   Y. Kitano, Y. Komura, H. Kajiwara, E. Watanabe: Acta Cryst. A **36**, 16 (1980)

3.39   Y. Komura: Bull. Jpn. Inst. Metals **22**, 626 (1983) (in Japanese)

3.40   A. Kelly, R.B. Nicolson: *Prog. Mater. Sci.* **10**, 151 (Pergamon, London 1963)

3.41   R.B. Nicolson, G. Thomas, J. Nutting: J. Inst. Metals **87**, 429 (1958/59)

3.42   H. Yoshida, D.J.H. Cockayne, M.J. Whelan: *Electron Microscopy*, Vol. 1, (Australian
Academy Sci. 1974) p. 430; Phil. Mag. **34**, 89 (1976)

3.43   H. Yoshida, H. Hashimoto, Y. Yokota, N. Ajika: Trans. Jpn. Inst. Met. **24**, 378 (1983)

3.44   X. Auvray, P. Georgopoulos, J.B. Cohen: Acta Met. **29**, 1061 (1981);
J.B. Cohen, P. Georgopoulos: Scripta Met. **16**, 1107 (1982)

3.45   T. Abe, K. Miyazaki, K. Hirano: Acta Met. **30**, 357 (1982)

3.46   T. Ishimasa, H.-U. Nissen, K. Fukano: Workshop on Physics of Small Particles
(Gwatt, Switzerland Oct. 1984); Phys. Rev. Lett. **55**, 511 (1985)

3.47   T. Ishimasa, Y. Fukano: 3rd Int. Sym. on Small Particles and Inorganic Clusters,
Berlin, 1984; to appear in Surface Sci.
D. Shechtman, I. Blech, D. Gratias, J.W. Cahn: Phys. Rev. Lett. **53**, 1951 (1984)

3.48   A.L. Mackay: Phys. Bull., Nov., 495 (1976); Surf. Sci. **156**, 241 (1985)

3.49   K. Hiraga, M. Hirabayashi, A. Inoue and T. Masumoto: Sci. Rep. RITU **A-32**, 309
(1985)

3.50   H.-J. Güntherodt, H. Beck (eds.): *Glassy Metals I* and *II*, Topics Appl. Phys. Vols.
46 and 53 (Springer, Berlin, Heidelberg 1981 and 1983)

3.51   M.L. Rudee, A. Howie: Phil. Mag. **25**, 1001 (1972)

3.52   S.R. Herd, P. Chaudhari: Phys. Stat. Sol. (a) **26**, 627 (1974)

3.53   Y. Ishida, H. Ichinose, H. Shimada, H. Kojima: Proc. 4th Intern. Conf. Rapidly
Quenched Metals, Sendai, Jpn. Inst. Met., Vol. 1, 421 (1982)

3.54   H. Ichinose, Y. Ishida: Trans. Jpn. Inst. Met. **24**, 405 (1983)

3.55   Y. Hirotsu, R. Akada: Jpn. J. Appl. Phys. **23**, L479 (1984)

3.56   F.E. Fujita: *Rapidly Quenched Metals*, ed. by S. Steeb, H. Warlimont (Elsevier,
Amsterdam 1985) p. 585

3.57   A. Howie, I.L. Krivanek, M.L. Rudee: Phil. Mag. **27**, 235 (1973)

3.58   A. Howie: J. Non-Cryst. Solids **31**, 41 (1978)

3.59   T. Hamada, F.E. Fujita: Jpn. J. Appl. Phys. **25**, 318 (1986)

3.60   T. Hamada, F.E. Fujita: Jpn. J. Appl. Phys. **24**, 249 (1985)

3.61   M. Hirabayashi: Trans. Jpn. Inst. Met. **24**, 317 (1983)

3.62   K. Hiraga, D. Shindo, M. Hirabayashi, O. Terasaki, D. Watanabe: Acta Cryst. A
36, 2550 (1980)

3.63   K. Hiraga, D. Shindo, M. Hirabayashi, O. Terasaki, D. Watanabe: Acta Cryst. A **38**,
269 (1982)

3.64   D. Watanabe, O. Terasaki: Mat. Res. Soc. Symp. Proc. **21**, 231 *Phase Transformations in Solids*, ed. by T. Tsakalakos (Elsevier, Amsterdam 1984)

3.65   S. Amelinckx, J. van Landuyt, G. van Tendeloo: *Modulated Structure Materials*, ed.
by T. Tsakalakos, NATO ASI Series (E. Nijhoff, Dordrecht 1984) p. 183

3.66   K. Hiraga, D. Shindo, M. Hirabayashi: J. Appl. Cryst. **14**, 185 (1981)

3.67   D. Shindo: Acta Cryst. A **38**, 310 (1982)

3.68   R. Portier, D. Gratias, M. Guymont, W.M. Stobbs: Acta Cryst. A **36**, 190 (1980);
J. Microscopy **119**, 163 (1980)

3.69   N. Kuwano, H. Mishio, M. Toki, T. Eguchi: phys. stat. sol. (a) **65**, 341 (1981)
3.70   M. Hirabayashi, K. Hiraga, D. Shindo: J. Appl. Cryst. **14**, 169 (1981)
3.71   D. Watanabe, O. Terasaki: *Modulated Structure Materials*, ed. by
        T. Tsakalakos, NATO ASI, Series E. (Martinus Nijhoff Pub., Dordrecht 1984) p.247
3.72   D. Shindo, K. Hiraga, M. Hirabayashi, Sci. Rep. Res. Inst. Tohoku Univ. **32**, 32
        (1984)
3.73   G. Van Tendeloo, J. Microscopy **119**, 125 (1980)
3.74   P.R. Okamoto, G. Thomas, Acta Met. **19**, 825 (1971)
3.75   W.M. Stobbs, J.P. Chevalier, Acta Met. **26**, 233 (1978)
3.76   L. Stenberg, Chemica Scripta **14**, 219 (1978-79)
3.77   L. Stenberg, S. Anderson, J. Solid State Chem. **28**, 269 (1979)
3.78   T. Ishimasa, Y. Kitano, Y. Komura: phys. stat. sol. (a) **66**, 703 (1981)
3.79   K. Hiraga, T. Yamamoto, M. Hirabayashi: Trans. Jpn. Inst. Met. **24**, 421 (1982)
3.80   P.D. Southwick, W.M. Stobbs, J. Microscopy **119**, 169 (1980)
3.81   S. Takeda, Y. Kitano, Y. Komura, J. Electron Microsc. **32**, 105 (1983)
3.82   H.Q. Ye, D.X. Li, K.H. Kuo: Phil. Mag. A**51**, 829,839,849 (1985)
3.83   S. Takeda, H. Horikoshi, Y. Komura: J. Microscopy **129**, 347 (1983)
3.84   K. Hiraga, M. Hirabayashi, N. Ishigaki: Proc. 8th European Cong. Electron Mi-
        croscopy, (Budapest 1984), Vol. 1, p. 215. Submitted to J. Microscopy
3.85   N. Kuwano, I. Hiroshig Be, Y. Tomokiyo, T. Eguchi: Proc. 7th Intern. Conf. High
        Voltage Electron Microscopy, (Berkeley, 1983), p. 291. See also N. Kuwano, S. Higo,
        K. Yamamoto, K. Oki, T. Eguchi, Jpn. J. Appl. Phys. **24**, L663 (1985)
3.86   Y. Kitano, H.U. Nissen, K. Kwasnitza: Cryogenics **22**, 635 (1982)
3.87   T. Onozuka, T. Yamamoto, M. Hirabayashi: Sci. Rep. Res. Inst. Tohoku Univ. A
        32, 21 (1984)
3.88   A.B. Lamine, M.J. Lahana, F. Reynaud, P. Stadelmann: J. Mat. Sci. Lett. **3**, 431
        (1984)
3.89   M. Takeda, H. Yoshida, H. Hashimoto: phys. stat. sol. (a)**87**, 473 (1985)
3.90   J.P. Zhang, H.Q. Ye, K.H. Kuo, S. Amelinckx: phys. stat. sol. (a) **88**, 475 (1985)
3.91   H.Q. Ye, S. Amelinckx: phys. stat. sol. (a) **88**, 483 (1985)
3.92   T. Onozuka, M. Hirabayashi: *Modulated Structures 1979*, ed. by J.M. Cowley, J.B.
        Cohen, M.B. Salamon, B.J. Wuensch, AIP Conf. Proc. **53**, Am. Inst. Phys., New
        York, 373 (1979)
3.93   T. Onozuka, M. Hirabayashi, H. Ota: Sci. Rep. Res. Inst. Tohoku Univ. A **32**, 54
        (1984)
3.94   R. Guan, H. Hashimoto, T. Yoshida: Acta Cryst. B **40**, 109 (1984)
3.95   R. Guan, H. Hashimoto, K.H. Kuo: Acta Cryst. B **40**, 560 (1984), **B 41**, 219 (1985)
3.96   S. Nagakura, Y. Nakamura: Trans. Jpn. Inst. Met. **24**, 329 (1983)
3.97   T. Sato, T. Takahashi: Transp. Jpn. Inst. Met. **24**, 386 (1983)
3.98   N. Ajika, H. Endoh, H. Hashimoto, M. Tomita, H. Yoshida: Phil. Mag. A **51**, 729
        (1985)
3.99   C.K. Wu: Trans. Jpn. Inst. Met. **26**, 7 (1985)
3.100   S. Nagakura, T. Suzuki, M. Kusunoki: Trans. Jpn. Inst. Met. **22**, 699 (1981)
3.101   R. Sinclair, H.A. Mohamed: Acta Met. **26**, 623 (1978)
3.102   M. Fukamachi, S. Kajiwara: Jpn. J. Appl. Phys. **19**, L479 (1980)
3.103   K.M. Knowles: Phil. Mag. A**45**, 357 (1982)
3.104   K.M. Knowles, J.M. Christian, D.A. Smith: J. Physique **43**, C-4, 185 (1982)
3.105   F.C. Lovey, W. Coene, D. van Dyck, G. van Tendeloo, J. van Landuyt, S. Amelinckx:
        Ultramicroscopy **15**, 345 (1984)
3.106   J.M. Cook, M.A. O'Keefe, D.J. Smith, W.M. Stobbs: J. Microscopy **129**, 295 (1983)
3.107   L.E. Tanner, A.R. Pelton, R. Gronsky: J. Physique **43**, C-4, 169 (1982)
3.108   M. Sugiyama, R. Ohshima, E.F. Fujita: Trans. Jpn. Inst. Met. **25**, 585 (1984)
3.109   D. Cherns: Phil. Mag. **30**, 549 (1974)
3.110   K. Takayanagi, K. Yagi: Trans. Jpn. Inst. Met. **24**, 337 (1983)
3.111   N. Osakabe, Y. Tanishiro, K. Yagi, G. Honjo: Surf. Sci. **109**, 353 (1981)
        R. Vanselow, R. Howe (eds.): *Chemistry and Physics of Solid Surfaces* IV, Springer
        Ser. Chem. Phys., Vol. 20 (Springer, Berlin, Heidelberg 1982) Chap.6
3.112   J.M. Cowley: In *Electron Microscopy of Materials*, ed. by W. Krakow, D.A. Smith,
        L.W. Hobbs, Mat. Res. Soc. Smp. Proc. Vol. 31 (Elsevier, New York 1984) p. 177

3.113   K. Takayanagi, K. Kobayashi, Y. Kodaira, Y. Yokoyama, K. Yagi: Proc. 7th Intern. Conf. , High Voltage Electron Microscopy (Berkeley 1983) p. 47

3.114   L.D. Marks, D.J. Smith, Nature **303**, 316 (1983)

3.115   D.J. Smith, L.D. Marks, Ultramicroscopy **16**, 101 (1985)

3.116   S. Iijima, T. Ichihashi, Jpn. J. Appl. Phys. **24**, L125 (1985)

3.117   W.A.T. Clark: In *Electron Microscopy of Materials*, ed. by W. Krakow, D.A. Smith, L.W. Hobbs Mat. Res. Soc. Smp. Proc., Vol. 31 (Elsevier, New York 1984) p.211

3.118   For example, O.L. Krivanek: Chemica Scripta **14**, 213 (1978–79), J. Microscopy **119**, 81 (1980)

3.119   Y. Ishida, H. Ichinose, M. Mori, M. Hashimoto: Trans. Jpn. Inst. Met., **24**, 349 (1983)

3.120   S. Ino: J. Phys. Soc. Jpn. **21**, 346 (1966)

3.121   L.D. Marks, D.J. Smith: J. Cryst. Growth **54**, 425 (1981)

3.122   J. Teuho, J. Mäki, M. Hirabayashi, K. Hiraga: J. Mat. Sci. Lett. **4**, 826 (1985)

3.123   K. Hiraga, M. Hirabayashi, M. Sagawa, Y. Matsuura: Jpn. J. Appl. Phys. **24**, L30 (1985)

3.124   R.S. Timsit, W.G. Waddington, C.H. Humphreys, J.L. Hutchison: Appl. Phys. Lett. **46**, 830 (1985)

3.125   T.S. Kuan: In *Electron Microscopy of Materials*, ed. by W. Krakow, D.A. Smith, L.W. Hobbs Mat. Res. Soc. Smp. Proc. 31 (Elsevier, New York 1984) p. 143

3.126   K. Kajiwara, H. Kawai, K. Kaneko, N. Watanabe: Jpn. J. Appl. Phys. **24**, L85 (1985)

3.127   F. Nizzoli, K.-H. Rieder, R.F. Willis (eds.): *Dynamical Phenomena at Surfaces, Interfaces and Superlattices*, Springer Ser. Surf. Sci., Vol. 3 (Springer, Berlin, Heidelberg 1985)

3.128   C.S. Baxter, W.M. Stobbs: Ultramicroscopy **16**, 213 (1985)

3.129   H. Hashimoto, H. Endo, T. Tanji, A. Ono, E. Watanabe: J. Phys. Soc. Jpn. **42**, 1073 (1977)

3.130   H. Formanek, M. Muller, M.H. Hahn, T. Koller: Naturwiss. **58**, 339 (1971)

3.131   H. Hashimoto, A. Kumao, K. Hino, H. Yotsumoto, A. Ono: Jpn. J. Appl. Phys. **10**, 1115 (1971)

3.132   F.P. Ottensmeyer, E.E. Schmidt, T. Jack, J. Powell: J. Ultrastructure Res. **46**, 546 (1972)

3.133   K. Mihama, A. Horata, R. Uyeda: Jpn. J. Appl. Phys. **12**, 746 (1973)

3.134   L. Reimer: *Scanning Electron Microscopy*, Springer Ser. Opt. Sci., Vol. 45 (Springer Berlin, Heidelberg 1985)

4. Field Ion Microscopy[1]

N. Ernst and G. Ehrlich[2]

With 26 Figures

The field ion microscope, invented by *Müller* [4.1] in 1951, was the first instrument to reveal surfaces on the atomic level. In the years since its invention, the field ion microscope and its derivatives have proved not only remarkably simple, but also versatile, allowing important new insights about atomic behavior in many different fields of study. Recently, other techniques, first electron microscopy and subsequently scanning tunneling microscopy, have developed the ability to probe solid surfaces on a scale approaching atomic dimensions. In the hands of *Crewe* and his colleagues [4.2,3], scanning electron microscopy has been shown capable of revealing individual metal atoms on light substrates. However, except for some initial exploratory studies [4.4], no consistent effort to apply these methods to the characterization of surface behavior has as yet appeared. The primary contribution of electron microscopy to surface studies is being made by fixed beam methods, which are beginning to reveal the atomic arrangement of the substrate [4.5,6]. Scanning tunneling microscopy (STM) [4.7] is much the newest technique and is still undergoing rapid development. The principal application has been to the examination of surface structure [4.8,9]: even the limited work available so far has clearly revealed the complex structural features to be expected on macroscopic crystals. Neither the operating principles nor the potential of this technique are well established, but it already is evident that STM has great promise for the examination of macroscopic surfaces.

Even though field ion microscopy is no longer alone in the ability to yield high-resolution spatial information about crystals, it is still first in the amount of incisive data generated about atomic behavior on and in crystals, and remains unique in two aspects: the ability to prepare highly perfect surfaces unmarred by thermal disorder, through field evaporation and desorption, and the capability of providing information about chemical constitution as well as location of individual atoms. The real strength of field ion microscopic methods is their ability to focus on individual features such as single atoms on crystals, as in Fig. 4.1, rather than in providing accurate information about atomic spacings or structures at an interface. Field ion microscopy has had quite a considerable impact on materials research [4.11,12] especially on the understanding of radiation damage in metals [4.13], alloy studies [4.14], as well as in

[1] Supported by the National Science Foundation under Grant DMR 84-20751.

[2] Guggenheim Fellow, 1984–1985.

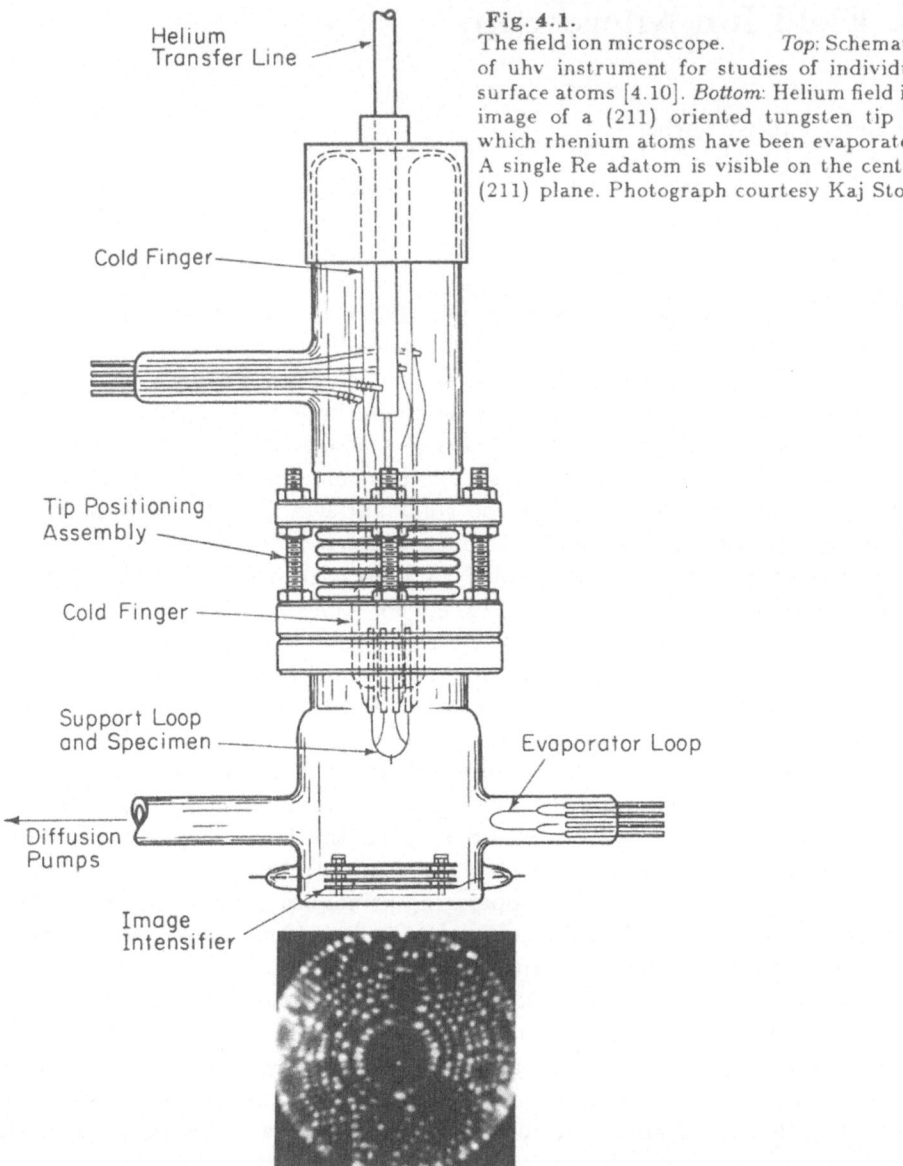

Fig. 4.1.
The field ion microscope. *Top*: Schematic of uhv instrument for studies of individual surface atoms [4.10]. *Bottom*: Helium field ion image of a (211) oriented tungsten tip on which rhenium atoms have been evaporated. A single Re adatom is visible on the central (211) plane. Photograph courtesy Kaj Stolt

Helium Transfer Line

Cold Finger

Tip Positioning Assembly

Cold Finger

Support Loop and Specimen

Diffusion Pumps

Image Intensifier

Evaporator Loop

the examination of the surface events on metals [4.15]. Rather than attempting a general overview of the many different areas on which field ion microscopy has been brought to bear [4.16], we shall emphasize here the physical processes important in the practical implementation of field-ion-microscopic techniques, processes about which much has been learned recently. In the last part of this review we provide examples of studies by these techniques, which illustrate how much can be learned through the ability to see individual surface atoms, and by the capability of analyzing in detail processes, such as evaporation, which occur in high fields.

4.1 Principles and Techniques

4.1.1 Magnification, Resolution, and Image Formation

The field ion microscope (FIM), shown schematically in Fig. 4.1, is a very simple instrument. It is essentially a point projection microscope in which the apex of the needle-shaped emitter is characterized by an average radius $\bar{r}$, as in Fig. 4.2. The average magnification $\overline{M}$ is given by

$$\overline{M} = R/(\bar{r}\beta), \tag{4.1}$$

where R is the tip-to-screen distance and β a numerical factor of about 1.6 accounting for the fact that in a real FIM the projection is not exactly radial [4.17–19]. Typical values of R and $\bar{r}$ are 8 cm and 3×10^{-6} cm, respectively; the magnification $\overline{M}$ is consequently of the order of 10^6. A real field ion emitter has regions with a variety of local tip radii, which may be determined by counting the number of rings appearing between prominent planes of a field ion micrograph [4.20]. The variation of the local radius, as well as the distribution of local electric field strength, make an exact calibration of the magnification factors for the field ion micrograph ambiguous. However, as discussed in Sect. 4.3.1b the surface unit cell can be mapped by recording the adsorption sites of single metal adatoms, so that interatomic separations even within densely packed crystal phases can be directly visualized.

Imaging of single surface atoms, on the fluorescent screen of an FIM as in Fig. 4.1, is done by ions from a neutral gas, usually helium, created by an intense local electric field at the surface. It was first pointed out be *Becker* [4.21]

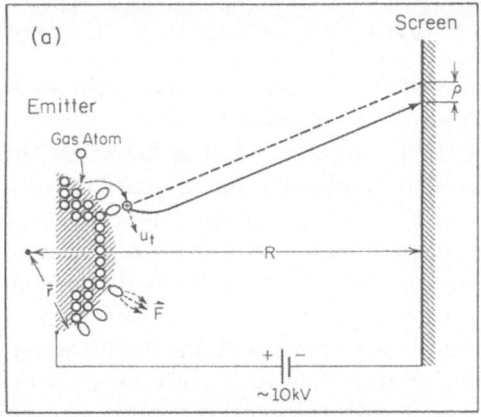
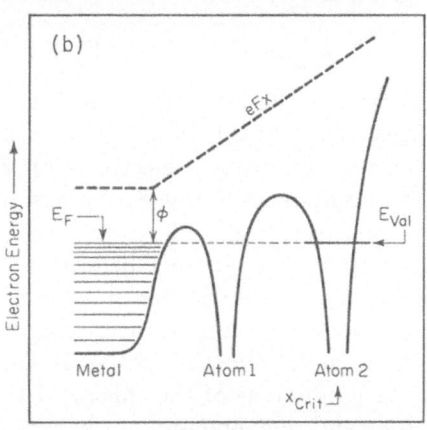

Fig. 4.2a,b. Operation of the field microscope. (a) FIM geometry. Incoming gas atoms are ionized in the high electric field **F** in front of protruding surface atoms, which are covered by a polarized layer of image gas. Due to the tangential velocity u_t of imaging ions, a scattering disk with radius ϱ is created on the phosphor screen. (b) Ionization mechanism. The applied electric field raises E_{val}, the valence electron level of atom 2. Tunneling of a valence electron from this atom, through field adsorbed atom 1, becomes possible at and beyond the critical distance x_{crit}, at wich E_{val} is aligned with the Fermi energy E_F of the metal emitter

that during imaging the surface under study is covered by a layer of adsorbed imaging gas due to the effect of the high local electric field [4.22-25]. The presence of field-adsorbed atoms was established by pulsed-field desorption of metal-helide ions using time-of-flight mass spectrometry [4.26]. Later, magnetic sector mass analysis of dc-field-desorbed ions revealed that the emitter-tip surface, including the densely packed planes, is indeed covered with field-adsorbed atoms when FIM is at optimum resolution [4.27-30]. In spite of some theoretical efforts [4.31-33], the effect of field-adsorbed gases on image formation is still not completely understood [4.34]. During imaging with helium gas the presence of small amounts of neon enhances the local brightness in the FIM [4.35-38]. All these findings favor an ionization mechanism in which the valence electron of the imaging gas atom tunnels through the adsorbed layer, as shown schematically in Fig. 4.2b.

Ions are generated at (and beyond) the critical distance at which tunneling of valence electrons from the image gas atom into empty electronic states above the Fermi level of the emitter becomes possible. The positively charged ions so formed are accelerated towards the counter electrode, which normally is a channel-plate intensifier combined with a phosphor screen (Sect. 4.1.4). The ion beam from a single surface atom is, however, not ideally straight, but is broadened due to three effects [4.17]: (i) The Heisenberg uncertainty of the tangential momentum of the imaging ion, (ii) the lateral velocity of the imaging ion coming from a surface at a finite temperature, and (iii) the lateral extension of the ionization zone at the critical distance. The optimum resolution of the FIM, defined by *Müller* and *Tsong* [4.17] as the minimum size of the image spot for a single atom on the screen divided by the magnification, has been given as [4.17,18].

$$\delta = [4\sqrt{\beta^2 r \hbar^2/(2kemF)} + 16\beta^2 r k_B T/(keF) + \delta_0^2]^{1/2}. \tag{4.2}$$

Here r is the local tip radius, m the mass of the imaging ion and F the local electric field; this is assumed to be related to the applied voltage V between emitter and screen electrode by $F = V/(kr)$. Equation (4.2) is based on the assumption that the gas atoms, which initially impinge on the surface with a relatively high polarization energy $(\alpha/2) \cdot F^2$ [4.17] are accommodated to the emitter temperature T before ionization. Experimental tests [4.39] have indeed shown a quadratic dependence of the image diameter of individual tungsten atoms on tip temperature, in agreement with (4.2), but the detailed results of *Chen* and *Seidman* [4.39] indicated incomplete accommodation. Nevertheless, the predictions of the theory, that resolution improves with decreasing tip radius and temperature, are in good agreement with the general experience that optimum resolution is obtained by proper sharpening and cryogenic cooling of the emitter tip.

Image contrast in FIM is governed by the electron tunneling rate above different atomic sites as well as by the flux of atoms to these sites. A detailed discussion of the physical processes involved is beyond the scope of this chapter and may be found in [4.34,39-43]. Both the (local) supply rate of image gas atoms as well as the effect of (local) field variations still remain to be investigated in detail. Recent theoretical estimates suggest that variations in the local electronic transition probabilities over the surface dominate the image contrast in FIM [4.44,45]. However, it is also clear from observations of the effect of adatoms on image contrast at long distances [4.46] that the gas supply must play an important role.

4.1.2 Field Evaporation and Desorption

For studies of well-defined surfaces, clean and smooth emitters are a prerequisite. In early work with the field emission microscope [4.40] these were obtained by resistively heating the wire supporting the emitter tip. However, heating to desorb impurities produces a thermally annealed end form, in which the tip is characterized by a relatively large radius, as well as large low-index faces [4.17,40,47]. With helium and neon as imaging gases, field ion microscopy of thermally cleaned emitter surfaces turns out to be difficult, as extremely high voltages have to be applied. It is of interest in this context, however, that field-ion-microscopic studies of metal surfaces after heating were the first to reveal the extensive thermal disorder characteristic of such surfaces [4.48], which is quite different from the high perfection of hard sphere models. Although resharpening of blunted tips by neon sputtering in situ is nowadays achieved routinely [4.49,50], the final preparation of sputtered or thermally treated emitter surfaces is almost always accomplished through electric-field-induced removal of surface atoms at cryogenic temperatures. Field desorption (FD) of adatoms and field evaporation (FEV) of substrate lattice atoms, first reported by *Müller* [4.51,52], produce a highly perfect end form, evident in Fig. 4.1.

Field evaporation and field desorption provide the basis for the application of field ion emission in areas like metallurgy [4.11], liquid-metal ion sources [4.53,54] and field-desorption mass spectrometry [4.55,56]. However, the underlying physical processes are still under continuing investigation. Field evaporation was originally viewed as involving the escape of an ion over a Schottky potential saddle, formed by the superposition of the image potential of the ion and the potential applied to the emitter [4.34,51,52,57]. This image hump model was recognized as oversimplified, but nevertheless had the considerable virtue of allowing quantitative predictions about the conditions for field evaporation [4.57]. A much more physically realistic model is due to *Gomer* and *Swanson* [4.58,59]. They proposed charge exchange, transforming a surface atom into an ion at some distance from the surface, as the limiting step in evaporation at high fields [4.60,61]. Neither of these models was able to account for a quite unexpected finding in the atom-probe – the appearance of highly charged ions in field evaporation [4.34,62–64].

The explanation soon offered for this phenomenon was post-field ionization [4.34]. The idea was this: in the initial act of field evaporation, ions are formed in some low charge state, but thereafter, during their sojourn in the high field, the ions lose additional electrons by tunneling into the emitter. However, quantitative calculations of the rate of field ionization for one-dimensional tunneling models [4.65], as well as three-dimensional estimates using the transfer Hamiltonian method for atoms with hydrogen-like wave functions [4.66], predicted negligible post-field ionization for systems in which high charge states were detected in the atom-probe. The mechanism of field evaporation was therefore left in uncertainty.

All this has been changed by studies of rhodium field evaporation, using mass and appearance spectrometry, to be discussed in more detail in Sect. 4.2.2d. These studies showed that the activation energy for desorption of singly and doubly charged rhodium ions was, in fact, identical, leading to the conclusion that post-ionization had to account for the appearance of the doubly charged entities [4.67]. Recent experiments on the energy distribution of field-

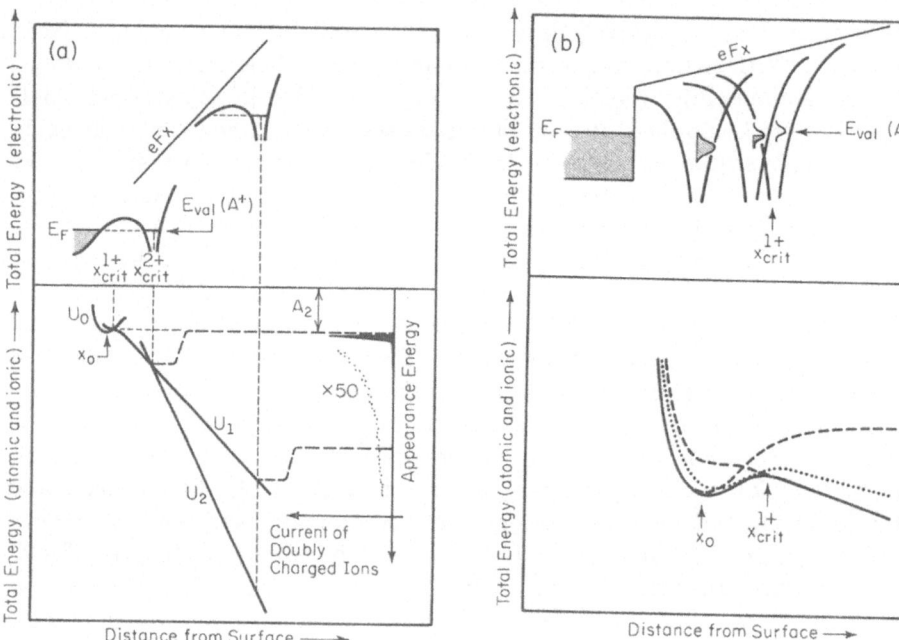

Fig. 4.3a,b. Field evaporation and post-field ionization. (a) Post-field ionization of an initially singly charged desorbing ion [4.68]. The potential energy of the valence electron of the singly charged ion A^+ is shown schematically at the top. x_0 : equilibrium position of the surface atom; x_{crit}^{1+} : critical distance for desorption of the singly charged ion; x_{crit}^{2+} : critical distance for post-field ionization of singly charged ion. At the lower left are drawn the potential energy U_0 of a neutral atom and of singly and doubly charged ions (U_1, U_2). The tailing in the energy distribution of doubly charged ions schematically indicated (*dotted points*) at the lower right reveals post-field ionization of desorbing singly charged ions. The appearance energy A_2 yields information about the binding energy of the field evaporating atom, as discussed in Sect. 4.2.2d. (b) Charge draining model for field evaporation of a singly charged ion [4.71]. Shown at the top is the valence level $E_{val}(A^0)$ of the evaporating atom, which is significantly broadened. At the critical distance for field evaporation x_{crit}^{1+}, the valence electron is completely drained out of the atom (A^0) into the Fermi sea of the emitter. At the bottom is indicated the corresponding potential energy of a singly charged ion, U_1. The dotted line is the potential energy of a hypothetical ion, with an effective charge $+0.5\,e$, the dashed curve gives the potential energy of a neutral atom, U_0

evaporated rhodium ions have provided direct evidence for the occurrence of post-field ionization of Rh^+ created in the initial desorption [4.68]. As is shown schematically in Fig. 4.3, the detection of a small abundance of low energy (higher appearance energy) Rh^{2+} establishes the post-field ionization of initially desorbed Rh^+ that has traveled $\approx 10\,\text{Å}$ or more into the gas phase. The same type of experiment has very recently been done on field-evaporated Re^{2+} and Re^{3+} using pulsed-laser field-desorption techniques [4.69]; the results again tend to support a post-field ionization mechanism for the generation of Re^{3+}. Measurements of field evaporation have also been carried out on tungsten [4.70], and show that the activation energies for appearance of W^{+3} and W^{+4} are the same; this suggests that for tungsten also post-field ionization is dominant in producing the higher charge states. Stimulated by all the experimental activity, there have also been new theoretical efforts. *Kingham* [4.71] has reexamined the existing theories, and proposed that field evaporation occurs by charge draining out of the 1st ionization level of the evaporating atom, creating a singly charged entity; this ion is subsequently field ionized to create higher charge states, as suggested in Fig. 4.3b. The process of post-field ionization has also been subjected to additional theoretical scrutiny [4.68,72,73]. In the most extensive of these examinations *Kingham* [4.73] used the semiclassical WKB approximation for tunneling probability calculations employing a model potential with an effective nuclear charge which is adjusted to fit experimental data. This procedure gives a field dependence of the ratio of ionic states in fair agreement with experiments for several systems [4.74–77]. However, a more incisive treatment, along the lines of *Chambers'* work [4.66] but allowing for the fact that the systems of interest cannot be adequately represented by hydrogen-like wave functions, is certainly still desirable. At present it appears that although a quantitative understanding of field evaporation is lacking, there are no severe conceptual problems confronting us.

4.1.3 Specimen Preparation

The actual field ion microscope is quite a simple affair. Despite that, field ion microscopy has been slow to catch on, in large measure due to some practical matters. There is no instrument commercially available, and that means considerable planning and study for anyone entering the field. The situation is somewhat better with the atom-probe, the combination FIM and time-of-flight mass spectrometer for determination of chemical constitution with atomic resolution (Sect. 4.1.7), in that a complete equipment can be ordered[3]. However, there is another barrier that must be overcome for any practitioner of the art – the preparation of suitable samples [4.17,18,78].

The conditions that must be met by the specimen are simple [4.79]. The emitter tip must be sharp enough so that the fields for evaporation and imaging are achieved at reasonable applied voltages, usually not exceeding 25 kV. The

---

[3] VG Scientific, Sussex, England.

sample shank must be smooth and symmetrical to avoid distorting the image. Furthermore, there cannot be any large defects or boundaries undercutting the tip, as otherwise the mechanical stresses due to the applied field may disrupt the specimen. The primary problem in finding an appropriate preparatory procedure is to discover a suitable electropolishing agent. A variety of recipes has appeared in the field emission literature [4.17,18]. The electron microscopy community [4.80] has also had to tackle the problem of producing thin specimens, and much useful information can be derived from perusing that literature. Once the polishing solution has been chosen, the remaining preparatory procedures are simple and are demonstrated by the techniques routinely used for producing tungsten emitters at Illinois. A small section of 0.125 mm wire is dipped in 2N sodium hydroxide solution, which has a ring-shaped platinum or nickel wire as counterelectrode, as in Fig. 4.4. Roughly 12 V DC is applied across the cell and the tungsten wire is attacked most rapidly at the air-liquid interface, forming a neck there, as illustrated in Fig. 4.4. That portion of the wire below the liquid level is thinned as polishing proceeds and finally pinches off at the narrowest place. When this happens, the polishing current suddenly drops, inasmuch as the area in contact with the solution is greatly reduced; at this moment the potential across the cell must be switched off. Any further dissolution at this stage instantly blunts the tip and must be avoided. This is accomplished automatically by a shutoff circuit [4.81]. The tungsten emitter tips obtained by these procedures ordinarily have a diameter of a few hundred angstrom and can be imaged at voltages of just a few kV. More inert materials, such as iridium, for example, may require the use of several different polishing agents [4.79], or prior thinning of the specimen. Nevertheless, after some exploration of the polishing behavior of the particular material, actual tip preparation can be carried out routinely in the simple apparatus in Fig. 4.4.

The high stresses imposed by the imaging field have not proved a severe obstacle to the utilization of the field ion microscope. Materials differing widely in their mechanical properties, and ranging from soft metals such as aluminum or gold [4.82] to refractories like rhenium have been successfully examined.

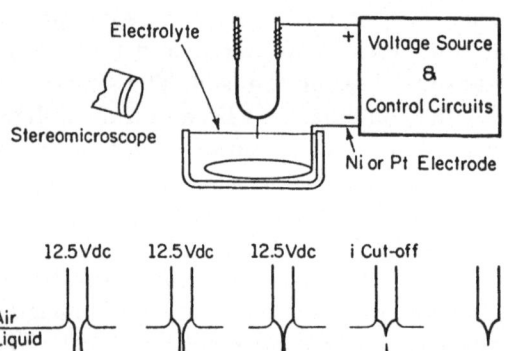

Fig. 4.4. Techniques for FIM specimen preparation [4.79]. *Top*: Electropolishing setup for tungsten emitters. *Bottom*: Tungsten emitter at various stages of electrochemical preparation

Techniques for preparing difficult materials continue to be developed [4.83] and alloys as well as pure metals are now routinely studied [4.84]. Success with semiconductors has been rather more difficult [4.85]. With silicon, for example, atomically smooth surfaces have only recently been imaged [4.86]. The problem here is in the field evaporation process, which under ordinary conditions does not remove single protruding surface atoms, but instead causes the evaporation of clusters [4.87], leaving an uneven surface. However, this problem can be overcome by field evaporation at temperatures $T > 150\,K$ [4.86,87]. Field ion microscope examination of overlayers has also been largely limited to metals and semiconductors. Electronegative layers appear to suffer extensive damage due to bombardment by low energy electrons formed by resonance ionization of helium (Sects. 4.1.7b and 4.2.2a) at some distance from the sample surface [4.88]. Despite this problem, biomolecular contours have been obtained by depositing the macromolecules on a blunt emitter at a low temperature, embedding them in benzene, and then creating an image by field desorbing the benzene from the surface [4.89]. It appears that special techniques can be found to deal even with quite difficult situations.

4.1.4 Image Detection

Images in first generation FIM tubes were generated on a phosphor screen deposited on a glass disk covered with a transparent conductive layer. The impact energy of imaging ions was directly converted into flashes of visible light, building up the FIM pattern which was then photographed using an externally mounted camera [4.17]. Because of the low ion current, typically $10^{-15}\,A$ from a single atom, relatively long exposure times on high-sensitivity film were required [4.90]. In practice, the sensitivity of the phosphor screen has restricted field ion microscopy to ions such as He^+ and H_2^+. Hydrogen, however, is capable of initiating modifications of the surface structure of the emitter, especially during imaging of less refractory metals [4.17,34]. The high electric field of about $4.5\,V/Å$ needed for He^+ imaging limits the range of materials accessible to routine field ion microscopy, as field evaporation starts below $4\,V/Å$ for most metals [4.17]. The experimental situation changed dramatically with the introduction of the Multi-Channel Plate (MCP), allowing the use of a wider range of imaging gases, in particular neon [4.34].

The use of the MCP in field ion microscopy has already been discussed in [4.34,91]. Briefly, however, the MCP consists of an array of something like 10^6 miniature electron multipliers, with a typical channel diameter of $10\,\mu m$ and a length of $0.5\,mm$ [4.92–94]. An ion (or electron) impinging on the front side of the MCP initiates a cascade of secondary electrons in one channel, with as many as 10^4 electrons leaving the back side of the MCP. These are then accelerated towards the phosphor screen, creating a highly amplified image of the emitter. Although the brighter image obtained with the MCP is a great advantage, there is also a significant drawback; the tiny capillaries of the channel plate introduce a huge amount of surface and therefore a highly undesirable source of contamination into the system [4.95]. For most FIM studies, but especially for observations of surface diffusion with hundreds of diffusion cycles, MCP has to be extensively outgassed through electron bombardment. This treatment may reduce the gain to 10^2, requiring sensitive photographic techniques or the use of light amplifying TV cameras [4.96,97] for image recording.

The gain can be raised up to 10^7 by sandwiching two channel plates into a chevron configuration [4.93]. In this way the ultrashort ($<100\,ps$) pulse of roughly 10^6 electrons arriving at the phosphor, which is initiated by one charged particle at the front, can be fed into a suitable amplifier and then registered by a counter [4.91]. A promising variant of this arrangement is the combination of a chevron plate with a position-sensitive detector[4] in place of the conventional screen [4.98]. In this assembly, drawn schematically in Fig. 4.5, the electron pulse created at the back of the chevron plate following the arrival of a single field ion, is accelerated towards a particular position of a resistive anode. Analog position signals, accumulated in a digital memory, are displayed on a video screen and the images can be stored on the hard disc of a com-

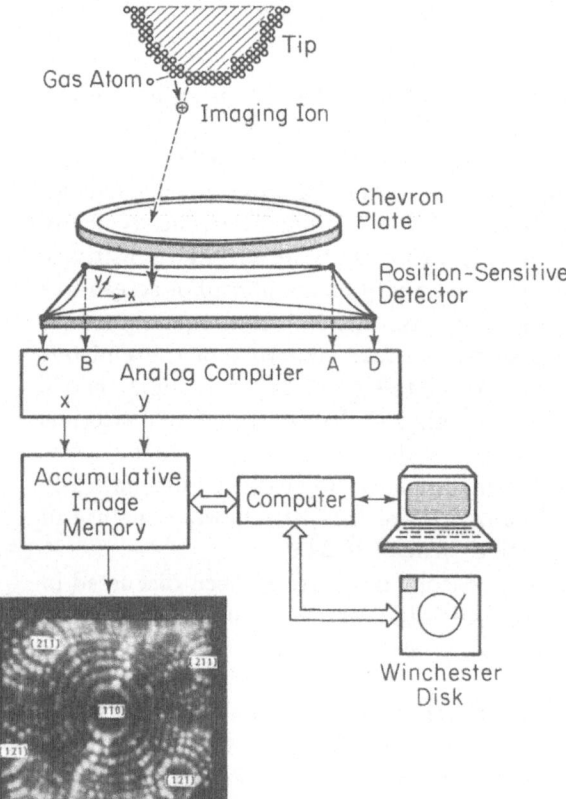

Fig. 4.5. Field ion microscope with a position sensitive detector [4.98]. Ions leaving the emitter generate a magnified image of the tip at the front side of a chevron channel plate. After charge amplification, the coordinates of each impinging ion are measured by an analog computer. The image shown has been built up by registering the arrival of single ions as a function of position

---

[4] Surface Science Lab., Inc. Mountain View, California USA.

puter, allowing further processing of the data. Presently, the recording of one FIM image, shown in Fig. 4.5, is accomplished in less than five minutes; but the data-recording time should be reduced by an order of magnitude after introduction of faster A/D converters [4.99]. Although FIM pictures taken with the position-sensitive detector have sharper contrast and better resolution than TV-camera pictures [4.99], the full power of the new technique, especially its single-particle sensitivity, is still to be explored [4.100].

4.1.5 Variants of the Field Ion Microscope

The field ion microscope provides tremendously detailed spatial information about adatoms at surfaces. With this data available, however, the desirability of more information about the properties of such well characterized structures becomes immediately apparent. Quite often some further hints about chemical composition can be obtained indirectly in the field ion microscope itself. For example, the field required for desorption may sometimes be used to discriminate between chemically different adatoms. Similarly, differences in the voltage at which metal and silicon atoms in a cluster image, have made it possible to distinguish between the two [4.101]. Especially for gas layers, however, further characterization is desirable and this provided the stimulus for development of instruments combining the high spatial resolution of the field ion microscope with additional analytical capabilities.

4.1.6 Field Emission Field Ion Microscopy

The field ion microscope evolved directly from the field emission microscope [4.40,102]. In this earlier instrument an image of a surface is formed on a phosphor screen by applying a high negative potential to the specimen, which is in the form of a sharp metal tip, causing electrons to tunnel through the surface. The resolution of the device is limited to roughly 20 Å by the considerable energy of the electrons parallel to the emitting surface. However, from the current versus voltage characteristics, that is from Fowler-Nordheim plots [4.40,102], it is possible to derive information about work functions, and therefore about the dipole moment in the adlayer [4.103,104]. The first effort to characterize changes in electron emission brought about by a single metal adatom was made by *Plummer* and *Rhodin* [4.105] who deposited a tungsten atom on a tungsten emitter in a field ion microscope and then carried out a Fowler-Nordheim analysis for the electrons emitted from the surface. This type of study was subsequently extended by *Kellogg* and *Tsong* [4.106], who reported dipole moments for various metal adatoms on W(110). In a separate examination, *Todd* and *Rhodin* [4.107] used the same approach to measure the work function of the (110) plane of a tungsten emitter itself, which had previously been characterized by field ion microscopy, in order to assess the possible contribution of surface imperfections to field emission measurements. These investigations have had to contend with a significant experimental problem: electron emission from low-index planes is swamped by emission from the surrounding rougher areas. It is therefore crucial to collect only electrons from the surface of interest. The standard way of doing this is to have a probe which intercepts the electron current at a small region of the fluorescent screen that is used for imaging in the field ion mode. The conditions to be met for collecting the electrons have been discussed by *Gadzuk* and *Plummer* [4.108]. Despite considerable early interest [4.109] there has so far been no detailed experimental study to establish how large the probe area can be relative to the size of the plane of interest without encountering contributions from the surrounding region. In a field emission study on emitters prepared by careful field evaporation, but without imaging by field ion microscopy, *Polizzotti* [4.110] found that only on a (110) with a characteristic dimension of at least 60 Å was the measured work function un-

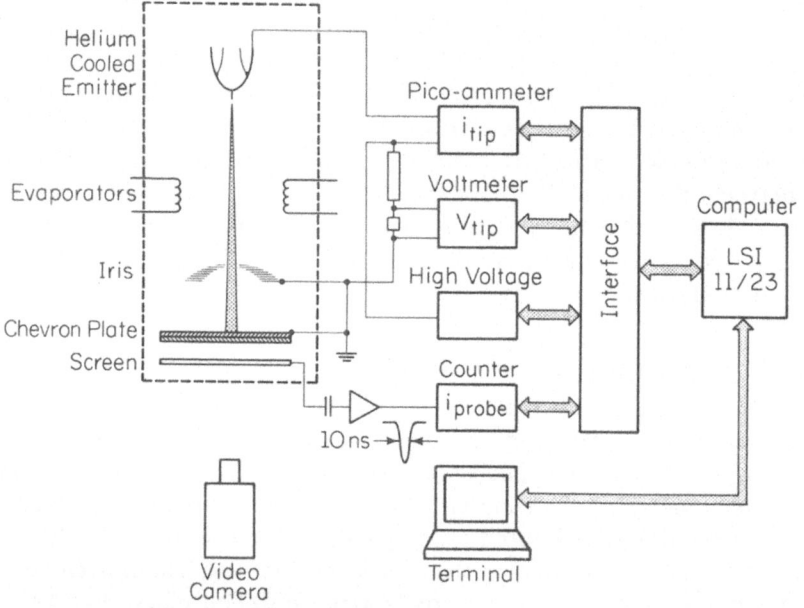

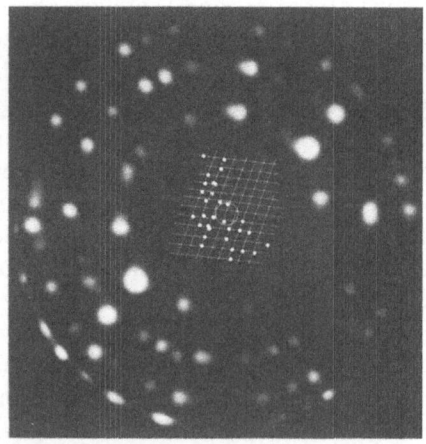

Fig. 4.6. Combined field electron and ion microscope system [4.111]. An externally adjustable iris diaphragm makes it possible to restrict the sample area, which is imaged either with ions or electrons. At the bottom is shown a field-ion image of the emitter. On this has been superposed the surface grid of binding sites on W(110), derived by observing the location of a diffusing adatom, as well as the circular outline of the probe-hole projected onto (110)

affected by emission from the periphery. However, this dimension could have been influenced by a non-uniform shank.

Recently, a combined field electron and field ion microscope has been reported [4.111] which should make it possible to establish the conditions appropriate for valid field emission measurements on single adatoms. A schematic of the instrument is shown in Fig. 4.6. The detector for both electrons and ions is a chevron channel plate. The surface area sampled in field emission is defined by an iris diaphragm, which can be externally adjusted to determine the effect of probe hole size upon the measured work function. Although definitive measurements on adatoms are yet to be reported, it does appear from introductory experiments with this equipment that not only the plane size but also the tip radius must be carefully considered in field electron emission, especially from very sharp (sputtered) tips.

4.1.7 The Atom-Probe

a) Pulsed-Voltage Atom-Probe

With the advent of the FIM it was for the first time possible to see individual atoms. The desirability of obtaining chemical information about surfaces with the same resolution became immediately obvious. In the 1950's, *Inghram* and *Gomer* [4.112] had already combined a field electron microscope with a mass spectrometer, characterized by the relatively low spatial resolution of 20 Å and a detection sensitivity limited by the electronic equipment available at the time. Chemical identification of single surface atoms and molecules was finally made possible by the atom-probe FIM, developed by *Müller* et al. [4.62,113] in 1968. In this instrument an atom of interest is field desorbed, mass analyzed and detected with high probability. In first generation atom-probes, ions emitted from selected sites of the specimen tip were allowed to pass through a small probe hole in the middle of the phosphor screen of an ordinary FIM. Mass analysis was performed by time-of-flight (TOF) measurement for the ions field desorbed by a short high-voltage pulse superimposed on the tip voltage. The desorbed ions travel through the probe hole, down a drift tube, and strike a high-gain detector that stops the timing device. In principle, the measured time interval, together with the operating parameters of the probe, then yield the mass-to-charge ratio of the particle.

The mass resolution $\Delta m/m$, with Δm being the full width at half maximum of a registered mass peak, is of the order of 1/100 for an atom-probe FIM equipped with a straight flight tube [4.113,114]. This relatively small value is a consequence of the anisochronous arrival of ions having the same mass-to-charge ratio, and arises from distortions during acceleration by the high-voltage pulse [4.115,116]. The resolution can be improved an order of magnitude by installing an energy focusing electrostatic deflector system, in which low-energy ions travel a shorter distance to the detector than high-energy ions, resulting in the simultaneous arrival of particles of the same mass-to-charge ratio [4.117,118].

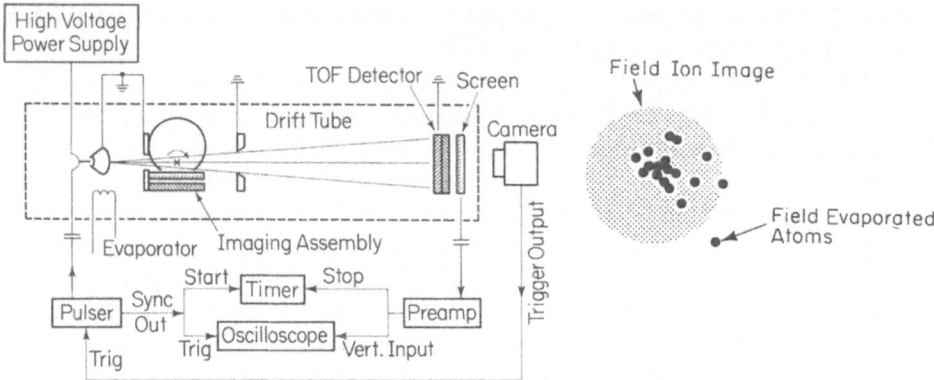

Fig. 4.7. Atom-probe field ion microscope [4.66,119]. *Left:* Schematic of experimental arrangement. *Right:* Location of rhodium ions, detected after field evaporation from rhodium deposit on W(110), in relation to field ion image of single adatom recorded on the same detector

The ability of the atom-probe FIM to provide spatial information is, up to a point, controlled by the size of the probe hole, at least for emitter tips having a regular end form. However, if the probe hole is so small that only one atom site is sampled, the probability that the selected atom will pass through the hole may become very small. The ability of the atom-probe FIM actually to identify the chemical nature of preselected atoms was demonstrated by *Chambers* and *Ehrlich* [4.119] for single rhodium atoms deposited on W(110). The alignment of the atom-probe shown in Fig. 4.7 was performed with a laser and a movable channel plate in order to circumvent limitations arising from conventional probe-hole operation. In this way the location of field evaporated rhodium ions on the detector screen could be recorded in relation to their field ion image on the same screen, demonstrating that the spatial resolution of the atom-probe is at least as good as the atomic resolution in the FIM. This feature has been quite useful in studying the mechanism of diffusion [4.120], as discussed in Sect. 4.2.1a. The TOF atom-probe has made significant contributions to a number of fields, but undoubtedly the greatest successes have been in metallurgical applications, which were surveyed in [4.11–14].

b) Other Atom-Probe Spectrometers

Several variations of the original time-of-flight atom-probe FIM appeared between 1970 and 1980: combinations with quadrupole and magnetic-sector spectrometers [4.121,122], the imaging atom-probe [4.91,123], and the pulsed-laser atom-probe [4.124,125]. The magnetic-sector instruments have been successful notably in combined mass and energy analyses of field ionization events. *Jason* [4.126], for example, measured well defined oscillating structures in the energy distribution of field ions like Ne^+, H_2^+, and CO^+, using a magnetic sector mass spectrometer equipped with a fluorescent screen for observing field electron emission patterns. His results were confirmed by *Müller* and coworkers [4.122,127], using TOF techniques and also the magnetic-sector atom-probe

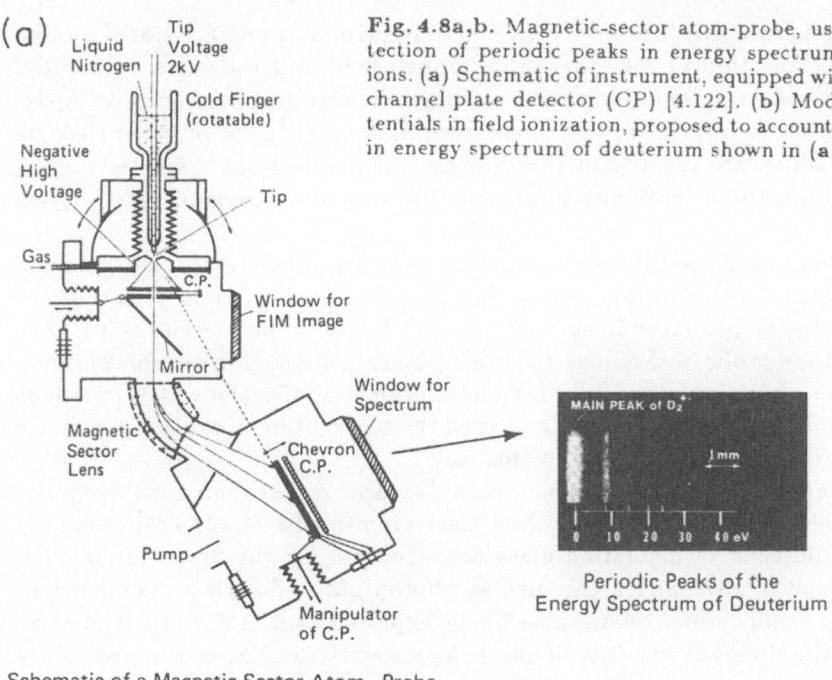

(a)

Liquid Nitrogen

Tip Voltage 2kV

Cold Finger (rotatable)

Negative High Voltage

Tip

Gas

C.P.

Window for FIM Image

Mirror

Magnetic Sector Lens

Window for Spectrum

Chevron C.P.

Pump

Manipulator of C.P.

Schematic of a Magnetic-Sector Atom—Probe

MAIN PEAK of D_2^+

1 mm

0   10   20   30   40 eV

Periodic Peaks of the Energy Spectrum of Deuterium

Fig. 4.8a,b. Magnetic-sector atom-probe, used in detection of periodic peaks in energy spectrum of field ions. (a) Schematic of instrument, equipped with chevron channel plate detector (CP) [4.122]. (b) Model of potentials in field ionization, proposed to account for peaks in energy spectrum of deuterium shown in (a) [4.126]

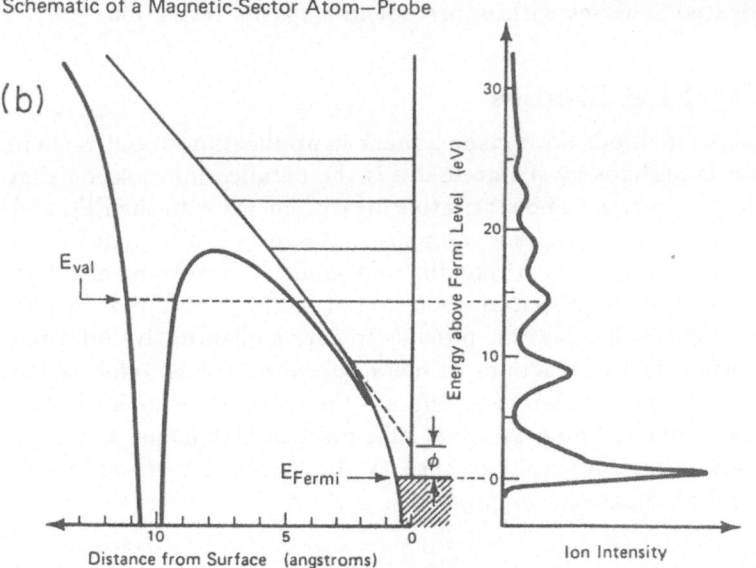

(b)

E_{val}

E_{Fermi}

ϕ

Energy above Fermi Level (eV)

30

20

10

0

Distance from Surface (angstroms)

10   5   0

Ion Intensity

shown schematically in Fig. 4.8a. Here, ions emitted from preselected sites of the emitter tip pass through the probe hole of a channel plate image intensifier and are subsequently mass and energy analyzed in a magnetic sector field. Single ions are detected by a chevron plate and fluorescent screen placed in the focal plane of the magnetic lens. In this way, the energy spectrum of D_2^+ field ions, shown in Fig. 4.8a, was photographically recorded. The observed oscillatory structure was interpreted as due to electronic resonance states formed by the superposition of the externally applied field and the surface potential [4.126–129], allowing field ionization at some distance from the surface. Apart from the intrinsic interest of this phenomenon, bombardment of the surface by electrons, which are created in this process with considerable kinetic energy, has turned out to be the limiting factor in probing some electronegative layers [4.88].

Imaging and pulsed-laser atom-probes both use TOF techniques for mass determination. In the voltage pulsed imaging-atom probe, time gating of the chevron detector produces images composed of specific ion species [4.91,123]. Imaging-atom probe techniques have also been combined with conventional atom probe operation, especially for metallurgical applications [4.11]. *Waugh* and *Southon* [4.130], for example, analyzed the segregation of oxygen at a single grain boundary in molybdenum in this way.

The nanosecond thermal pulse from a dye laser rather than a voltage pulse initiates field desorption in the pulsed-laser atom-probe [4.124,125]. Analysis of the composition of insulating glass was achieved for the first time in such an instrument [4.131], and areas such as photon-induced surface chemistry on metals and semiconductors are also being explored [4.86,132,133]. It is clear that with the different variants of the FIM presently available, the possibility exists for sophisticated analyses with unprecedented spatial resolution.

4.2 Illustrative FIM Studies

Field ion microscopic methods have made a mark in applications too diverse to review here. What is perhaps most impressive is the detailed information that has become available as a result of quantitative measurements with the FIM and its derivatives. In the next few pages we emphasize the role of such quantitative studies, by briefly surveying two entirely different kinds of measurements. Just by taking advantage of the high spatial resolution of the FIM, and its ability to depict individual atoms, it has become possible to derive quantitative information about the motion and interactions of metal atoms on solids, information never available before. Quite different in aim are the much more sophisticated measurements that probe the processes that take place in high fields. Although they require elaborate instrumentation, such studies promise to reveal much about chemical and physical surface processes, and are therefore examined in the last section.

4.2.1 Atomic Events on Solids

a) Surface Diffusion of Adatoms and Clusters

At the heart of any diffusion study is a measurement of the rate at which a deposit of the material under examination changes its spatial configuration. The tremendous resolution of the field ion microscope and the fact that it reveals surfaces on the atomic level has made this instrument ideal for establishing the

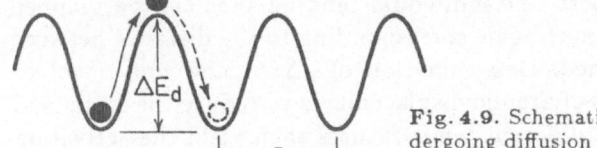

Fig. 4.9. Schematic of potential energy of adatom undergoing diffusion on a perfect surface with spacing a

surface transport characteristics of metal adatoms [4.15,134]. Basic to a determination of the diffusivity D is the Einstein relation, which in one dimension can be written as

$$\langle \Delta x^2 \rangle = 2\,Dt, \tag{4.3}$$

where $\langle \Delta x^2 \rangle$ is the mean-square displacement during time t. For an individual adatom the diffusion process can be modeled as in Fig. 4.9. The adatom usually is at one of the normal binding sites, where it is localized in its characteristic potential well. Every once in a while, the adatom is excited out of the well, and travels a short distance prior to deactivation. If the atom only jumps to nearest neighbor sites, at a distance a from the origin, then the diffusivity can be written as

$$D = \nu_0 a^2 \exp\left(-\Delta E_d / k_B T\right). \tag{4.4}$$

where ν_0 is the effective frequency at which an adatom attempts a jump in the diffusion direction, ΔE_d is the height of the barrier opposing atom motion, and k_B is as usual Boltzmann's constant. From the temperature dependence of the diffusivity, it is possible to deduce ΔE_d and therefore to gain insight into the energetics of surface interactions, whereas the diffusivity itself yields information about the kinetics of the transport process. The positions of an adatom on a surface between displacements are readily seen in the field ion microscope and are shown in Fig. 4.10 for a rhenium adatom [4.135] on W(211). By amassing

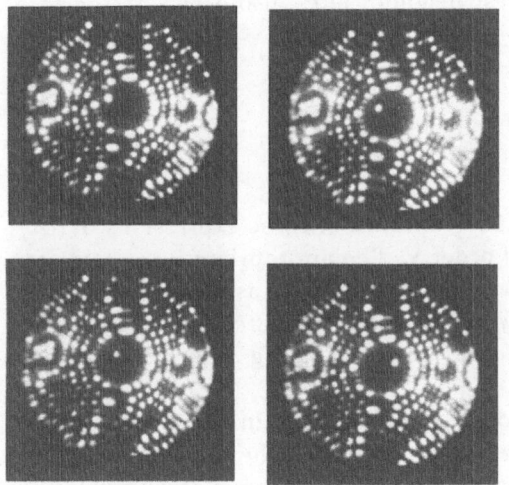

Fig. 4.10. Diffusion of a single rhenium adatom on the central (211) plane of a tungsten tip. Diffusion occurs during 30 s intervals at 351 K, between observations at T ≈20 K, and is always along direction of close packed atom rows on the surface. Courtesy Kaj Stolt

enough observations of this sort, the individual binding sites can be mapped out (Sect. 4.2.1b), so that a length scale corresponding to the distance between sites can be readily established. Determination of $\langle \Delta x^2 \rangle$ now only involves making sufficient observations of atomic displacements to reduce the statistical scatter, and measurements at different temperatures then yield the activation energy. This type of study has been done on a variety of metals and on different crystal facets, and has revealed quite varied behavior. On face-centered cubic surfaces, for example, self-diffusion has been found especially sensitive to the structure of the plane. On rhodium, surface diffusion of rhodium atoms on the close-packed plane, the (111), already occurs at roughly 50 K, whereas on (100), room temperatures are necessary for perceptible motion [4.136].

Such measurements of mean-square displacements serve to define the diffusivity. They do not necessarily prove that the picture of the diffusion process sketched above is applicable. One of the significant assumptions in the expression for the diffusivity D given by (4.4) is that atom jumps occur to nearest neighbor positions only. The validity of this picture, that is of the assumption that the jump distance is always small, cannot be deduced from measurements of the mean-square displacement alone – more information is required. This, it turns out, is readily available. The mean-square displacement is just the second moment of the distribution function for diffusion distances and, in principle, the entire distribution is accessible: it just requires many observations in order to establish the distribution without excessive scatter. The measured distribution function for displacements can then be compared with that for different models of the jump process. For example, if the adatom jumps to nearest neighbor sites only at a rate α, then the probability $p_x(t)$ of finding an atom at a distance x from the origin after time t is given by

$$p_x(t) = \exp(-\alpha t) I_x(\alpha t). \tag{4.5}$$

If, in addition, jumps to second nearest neighbor sites also occur, at the rate β, the more complicated relation

$$p_x(t) = \exp[-(\alpha + \beta)t] \sum_{j=-\infty}^{\infty} I_j(\beta t) I_{x-2j}(\alpha t) \tag{4.6}$$

describes the distance distribution [4.137]. In both of these expressions, $I_x(\alpha t)$ is just the modified Bessel function of order x. The appropriate jump rate can be found from a least-squares fit to the experimental data, as shown in Fig. 4.11 for the motion of tungsten atoms on W(211). From measurements available so far, it appears that on tungsten at least jumps spanning long distances are infrequent [4.137].

Though quite powerful, this approach is not sufficient in all cases to untangle the events in atom diffusion. Consider, for example, the diffusion of metal atoms on the (110) plane of the platinum metals. The atomic arrangement of

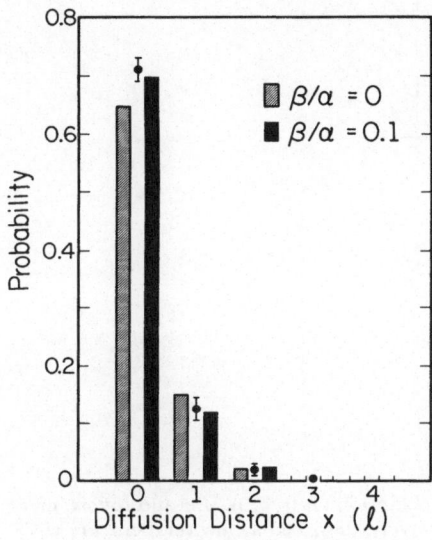

Fig. 4.11. Distribution of displacements for tungsten adatoms on W(211). Best fit is obtained assuming the ratio β/α of double to single jumps is $\approx 1/10$

the (110) plane in the bulk is reminiscent of the (211) plane in the bcc lattice, in that there are pronounced channels along which we expect the adatoms to move. On Rh(110), this is precisely what happens [4.136]. However, on platinum motion across the channels [4.138] is competitive with motion along the channels, and on Ir(110) movement is always across the channels [4.139]. What is the mechanism of this cross-channel diffusion? Two possibilities come to mind [4.138]. The adatom may just jump over the channel walls, possibly during a fluctuation in which the channel atoms are farther apart than usual. An alternative is a replacement mechanism, in which an adatom pushes a lattice atom into the adjacent channel and takes its place in the wall. These possibilities can be readily distinguished in diffusion of adatoms chemically different from the lattice. In the replacement mechanism the atom appearing in the adjacent channel is a lattice atom, whereas in the ordinary jump mechanism it is the original adatom that moves to the adjacent row. All that is required is a technique to reveal individual atoms as well as their chemical identity, that is, the atom probe. Experiments to examine cross-channel motion have been carried out by *Wrigley* [4.120,139], who looked at the behavior of tungsten on the (110) plane of iridium.

In experiments in which no diffusion was allowed to occur, iridium and tungsten atoms are readily distinguished in the TOF spectrum in Fig. 4.12. When a TOF analysis is done after cross-channel diffusion the atom that appears in the channel adjacent to the original site turns out to be iridium, that is a lattice atom. Furthermore, atom-by-atom analysis of the first substrate layer after diffusion frequently reveals a tungsten atom in the lattice, something not observed in the absence of diffusion. It is clear that the singular spatial resolution of the field ion microscope, combined where necessary with the ability to determine chemical identity in the atom-probe, has made it possible to obtain very detailed information about transport processes of individual adatoms.

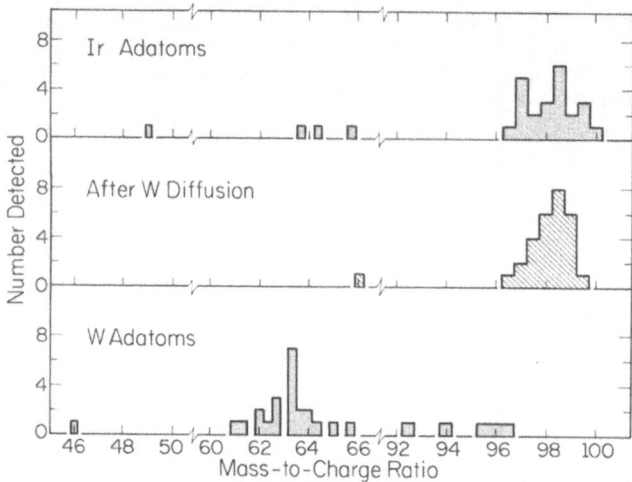

Fig. 4.12. Atom-probe observations of cross-channel motion on Ir(110). *Top* and *bottom*: measurements for iridium and tungsten adatoms respectively, desorbed at low temperature without any previous heating, to establish calibration. *Center*: observations on adatom subsequent to cross-channel motion [4.139]

These techniques have also proved useful in the examination of metal clusters [4.134,140,141]. When two adatoms are placed in adjacent channels of W(211), they associate into a dimer which moves as a unit as, for example, in Fig. 4.13. For dimers made up of rhenium adatoms, detailed studies have been carried out which reveal quite unexpected properties [4.81,135,142]. Cross-channel dimers diffuse more rapidly than the individual adatoms. A detailed kinetic analysis has revealed that this high mobility arises from the peculiar interactions between the rhenium atoms, which lower the barrier to motion for one atom when in close proximity to another. Recent studies by *Reed* [4.81] have also revealed a pronounced sensitivity to the arrangement of the cluster atoms on the surface. If two rhenium atoms are placed in the same rather than in adjacent channels, the resulting dimer (Fig. 4.14) behaves quite differently from cross-channel dimers. Migration is inhibited compared to that of a single rhenium atom, the barrier to diffusion of the dimer exceeding that of a single rhenium atom by 0.15 eV.

Although much less effort has been devoted to characterizing behavior of larger clusters, some intriguing properties of clusters on W(110) have already become evident. For overlayers of platinum, palladium and iridium [4.143], as well as for silicon [4.143,144], clusters can be formed with a very open spacing, which is apparent in Fig. 4.15. The stability of these structures depends strongly upon the temperature and surface concentration. This has been nicely demonstrated for iridium. At 460 K, clusters made up of 15 or more atoms are stable; clusters of 12 atoms, however, dissociate into chains of iridium atoms. Even large clusters have been found to have considerable mobility on W(110). For example, a palladium layer of more than 50 atoms sweeps over the surface

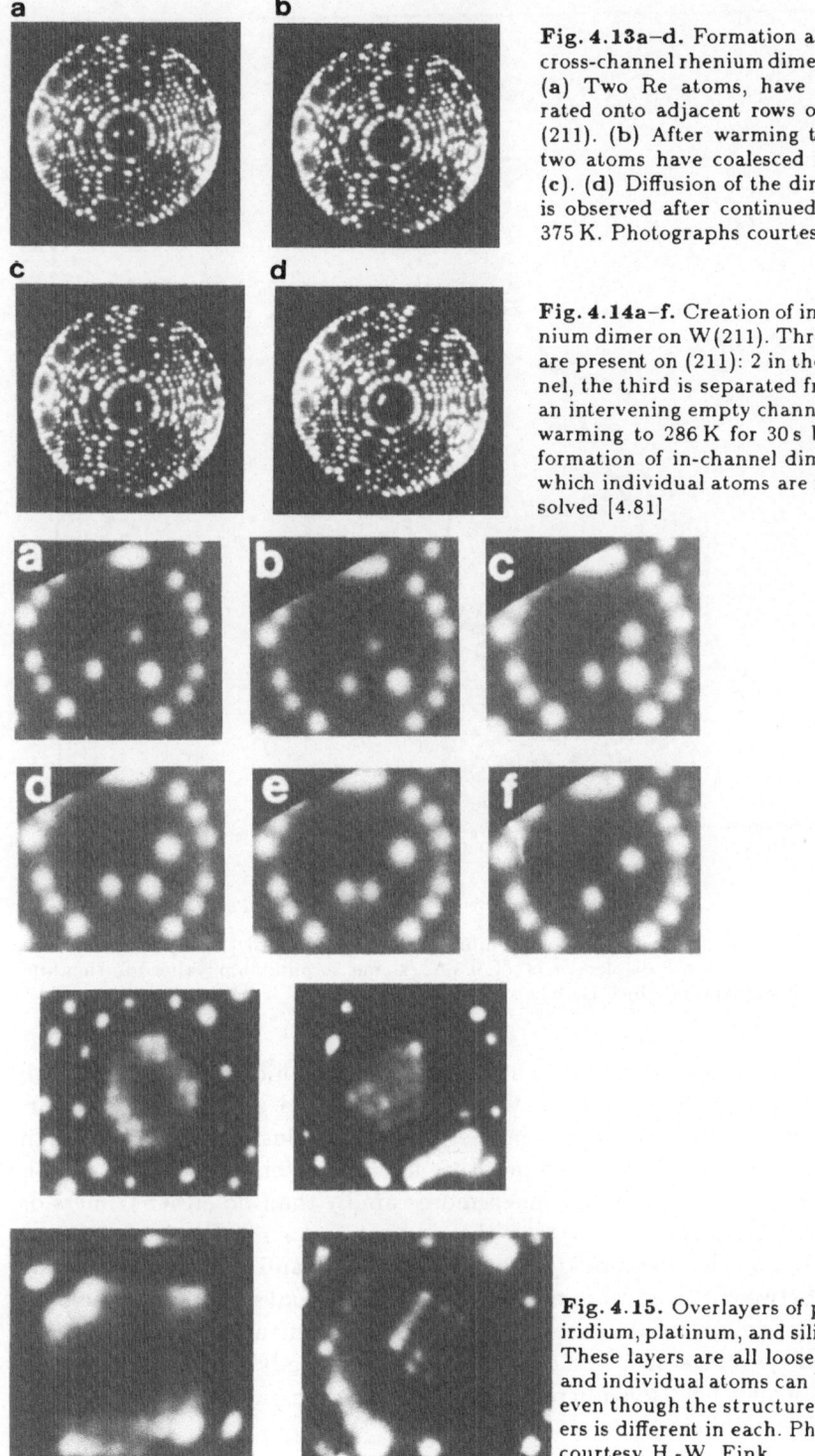

Fig. 4.13a–d. Formation and motion of cross-channel rhenium dimer on W(211). (a) Two Re atoms, have been evaporated onto adjacent rows of the central (211). (b) After warming to 375 K, the two atoms have coalesced into a dimer (c). (d) Diffusion of the dimer as a unit is observed after continued warming to 375 K. Photographs courtesy Kay Stolt

Fig. 4.14a–f. Creation of in-channel rhenium dimer on W(211). Three Re adatoms are present on (211): 2 in the same channel, the third is separated from these by an intervening empty channel. Repeated warming to 286 K for 30 s brings about formation of in-channel dimer (in f), in which individual atoms are no longer resolved [4.81]

Fig. 4.15. Overlayers of palladium, iridium, platinum, and silicon on W(110). These layers are all loosely packed, and individual atoms can be resolved, even though the structure of the layers is different in each. Photographs courtesy H.-W. Fink

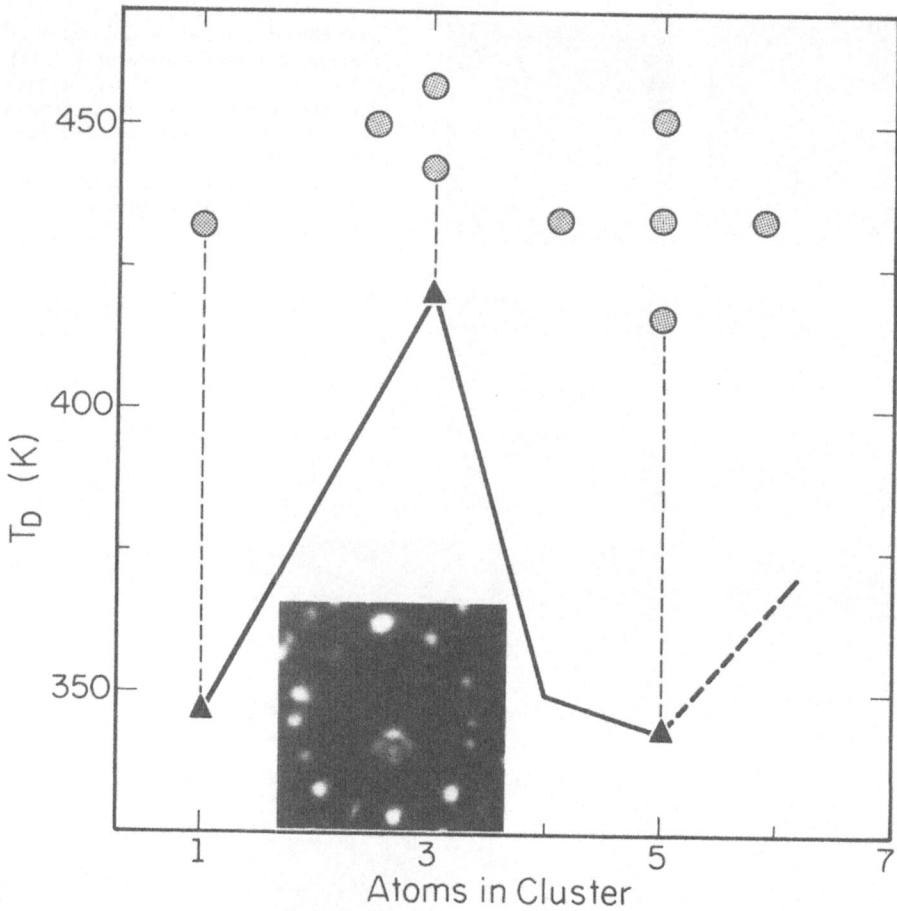

Fig. 4.16. Diffusion temperature T_D for rhenium clusters on W(110) [4.146]. T_D, the temperature for a mean-square displacement of $0.5\,\text{Å}^2/\text{s}$, has a minimum value for rhenium pentamers, a photograph of which is shown

at temperatures of 390 K [4.145]. The mechanism by which movement of such large clusters occurs has not been thoroughly explored and is not yet clear. However, there are unusual trends in the mobility as cluster size increases. In Fig. 4.16 is shown the diffusion temperature for Re clusters on W(110) [4.146]. The pentamers appear to diffuse much more rapidly than do either trimers or larger entities. It is also known that the pentamers have a much more open arrangement than do rhenium trimers on this plane, and there may be some correlation between the unusual mobility and the unusual structure of the pentameter. Much more needs to be done here to quantitatively map out the diffusion and structure of larger clusters, but it is already evident that this is one area in which field ion microscopic methods have radically changed the state of understanding.

b) Interactions Between Adatoms

The ability of the field ion microscope to visualize individual adatoms and therefore to establish a distance scale on the surface has also made it possible to examine interactions on surfaces in a very direct way. The idea behind such measurements is simple in the extreme. In an equilibrium system, the probability P_i of finding a specified configuration i is proportional to $\exp\left(-F_i/kT\right)$, F_i being the free energy of the system in that configuration. An experimental determination of P_i therefore affords an indication of the free energy F_i, which is the desired quantity for understanding interactions. Consider as an example a classical problem in crystal growth: the binding of a single atom at different sites on a surface. In the standard but obviously oversimplified view of crystals, atom binding should be strongest wherever the coordination between adatom and lattice is highest. The validity of this notion can be tested by examining the distribution of a single adatom [4.147] over the sites in a [111] channel on W(211). If the classical view is correct, sites in the center should be more heavily populated than sites at the ends, close to a descending lattice step. A test presupposes that it is possible to reproducibly locate atoms at different sites. That this is indeed the case is shown by the data (Fig. 4.17) on the actual positions observed for a tungsten adatom on W(211). There obviously is

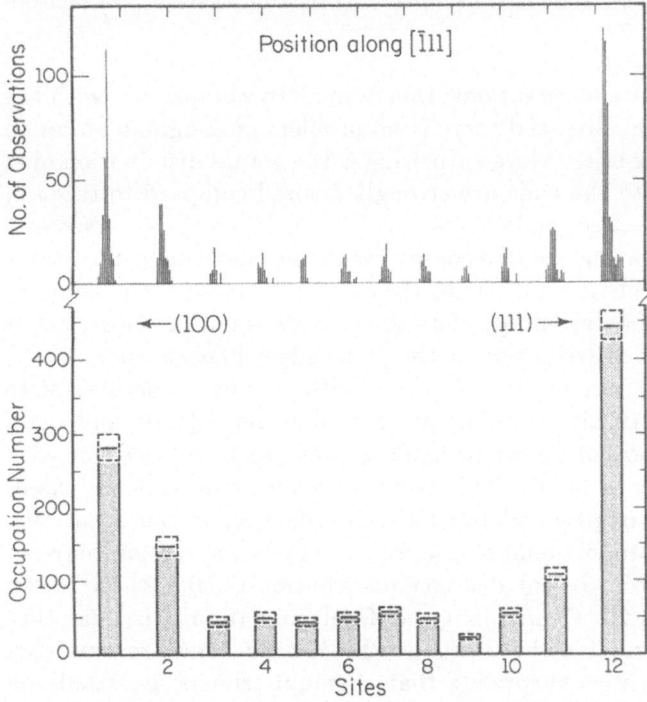

Fig. 4.17. Distribution of a single tungsten atom on W(211), after equilibration at 363 K [4.147]. *Top*: plot of positions at which adatom is observed in one diffusion channel: 12 clearly separated locations can be distinguished. *Bottom*: distribution of adatom over these 12 binding sites, derived from 1340 observations

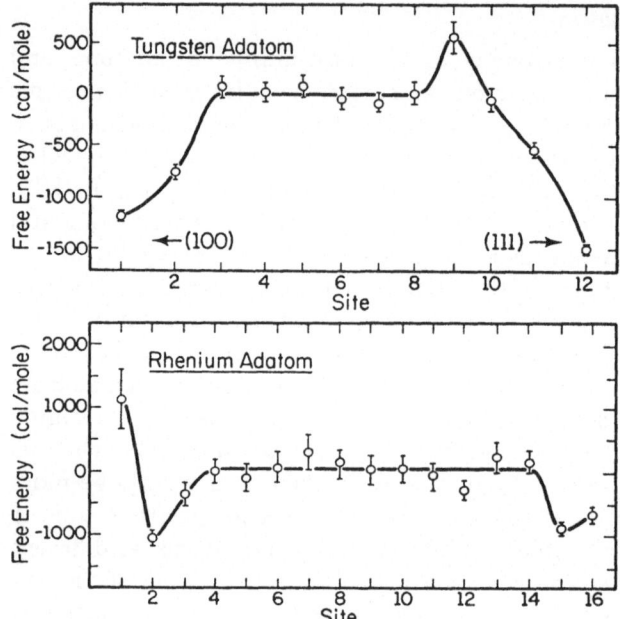

Fig. 4.18. Relative free energy of tungsten (*top*) and rhenium adatom (*bottom*) at different locations on W(211) [4.147]. Sites close to the ends of (211) show unusual behavior, but the details differ for the two adatoms

some scatter in the separate observations; this is small by comparison with the distance between the atom sites and there is no problem in assigning atoms to a finite number of binding sites. More surprising is the actual distribution over the sites. Positions close to the ends are strongly favored compard to those in the center of the channel. Also, end effects extend over several sites. As is evident from Fig. 4.18, comparing the free energy for rhenium and tungsten atoms at different sites, end effects are specific to the chemical nature of the adatom. The reason behind the low free energy close to the ends is not yet clear, but it seems likely that charge redistribution at the plane edges plays a role.

These techniques are not limited to planes with channel structures or to single adatoms. In Fig. 4.19, site mapping has been done on W(110), and again it is evident that allocation of atoms to binding sites can be carried out with a high degree of confidence [4.50]. This fact has made it possible to examine the probability of finding two adatoms at a specified separation from each other, which should be proportional to the free energy of interaction between the atoms. An example of a recent distance distribution [4.50,134] for Re on W(110) is given in Fig. 4.20. From this type of determination it is clear that at close separations, interactions between two rhenium adatoms are repulsive. In view of these results, it is surprising that rhenium trimers are stable on W(110); evidently many-atom interactions are responsible for the stability of rhenium clusters on this plane [4.148]. The ability of the field ion microscope to distinguish trimers from single adatoms has made it possible to look quan-

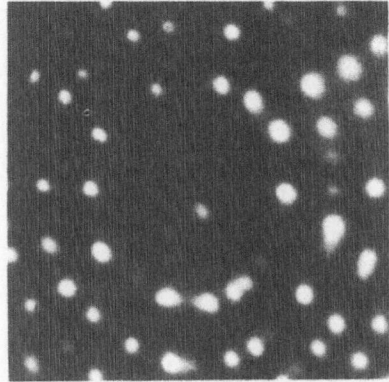

Fig. 4.19. Mapping of binding sites for rhenium adatom on W(110) [4.50]. A single Re adatom, at the left, is allowed to diffuse over the surface, visiting some sites several times. At right is shown a record of the positions actually observed. The scatter in the locations is small, and the grid of binding sites is clearly mapped out

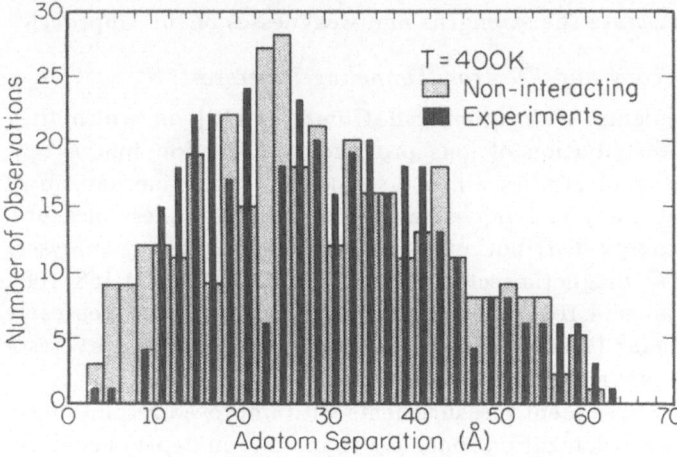

Fig. 4.20. Distance distribution of two rhenium adatoms on a W(110) plane of 64 Å diameter [4.148]. Equilibration at 400 K for 20 s. The data indicate repulsive interactions at distances < 10 Å

titatively at the kinetics of trimer dissociation, and therefore at the barrier to the dissociation process. The thermodynamics of the trimer-adatom equilibrium have also been examined [4.148]. The most important factor in limiting the accuracy of such determinations is the uncertainty in the number of sites available on the plane, but nevertheless the binding energy of rhenium in the trimer has been successfully estimated at 0.27 eV.

Although these experiments are tedious and time-consuming, they give a clear indication of the capability of field-ion microscopic methods to provide quantitative information about energetics at surfaces. One final point should be stressed about these determinations. In studies of the equilibrium distribution, or of diffusion, equilibration is done in the absence of any applied field;

imaging conditions are established only at low temperatures, where it can be shown that the act of observation has no significant effect upon the distribution. Properly done, such measurements therefore refer to an ordinary thermal system, unaffected by high fields.

4.2.2 Field Evaporation and Desorption Measurements

Although field ion images are by far the most immediately impressive demonstration of how powerful field ion microscopic techniques are, a great deal of information about the imaging process, and more important about events at the surface, has been obtained by analysis of the ions emitted from the surface under the influence of high fields. The level of instrumentation required for this, and the sophistication of the experiments, is much greater than in visual field ion microscopy. However, such measurements can make important contributions to the understanding of chemical and physical processes in high electric fields. Here we shall concentrate on a limited portion of these interesting studies, designed to illustrate the strengths and weaknesses of this approach.

a) Ion Energy Spectroscopy and Electron Tunneling Processes

The pioneering experiments of *Inghram* and *Gomer* [4.112], in which they measured the energy distribution of ions produced in field ion microscopy, were the first in a series of studies aimed at elucidating the mechanism of field ionization. A great many techniques have been utilized in these measurements of the field ion energy distribution: retarding potential energy analysers [4.17,22,34,121,149–157], magnetic-sector fields [4.56,68,122,126,151,158–160], electrostatic parallel plates [4.161] and TOF methods [4.69,127]. In their early study, *Müller* and *Bahadur* [4.149], using a retarding analyzer with 2 eV resolution, established the previously proposed existence of a critical distance for field ionization [4.112]. Subsequent measurements with improved resolution revealed an extremely narrow ionization zone, less than 0.2 Å in depth, i.e. about one-tenth the diameter of the imaging helium atom, during operation at best image field conditions [4.150]. These observations were in quantitative agreement with ionization at a critical distance of about 5 Å from the isopotential defining the surface [4.162]. The occurrence at higher fields of an oscillatory structure in the distribution of low-energy ions has been interpreted as resonance tunneling during space ionization of reflected or incoming atoms at separations beyond the critical distance [4.126–129]. An alternative explanation [4.163,164], namely surface plasmon creation, has not been confirmed by any experiment up to now [4.22,68,165].

Quite a different objective was pursued by *Utsumi* and *Smith* [4.154]: they attempted to probe the density of electronic states above the Fermi energy level by field ion energy analysis. Using a retarding analyzer with a resolution of less than 0.2 eV, they measured the kinetic energy distribution of He^+ and Ne^+ on both clean and nitrogen covered W(100) and W(110), and compared this to calculated results for the surface and bulk density of electronic states [4.166].

Hanson and *Inghram* [4.160] confirmed the existence of a second peak 2 eV beyond the main peak of He$^+$ field ionized at W(100) with a high-resolution magnetic sector. However, they were not able to duplicate the fine structure reported by Utsumi and Smith. *Hanson* and *Inghram* [4.167] came to the conclusion that the local arrangement of ion cores in the near-surface region in close proximity of the atom to be ionized is dominant in affecting the energy distribution, but such measurements have not been intensively pursued since then.

b) The Appearance Energy in Field Ionization

The appearance energy is by definition the energy required to remove an electron from molecular entities in their ground state to create ground state ions. The field ionization of He, for example, just amounts to the reaction

$$He \rightarrow He^+ + e^-,$$

occurring at or close to the emitter surface. As in ordinary mass spectrometry, the appearance energy for ionization processes occurring in high fields can, in principle, provide useful and important thermodynamic information, such as enthalpies of reactions. Especially for the evaporation of metals in high fields, this type of analysis has already yielded a surprising amount of information about atomic properties at the surface. The concept as well as the experimental determination of field ion appearance energies are not without their difficulties, and we shall therefore discuss these topics separately.

In a retarding potential experiment, appearance energies are measured by applying that voltage $\Delta \varsigma_0$ between emitter and collector which just suffices to keep the ion from being collected. For ions with a single charge e, the appearance energy A_1 is therefore given by

$$A_1 = e\Delta\varsigma_0 + \phi_c, \tag{4.7}$$

where ϕ_c is the work function of the collector. The physical significance of the appearance energy was examined in [4.150–153,156,157,159,161,168,169] but especially by *Forbes* [4.168], whose presentation we follow. He considered a thermionic cycle illustrated in Fig. 4.21a. For creation of singly charged ions in a retarding experiment, the following steps can be distinguished:

1. An electron is removed from the neutral atom, situated at the saddle position for field ionization, and placed at the Fermi level of the emitter.
2. In a step which requires no expenditure of work, the ion so formed is moved to a position just outside the collector.
3. The electron produced during ionization is taken around the circuit, across the retarding battery, to the collector. The amount of work in this step is $e\Delta\varsigma_0$.
4. The electron is then removed from the collector and allowed to recombine with the ion. An amount of work $\phi_c - I_1$ must be done, I_1 being the first ionization energy of the atom.

5. Finally the neutral atom is moved from just outside the collector back to its original position at the saddle.

In steps 1 and 5 the amount of work done equals $U_1^{sa} - U_0^{\infty}$ [4.152,168], the difference in the ion potential energy at the saddle and the potential energy of the neutral atom, infinitely far from the surface. The total amount of work done in this cycle is zero, as at the end the system is restored to its initial state; it follows that

$$A_1 = I_1 + U_0^{\infty} - U_1^{sa}. \tag{4.8}$$

For most gases accommodated to the temperature of a cryogenically cooled emitter, the difference in potential energies $U_0^{\infty} - U_1^{sa}$ is equal to $E_p^0(F)$, the polarization energy gained by the atom in moving from field-free space into the high-field region at the emitter surface [4.168]. In the field ionization of such gases

$$A_1 = I_1 + E_p^0(F). \tag{4.9}$$

For field evaporated or field desorbed ions, the appearance energy of singly charged species has been deduced by similar arguments as [4.168–171]

$$A_1 = I_1 + \Lambda(F) - Q(F) - zk_B T_s. \tag{4.10a}$$

Here Λ represents the binding energy of the atom at the surface, Q is the

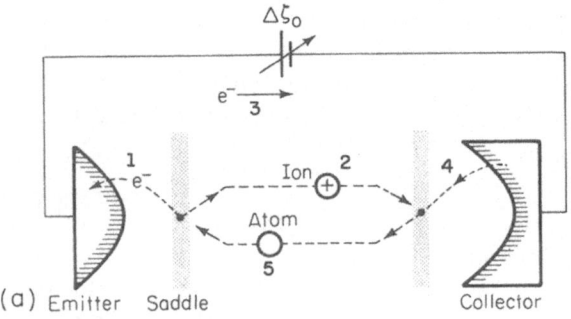

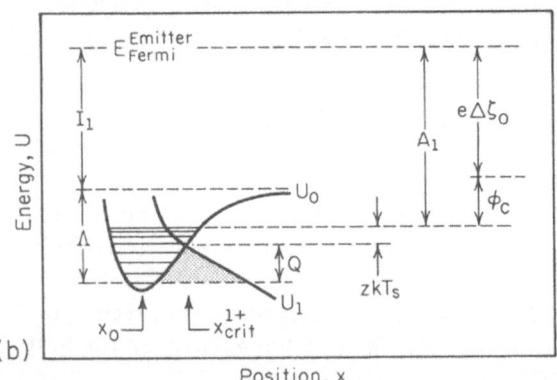

Fig. 4.21a,b. The appearance energy A_1 in field evaporation. (a) Thermodynamic cycle to determine the appearance energy [4.168]. (b) Potential diagram illustrating the quantities entering (4.10a) for the appearance energy of field desorbed ions

102

activation energy for field evaporation, and T_s the surface temperature. As illustrated in Fig. 4.21b, the last term zk_BT_s, z being a number around ten or less, accounts for the fact that the ion may desorb from a vibrationally excited state, exceeding the activation-energy barrier for field evaporation. For field evaporation of n-fold charged ions, (4.10a) can be generalized to [4.67,168]:

$$A_n = \sum_{i=1}^{n} I_i + \Lambda(F) - Q(F) - zk_BT_s, \tag{4.10b}$$

regardless of the mechanism of ion formation. The gas-phase ionization energies I_i are known or can be derived from appearance-energy measurements discussed in more detail in Sects. 4.2.2c and d. Since the activation energy for field evaporation $Q(F)$ can be obtained from the dependence of the evaporation rate upon the temperature of the surface, combined appearance and activation energy measurements together with experimental estimation of the zk_BT_s − term allow a determination of the binding energy $\Lambda(F)$. The experimental implementation of these ideas will be discussed next, together with selected examples of results obtained by appearance-energy analysis of field ionized and field desorbed molecules.

c) Experimental Studies of Field Ionization and Field Desorption

In the measurement of appearance energies the separation of species having different masses and charges is an experimental prerequisite. To achieve mass and charge separation, magnetic-sector spectrometers [4.151–153,155,156,172,173], as well as quadrupole mass filters [4.157,174] have been combined with retarding potential energy analyzers. Figure 5.22 shows a schematic of such an apparatus developed by one of the authors (N.E.) and his collaborators [4.172,175,176]. As in an ordinary atom-probe (Sect. 4.1.7), ions like He^+ and Ne^+ emitted from a selected surface site pass the probe hole of a channel plate intensifier, allowing the characterization of the emitter surface by field ion microscopy [4.175]. After mass separation in a magnetic mass spectrometer, ions are energy analyzed in a five-electrode energy filter equipped with two retarding gold meshes [4.169]. The ion rate is measured as a function of the voltage between emitter and retarder electrode. The onset value δ_0 of a retardation curve is ordinarily obtained by determining that voltage at which the ion signals first exceed the background rate. Under difficult circumstances, however, curve fitting procedures may eventually have to be applied [4.177].

Experiments carried out so far suggest that the considerations of Sect. 4.2.2b adequately describe the physical situation. As can be seen from the spectra displayed in Fig. 4.22, the onset voltage for $^{20}Ne^+$ generated by field ionization at the edge of a W(110) plane is significantly smaller than the value observed for $^4He^+$, as expected. In fact, it has been demonstrated that the differences in onset values, measured for rare gases, with the emitter at liquid nitrogen temperature, agree with the differences of free space ionization potentials within an accuracy of ± 0.1 V [4.153,156,157,169,172]. Also, onset voltages of rare-gas ion

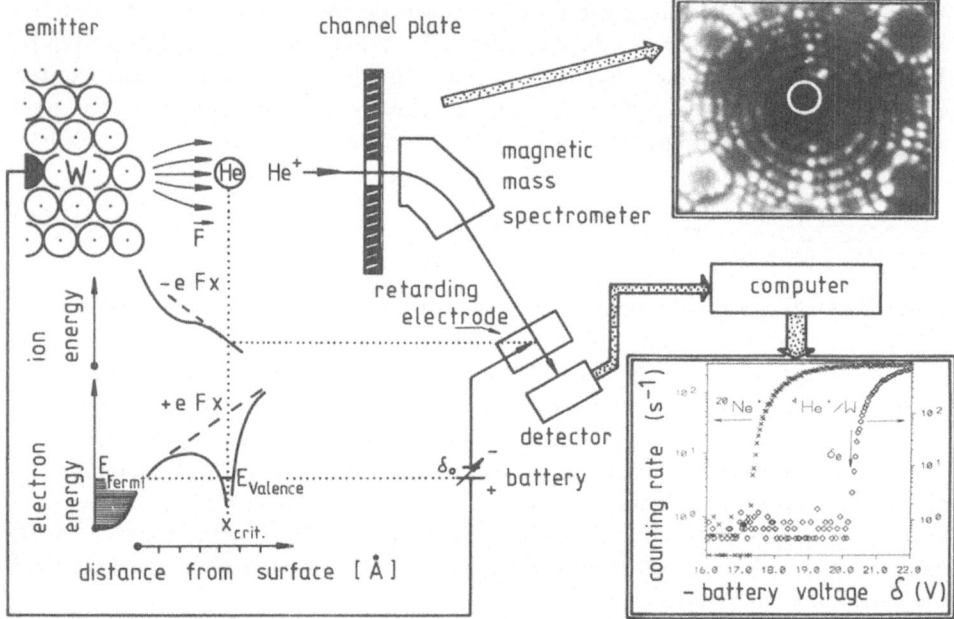

Fig. 4.22. Schematic of a field ion appearance spectrometer [4.172,176]. On the left is a schematic illustrating the field ionization of He. In the high local field in front of a step site, a helium atom is preferentially field ionized at and beyond the critical distance. x_{crit} is determined by the alignment of the Fermi energy of the metal with the valence-electron level of the atom. After passing through the probe hole of a channel plate intensifier, the He^+ ion beam is first mass and then energy analyzed in a retarding potential system. The white ring in the field ion micrograph (at right) indicates the size of the probe hole

retardation curves have been shown independent of the applied field strengths: the onsets did not show any upward shift, as might be expected from an increase in the polarization energy of the neutral atom [4.169,172,178]. Instead *Domke* et al. [4.157] measured decreasing onset values for Xe^+, Kr^+, and Ar^+ as the emitter temperature was raised up to 700 K. This behavior is explained by electron-tunneling processes, illustrated in Fig. 4.23a. The atom to be ionized can be located at distances smaller than the ordinary critical distance (shown, for example, in Fig. 4.22), due to the thermal depletion of electronic states below the Fermi level of the emitter. A further process, outlined schematically in Fig. 4.23b, involving the field ionization of high-energy particles, is also likely to occur [4.157]: a relatively small number of high-energy atoms having thermal energies $x k_B T_g$ may be ionized at the critical distance. Following the *Gomer* and *Swanson* theory of field desorption [4.58,59], the onset energy $e\delta_0$, together with the retarder work function ϕ_R shown in Fig. 4.23b, measure the total energy of field ionized particles. The appearance energy of ions originating from nonaccommodated high-energy atoms is thus given by [4.157]

$$A_1 = I_1 - x k_B T_g, \tag{4.11}$$

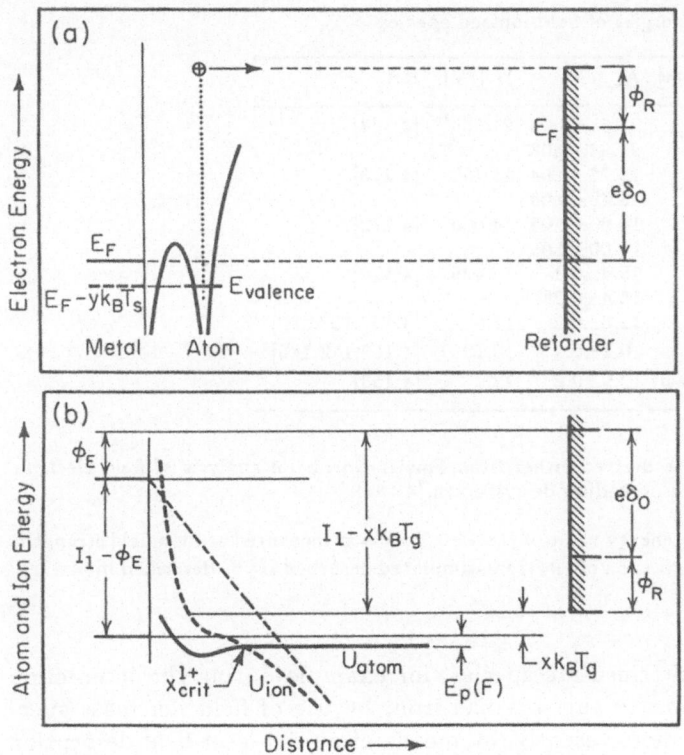

Fig. 4.23a,b. Potential diagrams illustrating different mechanisms involved in the decrease of appearance energy values [4.157]. (a) Electron tunneling into thermally depleted states below the Fermi level of a hot emitter. (b) Field ionization of hot neutrals at the critical distance

where $A_1 = e\delta_0 + \phi_R$. Similarly, for particles which are completely accommodated to the emitter temperature T_s, the relation

$$A_1 = I_1 + E_p(F) - y\,k_B T_s \tag{4.12}$$

has to be applied. Here, E_p represents, as before, the polarization energy of the atom [4.157]. So far, experiments, including measurements of ϕ_R by field electron emission, suggest that at least for rare-gas ions and molecular ions like H_2^+, H_2O^+ and NH_3^+ (4.11) is valid [4.157,178–181]. This is demonstrated in Table 4.1. The temperature term $xk_B T_g$ is negligibly small if measurements are performed at cryogenic temperatures. However, field ion appearance energies measured for CO^+ [4.177] and for paraffin ions like CH_4^+ [4.152,153], lie slightly above the adiabatic ionization energies. These results may possibly be ascribed to the influence of the electric field on Franck-Condon factors, which is still to be investigated [4.153,159].

Field fragmentation of molecules such as H_2 [4.158], as well as field induced surface reactions followed by desorption of product ions, are frequently observed phenomena during field ionization [4.34,56]. Attempts have been made to ob-

Table 4.1. Appearance energies of field ionized species

Ion/Metal	F[V/nm]	A_1 [eV]	I_1 [eV]	Ref.
$^{20}Ne^+$/W	30[a]	21.54±0.04	21.564[b]	[4.178]
	45	21.54±0.08		
$^{40}Ar^+$/W, Nb	14	15.78±0.04	15.759	[4.178]
	28	15.78±0.08		
$^{84}Kr^+$/W	9	14.00±0.05	14.000	[4.172]
	18	14.00±0.05		
H_2^+/W, Nb	19	15.4±0.1	15.425	[4.178]
	25	15.4±0.2[c]		
H_2O^+/W, Pt, Rh	7–14	12.6±0.2	12.6	[4.152,153,180]
NH_3^+/Pt, Rh	6–12	10.2±0.1	10.18	[4.152,172,180]
CH_4^+/Pt	unknown	12.9±0.2	12.7	[4.153]

[a] Field strength values were derived either from Fowler-Nordheim analysis of field electron emission currents or by the procedure described in [4.182].
[b] Data from [4.183].
[c] An additional appearance energy value of (14.4±0.2) eV was measured at high field strength; this arises from the field ionization of electron-stimulated desorbed H_2^+ as described in [4.178].

tain information on intermediate species, for example during the interaction of N_2 and H_2 with emitter surfaces like iron, by use of field ion mass spectrometry [4.184] and, more recently, by means of pulsed-laser field desorption [4.185,186]. Ionic species like H_3^+, H_3O^+, NH_4^+, even N_3^+ have been identified; but the reaction mechanisms remained obscure. *Anway* [4.151], *Goldenfeld* et al. [4.152] and *Heinen* et al. [4.153] have systematically measured appearance energy data for DC-field desorbed molecular ions; the data have been compared to calculated enthalpy values of model reactions. In this way the most probable surface reaction pathways have been derived, as demonstrated for a few examples in Table 4.2. For desorbed protonated water molecules H_3O^+, the measured appearance-energy value was found to depend on the emitter material. To explain this, a proton transfer reaction between adsorbed H_2O

Table 4.2. Appearance energies of field induced surface reaction products

Ion/Metal	A_1 [eV]	Proposed reaction	Ref.
H_3^+/Nb, Rh, W	12.1±0.1	$H_2(ad) + H(ad) \xrightarrow{-e^-} H_3^+$	[4.178,179]
H_3O^+/Pt, Rh /W	10.5±0.2 9.5±0.2	$2H_2O(ad) \xrightarrow{-e^-} {}^*OH + H_3O^+$	[4.153,180]
NH_4^+/Pt	9.5±0.1	$NH_3(ad) + H_2O(ad) \xrightarrow{-e^-} {}^*OH + NH_4^+$	[4.152]
$CH_3OH_2^+$/Pt	8.9±0.3	$2CH_3OH(ad) \xrightarrow{-e^-} {}^*OCH_3 + CH_3OH_2^+$	[4.153]

and subsequent attachment of a released OH radical to the emitter surface Me has been proposed, in which the binding energy of HO-Me contributes to the appearance energy $A_1(H_3O^+)$ [4.153,180]. It should be noted that during appearance-energy measurements only product ions and not surface reactants are analyzed. In special cases, additional experimental data about the temperature and field-strength dependence of the rate of ion formation have yielded information about reacting species. For example, in the formation of H_3^+ it has been possible to determine the lifetime and structure of an intermediate H_3 surface complex from an energy analysis of electron-stimulated field desorbed H_3^+ [4.178].

The effect of the high electric field on the kinetics and thermodynamics of surface reactions is, of course, an important consideration, as discussed in the book by *Beckey* [4.56] and the review by *Cocke* and *Block* [4.187]. Such effects can be eliminated, however. *Kruse* et al. [4.188] developed a TOF technique, in which no electric field is applied between two field-desorption pulses. Their recent measurements [4.189] have yielded data on the dependence of the adsorption time of NO on the temperature of a platinum emitter. In this way kinetic parameters have been derived for the field-free thermal desorption of NO from stepped surfaces of a platinum emitter.

d) Field Evaporation Studies

Field evaporation has already been introduced in Sect. 4.1.2. Recent studies relying on mass spectroscopic and appearance energy measurements provide a good indication of what can be accomplished by these methods. The spectrometry of field evaporated ions requires an extremely sensitive apparatus, since only a limited number of ions can be field evaporated in the course of an experiment in which the object under investigation is irrevocably altered by tip blunting. Mass spectrometric analysis of the low-temperature field evaporation of Be, Fe, Cu, and Zn was first achieved by *Barofsky* and *Müller* [4.190], who used a magnetic-sector mass spectrometer equipped with a high-gain resistance-strip detector. A change in the charge state of field evaporated beryllium from Be^{++} to Be^+ was observed as the temperature of the emitter was raised from 21 to 300 K. The measured relative abundance of differently charged beryllium ions strongly deviated from calculated data based on the image-potential hump model [4.52], especially at higher temperatures [4.190,191]. Qualitatively at least, the results of *Barofsky* and *Müller* [4.190] can be understood in terms of the post-ionization model: initially Be^+ desorbs, but loses a further electron by ordinary field ionization [4.73]. However, quantitative comparisons have not yet been made, due to the lack of an adequate field strength calibration.

Attempts to examine field evaporation processes by means of energy analysis were first made by *Müller* [4.22], and by *Waugh* and *Southon* [4.161]. These authors reported energy deficits of field evaporated multiply charged ions, such as Ta^{4+} and Ta^{3+} [4.22]. The results did not show the high deficits in kinetic energy predicted by the plasmon-loss theory of *Lucas* [4.163] but no details about the nature of the evaporation mechanism were given. Appearance spec-

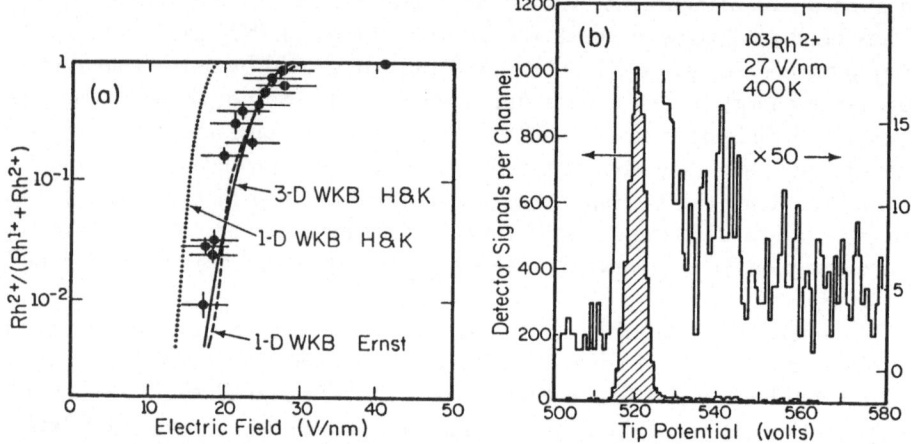

Fig. 4.24a,b. Field evaporation of rhodium. (a) Relative abundance of doubly charged rhodium as a function of field strength. Experimental points, obtained with a magnetic-sector mass spectrometer, are compared with different theoretical estimates of post-ionization [4.67,72]. (b) Differential energy distribution of Rh^{2+}; high tip potential corresponds to low kinetic energy. The detection of low energy Rh^{2+} provides direct evidence for the post-ionization of desorbing Rh^{1+} at some distance from the sample [4.68]

troscopic data on field evaporated and field desorbed Ag^+ have been reported [4.156,170]. These do not appear to be in agreement with expectations based on the image-potential theory; however, final conclusions on the desorption mechanisms could not be drawn.

The capabilities of appearance-energy analyses are demonstrated by detailed investigations of the field evaporation of singly and doubly charged rhodium ions, performed by one of the authors (N.E.) [4.67,68]. Four quantities were measured and compared to results of model calculations: (i) relative abundance of Rh^{2+} as a function of field strength, (ii) differential energy distributions, (iii) appearance energies and (iv) activation energies. Results of experiments of type (i) and (ii) shown in Fig. 4.24 provide strong support for the occurrence of post-field ionization of desorbing Rh^+. Relative abundances of Rh^{2+}, detected at different field strength values, are compared in Fig. 4.24a to calculated data based on the post-ionization model [4.67,72]. The post-field ionization of Rh^+ was strongly suggested by two further experimental results: (i) the detection of low-energy Rh^{2+}, as displayed in Fig. 4.24b, and (ii) the equality of activation energies for the field evaporation of Rh^{1+} and Rh^{2+} [4.67]. The appearance of a relatively small number of low-energy Rh^{2+}, under conditions where both ionic species are equally abundant, provides direct evidence for the post-ionization mechanism, since these low-energy Rh^{2+} must have been generated at a distance of more than about 1 nm (10 Å) from the surface [4.68]. For field strength values ranging between 15 and 40 V/nm, combined appearance- and activation-energy measurements have been performed on Rh^{2+} as well as on Rh^{1+}. Figure 4.25a shows retardation curves of singly and doubly charged rhodium; the onset voltage of the cut-off curves, together

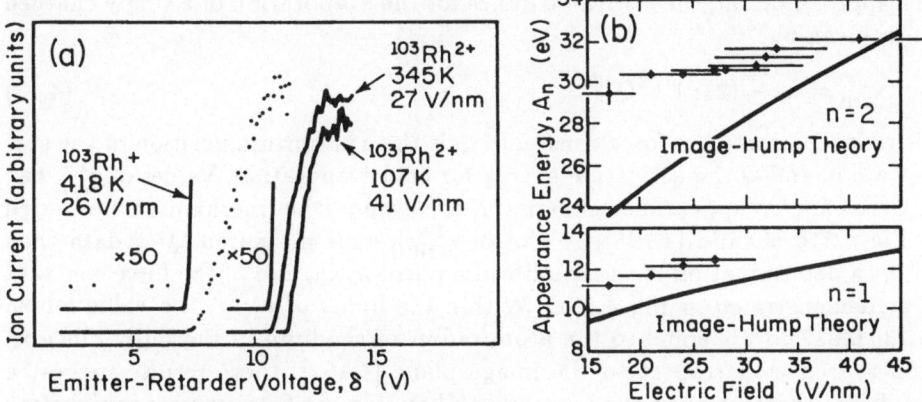

Fig. 4.25a,b. Analysis of rhodium field evaporation. (a) Retardation curves of singly and doubly charged field evaporated rhodium ions. (b) Appearance energies, derived from the onset of the retardation curves, as a function of field strength [4.67]. The predictions of the image-potential model [4.52] do not agree with experiment

with the actual value of the retarder work-function as well as the charge-number of the ion, defines the appearance energy A_n of n-fold charged ions [4.67]. It should be noted that $A_2 - A_1$, the difference in appearance energies of doubly and singly charged rhodium, derived from the data in Fig. 4.25b, equals the second ionization energy reported for gaseous Rh^{1+} in complete agreement with the occurrence of post-field ionization as well as with the theoretical consideration outlined in Sect. 4.2.2b. However, it is evident from Fig. 4.25b that the agreement between measured A_n data with calculated results, using the image-potential hump theory is very poor, especially for those Rh^{2+} field evaporated at lower field strength. The discrepancy between the activation energy for Rh^{2+} and Rh^{1+}, predicted by the image-potential model, and the measured values is even more severe. The image-potential model, which assumes the escape of an n-fold charged surface ion over a Schottky-type barrier, is consequently ruled out as the initial step in field evaporation of rhodium [4.67]. This conclusion is corroborated by *Kellogg*'s recent data [4.70] for the activation energy of field evaporated W^{3+} and W^{4+}. Alternative theoretical approaches to field evaporation have therefore been undertaken by *Chibane* and *Forbes* [4.192], *Kingham* [4.193] and most recently by *Tomanek* et al. [4.194].

Although more theoretical work is certainly necessary, some interesting physical quantities, like the equilibrium position, the vibrational frequency, the binding energy as well as the polarizability of the field evaporating atom, may be derived from data on the appearance and activation energy measured during field evaporation. This has successfully been done for rhodium. The procedure for obtaining this information has been outlined by *Forbes* et al. [4.195] and is based on three assumptions: (i) evaporation is thermally activated (ii) the atom is essentially neutral and bound in a potential trough, as shown in Fig. 4.21b, (iii) the atomic potential can be approximated as parabolic. In this approximation, the critical distance for the evaporation of a singly charged ion is given by

$$x_{crit}^{1+} = x_0 + (2/\kappa)^{1/2}Q^{1/2}. \tag{4.13}$$

where κ represents the force constant, x_0 is the equilibrium position of the surface atom, and Q the activation energy for field evaporation. Values of x_{crit}^{1+} can be derived from appearance energies, if the evaporation mechanism illustrated in Fig. 4.21b is valid [4.195]. A plot of x_{crit}^{1+} versus measured $Q^{1/2}$ data then allows a determination of the equilibrium position x_0, and of the force constant κ, as demonstrated in Fig. 4.26a. Within the limits of error, the value for x_0 (0.13±0.035 nm) is equal to the atom radius of rhodium in the bulk. Since x_0 is measured with respect to the image plane [4.162], these results suggest a significant contraction of the bond length between the field evaporating surface atom and the substrate lattice. From the force constant κ a vibration frequency of $(1.4_{+0.5}^{-0.3}) \cdot 10^{12} s^{-1}$ has been derived. This is comparable to, but significantly smaller than the Debye frequency values derived theoretically or from LEED experiments on macroscopic rhodium crystals [4.196,197].

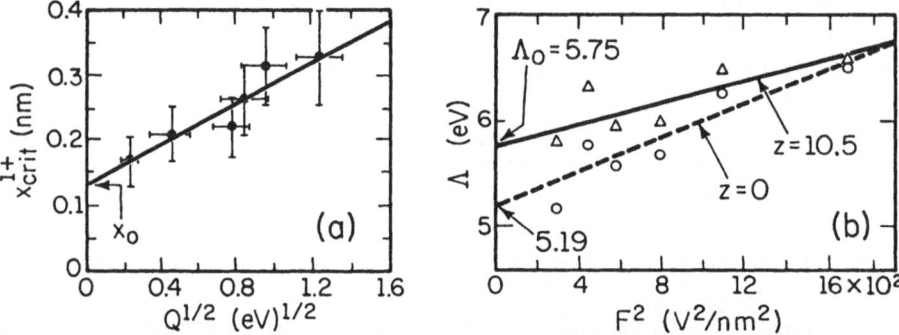

Fig. 4.26a,b. Derivation of atomic parameters from field evaporation data for rhodium. (a) Estimation of x_0, the equilibrium position of a surface atom, from data on the critical distance, in accord with (4.13). (b) Field-strength dependence of the binding energy of rhodium atoms, according to (4.15). The intercept yields the sublimation energy Λ_0, the slope gives the polarizability α [4.171,195]

The binding energy Λ of field evaporating rhodium atoms can be derived by applying (4.10b). For doubly charged ions this becomes

$$\Lambda(F) = A_2(F) + Q(F) + zk_BT_s - (I_1 + I_2). \tag{4.14}$$

The first and second ionization energies of the free atom, I_1 and I_2, have been determined by other spectroscopic methods [4.198]. Data for A_2 and Q are available from combined appearance- and activation-energy measurements. The quantity z in the temperature term zk_BT_s, which represents an experimental parameter involving the signal-to-noise ratio during the appearance-energy measurements (Fig. 4.21) [4.157], cannot be determined directly during field evaporation. However, *Forbes* and *Chibane* [4.171] have proposed a method in which z is derived by fitting experimentally determined Λ-values to a quadratic field-strength dependence

$$\Lambda(F) = \Lambda_0 + (\alpha/2)F^2,\tag{4.15}$$

as shown in Fig. 4.26b. In this model, α represents the polarizability of the field evaporating atom and Λ_0 is the sublimation energy. As demonstrated in Fig. 4.26b, a value of $z = 10.5$ yields a fitted line whose intercept corresponds to the known sublimation energy of rhodium (5.75 eV). The value of the polarizability α obtained from the slope of the curve is found to be (1.5 ± 0.5) Å^3. This is smaller than the gas-phase polarizability of Rh $(7.6\,\text{Å}^3)$, a result in general agreement with the findings of *Tsong* and *Kellogg* [4.199] who carried out experiments on the field-induced directional walk of single adatoms. The theoretical understanding of these results of two completely different experimental methods is still in an early stage in spite of efforts to apply quantum mechanical [4.199] as well as classical electrostatic [4.171] considerations on dipole formation in a single metal adatom.

The experimental estimate of binding energy values presented here is not sensitively affected by field-strength calibrations or by assumptions about evaporation mechanisms [4.200]. As such it is certainly an improvement over earlier methods which have relied on the measurement of the field-strength dependence of the desorption rate for metal atoms [4.201,202]. There is one important but difficult extension of these experiments that must still be made – the determination of the field-free adsorption energy of single selected adatoms. However, it is already clear from the existing material that combined appearance- and activation-energy measurements are capable of providing incisive information about surface properties on an atomic scale.

Acknowledgements. We want to express our appreciation to Prof. J.H. Block and the Fritz-Haber-Institut der Max-Planck-Gesellschaft for making this collaboration possible. In preparing this review, we have benefited from many discussions with colleagues in the department Grenzflächenreaktionen at the FHI, Berlin, and in the Coordinated Science Laboratories, Urbana. Special thanks are due to Gabriele Mehnert for undertaking the typing at a critical stage.

References

4.1   E.W. Müller: Z. Physik **131**, 136 (1951)
4.2   M. Isaacson, D. Kopf, M. Utlaut, N.W. Parker, A.V. Crewe: Proc. Nat. Acad. Sci. USA **74**, 1802 (1977)
4.3   M.S. Isaacson, J. Langmore, N.W. Parker, D. Kopf, M. Utlaut: Ultramicrosc. **1**, 359 (1976)
4.4   M. Utlaut: Phys. Rev. B**22**, 4650 (1980)
4.5   K. Takayanagi: J. Microsc. **136**, 287 (1984)
4.6   J.J. Metois, K. Takayanagi, Y. Tanishiro: Surf. Sci. **155**, 53 (1985)
4.7   G. Binnig, H. Rohrer: Helv. Phys. Acta **55**, 726 (1982)
4.8   G. Binnig, H. Rohrer: Physica B&C **127**, 37 (1984)
4.9   R.S. Becker, J.A. Golovchenko, B.S. Swartzentruber: Phys. Rev. Lett. **54**, 2678 (1985)

4.10   W.R. Graham, G. Ehrlich: Surf. Sci. **45**, 530 (1974)
4.11   R. Wagner: In *Crystals-Growth, Properties, and Applications,* ed. by H.C. Freyhard (Springer, Berlin, Heidelberg 1982), Vol. 6, p. 1, and references therein
4.12   R. Wagner: Phys. Blätt. **36**, 65 (1980)
4.13   D.N. Seidman: In *Radiation Damage in Metals,* ed. by N.L. Peterson and S.D. Harkness (Am. Soc. Metals, Metals Park. 1976) p. 28
4.14   S.S. Brenner, M.K. Miller: J. Metals **35**, 54 (1983)
4.15   G. Ehrlich: In Proc. 9th Intern. Vacuum Congress and 5th Intern. Conf. on Solid Surfaces. Invited Speakers Volume, ed. by J.L. de Segovia (ASEVA, Madrid 1983) p. 3
4.16   A recent review has been given by J.A. Panitz: J. Phys. E**15**, 1281 (1982)
4.17   E.W. Müller, T.T. Tsong: *Field Ion Microscopy, Principles and Applications* (American Elsevier, New York 1969)
4.18   K.M. Bowkett, D.A. Smith: In *Defects in Crystalline Solids,* Vol. 2, ed. by S. Amelinckx, R. Gevers, J. Nihoul (North-Holland, Amsterdam 1970)
4.19   T.J. Wilkes, G.D.W. Smith, D.A. Smith: Metallography **7**, 403 (1974)
4.20   M. Drechsler, P. Wolf: In Proc. 4th Intern. Conf. on Electron Microscopy, ed. by B.W. Bargmann, G. Möllenstedt, H. Niehrs, D. Peters, E. Ruska, C. Wolpers (Springer, Berlin 1960) p. 835
4.21   J.A. Becker: Solid State Physics **7**, 379 (1958)
4.22   E.W. Müller: Ber. Bunsenges. Phys. Chem. **75**, 979 (1971)
4.23   T.T. Tsong, E.W. Müller: Phys. Rev. Lett. **25**, 911 (1970); J. Chem. Phys. **55**, 2884 (1971)
4.24   K.D. Rendulic: Surf. Sci. **28**, 285 (1971); **34**, 581 (1973)
4.25   R.G. Forbes: Vacuum **31**, 567 (1981); J. Phys. D**18**, 473 (1985)
4.26   J.A. Panitz, E.W. Müller, S.B. McLane: Surf. Sci. **17**, 430 (1969)
4.27   R.J. Culbertson, T. Sakurai, G.H. Robertson: Phys. Rev. B**19**, 4427 (1979)
4.28   N. Igata, S. Sato: In *Proc. 27th Int. Field Emission Symp.,* ed. by Y. Yashiro, N. Igata (University of Tokyo, Tokyo 1980) p. 95
4.29   N. Ernst: Phys. Rev. Lett. **45**, 1573 (1980)
4.30   N. Ernst, J.H. Block: Surf. Sci. **117**, 561 (1982)
4.31   D.A. Nolan, R.M. Herman: Phys. Rev. B**8**, 4099 (1973)
4.32   H. Iwasaki, S. Nakamura: Surf. Sci. **49**, 664 (1975)
4.33   R.G. Forbes, M.K. Wafi: Surf. Sci. **93**, 192 (1980)
4.34   E.W. Müller, T.T. Tsong: In *Progress in Surface Science,* Vol. 4, ed. by S.G. Davison (Pergamon, Oxford 1974) p. 1
4.35   O. Nishikawa, E.W. Müller: J. Appl. Phys. **35**, 2805 (1964)
4.36   E.W. Müller: Surf. Sci. **8**, 462 (1967)
4.37   W. Schmidt, Th. Reisner, E. Krautz: Surf. Sci. **26**, 297 (1971)
4.38   A.P. Janssen, J.P. Jones: Surf. Sci. **33**, 533 (1972)
4.39   Y.C. Chen, D.N. Seidman: Surf. Sci. **26**, 61 (1971)
4.40   R. Gomer: *Field Emission and Field Ionization* (Harvard Univ. Press, Cambridge, MA 1961)
4.41   M.J. Southon: Ph.D. Thesis, University of Cambridge (1963)
4.42   H.A.M. van Eekelen: Surf. Sci. **21**, 21 (1970)
4.43   K.D. Rendulic, M. Leisch: Surf. Sci. **95**, L271 (1980)
4.44   H.H.H. Homeier, D.R. Kingham: J. Phys. D**16**, L115 (1983)
4.45   D.R. Kingham, N. Garcia: J. Physique **45**(C9), 119 (1984)
4.46   G. Ehrlich: Surf. Sci. **63**, 422 (1977)
4.47   V.T. Binh, M. Drechsler: J. Physique **45**(C9), 29 (1984) and references therein
4.48   E.W. Müller: J. Appl. Phys. **28**, 1 (1957)
4.49   A.P. Janssen, J.P. Jones: J. Phys. D**4**, 118 (1971)
4.50   H.-W. Fink, G. Ehrlich: J. Chem. Phys. **81**, 4657 (1984)
4.51   E.W. Müller: Naturwissen. **29**, 533 (1941)
4.52   E.W. Müller: Phys. Rev. **102**, 618 (1956)
4.53   R. Gomer: Appl. Phys. **19**, 365 (1979)
4.54   D.R. Kingham, L.W. Swanson: J. Physique **45**(C9), 133 (1984) and references therein
4.55   A.L. Suvorov, V.V. Trebukhovskii: Usp. Fiz. Nauk. **107**, 657 (1972) [English transl.: Sov. Phys. Usp. **15**, 471 (1973)]

4.56   H.D. Beckey: *Principles of Field Ionization and Field Desorption Mass Spectrometry* (Pergamon, Oxford 1977)

4.57   D.G. Brandon: In *Field-Ion Microscopy*, ed. by John J. Hren, S. Ranganathan (Plenum, New York 1968) p. 28

4.58   R. Gomer: J. Chem. Phys. **31**, 341 (1959)

4.59   R. Gomer, L.W. Swanson: J. Chem. Phys. **38**, 1613 (1963)

4.60   T.T. Tsong, E.W. Müller: Phys. Stat. Sol. (a) **1**, 513 (1970)

4.61   T.T. Tsong: J. Chem. Phys. **54**, 4205 (1971); Surf. Sci. **70**, 211 (1978)

4.62   E.W. Müller, J.A. Panitz, S.B. McLane: Rev. Sci. Instr. **39**, 83 (1968)

4.63   S.S. Brenner, J.T. McKinney: Appl. Phys. Lett. **13**, 29 (1968)

4.64   E.W. Müller, S.V. Krishnaswamy: Phys. Rev. Lett. **37**, 1011 (1976)

4.65   D.M. Taylor: Ph.D. Thesis, Cambridge University (1970)

4.66   R.S. Chambers: Ph.D. Thesis, University of Illinois at Urbana-Champaign (1975)

4.67   N. Ernst: Surf. Sci. **87**, 469 (1979)

4.68   N. Ernst, Th. Jentsch: Phys. Rev. B**24**, 6234 (1981)

4.69   T.T. Tsong, T.J. Kinkus: Phys. Rev. B**29**, 529 (1984)

4.70   G.L. Kellogg: Phys. Rev. B**29**, 4304 (1984)

4.71   D.R. Kingham: Vacuum **32**, 471 (1982)

4.72   R. Haydock, D.R. Kingham: Phys. Rev. Lett. **44**, 1520 (1980)

4.73   D.R. Kingham: Surf. Sci. **116**, 273 (1982)

4.74   M. Konishi, M. Wada, O. Nishikawa: Surf. Sci. **107**, 63 (1981)

4.75   G.L. Kellogg: Phys. Rev. B**24**, 1848 (1981)

4.76   G.L. Kellogg: Surf. Sci. **120**, 319 (1982)

4.77   H.O. Andrén, A. Henjered, D.R. Kingham: Surf. Sci. **138**, 227 (1984)

4.78   J.J. Hren, S. Ranganathan (eds.): *Field-Ion Microscopy* (Plenum, New York 1968) p. 231

4.79   R. Liu: Ph.D. Thesis, University of Illinois at Urbana-Champaign (1977) App. A

4.80   P. Hirsch, A. Howie, R.B. Nicholson, D.W. Pashley, M.J. Whelan: *Electron Microscopy of Thin Crystals* (R.E. Krieger Publishing, Malabar, FL 1977) App. 1

4.81   D.A. Reed, G.Ehrlich: Surf. Sci. **151**, 143 (1985)

4.82   E.D. Boyes, M.J. Southon: Vacuum **22**, 447 (1973)

4.83   J.J. Carroll, A.J. Melmed: Surf. Sci. **116**, 225 (1982)

4.84   B. Ralph, S.A. Hill, M.J. Southon, M.P. Southon, A.R. Waugh: Ultramicrosc. **8**, 361 (1982)

4.85   A.J. Melmed, R.J. Stein: Surf. Sci. **49**, 645 (1975)

4.86   G.L. Kellogg: Phys. Rev. B**28**, 1957 (1983)

4.87   T. Sakurai, R.J. Culbertson, A.J. Melmed: Surf. Sci. **78**, L221 (1978)

4.88   G. Ehrlich, F.G. Hudda: Phil. Mag. **8**, 1587 (1963)

4.89   J.A. Panitz: J. Microscopy **125**, 3 (1982)

4.90   E.W. Müller: Adv. Electron. Electron Phys. **13**, 83 (1960)

4.91   J.A. Panitz: Progress in Surface Science **8**, 219 (1978)

4.92   B. Leskovar: Phys. Today **30**, 42 (November 1977)

4.93   J.L. Wiza: Nucl. Instr. Methods **162**, 587 (1979)

4.94   M. Wulf: Physik in unserer Zeit **3**, 90 (1981)

4.95   A. van Oostrom: CRC Crit. Rev. Solid State Sci. **4**, 353 (1974)

4.96   J.A. Panitz: J. Vacuum Sci. Technol. **17**, 757 (1980)

4.97   H.-W. Fink.: Ph.D. Thesis, Technical University of Munich (1982)

4.98   Th. Schiller: Ph.D. Thesis, Technical University of Berlin (1985)

4.99   Th. Schiller: Private communication (1985)

4.100   Th. Schiller, U. Weigmann, S. Jaenike, J.H. Block: J. Physique **47** (C2), 479 (1985)

4.101   J.D. Wrigley, G. Ehrlich: In Proc. Materials Research Soc., San Francisco 1985, in press

4.102   R.H. Good, E.W. Müller: In *Handbuch der Physik*, Vol. 21, ed. by S. Flügge (Springer, Heidelberg 1956) p. 176

4.103   A. Modinos: *Field, Thermionic, and Secondary Electron Emission Spectroscopy* (Plenum, New York 1984)

4.104   L.W. Swanson, A.E. Bell: Adv. Electron. Electron Phys. **32**, 193 (1973)

4.105   E.W. Plummer, T.N. Rhodin: Appl. Phys. Lett. **11**, 194 (1967)

4.106   G.L. Kellogg, T.T. Tsong: Surf. Sci. **62**, 343 (1977)

4.107   C.J. Todd, T.N. Rhodin: Surf. Sci. **36**, 353 (1973)

4.108   J.W. Gadzuk, E.W. Plummer: Rev. Mod. Phys. **45**, 487 (1973)
4.109   R.D. Young, E.W. Müller: J. Appl. Phys. **33**, 91 (1962)
4.110   R.S. Polizzotti, G. Ehrlich: Surf. Sci. **91**, 24 (1980)
4.111   N. Ernst, G. Ehrlich: J. Physique **45**(C9), 293 (1984); **47** (C2), 47 (1985)
4.112   M.G. Inghram, R. Gomer: J. Chem. Phys. **22**, 1279 (1954); Z. Naturforsch. A**10**,869 (1955)
4.113   E.W. Müller: In *Methods of Surface Analysis*,Vol. 1, ed. by A.W. Czanderna (Elsevier, Amsterdam 1975) p. 329, and references therein
4.114   S.S. Brenner, J.T. McKinney: Surf. Sci. **23**, 88 (1970)
4.115   S.V. Krishnaswamy, E.W. Müller: Rev. Sci. Instrum. **45**, 1049 (1974)
4.116   G. Abend: Ph.D. Thesis, Freie Universität of Berlin (1979)
4.117   E.W. Müller, S.V. Krishnaswamy: Rev. Sci. Instrum. **45**, 1053 (1974)
4.118   O. Nishikawa, K. Kurihara, M. Nachi, M. Konishi, M. Wada: Rev. Sci. Instrum. **52**, 25 (1981)
4.119   R.S. Chambers, G. Ehrlich: J. Vac. Sci. Technol. **13**, 273 (1976)
4.120   J.D. Wrigley, G. Ehrlich: Phys. Rev. Lett. **44**, 661 (1980)
4.121   T. Utsumi, O. Nishikawa: J. Vac. Sci. Technol. **9**, 477 (1972)
4.122   E.W. Müller, T. Sakurai: J. Vac. Sci. Technol. **11**, 878 (1974)
4.123   J.A. Panitz: J. Vac. Sci. Technol. **11**, 206 (1974)
4.124   W. Drachsel, S. Nishigaki, J.H. Block: Int. J. Mass Spectrom. Ion Phys. **32**, 333 (1980)
4.125   G.L. Kellogg, T.T. Tsong: J. Appl. Phys. **5**, 1184 (1980)
4.126   A. Jason: Phys. Rev. **156**, 266 (1967)
4.127   E.W. Müller, S.V. Krishnaswamy: Surf. Sci. **36**, 29 (1972)
4.128   M.E. Alferieff, C.B. Duke: J. Chem. Phys. **46**, 938 (1967)
4.129   J. Appelbaum, E.G. McRae: Surf. Sci. **47**, 445 (1975)
4.130   A.R. Waugh, M.J. Southon: Surf. Sci. **68**, 79 (1977)
4.131   G.L. Kellogg: J. Appl. Phys. **53**, 6383 (1982)
4.132   S. Nishigaki, W. Drachsel, J.H. Block: Surf. Sci. **87**, 389 (1979)
4.133   C.F. Ai, T.T. Tsong: J. Chem. Phys. **81**, 2845 (1984)
4.134   A thorough review is given by D.W. Bassett: In *Surface Mobilities on Solid Materials*, ed. by V.T. Binh (Plenum, New York 1983) p. 63,83
4.135   K. Stolt, W.R. Graham, G. Ehrlich: J. Chem. Phys. **65**, 3206 (1976)
4.136   G. Ayrault, G. Ehrlich: J. Chem. Phys. **60**, 281 (1974)
4.137   G. Ehrlich: J. Vac. Sci. Technol. **17**, 9 (1980)
4.138   D.W. Bassett, P.R. Webber: Surf. Sci. **70**, 520 (1978)
4.139   J.D. Wrigley: Ph.D. Thesis, University of Illinois at Urbana-Champaign (1982)
4.140   G. Ehrlich: Phys. Today **34**, 44 (June 1981)
4.141   G. Ehrlich, K. Stolt: In *Growth and Properties of Metal Clusters*, ed. by J. Bourdon (Elsevier, Amsterdam 1980) p. 1
4.142   K. Stolt, J.D. Wrigley, G. Ehrlich: J. Chem. Phys. **69**, 1151 (1978)
4.143   H.-W. Fink, G. Ehrlich: Surf. Sci. **110**, L611 (1981)
4.144   T.T. Tsong, R. Casanova: Phys. Rev. Lett. **47**, 113 (1981)
4.145   H.-W. Fink, cited by G. Ehrlich: In *Chemistry and Physics of Solid Surfaces V*, ed. by R. Vanselow and R. Howe, Springer Ser. Chem. Phys., Vol. 35 (Springer Berlin, Heidelberg 1985) p. 293
4.146   H.-W. Fink, G. Ehrlich: Surf. Sci. **150**, 419 (1985)
4.147   H.-W. Fink, G. Ehrlich: Surf. Sci. **143**, 125 (1984)
4.148   H.-W. Fink, G. Ehrlich: Phys. Rev. Lett. **52**, 1532 (1984)
4.149   E.W. Müller, K. Bahadur: Phys. Rev. **102**, 624 (1956)
4.150   T.T. Tsong, E.W. Müller: J. Chem. Phys. **41**, 3279 (1964)
4.151   A.R. Anway: J. Chem. Phys. **50**, 2012 (1969)
4.152   I.V. Goldenfeld, I.Z. Korostyshevsky, G.B. Mischanchuk: Int. J. Mass Spectrom. Ion Phys. **13**, 297 (1974)
4.153   H.J. Heinen, F.W. Röllgen, H.D. Beckey: Z. Naturforsch. **29a**, 773 (1974)
4.154   T. Utsumi, N.V. Smith: Phys. Rev. Lett. **33**, 1294 (1974)
4.155   J.H. Block, L. Ernst, N. Ernst: Jap. J. Appl. Phys. **14**, 1813 (1975)
4.156   T.T. Tsong, W.A. Schmidt, O. Frank: Surf. Sci. **65**, 109 (1977)
4.157   M. Domke, E. Hummel, J.H. Block: Surf. Sci. **78**, 307 (1978)
4.158   G.R. Hanson: J. Chem. Phys. **62**, 1161 (1975)

4.159 A.J. Jason, A.C. Parr: Int. J. Mass Spectrom. Ion Phys. **22**, 221 (1976)
4.160 G.R. Hanson, M.G. Inghram: Surf. Sci. **55**, 29 (1976)
4.161 A.R. Waugh, M.J. Southon: J. Phys. D**9**, 1017 (1976)
4.162 N.D. Lang, W. Kohn: Phys. Rev. B**7**, 3541 (1973)
4.163 A.A. Lucas: Phys. Rev. Lett. **26**, 813 (1971)
4.164 A.A. Lucas, M. Sunjic: J. Vac. Sci. Technol. **9**, 729 (1972)
4.165 A.J. Jason, A.C. Parr, M.G. Inghram: Phys. Rev. B**7**, 2883 (1973)
4.166 N.E. Christensen, B. Feuerbacher: Phys. Rev. B**10**, 2349 (1974)
4.167 G.R. Hanson, M.G. Inghram: Surf. Sci. **64**, 305 (1977)
4.168 R.G. Forbes: Surf. Sci. **61**, 221 (1976)
4.169 N. Ernst: Ph.D. Thesis, Freie Universität of Berlin (1976)
4.170 E. Hummel, M. Domke, J.H. Block: Z. Naturforsch. **34a**, 47 (1978)
4.171 R.G. Forbes, K. Chibane: Surf. Sci. **121**, 275 (1982) and references therein
4.172 N. Ernst, G. Bozdech, J.H. Block: Int. J. Mass Spectrom. Ion Phys. **28**, 33 (1978)
4.173 R.J. Culbertson, T. Sakurai: J. Vac. Sci. Technol. **15**, 1752 (1978)
4.174 R.-G. Abitz: Ph.D. Thesis, Freie Universität of Berlin (1978)
4.175 N. Ernst, G. Bozdech: Unpublished work (1983)
4.176 N. Ernst, G. Bozdech, S. Kato, J.H. Block: J. Physique **45**(C9), 231 (1984)
4.177 O. Frank, W.A. Schmidt: Int. J. Mass Spectrom. Ion Phys. **29**, 117 (1979)
4.178 N. Ernst, J.H. Block: Phys. Rev. B**12**, 7092 (1984)
4.179 N. Ernst, J.H. Block: Surf. Sci. **126**, 397 (1983)
4.180 N. Ernst, G. Bozdech, J.H. Block: Ber. Bunsenges. Phys. Chem. **82**, 756 (1978)
4.181 N. Ernst, G. Bozdech, J.H. Block: Int. J. Mass Spectrom. Ion Phys. **28**, 27 (1978)
4.182 T. Sakurai, E.W. Müller: Phys. Rev. Lett. **30**, 532 (1973)
4.183 R.D. Levin, S.G. Lias: *Ionization Potential and Appearance Potential Measurements 1971–1981*, Natl. Stand. Ref. Data Ser., Natl. Bur. Stand, USA (1982)
4.184 W.A. Schmidt: Z. Angew. Chem. **4**, 151 (1968)
4.185 J.H. Block, W. Drachsel, N. Ernst, Th. Jentsch, S. Nishigaki: In *Ion Formation from Organic Solids*, ed. by A. Benninghoven, Springer Ser. Chem. Phys., Vol. 25 (Springer Berlin, Heidelberg 1983) p. 211
4.186 W. Liu, T.T. Tsong: Surf. Sci. **151**, 251 (1985)
4.187 D.L. Cocke, J.H. Block: Surf. Sci. **70**, 363 (1978), and references therein
4.188 N. Kruse, G. Abend, W. Drachsel, J.H. Block: In Proc. 3rd Intern. Congr. on Catalysis, Berlin (1984) p. 105
4.189 N. Kruse, T. Kessler, G. Abend, J.H. Block: J. Physique **45**(C9), 227 (1984)
4.190 D.F. Barofsky, E.W. Müller: Surf. Sci. **10**, 117 (1968)
4.191 T.T. Tsong: Surf. Sci. **9**, 31 (1968)
4.192 K. Chibane, R.G. Forbes: J. Physique **45**(C9), 99 (1984)
4.193 D.R. Kingham: In Proc. 29th Intern. Field Emission Symp., ed. by H.O. Andrén, H. Nordén (Almqvist & Wiksell Int., Stockholm 1982) p. 27; and J. Physique C**47** (C2),11 (1985)
4.194 D. Tomanek, H.J. Kreuzer, J.H. Block: Surf. Sci. **157**, L315 (1985)
4.195 R.G. Forbes, K. Chibane, N. Ernst: Surf. Sci. **141**, 319 (1984)
4.196 D.G. Castner, G.A. Somorjai, J.E. Block, D. Castiel, R.F. Wallis: Phys. Rev. B**24**, 1616 (1981)
4.197 C.M. Chan, P.A. Thiel, J.T. Yates, W.H. Weinberg: Surf. Sci. **76**, 296 (1978)
4.198 C.E. Moore: Ionization Potentials and Ionization Limits Derived from the Analyses of Optical Spectra, Natl. Stand. Ref. Data Ser., Natl. Bur. Stand, USA **34**, 1ff. (1970)
4.199 T.T. Tsong, G.L. Kellogg: Phys. Rev. B**12**, 1343 (1975)
4.200 M. Vesely, G. Ehrlich: Surf. Sci. **34**, 547 (1973)
4.201 G. Ehrlich, C.F. Kirk: J. Chem. Phys. **48**, 1465 (1968)
4.202 E.W. Plummer, T.N. Rhodin: J. Chem. Phys. **49**, 3479 (1968)

5. X-Ray and Neutron Diffraction

C.N.J. Wagner

With 16 Figures

X-ray diffraction has been a powerful tool to elucidate the crystal and defect structure of metallic alloys. Shortly after the discovery of x-ray diffraction
by M. von Laue in 1912, the simple description of the scattering of x-rays by
lattice planes

$$2d \sin \theta = \lambda$$

was given by W.L. Bragg, where d is the interplanar spacing of the diffracting
hkl planes, θ is the Bragg angle (2θ is the scattering angle, i.e., the angle
between the incident and the diffracted beams), and λ is the wavelength. The
powder method was introduced by P. Debye and P. Scherrer in 1915, which
became the standard method to study the structure of poly-crystalline alloys.
The neutron, which was discovered in the early 30's, has been used in diffraction
experiments on crystalline and non-crystalline materials since 1946 when E.O.
Wollan and C.G. Shull began their pioneering neutron scattering studies.

The early scattering experiments have been performed with relatively weak
sources and rather inefficient detectors. The past two decades have seen great
advances in new sources and detectors for neutron and x-ray scattering experiments. The development of pulsed x-ray synchrotron [5.1] in the 60's and
neutron spallation [5.2] sources in the 70's has not only increased their respective fluxes, but also made possible experiments which take advantage of the
time structure of these sources. The advent of Si and Ge solid-state detectors
for x-rays, and position-sensitive proportional and scintillation detectors for
x-rays and neutrons has considerably improved the utilization of conventional
radiation sources (sealed and rotating x-ray tubes, and nuclear reactors for
neutrons).

In this chapter, a brief survey will be presented of new techniques and
methods to analyze and characterize the structure of poly- and non-crystalline
metallic alloys. Most engineering alloys are used in poly-crystalline form with
grain or particle sizes in the micrometer range (micro-crystalline metals and alloys) or subgrain and sub-particle sizes in the nanometer range (nano-crystalline
alloys). These materials might exhibit random or preferred orientation of the
grains or sub-grains in their solid form. Non-crystalline metallic alloys have
been prepared by rapid quenching from the liquid and the vapor phases, by ion
implantation and by solid-state diffusion. These alloys, sometimes called metallic glasses or glassy metals, exhibit interesting structural, magnetic, electrical
and chemical properties [5.3].

In any scattering or diffraction experiment on poly- or non-crystalline alloys, the goal is to measure the scattered intensity as a function of the length of the diffraction vector $K = Q = k_1 - k_0$, where k_0 and k_1 are the incident and the scattered wave-vectors, respectively. In x-ray diffraction, $k_1 = k_0$ and it follows that the length of the diffraction vector is given by

$$K = Q = (4\pi/\lambda) \sin\theta \qquad (5.1)$$

where 2θ is the angle between the incident beam k_0 and the scattered beam k_1. This can be accomplished by varying (i) the scattering angle 2θ using a single wavelength (variable 2θ-method), or (ii) the wavelength λ of the incident beam at a fixed scattering angle 2θ (variable λ-method) [5.4]. Both techniques have been applied to the study of the bulk structure of micro-, nano-, and non-crystalline metals and alloys. Recently, the grazing-incidence scattering of x-rays has been used to investigate the structure of poly- and non-crystalline surfaces and thin films [5.5,6].

Scattering experiments permit us to determine the topological and the chemical short-range order in multi-component metallic alloys, whether the structure is poly- or non-crystalline [5.7,8]. It will be shown in this chapter, that the topological order can be deduced from the number (or density) structure factor $S_{nn}(Q)$, whereas the chemical short-range order can be evaluated from the concentration structure factor $S_{cc}(Q)$ [5.4]. In crystalline alloys, $S_{nn}(Q)$ corresponds to the Bragg reflections of the powder pattern. $S_{cc}(Q)$ describes the modulation of the Laue monotonic background scattering, which is the precursor of the superstructure reflections in perfectly ordered alloys, or leads to small-angle scattering when precipitation occurs [5.9].

The broadening and the displacement of the peaks in the powder pattern of crystalline alloys can be used to evaluate the size of the coherently diffracting domains (particle or sub-grain size) [5.8,10], the microstrains within these domains [5.10], and the macrostrains due to applied or residual stresses [5.11,12].

The intensity of the individual reflections in the powder pattern of a multiphase alloy allows us to evaluate the amount of each phase in the mixture [5.13]. If the grains or particles in the irradiated volume of the poly-crystalline alloy do not have a random distribution in orientation, the intensity of the Bragg peaks will depend on the orientation of the normal to the reflecting planes with respect to the laboratory coordinates, usually taken as the normal to the specimen surface and two perpendicular axes in the specimen surface [5.13].

5.1 Diffraction of Neutrons and X-Rays by Poly- and Non-Crystalline Alloys

5.1.1 Neutron and X-Ray Scattering

In a diffraction experiment, a beam of neutrons or x-rays defined by the wave-vector k_0 and the energy E_0 is incident upon a sample, as shown in Fig. 5.1. A diffracted (or scattered) beam with wave-vector k_1 and energy E_1 emerges from the sample, making the angle 2θ with the incident beam.

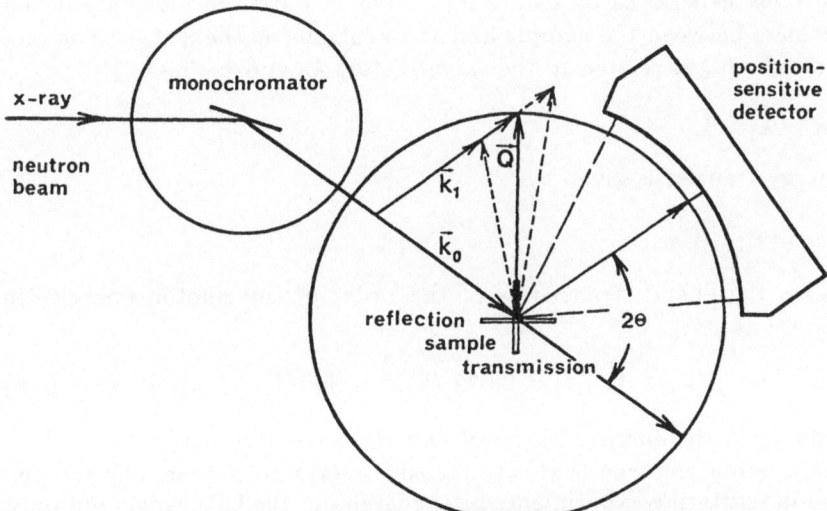

Fig. 5.1. Diffraction geometry for x-rays and neutrons, using the variable 2θ-technique with a monochromator in the primary beam. In x-ray diffraction, the incident wave-vector k_0 and the scattered wave-vector k_1 have equal lengths, i.e., $k_0 = k_1$, but in neutron diffraction, a range of k_1 is observed, because of the considerable momentum transfer which can take place in the neutron scattering process

The wave-vector k of the neutron particle or the x-ray photon is related to the momentum p through the de Broglie equation, i.e., $k = p/\hbar$, with $\hbar = h/(2\pi)$, h being Planck's constant. The momentum of the neutron is $p = mv$, m and v being the mass and the velocity of the neutron, respectively. The momentum of the x-ray photon can be expressed as $p = \hbar\omega/c$, where ω is the angular frequency and c is the velocity of light.

Consequently, we find the following relations for the wave-vector k:

neutron particle: $k = mv/\hbar$ (5.2)

x-ray photon: $k = \omega/c$. (5.3)

The kinetic energy of the thermal neutrons is given by

$$E = mv^2/2 = p^2/(2m) = \hbar^2 k^2/(2m) = [h^2/(2m)]/\lambda^2.$$

119

If we express the energy E of the neutrons in meV, the velocity v in m/s and the wavelength λ in Å, we obtain the following relations

$$E = 5.2276 \times 10^{-6} v^2 = 81.787/\lambda^2. \tag{5.4}$$

The energy of the x-ray photon can be written as

$$E = \hbar\omega = hc/\lambda = 12.40/\lambda \tag{5.5}$$

if we express the energy E in keV and the wavelength λ in Å.

In the most general case, we are interested in both the momentum and energy transfers between the sample and the neutrons or the x-rays. The momentum transfer $\hbar Q$ is related to the wave-vectors k_0 and k_1, i.e.,

$$\boldsymbol{Q} = \boldsymbol{K} = \boldsymbol{k}_1 - \boldsymbol{k}_0$$

and the energy transfer is given by

$$E = \hbar\omega = E_1 - E_0.$$

For neutrons, the energy transfer is of the order of the photon energies in metallic alloys, i.e.,

$$E = [\hbar^2/(2m)][(k_1)^2 - (k_0)^2] = 2.072[(k_1)^2 - (k_0)^2] \tag{5.6}$$

if we express again the energy E in meV and the wave-vector k in Å^{-1}. Thus, an incoming thermal neutron is able to transfer energy to the metallic sample. X-rays used in scattering experiments have energies in the keV range, but only shifts of about 1 eV can be readily measured, which are larger than the phonon energies in solids and liquids.

In the scattering process, we distinguish two cases: (i) elastic scattering, where the incident and scattered radiation have equal energies, i.e., $E_0 = E_1$, and (ii) inelastic scattering, in which measures are taken to define both incoming and scattered energies [5.7], because $E_0 \neq E_1$.

In this chapter, we will discuss only the elastic scattering of x-rays and neutrons, i.e., we will employ the *static approximation of the scattering*. In this case, $k_0 = k_1 = k = 2\pi/\lambda$. Thus, $Q = K = 2k \sin\theta = (4\pi/\lambda) \sin\theta$. The wavelength λ [Å] is related to the energy E [keV] of the x-ray photon or the velocity v [cm/μs], the length L [cm] of the flight path, and the time-of-flight t [μs] of the neutron, i.e.,

$$\lambda_{photon} = hc/E = 12.40/E, \tag{5.7}$$

$$\lambda_{neutron} = h/(mv) = (h/m)(t/L) = 0.3955(t/L). \tag{5.8}$$

5.1.2 General Scattering Theory for Solid and Liquid Solutions

Let us consider an alloy consisting of n elements $1, 2, 3, \ldots, n$. There are a total number of N atoms in volume V, and N_i of these atoms belong to element i.

$\varrho_0 = N/V$ is the average atomic density of the alloy, $\varrho_i = N_i/V$ is the average atomic density of element i, and $c_i = N_i/N = \varrho_i/\varrho_0$ is the average concentration of element i. As a consequence, we have the following sum rules:

$$\sum N_i = N; \quad \sum \varrho_i = \varrho_0; \quad \text{and} \quad \sum c_i = 1.$$

The actual atomic density $p_i(r)$ of element i can be expressed as a series of delta functions, i.e.,

$$p_i(r) = \sum_k \delta(r - r_{ik}). \tag{5.9}$$

a) Amplitude of Scattered Radiation

The amplitude $A_n(Q)$ of the scattered radiation (x-rays or neutrons) is given in the static approximation by [5.4]:

$$A_n(Q) = A_e(Q) \sum_i f_i(Q)P_i(Q) \tag{5.10}$$

where $A_e(Q)$ is the amplitude of scattering by an isolated scatterer (electron or nucleus), $f_i(Q)$ is the complex scattering amplitude (or length) of element i, and $P_i(Q)$ is the Fourier transform of $p_i(r)$, i.e.,

$$P_i(Q) = \int p_i(r) \exp(-iQr)dr. \tag{5.11}$$

It is advantageous to introduce the shape function $b(r)$ of the scattering object of volume V, defined as $b(r) = 1$ inside V and $b(r) = 0$ outside V [5.14]. Thus, we can write: $V = \int b(r)dr$. The Fourier transform of $b(r)$ is called the shape amplitude $B(Q)$, i.e.,

$$B(Q) = \int b(r) \exp(-iQr)dr. \tag{5.12}$$

b) Intensity of Scattered Radiation (Static Approximation)

The intensity $I_n(Q)$, scattered by N atoms in the sample of volume V, i.e., the coherent scattering cross-section, is given by [5.4]

$$I_n(Q) = A_n(Q)A_n^*(Q) = I_e(Q) \sum \sum f_i(Q)f_j^*(Q)P_i(Q)P_j^*(Q) \tag{5.13}$$

where $A^*(Q)$, $f^*(Q)$, and $P^*(Q)$ are the complex conjugate quantities of $A(Q)$, $f(Q)$, and $P(Q)$, respectively. Employing the convolution theorem of the Fourier transform [5.14], it is readily seen that the product $P_i(Q)P_j^*(Q)$ is the Fourier transform of the convolution $q_{ij}(r)$, i.e.,

$$P_i(Q)P_j^*(Q) = \int q_{ij}(r) \exp(-iQr)dr \quad \text{and} \tag{5.14}$$

$$q_{ij}(r) = P_i(r)*p_j(r) = \int p_i(u)p_j(u + r)du.$$

The function $q_{ij}(r)$ is the generalized Patterson function, also called the Hosemann q-function.

If we define $\varrho_{ij}(r)$ as the number of j-type atoms per unit volume at the distance r from an i-type atom, we can write

$$q_{ij}(r) = N_i\{\delta_{ij}\delta(r) + [\varrho_{ij}(r) - \varrho_j]\}v(r) + N_i\varrho_j v(r) \tag{5.15}$$

where δ_{ij} is the Kronecker symbol, $\varrho_j = c_j\varrho_0$, $N_i = c_i N$, and

$$v(r) = (1/V)\int b(\mathbf{u})b(\mathbf{u} + \mathbf{r})d\mathbf{u} = b(\mathbf{r})*b(-\mathbf{r})/V \tag{5.16}$$

is the common volume of the object of volume V and its ghost displaced by the vector $\mathbf{r}$. Combining (5.14 and 15), we can write immediately

$$P_i(\mathbf{Q})P_j(\mathbf{Q}) = N\{c_i\delta_{ij} + c_ic_j[I_{ij}(\mathbf{Q}) - 1] + c_ic_j\varrho_0 V(\mathbf{Q})\} \tag{5.17}$$

where the shape factor $V(\mathbf{Q})$ is related to the volume small-angle scattering, i.e.,

$$V(\mathbf{Q}) = \int v(\mathbf{r}) \exp(-i\mathbf{Qr})d\mathbf{r} = [B(\mathbf{Q})]^2/V \tag{5.18}$$

and $J_{ij}(\mathbf{Q}) = [I_{ij}(\mathbf{Q}) - 1]$ is the minor partial interference function or structure factor, i.e.,

$$J_{ij}(\mathbf{Q}) = I_{ij}(\mathbf{Q}) - 1 = \int v(\mathbf{r})[(\varrho_{ij}(\mathbf{r})/c_j) - \varrho_0] \exp(-i\mathbf{Qr})d\mathbf{r}. \tag{5.19}$$

c) Isotropic Distribution of Units of Structure in the Alloy

If we assume that the units of structure or building blocks in the alloys (for example, atoms, molecules, or small grains) are distributed isotropically, we can replace the volume integral in the Fourier transform by a single integral over the radial distance r. It is easily seen that the expressions containing the shape factor $V(\mathbf{Q})$ and the partial structure factor $J_{ij}(\mathbf{Q})$ can be written as [5.4]

$$V(Q) = \int 4\pi r^2 v(r)[(\sin Qr)/(Qr)]dr \quad \text{and} \tag{5.20}$$

$$J_{ij}(Q) = \int 4\pi r^2 v(r)\{[\varrho_{ij}(r)/c_j] - \varrho_0\}[(\sin Qr)/(Qr)]dr. \tag{5.21}$$

d) Scattered Intensity per Atom

The scattered intensity per atom $I_a(Q)$ can then be written as [5.4]:

$$I_a(Q) = I_n(Q)/[NI_e(Q)] = \langle[f(Q)]^2\rangle + \langle f(Q)\rangle^2\varrho_0 V(Q)$$
$$+ \sum\sum c_if_i(Q)c_j[f_j^*(Q)]J_{ij}(Q) \quad \text{where} \tag{5.22}$$

$\langle[f(Q)]^2\rangle = \sum c_if_i(Q)f_i^*(Q)$, $\langle f(Q)\rangle^2 = \langle f(Q)\rangle\langle f^*(Q)\rangle$ and $\langle f(Q)\rangle = \sum c_if_i(Q)$.

e) Total Structure Factor (or Interference Function)

Since the scattering amplitude $f_i(Q)$ is a function of Q for x-rays, it has been customary to introduce the total structure factor or interference function, defined in such a way that it will modulate about a constant value, usually chosen

122

to be unity. As can be readily seen from (5.22), the following definitions will satisfy this requirement [5.16]:

$$I(Q) = \{[I_a(Q) - (\langle f^2 \rangle - \langle f \rangle^2)]/\langle f \rangle^2\} - \varrho_0 V(Q) \tag{5.23}$$

and

$$S(Q) = [I_a(Q)/\langle f^2 \rangle] - [\langle f \rangle^2/\langle f^2 \rangle]\varrho_0 V(Q). \tag{5.24}$$

Consequently,

$$J(Q) = I(Q) - 1 = \sum_i \sum_j W_{ij}(Q)[I_{ij}(Q) - 1] \tag{5.25}$$

and

$$T(Q) = S(Q) - 1 = \sum_i \sum_j W_{ij}(Q)[\langle f \rangle^2/\langle f^2 \rangle][I_{ij}(Q) - 1] \tag{5.26}$$

where

$$W_{ij}(Q) = c_i f_i(Q) c_j [f_j^*(Q)]/\langle f(Q) \rangle^2. \tag{5.27}$$

The functions $I(Q)$ and $S(Q)$ are shown in Fig. 5.2 for an amorphous Be-Ti alloy, determined with x-rays and neutrons, respectively [5.15]. In the neutron case, $I(Q)$ would be ill-conditioned because $\langle f \rangle$ is small for the $Be_{37.5}Ti_{62.5}$ alloy (a so-called "zero-alloy"). Thus $S(Q)$ should be preferred, since it is applicable for all scattering experiments.

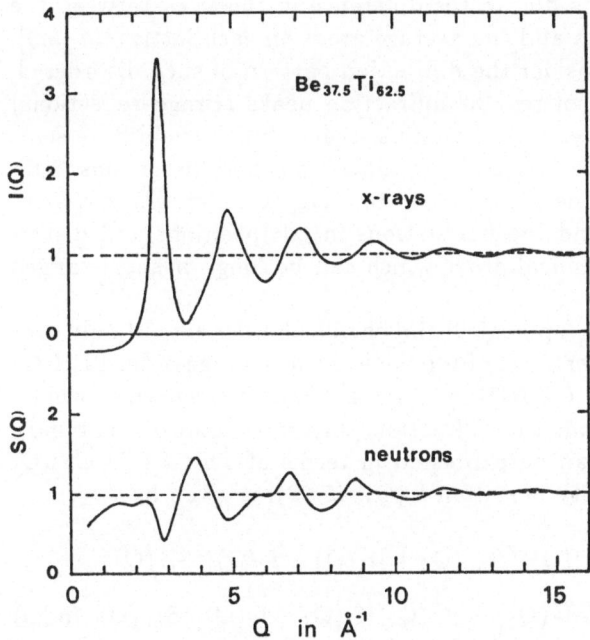

Fig. 5.2. The x-ray structure factor $I(K)$, (5.23), and the neutron structure factor $S(K)$, (5.24), for the amorphous $Be_{37.5}Ti_{62.5}$ alloy [5.15]

5.1.3 Binary Alloys

The total scattered intensity per atom reduces to the following expression for binary alloys.

$$I_a(Q) = \langle [f(Q)]^2 \rangle + \langle f(Q) \rangle^2 \varrho_0 V(Q) + c_1 f_1(Q) c_1 f_1^*(Q) [I_{11}(Q) - 1]$$
$$+ c_2 f_2(Q) c_2 f_2^*(Q) [I_{22}(Q) - 1]$$
$$+ \{ c_1 f_1(Q) c_2 f_2^*(Q) + c_1 f_1^*(Q) c_2 f_2(Q) \} [I_{12}(Q) - 1] \qquad (5.28)$$

because $I_{12}(Q) = I_{21}(Q)$ since $\varrho_{12}(r)/c_2 = \varrho_{21}(r)/c_1$, i.e., the number of 1–2 pairs must be equal to the number of 2–1 pairs in the binary alloy.

It is of interest to learn about the distribution of two kinds of atoms 1 and 2 in the alloy. In an ideal gas, the atoms 1 and 2 will be distributed in a completely random fashion. If such a gas is condensed, a departure from randomness will be observed in most case, because atoms 1 and 2 usually have different sizes and interact chemically with each other. Only if all atoms in the condensed matter are isotopes of the same element can the distribution be assumed to be random.

a) Topological and Chemical Order in Binary Solutions

A crystalline, disordered alloy, consisting of 1 and 2 type atoms, may be regarded as an ideal crystal consisting of "average" atoms with superimposed "inhomogeneity" represented by variations in the atomic scattering amplitudes and in the static displacements due to the difference in the sizes between the real atoms of the solid solution and the average atom on each lattic site [5.9]. It is often advantageous to consider the diffraction pattern of such disordered, crystalline solution to consist of regular diffraction peaks (Bragg reflections) produced by the average lattice, and of a diffuse scattering (Laue diffuse scattering) produced by the fluctuations of the compositions and distortions [5.8]. The average lattice describes the positional or topological order which is long-ranged in crystalline solids, and the fluctuations in compositions and distortions are a measure of the chemical order which can be long- or short-ranged in crystalline solid solutions.

In amorphous solids, both topological and chemical order extend only over short ranges, and we speak, therefore, of topological short-range order (TSRO) and chemical short-range order (CSRO) in such materials. It has been shown by *Bhatia* and *Thornton* [5.17] that the diffraction pattern of amorphous binary (or multi-component) alloys can be expressed in terms of TSRO and CSRO, i.e., the total, scattered intensity per atom $I_a(Q)$, (5.28), is given by

$$I_a(Q) = \langle f(Q) \rangle^2 \varrho_0 V(Q) + \langle f(Q) \rangle^2 S_{nn}(Q) + [f_1(Q) - f_2(Q)]^2 S_{cc}(Q)$$
$$+ \{ \langle f(Q) \rangle [f_1^*(Q) - f_2^*(Q)] + \langle f^*(Q) \rangle [f_1(Q) - f_2(Q)] \} S_{nc}(Q) \quad (5.29)$$

where
$$S_{nn}(Q) = 1 + \sum \sum c_i c_j J_{ij}(Q) = 1 + \sum c_i J_i(Q), \qquad (5.30)$$

$$S_{cc}(Q) = c_1 c_2 \{1 + c_1 c_2 [J_{11}(Q) + J_{22}(Q) - 2J_{12}(Q)]\},$$
$$= c_2 J_1(Q) + c_1 J_2(Q) - J_{12}(Q) \tag{5.31}$$

$$S_{nc}(Q) = c_1 c_2 \{c_1 [J_{11}(Q) - J_{12}(Q)] - c_2 [J_{22}(Q) - J_{21}(Q)]\}$$
$$= c_1 c_2 [J_1(Q) - J_2(Q)], \tag{5.32}$$

with

$$I_i(Q) = \sum c_j I_{ij}(Q) \tag{5.33}$$

which follows immediately from the definition of the atomic distribution function $\varrho_i(r)$, i.e.,

$$\varrho_i(r) = \sum \varrho_{ij}(r). \tag{5.34}$$

The number-number structure factor $S_{nn}(Q)$ describes the topological short-range order (TSRO), the concentration-concentration structure factor $S_{cc}(Q)$ the chemical short-range order (CSRO), and the number-concentration structure factor $S_{nc}(Q)$ the atomic size effect. At large values of Q, $S_{nn}(Q)$ modulates about one, $S_{cc}(Q)$ about $c_1 c_2$, and $S_{nc}(Q)$ about zero [5.4,9,17].

b) *Atomic Distribution and Correlation Functions*

Since amorphous and **poly-crystalline** alloys possess an isotropic distribution of their "units of structure" (atoms, molecules, small crystalline domains), we can evaluate the reduced, partial atomic distribution functions $G_{ij}(r)$ as the Fourier transforms of $J_{ij}(Q) = I_{ij}(Q) - 1$, i.e.,

$$G_{ij}(r) = 4\pi r \{[\varrho_{ij}(r)/c_j] - \varrho_0\} = (2/\pi) \int Q J_{ij}(Q) \sin Qr \, dQ. \tag{5.35}$$

The partial correlation function $T_{ij}(r)$ is defined as

$$T_{ij}(r) = 4\pi r \varrho_0 + G_{ij}(r) \tag{5.36}$$

and the partial radial distribution function $R_{ij}(r)$ is given by

$$R_{ij}(r) = r T_{ij}(r) = 4\pi r^2 \varrho_0 + r G_{ij}(r). \tag{5.37}$$

Similarly, we can define the number-concentration correlation functions $T_{nn}(r)$, $T_{cc}(r)$, and $T_{nc}(r)$ [5.4], i.e.,

$$T_{nn}(r) = 4\pi r \varrho_{nn}(r) = 4\pi r \varrho_0 + (2/\pi) \int Q[S_{nn}(Q) - 1] \sin Qr \, dQ, \tag{5.38}$$

$$T_{cc}(r) = 4\pi r \varrho_{cc}(r) = (2/\pi) \int Q\{[S_{cc}(Q)/(c_1 c_2)] - 1\} \sin Qr \, dQ, \tag{5.39}$$

$$T_{nc}(r) = 4\pi r \varrho_{nc}(r) = (2/\pi) \int Q S_{nc}(Q) \sin Qr \, dQ, \tag{5.40}$$

which describe the TSRO, the CSRO, and the size effect, respectively, in object or r-space.

It follows readily from (5.35–40) that

$$\varrho_{nn}(r) = c_1\varrho_{11}(r) + c_2\varrho_{22}(r) + 2c_1\varrho_{12}(r) = c_1\varrho_1(r) + c_2\varrho_2(r), \tag{5.41}$$

$$\varrho_{cc}(r) = c_1c_2\{[\varrho_{11}(r)/c_1] + [\varrho_{22}(r)/c_2] - 2[\varrho_{12}(r)/c_2]\},$$
$$= c_2\varrho_1(r) + c_1\varrho_2(r) - \varrho_{12}(r)/c_2 \tag{5.42}$$

$$\varrho_{nc}(r) = c_1c_2\{[\varrho_{11}(r) + \varrho_{12}(r)] - [\varrho_{22}(r) + \varrho_{21}(r)]\}$$
$$= c_1c_2[\varrho_1(r) - \varrho_2(r)]. \tag{5.43}$$

It should be realized that the Fourier transforms of $QJ(Q) = Q[I(Q) - 1]$, (5.25), and $QT(Q) = Q[S(Q) - 1]$, (5.26), yield the reduced atomic distribution functions $G_I(r)$ and $G_S(r)$, respectively, which are weighted sums of the partial atomic distribution functions $G_{ij}(r)$, (5.35), i.e.,

$$G_I(r) = (2/\pi)\int QJ(Q)\ \sin\ Qr\,dQ = \sum\sum W_{ij}(0)G_{ij}(r), \tag{5.44}$$

$$G_S(r) = (2/\pi)\int QT(Q)\ \sin\ Qr\,dQ$$
$$= \sum\sum W_{ij}(0)[\langle f(0)\rangle^2/\langle f^2(0)\rangle]G_{ij}(r). \tag{5.45}$$

These relations are exact for neutron scattering, but they are approximations for x-ray scattering since $W_{ij}(Q)$, (5.27), is a slowly varying function of Q.

It is readily seen that the differential correlation $D(r)$ is given by

$$D(r) = \langle f(0)\rangle^2 G_I(r) = \langle f^2(0)\rangle G_S(r). \tag{5.46}$$

The total correlation function $T(r)$ is defined as

$$T(r) = 4\pi r\varrho_0 + D(r) \tag{5.47}$$

and the radial distribution function $R(r)$ can be expressed as

$$R(r) = rT(r) = 4\pi r^2\varrho_0 + rD(r) = \sum\sum c_if_i(0)c_jf_j(0)\varrho_{ij}(r)/c_j. \tag{5.48}$$

c) Coordination Numbers

Since the reduced, partial atomic distribution functions $G_{ij}(r)$, and $G_{n-c}(r)$ of amorphous alloys exhibit only few well-defined peaks, it has been customary to concentrate on the first peak which is used to evaluate the coordination numbers N_{ij} and N_{n-c}, respectively, i.e.,

$$N_{ij} = \int 4\pi r^2\varrho_{ij}(r)dr = \int rT_{ij}(r)dr = \int R_{ij}(r)dr, \tag{5.49}$$

$$N_i = \int 4\pi r^2\varrho_i(r)dr = \sum_j N_{ij}, \tag{5.50}$$

$$N_{n-c} = \int 4\pi r^2\varrho_{n-c}(r)dr = \int rT_{n-c}(r)dr = \int R_{n-c}(r)dr. \tag{5.51}$$

5.1.4 Chemical Short-Range Order in Binary Alloys

It is possible to define a chemical short-range order (CSRO) parameter α_p for the pth coordination shell by rewriting the expression for the concentration coordination number N_{cc} using (5.42,49-51) [5.4,18,19], i.e.,

$$(N_{cc})_p = [c_2(N_1)_p + c_1(N_2)_p]\alpha_p \quad \text{where} \tag{5.52}$$

$$\alpha_p = (N_{cc})_p/[c_2(N_1)_p + c_1(N_2)_p]$$
$$= 1 - (N_{12})_p/\{c_2[c_2(N_1)_p + c_1(N_2)_p]\}. \tag{5.53}$$

The CSRO parameter α_1 for the first coordination shell can be evaluated from [5.19]:

$$\alpha_1 = 1 - N_{12}/\{c_2 N_{nn}[1 - (c_2 - c_1)(N_1 - N_2)/N_{nn}]\} \tag{5.54}$$

which to a first approximation is equal, but opposite in sign, to the Cargill-Spaepen CSRO parameter η_{12} [5.20], i.e.,

$$\eta_{12} = \{(N_{12}/c_2)[N_{nn}/(N_1 N_2)]\} - 1 = -\alpha_{cs} = -\alpha_1. \tag{5.55}$$

The atomic distribution function $\varrho_{nn}(r)$, which describes the topological order in the binary solution, yields the global coordination number N_{nn}. If the sizes of the atoms 1 and 2 are similar, we can assume that the number of atoms $\varrho_1(r)$ about a "one" atom is the same as the number of atoms $\varrho_2(r)$ about a "two" atom, i.e.,

$$\varrho_{nn}(r) = \varrho_1(r) = \varrho_2(r).$$

In this case, the number-concentration correlation function $\varrho_{nc}(r)$ is zero. This condition might be satisfied in crystalline, solid solutions, for which we can define the Warren-Cowley short-range order parameter α_{wc} [5.8], i.e.,

$$(\alpha_p)_{wc} = 1 - (N_{12})_p/[c_2(N_{nn})_p]. \tag{5.56}$$

With this definition of the Warren-Cowley parameter $(\alpha_p)_{wc}$, we can express the Laue diffuse scattering $I_{LDS}(Q)$ for a poly-crystalline sample [5.8] as

$$I_{LDS}(Q)/N = (f_1 - f_2)^2 S_{cc}(Q)$$
$$= c_1 c_2 (f_1 - f_2)^2 [1 + \sum_p (N_{nn})_p (\alpha_p)_{wc}(\sin Qr_p)/(Qr_p)] \tag{5.57}$$

which follows directly from (5.21,29 and 31), when $\langle f \rangle = 0$.

5.1.5 Topological Order in Crystalline Solid Solutions

In crystalline solid solutions, it is advantageous to express the atomic density $p_n(r) = \sum p_i(r) = \sum \delta(r - r_k)$ of a small crystal of volume dV directly as the product of the density $z_n(r)$ of an infinite crystal and the shape function $b(r)$ [5.14], i.e.,

$$p_n(r) = b(r)z_n(r) \quad \text{where} \tag{5.58}$$

$$z_n(r) = \sum_p \sum_q \sum_r \delta(r - r_k) \quad \text{and} \tag{5.59}$$

$$r_{pqr} = pa + qb + rc, \tag{5.60}$$

a, b, c being the translation vectors of the unit cell with volume $V_c = a \cdot b \times c$. The scattered amplitude can then be written as

$$A_n(Q) = F_t(Q)P_n(Q) = F_t(Q)[Z_n(Q) * B(Q)]$$

where $Z_n(Q)$ is the Fourier transform of $z_n(r)$ defined in (5.59), $B(Q)$ is the shape amplitude (5.12), and $F_t(Q)$ is the temperature corrected structure factor of the unit cell of the crystal [5.8], i.e.,

$$F_t(Q) = \sum f_n \exp(-M_n) \exp(hx_n + ky_n + lz_n) \tag{5.61}$$

where $f_n(Q)$ is the atomic scattering amplitude, M_n is the Debye-Waller temperature factor, and x_n, y_n, and z_n are the position of atom n in the unit cell. It has been shown [5.14] that

$$Z_n(Q) = \int z_n(r) \exp(-iQr)dr = (1/V_c) \sum_h \sum_k \sum_l \delta(Q - g_{hkl}) \tag{5.62}$$

where g_{hkl} is the reciprocal lattice vector, i.e.,

$$g_{hkl} = ha^* + kb^* + lc^*, \tag{5.63}$$

a^*, b^*, c^* being the translation vectors of the reciprocal unit cell.

The structure factor or interference function $I(Q) = I_n(Q)/\{I_eN[F_t(Q)]^2\}$ for a crystalline solution can then be written as

$$I(Q) = [Z_n(Q)]_*^2[B(Q)]^2/N = \sum \sum \sum V(Q - g_{hkl})/V_c. \tag{5.64}$$

In contrast to the amorphous solid solutions, the crystalline alloys exhibit peaks only at the reciprocal lattice points hkl, which are broadened by the shape factor $V(Q)$.

5.1.6 Integrated Intensity

In a crystalline powder, there are M small crystals, each with volume dV and N unit cells. Therefore, the total diffracted power about the reciprocal lattice point hkl is given by the Laue theorem [5.8]:

$$P_{hkl} = \int P(2\theta)d2\theta = [Mm_{hkl}\lambda^3R^2/(4 \sin\theta)](2\pi)^{-3} \int I_n(Q - g_{hkl})dQ,$$

$$P_{hkl} = [Mm_{hkl}\lambda^3R^2/(4 \sin\theta)]I_e(q)[F_t(q)]^2dV/(V_c)^2, \tag{5.65}$$

128

because at $q = Q - g_{hkl}$ the integral $\int I_n(Q - g_{hkl})dQ$ is equal to $\int I_n(Q)dq = NI_e(q)[F_t(q)]^2 \int I(q)dq = I_e(q)[F_t(q)]^2(2\pi)^3 dV/(V_c)^2$.

The total diffracted power per unit height of the Debye-Scherrer ring is then

$$P' = P/(2\pi R \ \sin 2\theta)$$
$$= I_0[(r_e)^2/(32\pi R)]\lambda^3[V/(V_c)^2][F_t(q)]^2 m_{hkl}(1 + \cos^2 2\theta)/(\sin^2\theta \ \cos\theta)$$

$$(5.66)$$

where R is the distance between the specimen and the detector, I_0 is the intensity of the primary beam, $r_e = 2.818 \times 10^{-13}$cm is the scattering length of a single electron for x-rays, $V = M dV = MNV_c$ is the volume of the powder, and m_{hkl} is the multiplicity factor of the (hkl) reflection.

5.2 Experimental Techniques

5.2.1 X-Ray and Neutron Sources

For many years, x-rays to be used in scattering experiments have been produced in sealed and open tubes, both using an electron beam, accelerated to a maximum of 100 keV and yielding a continuous spectrum and a series of characteristic x-ray wavelengths depending on the anode material. These x-ray tubes provide a beam with a brillance of 10^7 photons/(s mm^2 mrad^2 0.1%BW), where BW is the band-width of the spectrum [5.21]. When introducing the rotating anode tube in the early 60's, this brilliance could be improved by a factor of 10 or so.

A tremendous increase in brilliance occurred when the synchrotron radiation became available in the early 70's [5.1]. This radiation is produced by relativistically fast electrons. If the path of a bunch of such fast moving electrons with the energy $E = \gamma mc^2$ circulating in a storage ring is bent by a magnetic field into a circle of radius R, it will emit radiation into a cone of width $1/\gamma$ radians. The radiation will have a critical wavelength λ_c, given by $\lambda_c = R/\gamma^3$. For a 3 GeV ring, with $\gamma = 1.957 \times 10^3 E$ [GeV] $= 6 \times 10^3$ and a typical radius of 20 m, λ_c is approximately 1 Å, or the critical energy E_c is about 12 keV. The opening angle of the cone of radiation is about 2×10^{-4} radians or 40 s of arc. The brilliance of such a source of synchrotron radiation, produced by the bending magnet, is of the order of 10^{12} photons/(s mm^2 mrad^2 0.1BW) [5.1b].

A further increase in brilliance can be accomplished by the use of insertion devices in the straight sections of the storage ring for the electrons. One of these devices, the so-called wiggler, produces a beam with a brilliance of 10^{15} photons/(s mm^2 mrad^2 0.1%BW). The most recent insertion device, the undulator, is projected to yield a brilliance of as high as 10^{18} photons/(s mm^2 mrad^2 0.1BW) in the new generation of synchrotron sources presently on the drawing board [5.22].

Neutron sources have always been major installations. With the development of nuclear reactors, neutrons have become available in sufficient numbers to permit the performance of neutron-scattering experiments [5.7,23]. The neutrons produced in the reactor must first be moderated (or slowed down) to velocities which yield wavelengths around 1 Å (thermal neutrons). Since the absorption of thermal neutrons in most materials is much smaller than that of x-rays, it was possible to use large samples (10 mm in diameter, 50 mm in height), and consequently neutron beams with a cross-section of $10 \times 50 \, mm^2$ could be employed. This is important because the average flux of neutrons from the most powerful reactors is only 10^{14} neutrons/s cm^2.

During the past decade, accelerators have also been used to generate neutrons. An intense, pulsed beam of high-energy protons (or electrons) bombards a heavy-metal target and produces spallation neutrons, which after suitable moderation also contain a large number of neutrons with energies above 200 meV (or wavelengths below 0.7 Å), the so-called epithermal neutrons [5.2]. The peak flux of the pulsed sources is about 10^{15} neutrons/s cm^2, and the pulse duration is of the order of μs and the repetition rate is usually less than 100 Hz.

5.2.2 Instrumentation

The structure of poly- and non-crystalline alloys can be determined from the x-ray and neutron diffraction patterns. The basic requirement consists of the efficient measurement of the scattered intensity as a function of the length of the diffraction vector Q = K. Combining (5.1,5 and 8), we obtain

$$Q = K = (4\pi/\lambda) \sin\theta = [4\pi/(hc)](\sin\theta)E = (4\pi m/h)(\sin\theta)(L/t) \quad (5.67)$$

where E is the energy of the photons, and L and t are the path-length and the time-of-flight of the neutrons, respectively. It is readily seen that two experimental conditions can be employed [5.4]:

1. Variable 2θ-Method: Q can be varied by changing the angle 2θ between the incident and scattered beams using monochromatic x-rays or neutrons.
2. Variable λ-Method: Q can be varied by using polychromatic x-rays or neutrons and measuring the energy E of the photons or the time-of-flight t of the neutrons at a fixed scattering angle 2θ.

a) Variable Scattering Angle Method

Any modern scattering experiment, based on the measurement of the scattered intensity as a function of the scattering angle 2θ, consists of a radiation source, a monochromator to select a narrow band of wavelengths and to suppress unwanted radiation, a diffractometer with sample holder, and a radiation detector with associated electronic equipment, as shown in Fig. 5.1.

In the conventional diffractometer technique, a single detector (proportional, scintillation, or solid state detector) measures the scattered intensity through narrow collimators sequentially as a function of 2θ, either continuously or in steps, and controlled by a microprocessor or microcomputer. This technique is still very useful when using x-rays from powerful sources coupled with a sufficient amount of material because it provides the best peak to background ratio due to the fact that the detector sees only a very small solid angle ($<0.2°$) of the scattered radiation.

A great improvement in sensitivity and counting efficiency could be accomplished when it became possible to detect simultaneously all the radiation scattered by the sample (similar to the old Debye-Scherrer film technique). The development of curved position-sensitive detectors for x-rays and neutrons [5.24–26] had advanced to the state that permits the registration of the diffracted radiation over a large angular range in 2θ, as large as 120°. It must be emphasized that these detectors should be used with the sample and the flight paths in an evacuated chamber to reduce the parasitic scattering, in conjunction with a monochromatic radiation source. The overall gain in efficiency of such a system over the conventional scanning 2θ-diffractometer should be at least a factor of 100.

As long as one dimension of the cross-section of the radiation source can be made small (line focus of the conventional x-ray tube), it is possible to use focussing geometries with the scanning 2θ diffractometer, either in transmission or in reflection (the Bragg-Brentano geometry), which allow the use of a relatively large divergence of the primary beam [5.13]. If the cross-section of the source is large (point focus of the x-ray tube or beam line of the neutron source) and/or a position-sensitive detector is employed, the parallel beam geometry is preferred.

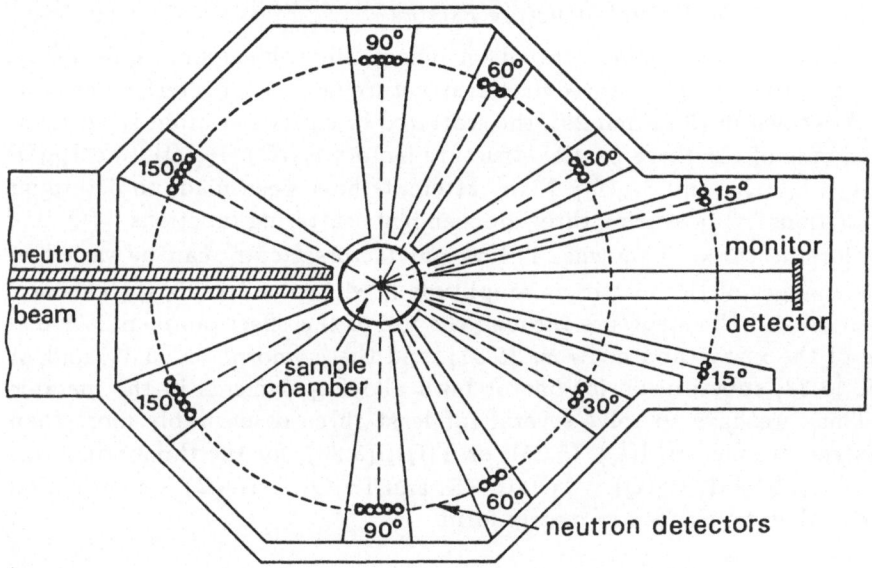

Fig. 5.3. Variable λ-technique: Time-of-flight neutron diffractometer

131

b) Variable Wavelength Method

With the development of intense sources of continuous radiation for x-rays and neutrons, the measurement of the scattered intensity at fixed scattering angle 2θ becomes very attractive. In the case of x-rays, the use of a solid-state detector with an energy resolution of $150\,\text{eV}$ at $6\,\text{keV}$ makes the technique of energy-dispersive diffraction possible [5.13]. The time structure of the neutrons produced in the pulsed spallation source allows the application of the time-of-flight technique [5.2], which is shown in Fig. 5.3.

In order to make absolute measurements of the scattered intensity, it is necessary to know the wavelength- or energy-dependence of the primary spectrum $I_0(E)$. In the case of neutron scattering, it is possible to use a vanadium sample to determine the wavelength spectrum [5.27,28]. In the energy-dispersive x-ray measurements, the precise determination of the primary beam spectrum is rather difficult [5.29]. Recently, the low-angle scattering from a metallic glass sample has been successfully used to evaluate the wavelength spectrum of the continuous radiation generated by a conventional x-ray tube [5.30].

5.3 Applications

5.3.1 Structure of Metallic Glasses and Liquids

Metallic glasses have been studied extensively during the past decade [5.3]. Because of their interesting physical and chemical properties, there has been a great interest in the determination of their structure, i.e., the topological and chemical short-range order.

a) Evaluation of the Partial Structure Factors

Most efforts have been concentrated on binary amorphous and liquid alloys, which are characterized by three partial structure factors (or interference functions). As shown in (5.22 and 29), the scattered intensity per atom $I_a(Q)$ is the weighted sum of the three partial structure factors $I_{11}(Q)$, $I_{22}(Q)$, and $I_{12}(Q)$ or $S_{nn}(Q)$, $S_{cc}(Q)$, and $S_{nc}(Q)$. Many attempts have been made to determine these functions by measuring three independent scattering functions $I_a(Q)$ and solving for the three unknowns. The weight factors $c_i c_j f_i f_j^*$ can be varied by suitable changes of the scattering amplitude f_i without affecting the chemical composition and the structure through the use of the effect of anomalous dispersion of the x-rays and neutrons [5.31], and the isomorphous [5.32] and/or isotopic [5.33] substitution for one or both alloying elements in the metallic glass. Thus, we have to solve several (at least three or desirably more than three) structure factors $J(Q)$, (5.25), or $T(Q)$, (5.26), for the three unknowns $P_1(Q) = I_{11}(Q) - 1$, $P_2(Q) = I_{22}(Q) - 1$, and $P_3(Q) = I_{12}(Q) - 1$. Equation (5.25) can then be written in matrix form

$$[J(Q)] = [W(Q)][P(Q)] \tag{5.68}$$

where the matrix elements $W_{ij}(Q)$ are defined in (5.27). The solution of (5.68) is found by applying the least-square method [5.34]. However, (5.68) is often ill-conditioned, i.e., the normalized determinant $|W_{ij}(Q)|_n$ of the matrix [W] is usually very small [5.34], or the figure of merit T (also called Turing's number) [5.35] is relatively large (>100), i.e.,

$$T = \| W(Q) \|_E \| W^{-1}(Q) \|_E \qquad (5.69)$$

where $\| W(Q) \|_E$ is the Euclidian norm of the matrix [W(Q)]. As a consequence, (5.68) does not yield reliable solutions for the individual partial structure factors $I_{ij}(Q) - 1$.

b) Isomorphous Substitution

Several experiments have been performed using isomorphous substitution with x-rays alone or in combination with neutron data. Elements which belong to the same column in the periodic table are suitable for isomorphous substitution. Examples are (Zr-Hf) and (Al-Ga) which have been employed to evaluate the partial functions in Ni-Zr [5.36,37] and La-Al [5.38] glasses, respectively. In this approach it is assumed that the structure is not affected by the substitution, a fact which is strongly supported by the occurrence of identical crystal structures in the corresponding crystalline alloys.

c) Isotopic Substitution

In neutron diffraction, it is possible to vary the scattering amplitude of an element by substitution with an isotope when available. The most suitable candidates are elements which contain isotopes with negative scattering lengths. Examples are Li, Ti, Ni, Dy and W. In such a case, it is possible to prepare a mixture of isotopes such that $f_i = b_i = 0$, a so-called "null-element". This has been done by *Lamparter* et al. [5.39,40] in their evaluation of the partial structure factors $I_{ij}(Q)$ of $Ni_{81}B_{19}$ and $Ni_{80}P_{20}$, which are shown in Fig. 5.4.

The reduced atomic distribution functions $G_{ij}(r)$ of the $Ni_{80}P_{20}$ alloy, which are the Fourier transforms of the partial structure factors $Q[I_{ij}(Q) - 1]$, are shown in Fig. 5.5. It is clearly seen that the metalloid atoms P are not nearest neighbors in this amorphous alloy.

Because ^7Li, and natural Ti and Mn have negative scattering lengths, it becomes possible to prepare so-called "zero-alloys" in Li, Ti or Mn containing glasses and liquids, where $\langle f \rangle = \langle b \rangle = 0$. In this case, one can determine the concentration correlation function $\varrho_{cc}(r)$ by Fourier transformation of the reduced structure factor $Q\{[S_{cc}(Q)/(c_1c_2)] - 1\}$ which is the only remaining term in the expression for the scattered intensity in (5.39). This approach has been successfully employed in amorphous Ni-Ti [5.41,42], Cu-Ti [5.43], Be-Ti [5.15], and (Fe-Mn)-P-C [5.44] alloys, and liquid Li alloys [5.45,46].

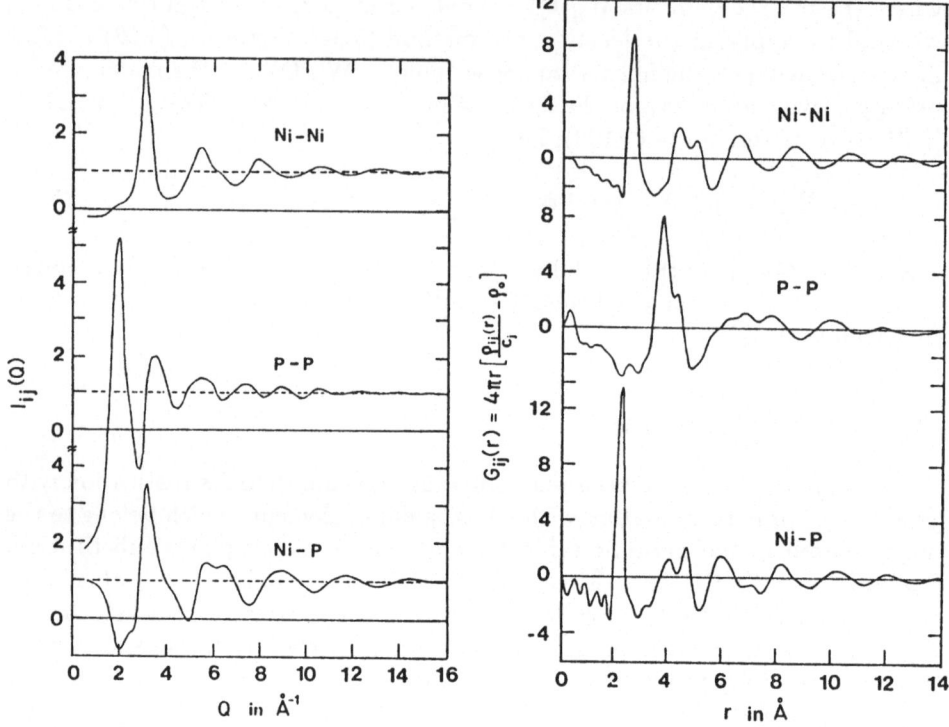

Fig. 5.4. The partial structure factors $I_{ij}(K)$ for the amorphous $Ni_{80}P_{20}$ alloy [5.41]

Fig. 5.5. The partial atomic distribution functions $G_{ij}(r) = 4\pi r\{[\varrho_{ij}(r)/c_j] - \varrho_0\}$ for the amorphous $Ni_{80}P_{20}$ alloy [5.41]

d) Anomalous Dispersion

The use of anomalous x-ray scattering has been suggested for the determination of the partial structure factors in amorphous alloys [5.31]. This technique utilizes the relatively sharp change in the complex scattering amplitude f_i of element i close to its absorption edge. In general, the atomic scattering amplitude must be written as

$$f(Q, E) = f_0(Q) + f'(Q, E) + if''(Q, E) \tag{5.70}$$

where $E = h\nu = hc/\lambda$ is the photon energy. $f_0(Q)$ is the Fourier transform of the electron density. $f''(Q, E)$ is directly related to the absorption coefficient $\mu(E)$, and is a slowly varying function of E below the edge. $f'(Q, E)$ has a steep negative peak at the edge and is relatively small at all other energies E, as shown in Fig. 5.6 [5.47].

With the availability of the highly intense synchrotron radiation, it became possible to select x-ray energies close to the absorption edge of element i. Unfortunately, the change in f_i due to this anomalous dispersion is not very large (of the order of 6–8 electron units at the $K - $ edge), and as a consequence

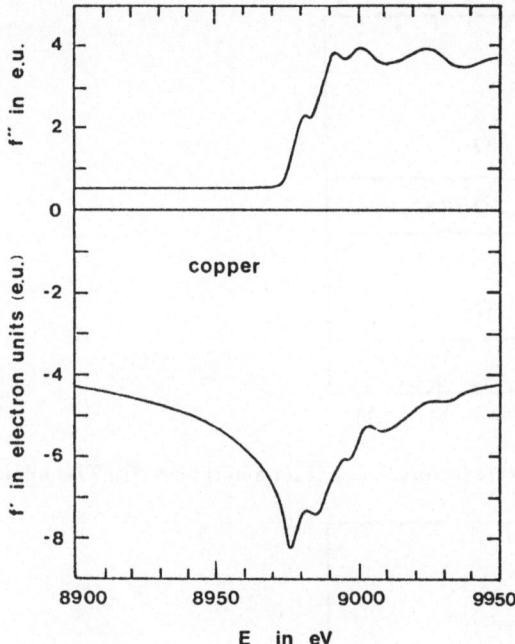

Fig. 5.6. Anomalous dispersion corrections f' and f'' of the x-ray scattering amplitude f for Cu [5.47]

the change in the weight factors $W_{ij}(Q)$, (5.27), is so small that the normalized matrix [W] is very small, less than 10^{-3}. *Shevchik* [5.48] has suggested the use of the differential anomalous scattering (DAS) technique. Some success was reported in the analyses of Ge-Se [5.6], Mo-Ge [5.49], and Mo-Ni glasses [5.50], respectively.

e) Chemical Short-Range Order in Binary Liquid and Amorphous Alloys

The first clear evidence for the existence of chemical short-range order in liquid metals was given by *Ruppersberg* and *Egger* [5.45], while studying a series of liquid Li-Pb alloys by neutron diffraction. They observed that the scattered intensity from a $Li_{80}Pb_{20}$ alloy was entirely governed by the $S_{cc}(Q)$ term in the total structure factor $S(Q)$, because $\langle f \rangle$ is zero for this composition. Assuming that the $S_{nc}(Q)$ function is small, the Warren-Cowley short-range order parameter α_{wc}, (5.52), was found to be -0.25, indicating a preference for unlike nearest neighbors. Similar observations were made in liquid and poly-crystalline Li-Ag alloys [5.51].

Since many metallic glasses have been prepared by rapid quenching from the liquid state, it is natural to assume that any chemical order which might be present in the liquid must be quenched into the amorphous state [5.9]. Evidence for such chemical order was found in many metal-metal and metal-metalloid glasses, whose partial structure factors were determined by the methods of isomorphous and isotopic substitution [5.15,36–41,52–55].

The chemical short-range order parameter α_1, (5.54), can be readily determined from the number-concentration correlation functions $T_{n-c}(r)$, which are

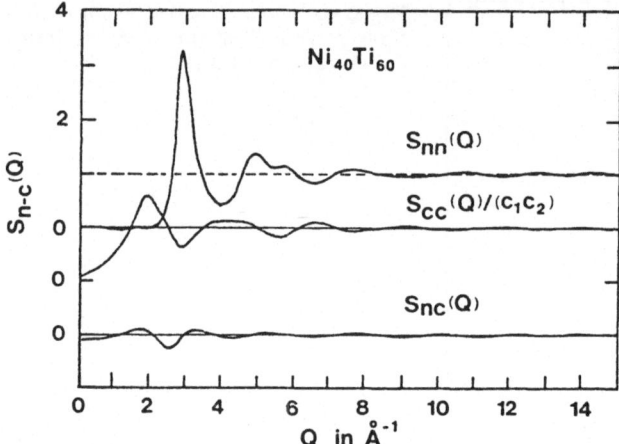

Fig. 5.7. The number-concentration structure factors $S_{n-c}(K)$ for amorphous $Ni_{40}Ti_{60}$ alloy [5.40]

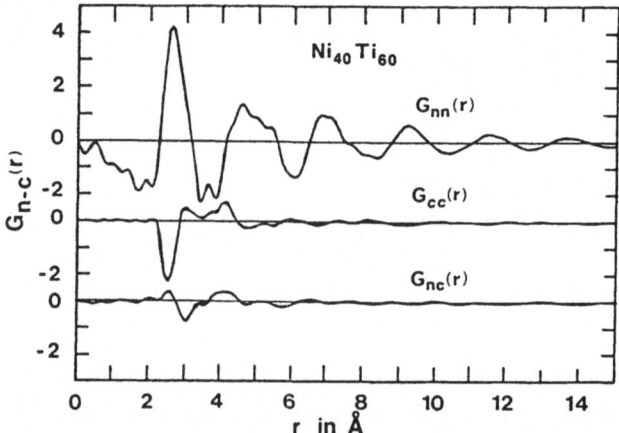

Fig. 5.8. The reduced, atomic correlation functions $4\pi r \varrho_{n-c}(r)$ for the amorphous $Ni_{40}Ti_{60}$ alloy [5.40]

the Fourier transforms of the number-concentration structure factors $S_{n-c}(Q)$, (5.38–40). An example of the functions $S_{n-c}(Q)$ is given in Fig. 5.7 for the amorphous $Ni_{40}Ti_{60}$ alloy [5.41]. The modulation of the function $S_{cc}(Q)/(c_1 c_2)$ about one clearly indicates that there exists chemical short-range order in this alloy. The concentration correlation function $T_{cc}(r) = 4\pi r \varrho_{cc}(r)$ exhibits a negative first peak which is the consequence of the preference for unlike nearest neighbors, as shown in Fig. 5.8. Using the values of the partial coordination numbers $N_{NiNi} = 2.3$, $N_{NiTi} = 7.9$ and $N_{TiTi} = 8.0$, (5.49), one obtains a value of $\alpha_1 = -0.15$ for the amorphous $Ni_{40}Ti_{60}$ alloy. A similar value of $\alpha_1 = -0.17$ has been determined for the amorphous $Ni_{80}P_{20}$ alloy from the partial co-ordination numbers $N_{NiNi} = 9.4$, $N_{NiP} = 2.3$ and $N_{PP} = 0$, which are the areas under the first peak of $c_j R_{ij}(r) = 4\pi r^2 c_j \varrho_0 + c_j r G_{ij}(r)$. The partial atomic distribution functions $G_{ij}(r)$ of the alloy $Ni_{80}P_{20}$ are shown in Fig. 5.5.

5.3.2 Phase Analysis of Poly-Crystalline Mixtures

The analysis of x-ray and neutron powder patterns is the perfect technique to determine quantitatively the amount of crystalline phases in powder mixtures, since each component produces a characteristic pattern independent of the others. The quantitative analysis by diffraction is based on the fact that the integrated intensity of a powder pattern peak of a particular phase α in a mixture of phases depends only on the volume fraction of that phase in the mixture [(5.13,56,57] . The relation between the integrated intensity P'_α and the volume V_α of phase is given in (5.66). Thus, we can write

$$P'_\alpha = I_\alpha = CR_\alpha V_\alpha \tag{5.71}$$

where C is a constant and R_α is the so-called R − factor of phase α, i.e.,

$$R_\alpha = (F_t)^2 m_{hkl}(1/V_c)^2(1 + \cos^2 2\theta)/(\sin^2 \theta \ \cos \theta). \tag{5.72}$$

For a flat, infinitely thick sample, wider than the x-ray beam, the volume V_α can be replaced by

$$V_\alpha = v_\alpha A_0/(2\mu_m) \tag{5.73}$$

where $v_\alpha = V_\alpha/V$ is the volume fraction of the phase α, A_0 is the cross-section of the primary beam, and μ_m is the linear absorption coefficient of the mixture. Consequently, (5.71) can be written as

$$I_\alpha = C_1 A_0 R_\alpha v_\alpha / \left(\sum v_\alpha \mu_\alpha \right) \tag{5.74}$$

where μ_α is the mass absorption coefficient of phase α. The value of the constant $C_1 A_0$ is unknown, because I_0 is generally not known. However, this constant will cancel out if we measure the ratio of I_α to the intensity of some standard reference line. The volume fraction v_α can then be found from this ratio.

The three main methods of analysis differ in what is used as a reference line: (i) the external standard method (using a diffraction peak from a pure phase), (ii) the direct comparison method (using a peak from another phase in the mixture), and (iii) the internal standard method (using a line from a foreign material mixed with the specimen).

In all methods, the absorption coefficient μ of the mixture is itself a function of v_α and can have a large effect on the measured intensity I_α. In addition, there are other practical difficulties in determining the intensities of the powder pattern peaks of all phases in the mixture. The most important of these complicating factors are:(i) preferred orientation of the grains in the poly-crystalline mixture, (ii) grain or particle size, and (iii) overlap of the peaks from different phases.

The effect of preferred orientation can be minimized in the direct comparison method by measuring several peaks for each phase. If only two phases are present in the sample (e.g., the retained austenite problem in steels), we can write (5.74) as follows:

$$X_\alpha = I_\alpha/R_\alpha = C_2 v_\alpha \tag{5.75}$$

where $C_2 = C_1 A_0 / \sum v_\alpha \mu_\alpha$. In the case of a textured sample, one can assume [5.58] that

$$X_\alpha = (1/n) \sum [(I_\alpha/R_\alpha)_i]_{\text{textured}}$$
$$= (1/n) \sum [(I_\alpha/R_\alpha)_i]_{\text{random}} = C_2 v_\alpha \tag{5.76}$$

where n is the number of the available reflections of phase α (n>4). To eliminate the constant C_2, we form the ratio of X_α due to the phase α and X_β due to phase β, i.e., $v_\alpha/v_\beta = X_\alpha/X_\beta$. In addition, we know that $v_\alpha + v_\beta = 1$. Thus we obtain the volume fraction of phase α in the mixture of phases α and β with preferred orientation

$$v_\alpha = 1/(1 + X_\beta/X_\alpha). \tag{5.77}$$

An example of a retained austenite measurement in an alloy steel is shown in Fig. 5.9. The powder pattern was obtained with $\mathrm{Mo\,K_\alpha}$ radiation, and the pairs of the bcc α (ferrite) peaks (200) and (211) and the fcc γ (austenite) peaks (220) and (311) were used to determine the volume fraction $v_\gamma = 0.16$.

The problem of peak overlap can be overcome by the following analysis. Suppose that there are N phases α present in the mixture and that the diffraction peak i is a composite line with contributions from n peaks i. Then we can rewrite (5.74) as follows:

$$I_i = \sum I_{i\alpha} = \sum C_{i\alpha} v_\alpha \tag{5.78}$$

where $C_{i\alpha} = CA_0 R_{i\alpha} / \sum v_\alpha \mu_\alpha$. Equation (5.78) can be readily written in matrix notation, i.e.,

$$[I] = [C][v].$$

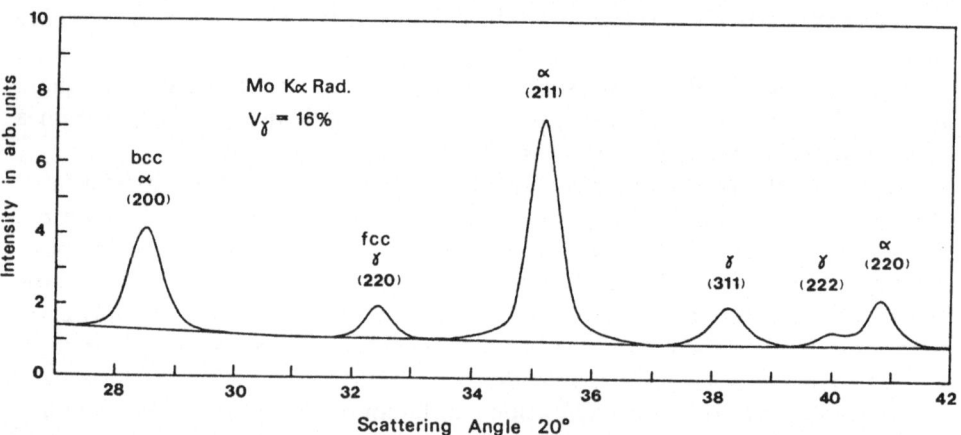

Fig. 5.9. Retained austenite in an alloy steel. (Diffraction pattern, taken with $\mathrm{Mo\,K_\alpha}$ radiation)

Assuming that $n>N$, we can use the least-square method to determine $[v]$, i.e.,

$$[v] = [[C]^T[C]]^{-1}[C]^T[I] \qquad (5.79)$$

where $[A]^T$ and $[A]^{-1}$ are the transpose and inverse, respectively, of the matrix $[A]$.

5.3.3 Small-Angle Scattering

For a homogeneous sample of volume V, containing N atoms, the small-angle scattering can be expressed as

$$I_{sa}(Q) = V\varrho_0\langle f\rangle^2 \int v(r) \exp(-iQr)dr$$
$$= V\varrho_0\langle f\rangle^2 V(Q) = \varrho_0\langle f\rangle^2[B(Q)]^2 \qquad (5.80)$$

where $v(r)$ is the size function, defined in (5.16), which is the convolution product of the shape function $b(r)$, i.e., $b(r) = 1$ inside the volume V of the sample and $b(r) = 0$ outside V. $B(Q)$ is the Fourier transform of $b(r)$. It is readily shown that $B(0) = V$, $V(0) = V$, and $v(0) = (2\pi)^{-3} \int V(Q)dQ = 1$.

In an inhomogeneous sample, any variation in the atomic density or concentration over regions larger than the atomic sizes of the alloying elements should produce appreciable scattering at small angles [5.59]. If there are segregated, well separated regions with N_r atoms in volume V_r, with a scattering power per unit volume $p_r = (N_r/V_r)\langle f_r\rangle = \varrho_r\langle f_r\rangle$ in the alloy with an electron density $p = \varrho_0\langle f\rangle$, we have to replace the term $(\varrho_0\langle f\rangle)$ in (5.80) with the expression $(p_r - p)$ and obtain

$$I_{sa}(Q) = V_r(p_r - p)^2 \int v(r) \exp(-iQr)dr = V_r(p_r - p)^2 V(Q). \qquad (5.81)$$

It follows immediately that $I_{sa}(0) = (V_r)^2(p_r - p)^2$ because $V(0) = \int v(r)dr = V_r$.

The size function $v(r)$ of the segregated domains can, in principle, be directly determined by Fourier transformation of the small-angle scattering $I_{sa}(Q)$. If we assume that the domains are distributed isotropically, we can replace $V(Q)$ by (5.20), and obtain

$$I_{sa}(Q) = V_r(p_r - p)^2 \int 4\pi r^2 v(r)[\sin Qr/(Qr)]dr. \qquad (5.82)$$

Thus,

$$V_r(p_r - p)^2[rv(r)] = [1/(2\pi)] \int QI_{sa}(Q)(\sin Qr)dQ. \qquad (5.83)$$

Since $v(0) = 1$, it follows that

$$V_r(p_r - p)^2 = [1/(2\pi)] \int Q^2 I_{sa}(Q)dQ. \qquad (5.84)$$

Combining (5.83 and 84), we obtain the final expression for $v(r)$:

$$v(r) = \int Q^2 I_{sa}(Q)[\sin Qr/(Qr)]dQ/\int Q^2 I_{sa}(Q)dQ. \tag{5.85}$$

Rather than using (5.85) to evaluate the size function $v(r)$ itself, it has been common practice to evaluate the radius of giration R_g of the segregated domains [5.14], defined as:

$$(R_g)^2 = \int 4\pi r^2 v(r)r^2\,dr/\int 4\pi r^2 v(r)dr. \tag{5.86}$$

In the Guinier approximation, we can write the small-angle scattering $I_{sa}(Q)$ in the following form [5.14]:

$$I_{sa}(Q) = I_{sa}(0)\,\exp\left[-(QR_g)^2/3\right]. \tag{5.87}$$

The slope of the function $\ln[I_{sa}(Q)]$ plotted versus Q^2 is proportional to the radius of giration R_g of the segregated regions in the poly- or non-crystalline solids.

It must be emphasized that the foregoing discussion has been based on the assumption that the segregated domains are relatively large and well separated in the otherwise homogeneous alloy. If the regions are of the order of atomic dimensions, we must apply the Bhatia-Thornton [5.17] formalism to interpret correctly the scattering at small angles.

5.3.4 Line Profile Analysis of Powder Pattern Peaks

The analysis of the changes in the profile and the position of a particular hkl reflection in a powder pattern provides a valuable method for the investigation of the structure and properties of crystalline materials. In particular, the effect of deformation on the position and the broadening of the peaks in the powder patterns of metals and alloys has been studied extensively over the past three decades.

It is usually assumed that the broadening of the peak profiles is produced by a reduction in the size of the coherently diffracting domains (crystallite size), by faulting on certain (hkl) planes, and by microstrains within the coherently diffracting domains. The changes in position might be due to residual stresses in bulk specimens, to faulting, and to lattice parameter changes produced by dislocations and segregation of solute atoms [5.10].

With modern diffractometer techniques, it is possible to separate the causes of peak broadening and peak shifts. The broadening produced by small crystallite sizes and faulting is independent of the order of reflection, whereas the strain broadening depends on the order of reflection. The peak shifts produced by faulting and residual stresses vary with the crystallographic orientation of the reflecting planes.

a) Particle Size and Strain Broadening

In practice, it is advantageous to express the diffracted intensity $P'(Q)$ (actually the diffracted power per unit height of the Debye-Scherrer ring) as a Fourier series [5.8,10]:

$$P'(Q) = K(\theta)(F_t)^2 \sum C_n \exp[-iL(Q - Q_0)] \qquad (5.88)$$

where $K(\theta)$ is slowly varying function of θ, i.e.,

$$K(\theta) = I_0(r_e)^2 M\lambda^3 m_{hkl}(1 + \cos^2 \theta)/(32\pi RV_{cc}^* \cos \theta \sin^2 \theta)$$

and $L = nd_{hkl}$ is a distance normal to the reflecting planes with interplanar spacing d_{hkl}. The Fourier coefficients C_n are given by

$$C_n = (1/\Delta Q) \int P'(Q) \exp[i(Q - Q_0)L]dQ \qquad (5.89)$$

where ΔQ is the integration interval, defined by

$$d_{hkl}\Delta Q = 2\pi. \qquad (5.90)$$

The Fourier coefficients C_n are usually complex quantities, i.e., $C_n = A_n + iB_n$. Therefore, we can write (5.92) as

$$P'(Q) = K(\theta)(F_t)^2 \sum [A_n \cos L(Q - Q_0) + B_n \sin L(Q - Q_0)]. \qquad (5.91)$$

The Fourier coefficients $|C_n|$ are the product of the particle size coefficients $(A_n)_{PF}$ and the strain or distortion coefficients $|(C_n)_D|$, i.e., $|C_n| = (A_n)_{PF}|(C_n)_D|$. The particle size coefficients $(A_n)_{PF}$ are independent of the order of reflection and can be approximated for small values of $L = nd_{hkl}$:

$$(A_L)_{PF} = 1 - L/D_e = \exp(-L/D_e) \qquad (5.92)$$

where D_e contains the average size of the coherently diffracting domains $D(hkl)$ and the fictitious size $D_F(hkl)$ due to faulting [5.8,10]. The strain coefficients $|(C_L)_D|$ can be written for small values of $L = nd_{hkl}$, $\varepsilon_L = (\Delta L/L)$, and $Q_0 = 2\pi s_0 = 4\pi \sin \theta_0/\lambda = 2\pi/d_{hkl}$, where θ_0 is the Bragg angle of the peak maximum:

$$\begin{aligned}
|(C_L)_D| &= 1 - 2\pi^2 L^2(\langle(\varepsilon_L)^2\rangle - \langle\varepsilon_L\rangle^2)(s_0)^2 \\
&= \exp[-2\pi^2 L^2(\langle(\varepsilon_L)^2\rangle - \langle\varepsilon_L\rangle^2)(s_0)^2].
\end{aligned} \qquad (5.93)$$

The diffracted intensity $I(Q)$ can then be written as

$$\begin{aligned}
I(Q) &= P'(Q)/[K(\theta)(F_t)^2] \\
&= \sum (A_L)_{PF}|(C_L)_D| \exp[-iL(Q - Q_0 - \langle\varepsilon_L\rangle Q_0)].
\end{aligned} \qquad (5.94)$$

To separate the particle size term $(A_L)_{PF}$ from the strain term $|(C_L)_D|$, the Warren-Averbach method is applied [5.8]:

$$\ln(|C_L|) = \ln(A_L)_{PF} - 2\pi^2 L^2(\langle(\varepsilon_L)^2\rangle - \langle\varepsilon_L\rangle^2)(s_0)^2. \qquad (5.95)$$

By plotting $\ln(|C_L|)$ as a function of the order $(s_0)^2[= (h^2 + k^2 + l^2)/a^2$ for cubic lattices] of reflection and extrapolating to $(s_0)^2 = 0$, one finds the value

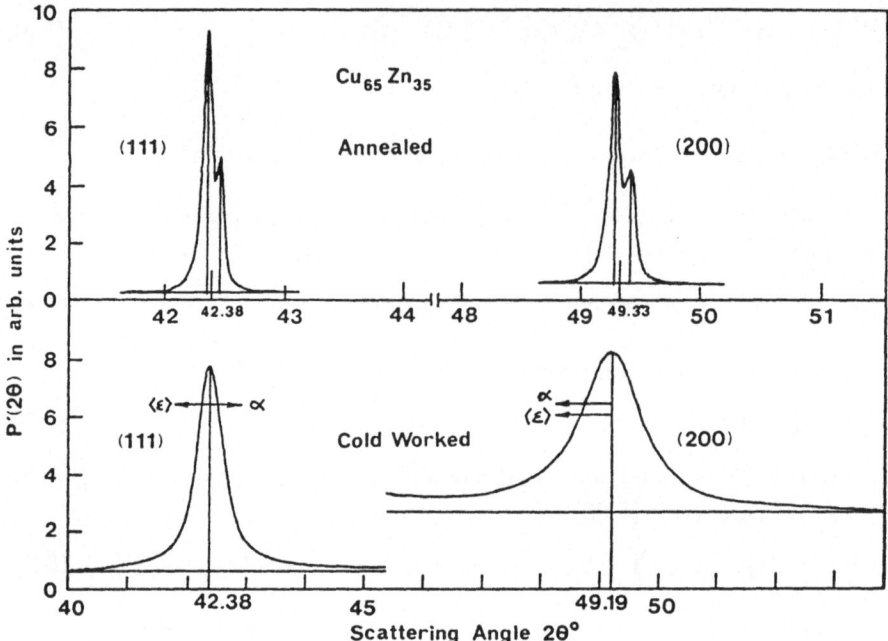

Fig. 5.10. Line profiles of annealed and cold-worked α-brass (Cu$_{65}$Zn$_{35}$; Cu K$_\alpha$)

of $(A_L)_{PF}$ for each L. The slopes of the curves $\ln(|C_L|)$ plotted versus $(s_0)^2$ are a measure of the mean-square strain deviation $[\langle(\varepsilon_L)^2\rangle - \langle\varepsilon_L\rangle^2]$ for different values of the distance L normal to the reflecting planes. The initial slope of the plot $(A_L)_{PF}$ vs. L is a measure of the effective particle size D_e. It follows from (5.92) that

$$[d(A_L)_{PF}/dL]_{L\to 0} = -1/D_e. \tag{5.96}$$

A detailed analysis of the effect of experimental errors in the determination of the peak profiles on the Fourier coefficients has been made by *Delhez* et al. [5.60].

The profiles of the powder pattern peaks (111) and (200) from annealed and cold-worked samples of 65/35 α-brass (Cu$_{65}$Zn$_{35}$) are shown in Fig. 5.10. The pattern was obtained with Cu K$_\alpha$ radiation in the focussing Bragg-Brentano reflection geometry. The peaks from the annealed standard were used to correct for instrumental broadening [5.8,10]. The cosine Fourier coefficients $A_{hkl}(L)$ of the filings of Cu$_{65}$Zn$_{35}$, prepared at liquid nitrogen temperature, are shown in Fig. 5.11. Using (5.95), the particle size and strain coefficients A_L^P and $|C_L^D|$, respectively, were separated. The initial slope of A_L^P yielded the effective particle size $D_e = 65$ Å, as shown in Fig. 5.11. The strain coefficients allowed the evaluation of the root-mean-square strains $(\langle\varepsilon^2\rangle - \langle\varepsilon\rangle^2)^{1/2}$, which are also shown in Fig. 5.11. It should be realized that $\langle\varepsilon_L\rangle$ is assumed to be zero in filings.

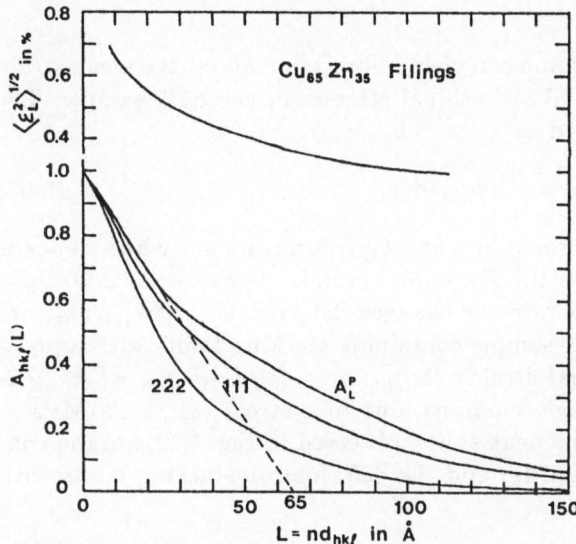

Fig. 5.11. Fourier coefficients of the line profiles of $Cu_{65}Zn_{35}$ filings, prepared at liquid nitrogen temperature

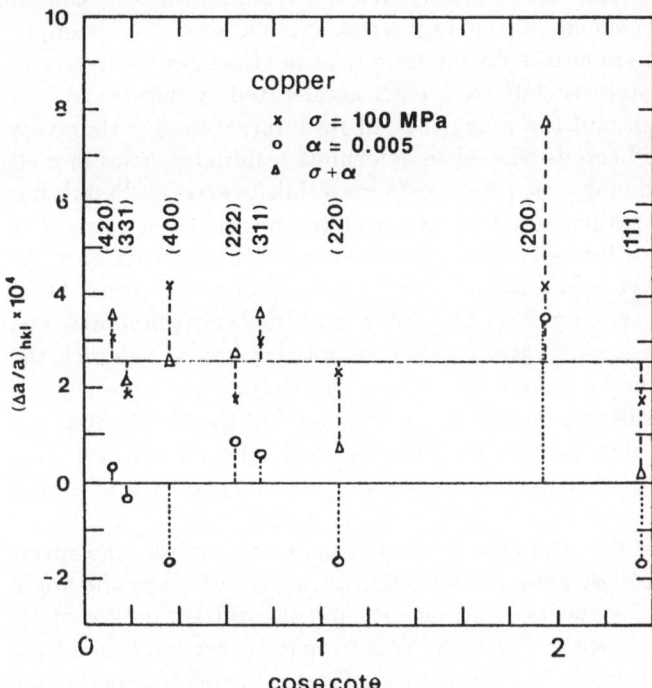

Fig. 5.12. Relative lattice parameter changes $\Delta a_{hkl}/a_{hkl}$ in copper, due to stacking faults (with the probability $\alpha = 0.005$), and residual strain $\varepsilon = -(\nu/E)\sigma$ (with $\sigma = 100\,\text{MPa} \simeq 10\,\text{kg/mm}^2 \simeq 14\,\text{Ksi}$) [5.10]

b) Peak Shift

The position of the peak maximum can yield information about the mean strain $\langle \varepsilon_L \rangle$, which is usually produced by residual stresses in the bulk sample. The peak shift ΔQ can be expressed as

$$\Delta Q / Q_0 = -\langle \varepsilon_L \rangle - G_{hkl} \alpha = -\Delta d_{hkl} / d_{hkl} \tag{5.97}$$

where α is the stacking fault probability and G_{hkl} is a constant which depends on the reflecting planes (hkl) [5.10]. For cubic crystals, we can write $\Delta Q / Q_0 = -\Delta a / a$. The relative lattice parameter changes $\Delta d_{hkl} / d_{hkl} = \Delta a_{hkl} / a_{hkl}$ are shown in Fig. 5.12 for a copper sample containing stacking faults with a probability $\alpha = 0.005$ and residual strains $\langle \varepsilon_L \rangle_{hkl} = -(\nu/E) \langle \sigma \rangle_{hkl}$, where ν is Poisson's ratio and E is Young's modulus, due to a stress $\langle \sigma \rangle = 100\,\text{MPa} \simeq 10\,\text{kg/mm}^2 \simeq 14\,\text{Ksi}$ [5.10]. The peak shifts, observed in Fig. 5.10, are the consequence of the effect of a strain $\langle \varepsilon \rangle$ and the deformation stacking faults with the probability α [5.61].

5.3.5 Residual Stress Measurements

Residual stresses are stresses which remain in a material when no force is applied. They can be introduced into metals and alloys by any mechanical, chemical, or thermal process. The build-up of such residual stresses can be either detrimental or beneficial during industrial applications. For example, initiation and propagation of cracks during fatigue or in stress corrosion can be impeded by compressive stresses, but are greatly accelerated by tensile stresses. The knowledge of the sign and the magnitude of residual stresses is thus very important. Methods have been developed to determine residual stresses in metals, based on acoustic and magnetic response of materials to stresses [5.62], but these have proved to be rather sensitive to variations in microstructure. The x-ray and neutron methods for residual stress measurements have been successfully employed for many years [5.2,13,63]. They are based on the determination of changes in interplanar spacings d_{hkl} by standard diffraction techniques, i.e., the d-spacing serves as an internal strain gauge. In poly-crystalline samples, the conventional powder method can be applied when the grains are randomly oriented and their size is less than 0.05 mm in diameter so that the Debye-Scherrer rings consist of uniformly distributed diffraction spots. Under these conditions, it is possible to apply isotropic elasticity theory to convert the measured strain into stresses.

In order to determine the complete strain tensor in the sample, the specimen must be rotated about two axes. One rotation is characterized by the angle ψ between the normal to the specimen surface P_3 and the diffraction vector Q, which is parallel to L_3' as shown in Fig. 5.13. This tilt can be accomplished by a rotation about an axis L_2' lying in the specimen surface and being perpendicular to the diffraction plane, defined by the incident and diffracted beams k_0 and k_1, respectively, i.e., the Ω-diffractometer (Fig. 5.13), or by a rotation about an axis lying in the specimen surface and in the diffraction plane, but being

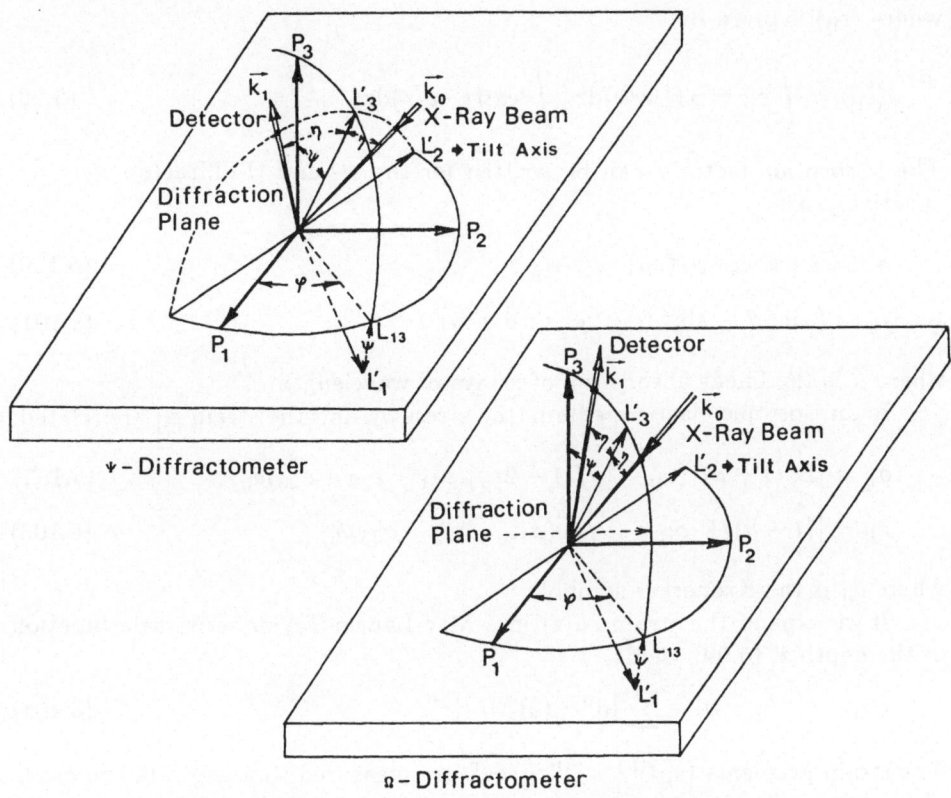

Fig. 5.13. Diffraction geometry for the Ψ- and Ω-diffractometers

perpendicular to the diffraction vector Q, i.e., the Ψ-diffractometer (Fig. 5.13). The second rotation is characterized by the angle ϕ and is carried out about an axis parallel to the specimen normal P_3 (Fig. 5.13). A useful guide for x-ray stress evaluation has recently been published by *Hauk* and *Macherauch* [5.64].

The residual or applied strain can be determined from a change in interplanar spacing $d_{\phi\psi}$ of lattice planes whose normal L_3' forms the angle ψ with the normal P_3 to the specimen surface, and its projection L_{13} onto the specimen surface P_1P_2 forms the angle ϕ with the P_1 axis as shown in Fig. 5.13. The strain $\varepsilon_{\phi\psi}$ is then given by the relation:

$$\varepsilon_{\phi\psi} = (d_{\phi\psi} - d_0)/d_0$$

where d_0 is the interplanar spacing of the unstrained material. Since x-rays penetrate the specimen surface, we measure the strain $\langle \varepsilon_{\phi\psi} \rangle$ averaged over the penetration depth t, i.e. [5.11],

$$\langle \varepsilon_{\phi\psi} \rangle = \langle \varepsilon_{3'3'} \rangle = \langle \varepsilon_{11} \rangle \cos^2 \phi \, \sin^2 \psi + \langle \varepsilon_{12} \rangle \sin 2\phi \, \sin^2 \psi + \langle \varepsilon_{22} \rangle \sin^2 \phi \, \sin^2 \psi$$

$$+ \langle \varepsilon_{13} \rangle \cos \phi \, \sin 2\psi + \langle \varepsilon_{23} \rangle \sin \phi \, \sin 2\psi + \langle \varepsilon_{33} \rangle \cos^2 \psi \qquad (5.98)$$

where $\langle \varepsilon_{ij} \rangle$ is given by

$$\langle \varepsilon_{ij} \rangle = \int \varepsilon_{ij} \exp(-z/\tau)dz / \int \exp(-z/\tau)dz. \tag{5.99}$$

The absorption factor τ can be written for the Ψ- and Ω-diffractometers, respectively, as:

$$\tau_\psi = \sin\theta \, \cos\psi/(2\mu) \tag{5.100}$$

$$\tau_\omega = (\sin^2\theta - \sin^2\psi)/(2\mu \, \sin\theta \, \cos\psi) \tag{5.101}$$

where μ is the linear absorption of x-rays of wavelength λ.

In an isotropic elastic medium, the stress σ_{ij} and the strain ε_{ij} are related:

$$\sigma_{ij} = [E/(1+\nu)]\{\varepsilon_{ij} - [\nu/(1-2\nu)](\varepsilon_{11} + \varepsilon_{22} + \varepsilon_{23})\delta_{ij}\}, \tag{5.102}$$

$$\varepsilon_{ij} = [(1+\nu)/E]\sigma_{ij} - (\nu/E)(\sigma_{11} + \sigma_{22} + \sigma_{33})\delta_{ij} \tag{5.103}$$

where δ_{ij} is the Kronecker symbol.

If we expand the strain $\varepsilon_{ij}(z)$ in a Mac-Laurin-Taylor series as a function of the depth z, (5.99) yields

$$\langle \varepsilon_{ij} \rangle = \varepsilon_{ij}(z = 0) + \sum[d^n \varepsilon_{ij}(0)/dz^n]\tau^n. \tag{5.104}$$

The strain gradients $(\varepsilon_{ij})^{(n)} = d^n \varepsilon_{ij}/dz^n$ are measured to a depth of the order of the absorption factor τ.

Since $\langle \varepsilon_{\phi\psi} \rangle$ depends on the angles ψ and ϕ, (5.98), there are two basic choices in the experimental methods which was pointed out by *Lode* and *Peiter* [5.65]. In the conventional method, $\langle \varepsilon_{\phi\psi} \rangle$ is measured as a function of the angle ψ for fixed values of ϕ, which is called the ψ-method. It is also possible to determine $\langle \varepsilon_{\phi\psi} \rangle$ as a function of ϕ at fixed angles ψ, which is called the ϕ-method.

It follows from (5.98, 99, and 104), that $\langle \varepsilon_{\phi\psi} \rangle$ is a function of powers of trigonometric functions in ϕ and ψ. It becomes possible to express $\langle \varepsilon_{\phi\psi} \rangle$ as a function of $\sin^n\psi$, or $\sin(n\phi)$ and $\cos(n\phi)$ which yield the differential and integral methods, respectively. The differential ψ-method, which includes the well-known $\sin^2\psi$ – method as a special case, was described by *Cohen* et al. [5.11], whereas the integral method was introduced by *Lode* and *Peiter* [5.65]. Extensions of the ψ-differential and ϕ-integral methods [5.12] permit the evaluation of the strains ε_{ij}^0 and their gradients $(\varepsilon_{ij})^{(n)}$ in the surface of the specimen [5.12].

a) The ψ-Differential Method

Equation (5.98) can be written in the following form

$$\langle \varepsilon_{\phi\psi} \rangle = \alpha + \beta \, \sin^2\psi + \gamma \sin 2\psi \quad \text{where} \tag{5.105}$$

$$\alpha = \langle \varepsilon_{33} \rangle, \tag{5.106}$$

$$\beta = \langle \varepsilon_{11} \rangle \cos^2 \phi + \langle \varepsilon_{22} \rangle \sin^2 \phi + \langle \varepsilon_{12} \rangle \sin 2\phi - \langle \varepsilon_{33} \rangle, \tag{5.107}$$

$$\gamma = \langle \varepsilon_{13} \rangle \cos \phi + \langle \varepsilon_{23} \rangle \sin \phi. \tag{5.108}$$

If we measure $\langle \varepsilon_{\phi\psi} \rangle$ for positive and negative values of ψ, say $-45° < \psi < +45°$, we can form the following expression:

$$(a_+)_{\phi\psi} = [\langle \varepsilon_{\phi\psi} \rangle_{\psi>0} + \langle \varepsilon_{\phi\psi} \rangle_{\psi<0}]/2 = \alpha + \beta \sin^2 \psi, \tag{5.109}$$

$$(a_-)_{\phi\psi} = [\langle \varepsilon_{\phi\psi} \rangle_{\psi>0} - \langle \varepsilon_{\phi\psi} \rangle_{\psi<0}]/2 = \gamma \sin |2\psi|. \tag{5.110}$$

The coefficients α, and β can be evaluated from the least-square fit of the values of $(a_+)_\psi$ when plotted as a function of $\sin^2 \psi$. The coefficient γ can likewise be obtained from the plot of $(a_-)_\psi$ vs. $\sin |2\psi|$.

Values of the strain $\langle \varepsilon_{\phi\psi} \rangle$ are shown in Fig. 5.14 for a cast and ground rail-steel (with 0.75wt.% carbon). The measurements were made for values of $\psi = 0, 9, 18, 27, 36$ and $45°$, and $\phi = 0°, 45°$ and $90°$, using the (310) reflection determined with $\mathrm{Co\,K_\alpha}$ radiation and a position-sensitive detector. Also

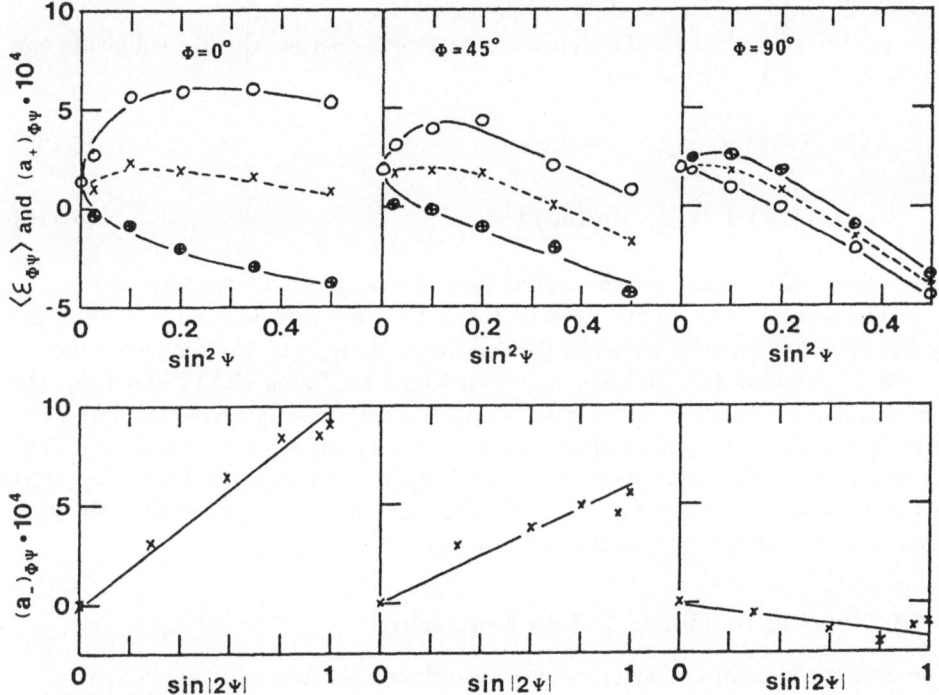

Fig. 5.14. Values of $< \varepsilon_{\phi\psi} >$ for a cast and ground rail-steel (plain carbon steel with 0.75 wt.% carbon). The x-ray measurements were made with $\mathrm{Co\,K_\alpha}$ radiation, using the (310) reflection. The values of $< \varepsilon_{\phi\psi} >$, $(a_+)_{\phi\psi}$, (5.109), and $(a_-)_{\phi\psi}$, (5.110), were plotted as function of $\sin^2 \psi$ and $\sin |2\psi|$, respectively

shown are the plots of $(a_+)_{\phi\psi}$ versus $\sin^2\psi$ and $(a_-)_{\phi\psi}$ vs. $\sin|2\psi|$. The ψ-differential method yielded the strain tensor: $\sigma_{11} = -63$, $\sigma_{12} = -4$, $\sigma_{13} = -93$, $\sigma_{22} = -220$, $\sigma_{23} = -15$, and $\sigma_{33} = -33$ MPa. The error is estimated to be ± 15 MPa.

b) The ϕ-Integral Method

Since the absorption factor τ, (5.100,101), depends only on the tilt angle ψ, it becomes advantageous to develop $\langle \varepsilon_{\phi\psi} \rangle$ in (5.98) as a function of ϕ [5.12], i.e.,

$$\langle \varepsilon_{\phi\psi} \rangle = A_0/2 + A_1 \cos\phi + A_2 \cos 2\phi + B_1 \sin\phi + B_2 \sin 2\phi \qquad (5.111)$$

where

$$A_0 = [\langle\varepsilon_{11}\rangle + \langle\varepsilon_{22}\rangle] \sin^2\psi + 2\langle\varepsilon_{33}\rangle \cos^2\psi, \qquad (5.112)$$

$$A_1 = \langle\varepsilon_{13}\rangle \sin 2\psi, \qquad (5.113)$$

$$A_2 = (1/2)(\langle\varepsilon_{11}\rangle - \langle\varepsilon_{22}\rangle) \sin^2\psi, \qquad (5.114)$$

$$B_1 = \langle\varepsilon_{23}\rangle \sin 2\psi, \qquad (5.115)$$

$$B_2 = \langle\varepsilon_{12}\rangle \sin^2\psi. \qquad (5.116)$$

It is readily seen that (5.111) represents a Fourier series whose coefficients can be determined by the relations

$$A_n = (1/2\pi) \int \langle\varepsilon_{\psi\psi}\rangle \cos(n\phi)d\phi, \qquad (5.117)$$

$$B_n = (1/2\pi) \int \langle\varepsilon_{\phi\psi}\rangle \sin(n\phi)d\phi, \qquad (5.118)$$

when the strains $\langle\varepsilon_{\phi\psi}\rangle$ are measured over the range of ϕ from 0 to 360°.

The strains $\langle\varepsilon_{\phi\psi}\rangle$, presented in Fig. 5.14, were also plotted as a function of the rotation angle ϕ between 0° and 360° in steps of 15° for the values of $\psi = 9, 27, 36$, and 45°, and are shown in Fig. 5.15. Using (5.117 and 118), the cosine and sine Fourier coefficients A_n and B_n, respectively, were calculated and used to evaluate the strain tensor and the corresponding stress tensor $\langle\sigma_{ij}\rangle$. The following values were obtained: $\sigma_{11} = -65$, $\sigma_{12} = -15$, $\sigma_{13} = -80$, $\sigma_{22} = -220$, $\sigma_{23} = 15$, and $\sigma_{33} = -25$ MPa. These values are in good agreement with those obtained with the ψ-differential method, shown in Fig. 5.14.

5.3.6 Grazing Incidence X-Ray Scattering

The grazing incidence scattering was introduced by *Marra* et al. [5.5] to determine the structure of thin films on crystalline or amorphous substrates. In this geometry (Fig. 5.16), the primary x-ray beam hits the sample at a shallow grazing angle ϕ. Total reflection will occur if the specimen is sufficiently smooth and flat, and if the angle ϕ is smaller than the critical angle θ_c, approximately

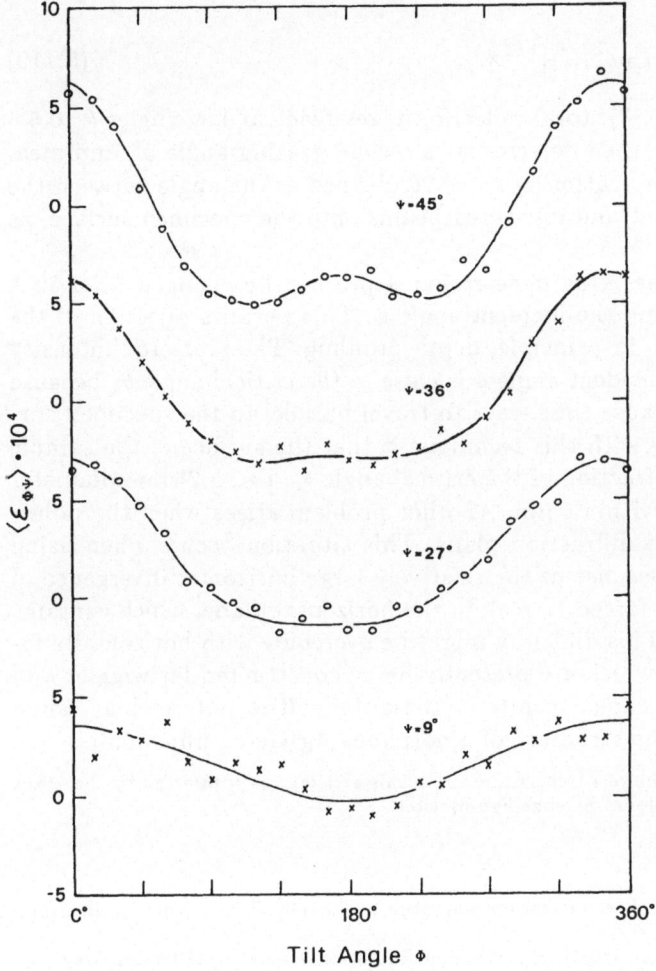

Fig. 5.15. Values of $< \varepsilon_{\phi\psi} >$ for a plain carbon steel with 0.75% carbon, plotted as a function of ϕ for $\psi = 9, 27, 36$, and $45°$

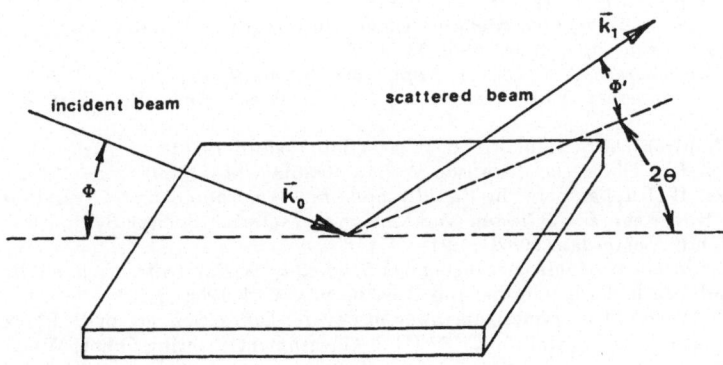

Fig. 5.16. Geometry of the grazing incidence scattering [5.6]

given by

$$\theta_c = (2\beta)^{1/2} = [e^2/(mc^2)\langle f\rangle]^{1/2}\lambda. \qquad (5.119)$$

Because β is small ($\sim 10^{-6}$), total reflection takes place at low angles $\theta < 0.5°$. The scattered x-rays are then detected at a second grazing angle ϕ', and measured as a function of the scattering angle 2θ, defined as the angle between the projections of the incident and diffracted beams onto the specimen surface, as shown in Fig. 5.11.

In this geometry, the x-ray penetration depth can be changed from 30 Å to several μm by changing the incident angle ϕ. This permits rejection of the substrate scattering and, in principle, depth profiling. The scattered intensity is maximized when the incident angle ϕ is close to the critical angle θ_c because refractive index effects cause the x-rays to travel parallel to the specimen surface. The basic difficulty with this technique is that the surface of the sample must be flat to within a fraction of the critical angle θ_c, i.e., a 25 mm diameter sample must be flat to within 5 μm. Another problem arises when the polarization vector lies in the diffraction plane. This situation occurs when using synchrotron radiation. Because of the relatively large horizontal divergence of its primary beam, one is forced to scan in the horizontal plane, which contains the polarization vector. This difficulty might be overcome with horizontally focusing monochromators which are presently being constructed for wiggler and undulator synchrotron beams. Inspite of these difficulties, interesting results have been obtained on the structure of amorphous Ag-Ge-Se films [5.6].

Acknowledgements. Part of the research, discussed in this article, was supported by the grant DMR 83-10025 from the National Science Foundation.

References

5.1   C. Kunz (ed.): *Synchrotron Radiation*, Topics Current Phys., Vol. 10 (Springer, Berlin, Heidelberg 1979)
        H. Winick, S. Doniach (eds.): *Synchrotron Radiation Research*, (Plenum, New York 1980)
5.2   J.M. Carpenter, G.H. Lander, C.G. Windsor: Rev. Sci. Instrum. **55**, 1019 (1984)
5.3   H. Beck, H.-J. Güntherodt: *Glassy Metals I* and *II*, Topics Appl. Phys., Vols. 46 and 53 (Springer, Berlin, Heidelberg 1981 and 1983)
5.3   F.E. Luborsky, (ed.): *Amorphous Metallic Alloys* (Butterworths, London 1983)
5.4   C.N.J. Wagner: J. Non-Crystalline Solids **31**, 1 (1978)
5.5   W. Marra, P. Eisenberger, A.Y. Cho: J. Appl. Phys. **50**, 6279 (1979)
5.6   P.H. Fuoss, P. Eisenberger, W.K. Warburton, A. Bienenstock: Phys. Rev. Lett. **46**, 1537 (1981)
5.7   G.E. Bacon: *Neutron Diffraction*, 3rd ed. (Clarendon, Oxford 1975)
5.8   B.E. Warren: *X-Ray Diffraction* (Addison-Wesley, Reading, MA 1969)
5.9   C.N.J. Wagner, H. Ruppersberg: In *Rapidly Solidified Amorphous and Crystalline Alloys*, ed. by B.H. Kear, B.C. Giessen, and M. Cohen, Materials Society Symp., Vol. 8 (North-Holland, Amsterdam 1982) p. 91
5.10   C.N.J. Wagner: In *Local Atomic Arrangements Studied by X-Ray Diffraction*, ed. by J.B. Cohen and J.E. Hilliard (Gorden and Breach, New York 1966) p. 219
5.11   J.B. Cohen, H. Doelle, M.R. James: *Accuracy in Powder Diffraction*, ed. by S. Block and C.R. Hubbard, NBS Special Publ. 567 (U.S. Government Printing Office, Washington 1980) p. 453
5.12   C.N.J. Wagner, M.S. Boldrick, V. Perez-Mendez: Adv. X-Ray Analysis **26**, 275 (1983)

5.13   B.D. Cullity: *Elements of X-Ray Diffraction*, 2nd ed. (Addison-Wesley, Reading, MA 1978)

5.14   A. Guinier: *X.Ray Diffraction* (Freeman, San Francisco 1963)

5.15   A.E. Lee, G. Etherington, C.N.J. Wagner: In *Proc. 5th Intern. Conf. Rapidly Quenched Metals*, ed. by S. Steeb and H. Warlimont (North-Holland, Amsterdam 1984) p. 491

5.16   C.N.J. Wagner: In *Amorphous Metallic Alloys*, ed. by F.E. Luborsky (Butterworth, London 1983) p.58

5.17   A.B. Bhatia, D.E. Thornton: Phys. Rev. B **2**, 3004 (1970)

5.18   C.N.J. Wagner, H. Ruppersberg: In *Application of Nuclear Techniques to the Studies of Amorphous Metals*, ed. by U. Gonser, Atomic Energy Review, Supplement No.1, 1981, p. 101

5.19   C.N.J. Wagner: In *Proc. 5th Intern. Conf. Rapidly Quenched Metals*, ed. by S. Steeb and H. Warlimont (North-Holland, Amsterdam 1984) p. 405

5.20   C.S. Cargill, F. Spaepen: J. Non-Crystalline Solids **43**, 91 (1981)

5.21   D.E. Moncton: *Major Facilities for Materials Research and Related Disciplines*, National Research Council (National Academy Press, Washington 1984) p. 15

5.22   J.E. Spencer, H. Winick: In *Synchrotron Radiation Research*, ed. by H. Winick and S. Doniach (Plenum, New York 1980) p. 662

5.23   R. Pynn: Rev. Sci. Instrum. **55**, 837 (1984)

5.24   R.C. Hamlin (ed.): Trans. Am. Cryst. Assoc. **18** (1982)

5.25   B. Sleaford, V. Perez-Mendez, C.N.J. Wagner: Adv. X-Ray Analysis **26**, 269 (1983)

5.26   E.R. Woelfel: J. Appl. Cryst. **16**, 341 (1983)

5.27   R.N. Sinclair, D.A.G. Johnson, J.C. Dore, H.H. Clarke, A.C. Wright: Nucl. Instrum. Meth. **117**,445 (1974)

5.28   K. Suzuki, M. Misawa, K. Kai, N. Watanabe: Nucl. Instrum. Meth. **147**, 519 (1977)

5.29   T. Egami: In [Ref. 5.3a, p. 25]

5.30   G. Fritsch, C.N.J. Wagner: Z. Physik B **62**, 189 (1986)

5.31   J.G. Ramesh, S. Ramaseshan: J. Phys. C **4**, 3029 (1971)

5.32   C.N.J. Wagner, D. Lee: J. Physique **41**, C8–242 (1980)

5.33   J.E. Enderby, D.M. North, P.A. Egelstaff: Phil. Mag. **14**, 961 (1966)

5.34   F.G. Edwards, J.E. Enderby, R.A. Howe, D.I. Page: J. Phys. C **9**, 3483 (1975)

5.35   A.K. Livesey, P.H. Gaskell: *Proc. 4th Intern. Conf. Rapidly Quenched Metals*, ed. by T. Masumoto and K. Suzuki, Vol. 1 (Jap. Inst. Metals 1982) p.335

5.36   D. Lee, A. Lee, C.N.J. Wagner, L.E. Tanner, A.K. Soper: J. Physique **43**, C-9–19 (1982)

5.37   A.E. Lee, G. Etherington, C.N.J. Wagner: J. Non-Crystalline Solids **61/62**, 349 (1984)

5.38   A. Williams: J. Non-Crystalline Solids **45**, 183 (1981)

5.39   P. Lamparter, W. Sperl, S. Steeb, J. Bletry: Z. Naturforsch. **37a**, 1223 (1982)

5.40   T. Fugunaga, N. Watanabe, K. Suzuki: J. Non-Crystalline Solids **61/62**, 343 (1984)

5.41   P. Lamparter, S. Steeb: In *Proc. 5th Intern. Conf. Rapidly Quenched Metals*, ed. by S. Steeb and H. Warlimont (North Holland, Amsterdam 1984) p. 459

5.42   H. Ruppersberg, D. Lee, C.N.J. Wagner: J. Phys. F **10**, 1645 (1980)

5.43   M. Sakata, N. Cowlam, H.A. Davies: J. Physique **41**, C8–190 (1980)

5.44   C. Janot, B. George, C. Tete, A. Chamberod, L. Laugier: J. Physique **46**, 1233 (1985)

5.45   H. Ruppersberg, H. Egger: J. Chem. Phys. **63**, 4095 (1975)

5.46   P. Chieux, H. Ruppersberg: J. Physique **41**, C8-145 (1980)

5.47   J.H. Hoyt, D. de Fontaine, W.K. Warburton: J. Appl. Cryst. **17**, 344 (1984)

5.48   N. Shevchik: Phil. Mag. **35**, 805 (1977)

5.49   J. Kortright, A. Bienenstock: J. Non-Crystalline Solids **61/62**, 273 (1984)

5.50   S. Aur, D. Kofalt, Y. Waseda, T. Egami, H.S. Chen, B.K. Teo, R. Wang: J. Non-Crystalling Solids **61/62**, 331 (1984)

5.51   R. Reiter, H. Ruppersberg, W. Speicher: *Liquid Metals 1976*, Inst. Phys. Series No. 30, 133 (1977)

5.52   M. Maret, A. Soper, G. Etherington, C.N.J. Wagner: J. Non-Crystalline Solids **61/62**, 313 (1984)

5.53   P. Lamparter, S. Steeb, E. Grallath: Z. Naturforsch. **38a**, 1210 (1983)

5.54   N. Cowlam, W. Guoan, P.P. Gardner, H.A. Davies: J. Non-Crystalline Solids **61/62**, 337 (1984)

5.55 E. Nold, P. Lamparter, H. Olbrich, G. Rainer-Harbach, S. Steeb: Z. Naturforsch. **36a**, 1032 (1981)

5.56 H.P. Klug, L.E. Alexander: *X-Ray Diffraction Procedures for Polycrystalline and Amorphous Materials*, 2nd ed. (Wiley, New York 1974)

5.57 E. R. Woelfel: J. Appl. Cryst. **14**, 291 (1981)

5.58 R. Gullberg, R. Lagneborg: Trans. Met. Soc. AIME **236**, 1482 (1966)

5.59 C.S. Cargill, III: In *Solid State Physics*, **30**, 227 (Academic, New York 1975)

5.60 R. Delhez, T.H. de Keijser, E.J. Mittemeijer: *Accuracy in Powder Diffraction*,, ed. by S. Block and C.R. Hubbard, NBS Special Publ. 567 (U.S. Government Printing Office, Washington 1980) p. 213

5.61 R.P.I. Adler, H.M. Otte, C.N.J. Wagner: Met. Trans. **1**, 2377 (1970)

5.62 M.R. James, O. Buck: *CRC Critical Rev. in Solid State Sci.*, Vol. 9 (CRC Press, Boca Raton, Florida 1980) p. 61

5.63 M.R. James, J.B. Cohen: *Experimental Methods in Materials Science* , Treatise on Materials Science and Technology **19A**, 20 (1980)

5.64 V.M. Hauk, E. Macherauch: Adv. X-Ray Analysis **27**, 81 (1984)

5.65 W. Lode, A. Peiter: Metall **35**, 758 (1981)

6. Extended X-Ray Absorption Fine Structure

D.G. Stearns and M.B. Stearns
With 5 Figures

The x-ray absorption spectrum of an atom exhibits sharp edges which correspond to the excitation threshold of the core electrons. If the atom is surrounded by other atoms in a condensed phase, then the absorption cross section above the absorption edge is observed to oscillate, exhibiting an often complex structure that can extend for hundreds of electron volts. These oscillations are aptly called the extended x-ray absorption fine structure (EXAFS). The fundamental understanding of EXAFS has advanced in the last two decades to the point where quantitative theoretical formulations can accurately account for the fine structure. The oscillations of the absorption cross section are due to the interference between the outgoing photoelectron wave and the incoming wave that is generated when the photoelectron scatters off of the atoms that surround the excited atom. Hence, EXAFS contains detailed information about the local environment of a particular type of atom. With the theoretical framework providing a relatively simple means of interpretation. EXAFS spectroscopy has developed into an important experimental tool for studying local atomic structure in solids.

In this chapter we present a review of the experiments that have used EXAFS spectroscopy to investigate the atomic structure of metal systems. The chapter begins with a general discussion of the theory, measurement and analysis of EXAFS, in which we highlight the advantages and limitations of the technique. This presentation is intended to provide a non-expert with an introduction to the field. The interested reader can find a more sophisticated and comprehensive discussion of EXAFS spectroscopy in several recent review articles [6.1–6]. The major emphasis and hence largest portion of this chapter is then devoted to cataloging the experiments in which EXAFS has been used to investigate the structure of metal systems. The increasing importance of EXAFS spectroscopy in materials science and metallurgy will be evident as we review the unique structural information that has been obtained with this technique.

6.1 Theory

6.1.1 Overview

The existence of EXAFS has been recognized for over sixty years, and throughout this period there were many attempts to explain the phenomenon.

We refer the reader to a nice historical account of the field in the review article by *Hayes* and *Boyce* [6.5]. It was not until the 1970's, however, that the origin of the EXAFS was placed on a firm theoretical footing. *Sayers* et al. [6.7a] presented a simple model, where the atoms surrounding the excited atom were treated as point scatterers, which could account for most of the features observed in K-absorption fine structure of Ge and Cu. The most important consequence of this model was that it showed that the fine structure could be related to the radial distribution of atoms that surround the absorbing atom through a simple Fourier transform [6.7b]. This interpretation of EXAFS, along with the development in the early 1970's of synchrotron radiation as an intense, broadband x-ray source, established EXAFS as a potentially powerful probe of atomic structure. In the mid-1970's the theory of EXAFS was refined to incorporate more realistic atomic potentials and such effects as the photoelectron scattering off of many atoms [6.8–10]. These more sophisticated treatments indicated that accurate structural information could be extracted from EXAFS if the atomic scattering factors were known. To this end the atomic scattering factors were calculated and tabulated for many elements [6.11,12]. It was hoped that these could be applied to any system of interest, making EXAFS the premier technique for studying short-range order. In the late 70's and early 80's it became apparent, however, that EXAFS spectroscopy has some distinct limitations. In particular, the atomic scattering factors are not universal, but depend somewhat on the chemical environment of the atoms (Sect. 6.3.2a). Another problem is that EXAFS is rather insensitive to broad features in the radial distribution function [6.14]. The recognition of these limitations in the last few years has represented the maturing of the field of EXAFS spectroscopy. What we are left with is a powerful structural probe that must be used with discretion, and preferably in conjunction with other techniques.

6.1.2 The Standard EXAFS Formula

In this subsection we discuss the basic principles and approximations that are involved in deriving the standard EXAFS formula. Rigorous and sophisticated derivations can be found in many places [6.2,5,8–10]. A typical example of EXAFS is presented in Fig. 6.1, which shows the absorption cross section above the K-edge for metallic Ni. EXAFS also occurs above L-edges, however the theory is slightly more complicated (see, for instance, [6.10]), so that we will consider here only excitation of the K-shell (1s) electrons.

The standard EXAFS formula is always derived using the single electron picture. We assume that when an atom (herein called the "central" atom) absorbs an x-ray with energy in excess of the K-edge threshold, then a 1s core electron is excited into the continuum. Due to the strong localization of the core state, we can think of the photoelectron as being emitted from the origin (defined as the center of the central atom). As the photoelectron wave expands into space, it is affected by the local potentials within the solid. In EXAFS theory the potentials are invariably treated in the muffin-tin approximation.

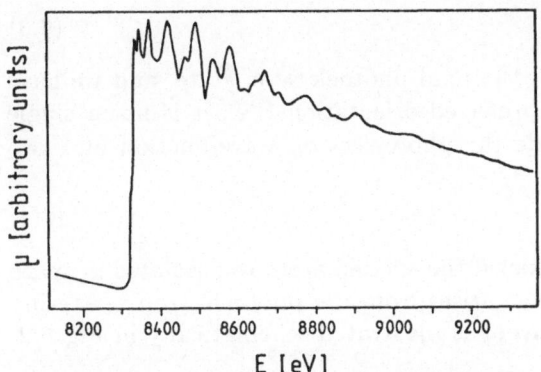

Fig. 6.1. The x-ray absorption cross section at the K-edge in Ni, showing the characteristic EXAFS oscillations

That is, the central atom and neighboring atom potentials are considered to be localized, spherically symmetric and surrounded by a region of constant potential. This formulation is critical to the success of the theory. It allows the photoelectron outside of the atomic potentials to be described as a free electron with a wavevector k given by

$$k^2 = 2m(E - E_0)/\hbar^2. \tag{6.1}$$

The energy E_0 is the value of the constant potential in between the atoms in the muffin-tin picture. In a metal, E_0 should correspond to the bottom of the conduction band.

The basic goal in EXAFS theory is to calculate the effect that the neighboring atomic potentials have on the outgoing photoelectron. If there were no neighboring atoms, then the absorption cross section would show no fine structure. The presence of EXAFS is precisely due to the scattering of the photoelectron off of the neighboring atom potentials. The scattering of the photoelectron is weak (particularly at high energies) so that the neighboring atom potentials are treated as a perturbation to the case of free atom absorption. On the other hand, the lifetime of the photoelectron is sufficiently short so that it will dissipate before it can get very far from the central atom. Hence, in EXAFS theory it is only necessary to include the scattering from the first few shells of atoms that surround the central atom.

EXAFS are represented by the quantity $\chi(E)$ which is defined as the normalized variation in the absorption cross section

$$\chi(E) = [\mu(E) - \mu_0(E)]/\mu_0(E). \tag{6.2}$$

Here $\mu(E)$ is the total absorption of the K-shell at energy E, which contains the EXAFS, and $\mu_0(E)$ is the free atom absorption that would occur if the central atom was isolated in space.

We now present some qualitative arguments that should help to motivate the formal expression for χ in terms of atomic parameters. The absorption cross section μ is proportional to the square of the dipole transition matrix element, i.e.,

$$\mu \sim (\langle \psi_c | r | \psi_p \rangle)^2. \tag{6.3}$$

Here $\langle \psi_c |$ is the core state, $| \psi_p \rangle$ is the final photoelectron state, and we have assumed that the sample has no preferred orientation (i.e., it is not a single crystal). The next step is to divide the photoelectron wavefunction into two parts

$$\psi_p = \psi_0 + \psi_{sc} \tag{6.4}$$

where ψ_0 is the wave that would exist if the excited atom was isolated in space and ψ_{sc} is the part of the wave that arises from the photoelectron scattering off of neighboring atoms. This concept is illustrated schematically in Fig. 6.2.

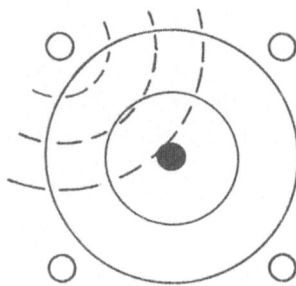

Fig. 6.2. A schematic illustration of the origin of EXAFS. The outgoing photoelectron wave (*solid line*) is backscattered by the neighboring atoms (*dashed line*). The absorption probability is modified by the amplitude of the scattered wave at the central atom

Remember that since the scattering is weak, ψ_{sc} is a small quantity. It can be seen from inserting (6.4) into (6.3) and expanding that the largest term corresponds to the free atom absorption cross section μ_0 and that the next largest terms, which are responsible for EXAFS, are given by

$$\mu - \mu_0 \sim \chi \sim \mathrm{Re} \{ \langle \psi_c | r | \psi_{sc} \rangle \langle \psi_0 | r | \psi_c \rangle \}. \tag{6.5}$$

Consider the first matrix element in (6.5) and recall that the core state is strongly localized. Since the scattered wave ψ_{sc} is well approximated by a plane wave over the region of the core state, then the matrix element is proportional to the amplitude of the scattered photoelectron wavefunction. The oscillations that characterize EXAFS can be understood as follows: as the wavelength of the photoelectron changes with energy, the amplitude of the scattered wave at the origin varies periodically. (Consider varying the distance between the rings in Fig. 6.2.)

In order to proceed further we must estimate the amplitude of the scattered part of the photoelectron wave. The localization of the core state guarantees that the photoelectron is created near the origin. As it leaves the central atom, the photoelectron accumulates a phase shift $\delta(k)$, and then propagates into space as an outgoing spherical wave. When the photoelectron reaches a neighboring atom at a distance r from the central atom, it has accumulated a total phase of $kr + \delta(k)$ and its amplitude has been reduced by a factor of $1/r$. Part of the photoelectron wave is scattered by the neighboring atom. Since we are interested in the amplitude of the scattered wave at the origin, then we will consider only that part of the photoelectron wave which is backscattered. This is called the single scattering approximation, in which we neglect the possibility of the photoelectron wave scattering off of more than one atom before returning to the origin.

The photoelectron wave is backscattered with a strength $f(k)$ and in scattering it experiences a phase shift $\phi(k)$. The scattering process is generally treated in the "small atom approximation" in which the spherical photoelectron wave is approximated as a plane wave. This is important because it makes the scattering factors $f(k)$ and $\phi(k)$ independent of the distance r between the origin and scattering atom. The scattered photoelectron propagates back to the excited atom as a spherical wave. When it gets back to the origin it has picked up another $kr + \delta(k)$ of phase and is attenuated by another factor of $1/r$. Hence the scattered wave arrives at the origin with an amplitude proportional to,

$$\psi_{sc} \sim [f(k)/r^2] \exp\{i[2kr + 2\delta(k) + \phi(k)]\}. \tag{6.6}$$

Of course, the total scattered wave is found by summing the contributions from all the neighboring atoms. The expression for the scattered wave is then inserted into (6.5) to derive $\chi(k)$. When this is done rigorously, and $\chi(k)$ is properly normalized, we arrive at the standard EXAFS formula

$$\chi(k) = -\sum_i [N_i f_i(k)/kr_i^2] \sin[2kr_i + 2\delta(k) + \phi_i(k)]$$
$$\times \exp[-2\sigma_i^2 k^2] \exp[-2r_i/\lambda_i(k)]. \tag{6.7}$$

Here N_i is the number of neighbor atoms in the ith shell located a distance r_i from the central atom. $f_i(k)$ and $\phi_i(k)$ are, respectively, the scattering amplitude and phase shift that characterize the interaction of the photoelectron with the neighboring atom potential. $\delta(k)$ is the phase shift that the photoelectron receives from the central atom potential. The quantity σ_i is the Debye-Waller parameter, and $\lambda_i(k)$ is the photoelectron mean-free path. The Debye-Waller factor and the attenuation factor $D_i(k)$

$$D_i(k) = \exp[-2r_i/\lambda_i(k)] \tag{6.8}$$

are separately incorporated into the theory. The Debye-Waller factor can be derived by averaging over an ensemble of χ's where the atoms in the ith shell are distributed normally about the mean distance r_i [6.15]. Then the Debye-Waller parameter σ_i characterizes the attenuation of EXAFS due to harmonic vibration of the neighbor atoms with respect to the excited atom. (Note that correlated motion between the excited atom and the neighbors can not effect the EXAFS.) The attenuation factor D_i is empirically introduced to account for the finite lifetime of the photoelectron due to inelastic scattering processes. In the case where the sample is a single crystal and the incident x-rays are polarized along a direction $\hat{z}$, (6.7) must be modified by including an angular factor $3(\hat{z} \cdot \hat{R}_i)^2$, $\hat{R}_i$ being a unit vector which points from the central atom to the ith neighbor atom.

The EXAFS formula (6.7), along with (6.1) that defines the photoelectron wavevector k, provide the basis for extracting structural information from x-ray absorption spectra. In principle, EXAFS is related to the atomic environment of the absorbing atom in a simple way: the fine structure is composed of a sum of sinusoids, each sinusoid corresponding to a shell of neighboring atoms. The frequency of the sinusoid depends mainly on the bond length r_i (since the phase $2\delta_i + \phi_i$ is always much less than $2kr_i$). The amplitude of the sinusoid is proportional to the coordination number N_i. It is clear, however, that even the simple expression (6.7) for EXAFS involves an abundance of parameters. This makes the analysis of EXAFS in practice a complicated business. If accurate structural information about bond lengths r_i and coordination numbers N_i is to be obtained from EXAFS, there must be some precursory knowledge of the other parameters such as the atomic scattering factors and phase shifts, the Debye-Waller parameter, the photoelectron mean-free path and the zero of photoelectron energy E_0. The techniques that have been developed for analyzing EXAFS will be discussed in Sect. 6.3.

In the case of amorphous systems, where the atomic shells may not be well defined, it is convenient to express the EXAFS equation (6.7) in a different form. We can let the distance r_i become a continuous variable r and introduce the radial distribution function $\varrho_{ab}(r)$, which represents the fraction of atoms of type b that exist a distance r from an atom of type a. The EXAFS function $\chi(k)$ can then be written as,

$$\chi_a = - \sum_b (1/k) f_b(k) \exp(-2\sigma_b^2 k^2)$$

$$\times \int_0^\infty dr \varrho_{ab}(r)(1/r^2) \sin[2kr + 2\delta_a(k) + \phi_b(k)] \exp[2r/\lambda_b(k)]. \quad (6.9)$$

X-ray and neutron diffraction have been widely used to obtain information about the radial distribution of atoms in a solid. It is interesting to compare EXAFS with diffraction as a structural tool [6.16]. The intensity of the diffracted beam is measured as a function of the momentum transfer wavevector q, which is defined as

$$q = (4\pi/\lambda)\sin\theta. \quad (6.10)$$

Here λ is the x-ray or neutron wavelength and 2θ is the scattering angle. The diffracted intensity I(q) is then given by

$$I(q) = \sum_a c_a \sum_b \int_0^\infty (dr/qr) \varrho_{ab}(r) \sin(qr) f_a(q) f_b^*(q). \quad (6.11)$$

In this equation c_a is the concentration of atoms of type a, and $f_a(q)$ and $f_b^*(q)$ are the (complex) scattering factors for atoms of type a and b, respectively. There are two important comparisons to be made. In the diffraction experiment the distances between all pairs of atoms are being measured at once. In this respect EXAFS has a very significant advantage over diffraction. The EXAFS experiment provides the radial distribution function about a single type of atom, which is selected by working at the appropriate K – or L – absorption edge. This advantage, however, is partly offset by a different limitation. For reasons that we presently discuss, the EXAFS analysis is only valid for values of the wavevector $k \gtrsim 3A^{-1}$. Comparison of the sinusoidal terms in (6.9 and 11) shows that the momentum wavevector q corresponds to 2k in the EXAFS experiment. This means that all of the information available from diffraction in the region $q \lesssim 6\,\text{Å}^{-1}$ is inaccessible to EXAFS. This results in an inherent limitation of EXAFS spectroscopy in studying systems with broad radial distributions, such as amorphous materials. The atomic shells as measured with EXAFS are distorted and artificially sharpened, since the low-frequency components in the radial distribution are missing. The obvious resolution of this limitation is that, whenever possible, both EXAFS and diffraction should be used to complement the structural investigation of an amorphous system.

6.1.3 Validity of the Theory

Many approximations have been made in order to derive the relatively simple expression (6.7) that describes EXAFS. Nevertheless, this formula can be used to interpret measured absorption cross sections in most systems. It is important, however, that the experimentalist be aware of the underlying approximations, as they determine the limitations of the theory. When necessary, the approximations can often be relaxed by modifying the theory, which inevitably makes the description more complicated and the analysis more cumbersome and uncertain. Instead it is preferable to choose to study systems that are amenable to the simple EXAFS theory.

We proceed to briefly review the approximations that are incorporated into the standard EXAFS theory, and discuss the range of validity of these approximations.

a) Multiple Scattering

The standard EXAFS formula (6.7) takes into account only single scattering events in which the photoelectron is backscattered from a neighbor. The simple theory neglects all of the multiple scattering contributions in which the photoelectron scatters off of several atoms before eventually returning to the central atom. The validity of the single scattering approximation has been extensively investigated and formal expressions for the multiple scattering contributions have been derived [6.9,10,17]. The theory indicates that the multiple scattering paths make contributions that mimic single scattering at a distance equal to the total path length. This is important as it implies that multiple scattering can never interfere with the determination of nearest-neighbor bond lengths. It also suggests that close-packed structures are more susceptible to contributions from multiple scattering than systems with open structures. It is found that for photoelectron energies greater than $\sim 50\,\text{eV}$ the multiple scattering is usually negligible because the photoelectron scattering is weak and the mean-free path is short $(\lambda = 5-10\,\text{Å}\ [6.18])$. An important exception are the multiple scattering paths in which the photoelectron is scattered in the forward direction, since the scattering factors can be strongly peaked in the forward direction [6.11,19]. Consequently, when two neighbor atoms are aligned with the central atom, such as the first and fourth shells in an fcc lattice, the scattering from the more distant neighbor will be amplified. In such cases, the multiple scattering must be taken into account in the analysis. It has even been suggested that this effect can be used to measure bond angles [6.20,21]. Multiple scattering becomes very important at low photoelectron energies $E \lesssim 30$ eV), as the mean-free path becomes longer [6.22] and the scattering becomes more isotropic. This is one of the reasons that the standard formula (6.7) is invalid near the absorption edge.

b) The Small Atom (or Plane Wave) Approximation

As the photoelectron is emitted from the central atom, it propagates as an outgoing spherical wave and is scattered by the neighboring-atom potentials. The theory is greatly simplified by making the small atom approximation, in which the curvature of the photoelectron wavefront is neglected at both the neighboring atom site and upon return to the central atom. By treating the photoelectron as a plane wave, the scattering parameters $f_i(k)$ and $\phi_i(k)$ become independent of the distances r_i. The validity of this approximation has been studied by *Lee* and *Pendry* [6.10] and they conclude that the approximation is valid for $k \gtrsim 5 \, \text{Å}^{-1}$. The approximation becomes better at larger neighboring atom distances r_i and at higher values of k, because the effective radius of the scattering potential decreases at higher photoelectron energies. The small atom approximation can lead to errors when the near-neighbor atoms are large. *Hayes* and *Boyce* [6.5] suggested that the approximation is invalid when the near neighbors have atomic number $Z \gtrsim 50$.

It should be mentioned that the small atom approximation can be avoided, at the expense of a much more complicated formulation of EXAFS than is given by (6.7). By properly treating the photoelectron wave as spherical and including multiple scattering terms, *Muller* and *Schaich* [6.23] claim to be able to reproduce the fine structure for Cu down to the absorption edge.

c) The Muffin-Tin Potential

A fundamental assumption in the theoretical formulation of EXAFS is that the neighbor atoms can be represented by a muffin-tin potential. In this picture the interatomic potential is constant, so that the photoelectron can be treated as a free electron in the interatomic region. However, the true interatomic potential deviates from a constant value due to the electron density variations associated with chemical bonding. This is particularly important in covalent systems where the bonding is strongly localized, and is less important in metallic systems where screening from the conduction electrons smoothes out the interatomic potential. The muffin-tin model is adequate at high k because the kinetic energy of the photoelectron is large compared to the chemical bonding energies. However, at low k the photoelectron wavefunction is sensitive to the detailed structure of the interatomic potential, so that the free electron model is unjustified. Hence the muffin-tin model is often not valid in the energy region near the absorption edge ($E \lesssim 25\,\text{eV}$).

d) Many-Body Effects

The standard EXAFS formula (6.7) is derived within the single electron formalism. Realistically, however, the atomic absorption of x-rays is a complex many-body problem. As the x-ray is absorbed, all of the electron states in the atom change. There are, in general, many excitation channels in which the energy of the x-ray is divided between the excited core electron and other electrons. These processes are called "shake-up" transitions. The photoelectrons that are associated with the "shake-up" processes have less energy than the "primary" photoelectrons (those that receive all of the x-ray energy), and hence contribute out of phase to the main EXAFS signal. Since the spectrum of "shake-up" electrons is fairly continuous, their contribution is mostly incoherent. It is found that for $k > 4\,\text{Å}^{-1}$ the overall effect of the "shake-up" transitions is to decrease the EXAFS amplitude by 20–30% [6.18,24].

Another important many-body effect is the creation and subsequent screening of the core hole. The EXAFS theory is based upon a system of static atomic potentials. In reality, the central atom potential that the photoelectron "sees" is dynamically changing as the core hole is screened by the valence or conduction electrons. There is considerable evidence that the chemical environment of the central atom affects the phase shift δ_i [6.25,26]. Furthermore, the dynamic screening of the central atom potential could make E_0 a k-dependent parameter.

It is also important to consider the finite lifetime of the photoelectron and core hole due to their interaction with the other electrons of the system. The photoelectron will tend to polarize the electrons around it. The exchange and correlation energies of the photoelectron within the neighboring atom potential, along with its decay through plasmon emission, can be incorporated into the backscattering amplitudes by using a local density approximation [6.11,12]. The finite lifetime of the photoelectron due to inelastic scattering outside of the neighboring atom potential is incorporated into the simple theory through the attenuation factor $\exp[-2r_i/\lambda_i(k)]$. *Stern* et al. [6.18] have evaluated this term in several systems. They found that the mean-free path is k-dependent, typically varying between ~ 5–$15\,\text{Å}$, and is quite sensitive to the chemical environment of the central atom. Finally, the finite lifetime of the core hole will significantly broaden the EXAFS, and should be taken into account for central atoms having $Z \gtrsim 40$ [6.27].

e) Anharmonic Disorder

The standard EXAFS formula contains a Debye-Waller factor $\exp(-2\sigma_i^2 k^2)$ that can account for either static or thermal disorder when the neighboring atoms are described by a Gaussian distribution about their mean distances r_i.

This formulation is invalid for systems having anharmonic atomic vibrations, or static disorder that can not be represented by shells with Gaussian broadening. The theory can be modified to account for asymmetric atomic distributions (Sect. 6.4.3) at the expense of adding more parameters. An arbitrary radial distribution can be modeled using the formulation of (6.9). Again it is emphasized, however, that since the standard EXAFS theory is not valid in the range $k \lesssim 3$ Å^{-1}, EXAFS can not provide accurate information about broadened features in the radial distribution function. This is why EXAFS is of limited value as a probe of very disordered systems [6.14,28].

6.2 Experimental Techniques

The various experimental techniques that are regularly used to measure EXAFS have been discussed in detail elsewhere [6.2,5]. Hence we present only a brief review.

The goal in an EXAFS experiment is to measure the x-ray absorption cross section of a sample over an energy range of 1000–2000 eV. The starting energy is defined by the position of the K–or L–absorption edge of the atomic constituent that is to be investigated. In particular, it is necessary to measure the absorption cross section from about two hundred electron volts below the absorption edge to 1000 eV or higher above the edge.

Normally the EXAFS signal amounts to a modulation in the absorption cross section of about 1% per near neighbor. The oscillations die out rapidly above the absorption edge due to the energy dependence of the atomic scattering factors and due to the incoherence induced by vibration of the atoms (as represented by the Debye-Waller factor). The attenuation at high k due to

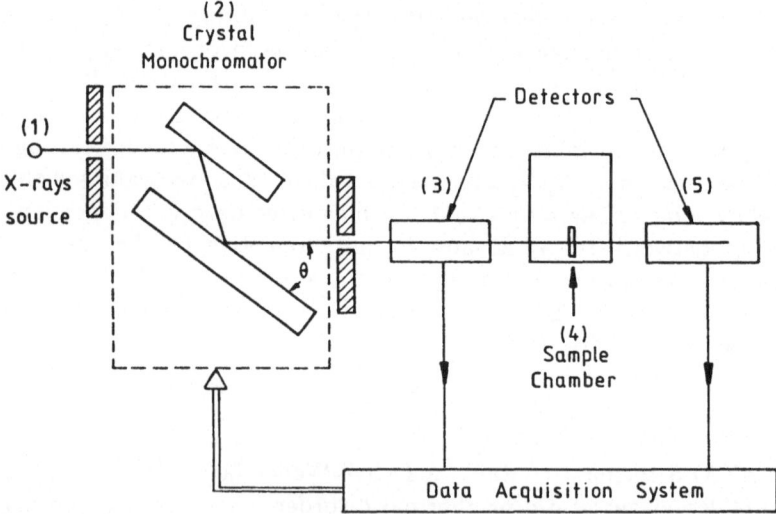

Fig. 6.3. The typical experimental configuration of a transmission EXAFS experiment

thermal vibration can be minimized by measuring the EXAFS spectra at low temperature. The measured spectrum must have very good statistics (better than 0.1%) in order to provide adequate signal in the important high k region.

The configuration of a typical EXAFS experiment is shown in Fig. 6.3. The standard constituents are (1) a broadband x-ray source, (2) a monochromator that selects the x-ray energy, (3) a monitor of the incident intensity, (4) the sample housed in a suitable environment, and (5) a monitor of the transmitted or fluorescent x-ray intensity or some other signal that is proportional to the amount of x-rays absorbed by the sample. In general, the monochromator and sample stage, as well as the data acquisition, are computer controlled.

The two types of x-ray sources that are commonly used in EXAFS experiments are the rotating anode x-ray generator and synchrotron radiation. The x-ray spectrum of a rotating anode source consists of bremsstrahlung radiation of energy up to the accelerating potential of the electrons, along with the characteristic emission lines of the target material and any impurities that happen to be present. This line emission can introduce undesirable, anomalous features in an EXAFS spectrum owing to the difficulty of normalizing to an incident x-ray intensity that is rapidly varying. The energy spectrum of x-rays generated as synchrotron radiation (SR) is smooth and continuous, extending from the infra-red to the hard x-ray region. The high-energy cutoff is determined by the energy of the circulating electron beam. The most attractive feature of SR is the high x-ray intensity. Devices inserted into the electron beam such as wigglers are used to enhance the SR emission, and can provide x-ray fluxes on the order of $\sim 10^{12}$ photons/s. At these intensities, an EXAFS spectrum can be measured in a few minutes. In comparison, the rotating anode source provides x-ray fluxes that are typically four or five orders of magnitude below SR fluxes, so that measuring an EXAFS spectrum can require days. Clearly SR is the x-ray source of choice in EXAFS experiments. Unfortunately, at this time SR is available at only a few user facilities throughout the world, and is relatively inaccessible to the average experimentalist. Therefore the convenience of a rotating anode as a laboratory source is an important consideration. As a general rule, EXAFS is concentrated systems could be measured on rotating anode sources for convenience, whereas EXAFS in dilute systems or other cases where the signal is expected to be weak, require the use of SR.

The x-ray monochromator typically consists of one or two single crystals of silicon or germanium. The incident continuous spectrum of x-rays is dispersed into Bragg reflections at different angles and a slit selects radiation within a narrow energy range ($\sim 1 - 5$ eV). Often the reflection from a second crystal is used to conveniently make the exit beam parallel to the entrance beam. The x-ray energy is changed by simply rotating the two crystals while simultaneously translating the slit. The crystals can be curved in order to focus the x-ray beam.

The EXAFS sample is a thin layer of the material of interest. It is usually housed in a chamber that provides environmental controls such as heating and cooling, and yet allows the x-ray beam relatively direct access to the sample. Windows of the chamber are made of transparent (i.e., low Z) material such as beryllium or kapton. Exotic sample chambers have been built for special applications, such as EXAFS studies of systems under high pressure [6.29].

The experimental features that we have thus far described are common to all EXAFS experiments. To go further, however, it is necessary to distinguish between the several different methods by which EXAFS are measured, as this can influence the sample geometry, as well as the choice of x-ray detectors. The best method for measuring EXAFS in concentrated systems is measuring the transmission of x-rays through the sample. Usually the initial and transmitted x-ray intensities are measured with gas ionization detectors. The logarithm of the ratio of the signals is proportional to the total absorption cross section of the sample. Signal-to-noise considerations show [6.2] that the sample should have an optimum thickness of 2.5 absorption lengths, which amounts to several microns for most metals. Furthermore, the detector that monitors the incident x-ray intensity should absorb ~20% of the incident beam [6.5].

When the atomic constituent that is being studied exists in the sample at a very low concentration, the EXAFS signal in a transmission measurement is swamped by the background absorption. In this case it is preferable to measure the EXAFS from the x-ray fluorescence that arises via the radiative decay of the core hole. X-ray fluorescence is an efficient process for medium and high Z elements and occurs at a well defined energy. Hence the intensity of the fluorescence emission line provides a sensitive measure of the number of x-rays absorbed by the atoms of interest, while in principle energy discrimination can minimize the background. In the fluorescence method, the absorption cross section is proportional to the ratio of the fluorescence intensity to the incident intensity. Measurement of the fluorescence signal is accomplished with some combination of metal foil filters and energy-sensitive detectors such as proportional counters, scintillation counters or semiconductor detectors. The detectors are arranged to collect as much solid angle as possible (the fluorescence is isotropic), with an emphasis on collection at 90° to the incident beam where scattered radiation is a minimum. In this case the sample should be thick enough to absorb essentially all of the incident x-rays. As a rule, signal-to-noise favors the fluorescence method when the atomic concentration of the species under investigation is less than ~0.5% [6.2]. Otherwise the transmission method is preferred.

Although the transmission and fluorescence methods are by far the most common, EXAFS has been measured by other techniques such as total photoelectron yield [6.30,31]. This technique is complementary to the fluorescence method in that it makes use of the non-radiative by-products of x-ray absorption. Owing to the short mean-free path of the photoelectrons, only the atoms within a few hundred angstroms of the surface are sampled, and hence the technique is well suited to studying the atomic structure near surfaces. Other methods that have been used to measure EXAFS include reflectivity [6.32] and luminescence [6.33].

Finally we mention a few of the experimental problems that have been discussed in the literature. Every EXAFS experiment is constrained to an energy resolution defined by the nature of the x-ray source and the monochromator. The finite resolution necessarily introduces a broadening of the fine structure, which will tend to decrease the amplitude. If the energy resolution function of the experimental apparatus is known, it can be accounted for in the data analysis [6.34]. Another problem particularly associated with crystal monochromators is the existence of higher-order reflections in the x-ray beam, known as harmonic contamination. If the monochromator consists of two separate crystals then the harmonics can be minimized by making the crystals slightly nonparallel, a process known as "detuning". Otherwise the harmonic contamination must be considered in the data analysis [6.35]. It has also been pointed out that non-uniformity of the sample thickness can cause an overall reduction in the EXAFS amplitude [6.36].

6.3 Analysis

The details of analysis are discussed in many articles, see for instance [6.2,5,13]. We now present a description, illustrated by an example, of the standard data manipulations used in reducing and analyzing EXAFS spectra.

6.3.1 Basic Manipulations

The starting point in EXAFS analysis is usually a spectrum, such as shown in Fig. 6.1, of the total absorption cross section μ measured as a function of x-ray energy E. The first step is to remove the absorption signal that corresponds to all of the higher shells of electrons, leaving only absorption due to the K-or L-shell of interest. This is accomplished by fitting the data below the absorption edge to a function such as the Victoreen formula,

$$\mu(E) = A/E^3 + B/E^4 \tag{6.12}$$

and then subtracting the extrapolated background absorption from the data.

Next the data is normalized to derive the EXAFS function $\chi(E)$, as given by

$$\chi(E) = [\mu(E) - \mu_0(E)]/\mu_0(E). \tag{6.13}$$

This step requires knowledge of the free atom absorption cross section $\mu_0(E)$. It is generally assumed that $\mu_0(E)$ can be generated from $\mu(E)$ by simply removing the fine structure by taking a running average, using spline fitting or other smoothing techniques [6.5,37]. We note, however, that this procedure will inevitably neglect the structure that has been observed in free atom absorption spectra [6.38]. The simulated free atom absorption μ_0 is shown as the dotted line in Fig. 6.4a.

In order to connect the experimental spectrum with the EXAFS theory it is necessary to convert the energy scale to the photoelectron momentum k.

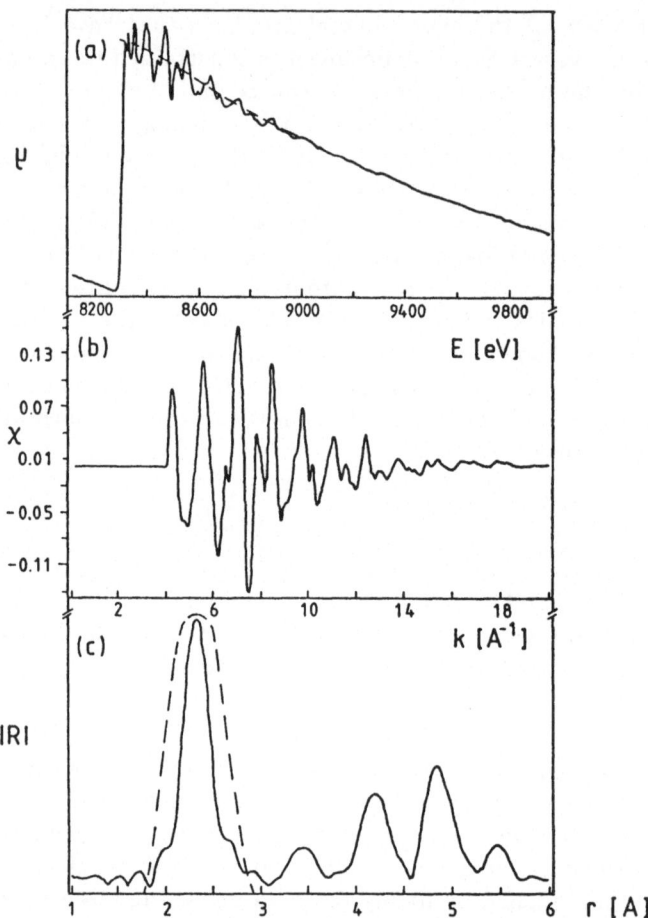

Fig. 6.4. (b) The x-ray absorption spectrum at the K-edge of Ni, showing the simulated free atom absorption (dashed line). (b) The corresponding EXAFS function $\chi(k)$. (c) The amplitude of the radial scattering function $R(r)$, which is the Fourier transform of $\chi(k)$. The dashed line represents a window function used to extract the contribution from the first neighbor shell

The two parameters are related by the equation,

$$k^2 = 2m(E - E_0)/\hbar^2 \tag{6.14}$$

where E_0 is the zero of photoelectron energy. In practice, it is impossible to determine a unique value for E_0, so that it is treated as an adjustable parameter. E_0 is generally chosen to be within $\sim 30\,\text{eV}$ of the absorption edge. Subtracting the free atom absorption and transforming the energy scale to k-space results in the EXAFS function $\chi(k)$ as shown in Fig. 6.4b.

The EXAFS spectrum has now been manipulated into a form that is compatible with the standard expression (6.7). The next step is to take a Fourier transform of the spectrum with respect to the variable 2k. At this stage the spectrum is multiplied by a function $k^n W(k)$, where n is an integer usually varying between one and three, and $W(k)$ is a window function. The k^n factor compensates for the k-dependence of the factor $f(k)/k$ in the standard formula (6.7). It weights the data at high-k where the EXAFS theory is on a firmer footing, and where the data is less sensitive to the particular choice of E_0. The function $W(k)$ is a rectangular window with smoothly decaying edges defined by a Gaussian or some similarly smooth function. The window defines an effective k range for the transform while suppressing termination errors. The lower cutoff should be chosen at $\sim 3 \text{ Å}^{-1}$ and the upper cutoff is determined by the maximum k at which the EXAFS signal is still above the noise level (preferably $k=15-20 \text{ Å}^{-1}$).

The Fourier transform yields the atomic scattering function $R(r)$ defined by

$$R(r) = -\frac{1}{2\pi} \int_{-\infty}^{\infty} d(2k)k^n W(k)\chi(k) \exp(i2kr). \tag{6.15}$$

$R(r)$ is a complex function, having real and imaginary parts. The magnitude $|R(r)|$, which is shown in Fig. 6.4c, is amenable to structural interpretation. The peaks in $|R(r)|$ correspond to the atomic shells in the radial distribution function $\varrho(r)$, except shifted to lower r by a few tenths of an Angstrom due to the phase shifts $2\delta(k) + \phi_i(k)$. The line shapes in $|R(r)|$, however, vary significantly from the true radial distribution function $\varrho(r)$ for two reasons. First, $R(r)$ represents a convolution of $\varrho(r)$ with a complicated function that corresponds to the Fourier transform of all the prefactors to the sine function in (6.7). Secondly, the window function $W(k)$ filters the low-frequency components out of $\varrho(r)$, which artificially sharpens up the radial distribution. Hence, simple inspection of $R(r)$ is only useful as a qualitative indication of the local atomic structure.

When $|R(r)|$ manifests well-separated peaks, the contribution to the EXAFS from the individual shells can be extracted. (Actually, only the nearest-neighbor peak is reliable.) The individual shells are obtained by multiplying $R(r)$ with a window function $W'(r)$ that selects the ith shell, as shown by the dotted line in Fig. 6.4c, and then backtransforming to get

$$\chi_i(k) = - \int_{-\infty}^{\infty} dr \, W'(r)R(r) \exp(-i2kr). \tag{6.16}$$

$\chi_i(k)$ represents the contribution to the EXAFS from the ith shell. If we neglect some broadening due to the window function W', the EXAFS $\chi_i(k)$ can be written as

$$\chi_i(k) = k^{n-1}W(k)N_i[f_i(k)/r_i^2] \exp[i2kr_i + i2\delta(k) + i\phi_i(k)]$$
$$\times \exp(-2\sigma_i^2 k^2) \exp[-2r_i/\lambda_i(k)]. \tag{6.17}$$

The function $\chi_i(k)$ has the form of $\varrho \exp(i\theta)$, so that the modulus and argument are easily separated. The argument yields the total phase shift $2kr_i + 2\delta(k) + \phi_i(k)$. The distance r_i can be determined from this quantity if the scattering phase shifts are known. Likewise, if the structure is known then the scattering phase shifts can be determined. The modulus of χ_i contains structural information about the coordination number N_i, the Debye-Waller parameter σ_i and the attenuation factor D_i.

6.3.2 Determination of Structural Parameters

a) The Assumption of Transferability

It is known that EXAFS contains detailed information about the local atomic structure that surrounds a particular atom type in a condensed phase system. The primary goal of structural analysis is to determine the partial radial distribution function, that is the distances, line shapes and coordination numbers of the different atomic shells. Inspection of the standard EXAFS expression (6.7) shows that these structural parameters are intimately correlated with the scattering phase shifts $\phi_i(k)$, $\delta(k)$ and the scattering amplitudes $f_i(k)$. Thus, in order to extract the desired structural information from the EXAFS, it is necessary to separately determine the atomic scattering factors. This problem is usually addressed by invoking the "assumption of transferability". The assumption asserts that the atomic potentials, and hence the scattering factors, are universal quantities which depends only on the atom types, i.e., they are independent of the chemical environment of the atoms. Then, in principle, the atomic scattering factors for all the elements are determined once; either calculated from first principles or measured in systems with known structures. It is assumed that these scattering factors can be universally applied to analyzing systems of unknown structure.

The atomic scattering amplitudes and phase shifts have been calculated and tabulated by *Teo* and *Lee* [6.12]. However, it is more desirable to directly measure the scattering factors in a system of known atomic structure. This is done by Fourier transforming the EXAFS spectrum and then backtransforming the separate peaks in $R(r)$ to get the EXAFS contribution $\chi_i(k)$ from each atomic shell as described above. When the structural parameters N_i, R_i and σ_i are known, it is straightforward to extract the scattering phase shifts and amplitudes [6.2,13].

The concept of transferability of atomic scattering factors between systems that have the same atomic constituents has been investigated by several groups [6.13,18,34,39,40]. It has generally been found that for nearest neighbors either the calculated or measured phase shifts can be successfully transferred to a bond-length accuracy of 0.01–0.02 Å. However, the most accurate structural determinations are achieved by transferring phases that have been measured in chemically similar systems of known structure. Furthermore, it has been observed that chemical effects cause variations in the scattering phases of atoms that are in more distant shells, especially for non-metals [6.13]. Hence it is desirable, whenever possible, to transfer phase between atom pairs having the same atomic shell [6.13]. The transferability of scattering amplitudes has been found to be much less successful. The scattering amplitudes are quite sensitive to chemical environment [6.2,18,34]. Another problem is that the small atom approximation is not strictly valid for $k \lesssim 5 - 10 \, \text{Å}^{-1}$ which makes the scattering amplitude, in reality, dependent on r_j. Consequently, scattering amplitudes should be transferred only between chemically similar systems and then only between shells at approximately the same distances.

b) Bond Length Determination

EXAFS spectroscopy is best suited for determining the distance of a nearest-neighbor shell which exhibits simple Gaussian broadening due to static or thermally induced disorder. In this case the position of the first peak in the radial scattering function $|R(r)|$ corresponds to the nearest-neighbor distance within an accuracy of 5–10%. The peak is shifted to a shorter distance than the true distance due to the scattering phase shift $2\delta(k) + \phi_i(k)$. In order to accurately determine the bond length from EXAFS, it is necessary to both know and properly account for the scattering phases. Several procedures have been presented in the literature for determining bond lengths [6.2,5,11,13,39]. All of these methods rely on the assumption of phase transferability. The bond length is best extracted from the EXAFS by using scattering phase shifts that have been measured in chemically similar systems of known structure. The zero of energy E_0 is an adjustable parameter in these procedures, which to some extent compensates for the chemical differences between systems. The analysis of EXAFS to determine bond lengths is particularly simplified when values of E_0 are chosen that make the phase shifts linear functions of k [6.13,41]. When the scattering phase shifts are carefully accounted for, bond lengths can be determined from EXAFS with accuracy of ~1%.

The procedures for extracting the shell distances r_i require that the contributions from the different shells can be separated by Fourier analysis. When the peaks in $|R(r)|$ overlap significantly the analysis is inaccurate. An even worse situation exists in the analysis of highly disordered systems in which the concept of Gaussian broadened shells is invalid. In these cases it is usually necessary to resort to modeling the EXAFS by assuming specific forms for the line shapes in the radial distribution function [6.1,5,14,34]. Structural information is inferred by least-squares fitting the model EXAFS to the measured spectra in either k-space or r-space.

c) Coordination Number and Debye-Waller Parameter

The amplitude of the EXAFS oscillations from a particular shell of neighbor atoms contains important structural information. In principle, if the scattering amplitude $f_i(k)$ is known, and if the distance r_i is previously determined, then the modulus of $\chi_i(k)$ as given in (6.17) will yield the coordination number N_i and the Debye-Waller parameter σ_i [6.2]. The analysis, however, relies on the assumption of scattering amplitude transferability. As discussed above, the scattering amplitudes for the same atom in chemically dissimilar systems have been found to be significantly different. Furthermore, the measurement of the EXAFS amplitude is susceptible to inaccuracy arising from complex physical phenomena such as many-body effects (Sect. 6.1.3d). In particular, the finite lifetime of the photoelectron modifies the amplitude through the attenuation factor $D_i(k) = \exp[2r_i/\lambda_i(k)]$. The mean-free path λ is usually treated as a free parameter and, for simplicity, is often assumed to be independent of k. We also note that the structural parameters contained in the amplitude are strongly correlated, which makes it difficult to uniquely determine N_i and σ_i. *Lengeler* and *Eisenberger* [6.34] claim that by using careful curve-fitting analysis, the coordination numbers and Debye-Waller parameters can be determined to an accuracy of 15% and 20%, respectively.

6.3.3 Guidelines for Using EXAFS Spectroscopy

The following is a list of guidelines that we feel should be observed if EXAFS spectroscopy is to be used as an accurate probe of atomic structure. This list has been generated by considering the limitations of the technique due to both theoretical and practical experimental constraints.

1. EXAFS is best suited for studying systems in which the absorbing atom is surrounded by well-defined neighboring shells that manifest simple Gaussian broadening due to either thermal or static disorder.
2. The EXAFS spectrum should be measured for as large a k-range as possible. The simple EXAFS theory is not valid at low-k so that the spectrum below $k \simeq 3\,\text{Å}^{-1}$ should not be used in the structural analysis. Preferably, the spectrum should extend to at least $k = 15\,\text{Å}^{-1}$.
3. When transferring scattering phase shifts and amplitudes to determine structural parameters, the scattering factors should be measured in a chemically similar system of known structure.
4. Phase shifts and amplitudes should be measured and transferred for each atomic shell, even when the shells consist of the same atom type.
5. Multiple scattering should be considered when analyzing distances longer than nearest-neighbor. This is especially important in close-packed systems or when two neighbor atoms are aligned with the central atom.
6. Whenever possible, structural investigations using EXAFS should be supplemented with other techniques.

6.4 Experimental Applications

Although the EXAFS technique has been used to investigate a large number of problems in metallic systems we shall survey only the three areas where it has been used most extensively:

The first problem, to which EXAFS is ideally suited, is the study of the local environment or static relaxation effects surrounding dilute solute atoms in metallic hosts.

The second application is in the determination of Debye-Waller factors for metals and the investigation of the temperature dependence of the mean square displacements. Here EXAFS works well for isotropic metals but is less satisfactory, requiring considerable modelling, for anisotropic materials.

The third application is the study of the structure of amorphous materials. In these investigations the credibility of the EXAFS results is often questionable since many of the conditions necessary for the validity of the EXAFS analysis are violated. For reliability the EXAFS analysis should be supplemented by other techniques such as neutron or x-ray scattering studies. In spite of the inappropriate application of the EXAFS analysis to amorphous materials these studies have been widely pursued because of the scarcity of other more appropriate techniques for examining their structure.

We shall not discuss the large volume of work done on catalysts since much of that work does not involve the metallic state.

6.4.1 Local Environment Surrounding Solute Atoms

One of the most straightforward applications of the ability of EXAFS to probe the local environment surrounding selected species is the study of solute-host bond lengths and clustering effects in dilute alloys. The comparison of the bond lengths with the change in lattice constants from the host materials and various theoretical models can then yield information on the forces and interactions between the atoms in metals. The measured bond lengths and variation of the lattice constants for substitutional impurities are listed in Table 6.1. Also

Table 6.1. Measured and calculated differences between solute-host and host-host bond lengths, Δr. Also listed is the measured relative lattice constant change per atomic fraction, $\Delta a/ac$

Host	Solute	At.%	EXAFS	Δr [Å] Elastic theories [6.42]	theories [6.54]	Discrete models	$\Delta a/ac$[a]	Ref.
Al	Mg	3.0	0.075±0.02	0.038	0.045	0.02[b]	0.11	[6.42]
Al	Cu	2,2.5	-0.13 ±0.02	-0.043	-0.05		-0.12	[6.43]
	Cu	0.5	-0.07 ±0.03[c]					[6.45]
Al	Zn	0.83	-0.02 ±0.01	-0.007	-0.008		-0.02	[6.46]
Ti	Cu	0.5	-0.05 ±0.045					[6.48]
Cr	Ge	1.0	0.05					[6.49]
Fe	Ge	1.2,3.0	0.06 ±0.02	0.02	0.04	0.07[d]	0.06	[6.50]
	As	1.0	0.03 ±0.02	0.03	0.06	0.07[d]	0.09	[6.50]
	Sn	0.5	0.06 ±0.02	0.07	0.15	0.095[d]	0.24	[6.50]
	Sb	1.0	0.10 ±0.03	0.17	0.35	0.12[d]	0.56	[6.50]
Cu	Fe	75ppm	-0.016±0.01	0.006	0.007		0.02[e]	[6.51]
Cu	Sn	0.3,1.0	0.08 ±0.02	0.085	0.105	0.07	0.28	[6.52]
Cu	Au	5	0.10 ±0.01	0.05	0.06		0.15	[6.53]
Au	Cu	5	-0.04 ±0.01	-0.03	-0.05		-0.09	[6.53]

[a] W.B. Pearson: *Handbook of Lattice Spacings and Structure of Metals*, Vols. 1 and 2 (Pergamon, New York 1958); [b] Ref.[6.55]; [c] This value is claimed to be in error due to the Cu not being in solid solution; [d] Scaled from a cluster calculation of the Xe-Fe bond length, as given in (6.20); [e] There is considerable disagreement about this value. The preferred value from Pearsons first volume is quoted here (footnote a)

listed are the bond lengths obtained from calculations made with continuous and discrete models.

Other local environment studies have also been made on structural defects; such as Guiner-Preston (G.P.) zones, which are planar extended defects, and radiation induced self-interstitials.

a) Al Host

The Mg K edge spectra (starting at 1303 eV) of alloys containing 3.0 and 7.3 at.% Mg in Al with about 5μ thickness were measured in transmission at liquid nitrogen temperatures [6.42]. The Al K-edge occurs 256 eV above that of Mg; limiting the Mg data to that range. This short range plus the energy resolution of the monochromator being only 4 eV in this energy range caused the quality of the data in this experiment to be rather poor. Using calculated phase shifts [6.12] it was concluded that the Al-Mg bond length in the 3 at.% alloy was 2.93 ± 0.02 Å an increase of 0.075 ± 0.02 Å over that of pure Al.

Foils of 150 μm thickness containing 2 at.% and 2.5 at.% Cu in Al were studied by transmission spectra at 15 K in three forms: a homogeneous solid solution, dispersed G.P. zones and dispersed small θ' precipitates [6.43]. The θ' samples were used as reference compounds to find the E_0 value which matched calculated phase shifts. It was found that the Al-Cu bond length was 2.725 Å which is contracted by 0.13 ± 0.02 Å of the host lattice bond length. Continual controversy exists as to the chemical composition and structure of the G.P. zones and their surroundings. From these EXAFS studies a model was proposed in which they are composed of an enriched Cu monolayer surrounded by a Cu depleted matrix. The Al-Cu bond length in the G.P. zone was found to be that of pure Al (2.86 Å) with the Al nearest-neighbor distance on each side of the plane of the zone being 0.23 ± 0.02 Å less, corresponding to a local contraction of the d_{100} spacing between the zone plane and the next (100) plane of $17\pm2\%$. However this model is inconsistent with recent x-ray data which concluded that the G.P. zones consisted of layers which were 100% Cu with the facing Al planes being collapsed and bowed toward the zone [6.44].

A supersaturated solid solution of 0.5% Cu in Al was measured in transmission at 77 K and analyzed with calculated phase and amplitude functions. This gave a Al-Cu bond length of 2.79 ± 0.03 Å which is 0.07 Å less than the bond length of Al in the pure host matrix [6.45]. This value does not agree with that measured in [6.43] and is claimed by those authors to be in error due to the Cu not being in solid solution [6.46].

The Al-Zn bond length in a 0.83 at.% alloy was measured at 15 K in transmission and analyzed using calculated phase and amplitude functions. It was found to be smaller than the host bond length by -0.02 ± 0.01 Å [6.46]. The effects of clustering in more concentrated alloys of 1.3, 2.6, 4.4 and 6.8 at.% were also studied. It was concluded that immediately after quenching the 6.8 at.% alloy to liquid nitrogen temperatures the state of this alloy is not an ideal solution but is made up of a majority of tetrahedron and bi-tetrahedron clusters.

Electron-irradiation-produced self-interstitials trapped near Ag atoms in a 400 ppm Ag in Al sample were preliminary studied using fluorescent detection. The effects on annealing were also observed [6.47].

b) Ti Host

Room temperature fluorescent measurements were made near the Cu edge on a 0.5 at.% Cu in Ti sample. Using semi-emperical phase shifts a Ti-Cu bond length of 2.875±0.045 Å was found which is 0.05±0.045 Å less than the average Ti-Ti bond length of hcp Ti (2.923 A) [6.48].

c) Cr Host

Measurements were made on 1 at.% Ge in Cr near the Ge edge and analyzed using calculated phase functions. They gave a Cr-Ge bond length which had an increase of 0.05 Å over that of Cr-Cr (2.498 Å) [6.49]. No errors were quoted.

d) Fe Host

Measurements at 80 K were made on 0.5 to 3 at.% binary alloys near the K edges of the 4 sp and 5 sp elements Ge, As, Sn and Sb in Fe. Using some phase functions determined from model compounds as well as calculated phase and amplitude functions [6.12] and a nonlinear least-squares fit to the back-transformed $\chi(k)$ gave the following bond length increases over that of Fe-Fe in pure Fe (2.480 Å): Fe-Ge, 0.06±0.02 Å; Fe-As, 0.03±0.02 Å; Fe-Sn, 0.06±0.02 Å; Fe-Sb, 0.10±0.03 Å [6.50].

e) Cu Host

Room-temperature fluorescence measurements on 75 ppm of Fe in Cu were analyzed with calculated phase and amplitude functions. They gave a Cu-Fe bond length which was 0.016±0.01 Å less than the Cu-Cu bond length in pure Cu (2.556 Å) [6.51].

Transmission measurements at 80 K on alloys containing 0.3 and 1 at.% Sn in Cu were analyzed using the phase function from a model compound. They gave a Cu-Sn bond length which was 0.08±0.02 Å larger than the Cu-Cu bond length in pure Cu [6.52].

Transmission measurements of 5% Au in Cu and 5% Cu in Au have also been reported but the details of the measurements are not yet published [6.53]. It was found that the Au-Cu bond length in Cu-Au was 0.10±0.01 Å larger than that of pure Cu and the Cu-Au bond length in Au-Cu was 0.04±0.01 Å less than that of pure Au.

6.4.2 Comparison to Theoretical Models

The elastic theory treats the matrix as a continuous homogeneous isotropic elastic medium [6.42,44] and relates the bond length change to the change in lattice constant with alloying, $\Delta a/ac$, where a is the lattice constant, as determined by x-ray scattering, and c the atomic concentration of the solute atoms. For many alloy systems this quantity is not very accurately determined. The relation is given by

$$\Delta r = (y/r)(\Delta a/ac)\{r_h[1 + (\gamma - 1)c/y]\} \qquad (6.18)$$

where Δr is the difference between the solute-host and host-host bond lengths, and r_h is the nearest-neighbor host distance; γ is an elastic coefficient given by $\gamma = 1 + 4\varsigma\kappa/3$ which is close to 1.5 for all metals, ς being the shear modulus and κ the compressibility coefficient. The quantity y is the ratio of the atomic volume to the spherical volume contained in the first shell; it is 0.169 for fcc and 0.184 for bcc lattices.

Another version of a continuum theory [6.54], in spirit similar to that described above, expresses the bond length change by

$$\Delta r = 3aI_1(\Delta a/ac)/8 \quad \text{for fcc} \quad \text{and}$$

$$\Delta r = 3aI_1(\Delta a/ac)/4 \quad \text{for bcc lattices} \qquad (6.19)$$

where I_1 is the length of a vector whose value can be obtained from graphs in [6.54].

As can be seen from Table 6.1 both of the continuous elastic theories give very similar results for the fcc lattice but differ by about a factor of two for the bcc host lattice. They underestimate the measured host-solute bond lengths for the Al alloys and give mixed agreement for solutes in Fe and Cu hosts.

There exist several versions of discrete models. Here the atoms are represented by potentials and the forces derived from their derivatives:

One such estimate for Mg in Al gave a first shell bond-length difference of 0.02 Å [6.55] which is much too small; even smaller than that given by the elastic theories. This is attributed to an incorrect treatment of the pairwise potential [6.42].

Another estimate made for impurities in Fe is based on a discrete crystallite calculation using two body potentials for Xe in Fe where an Fe-Xe bond length was obtained which was 0.22 Å greater than the Fe-Fe bond length [6.56]. Scaling this value by the difference in atomic radii, i.e.,

$$\Delta r(Z) = \Delta r(Xe)(r_Z - r_{Fe})/(r_{Xe} - r_{Fe}), \qquad (6.20)$$

where the atomic radii are derived from dividing the atomic weight by the density, gives the values listed in column 7 of Table 6.1. It is seen that the trends are quite well described and the magnitudes in reasonable agreement with the measured values.

A microscopic elastic theory taking into account the discrete positions of the host lattice atoms gave a calculated bond length for Cu-Sn of 0.07 Å in good agreement with the measurement [6.55].

Many of these experiments are on elements of only about 1 at.% abundance with the result that transmission measurements have a poor signal to noise ratio. In some cases the quoted accuracy of the measurements is questionable. That the bond lengths depend strongly on the electronic or chemical nature of the bonds has now been seen clearly in covalent and ionic compounds. It has been shown that the bond length of an impurity atom in a host strongly preserves the bond length corresponding to its own compound. E.g., the nearest-neighbor bond lengths surrounding dilute amounts of In in GaAs, or Ga in InAs, strongly retain the bond lengths they had in their respective compounds InAs or GaAs [6.57]. Furthermore Sr or Y in CaF_2 closely maintain the bond lengths corresponding to their respective compounds SrF_2 and YF_3 [6.58].

It is desirable to continue experiments of this type, repeating some of the questionable results and extending measurements to other alloy systems; especially using fluorescent techniques. Examining more systems will allow the determination of when and why the various theoretical models apply.

6.4.3 Debye-Waller Factors – Mean-Square Displacements

a) Isotropic, Harmonic Materials

Since EXAFS is sensitive to the local environment it gives information about the interatomic vibrations which are mainly dependent on short-range forces. Thus the Debye-Waller factor determined from EXAFS, $\exp(-2\sigma_i^2 k^2)$, contains the mean-square fluctuations in distances; $\sigma_i^2 = \langle [(u_i - u_0) \cdot \hat{r}_i]^2 \rangle$, where r_i is a vector from the central atom to the neighboring atom and u_i, u_0 are, respectively, the displacement vectors at the lattice point r_i and the central atom and the brackets indicate a thermal average. This is different from the mean-squared vibrational amplitudes $u_i^2 = \langle (u_i \cdot \hat{r}_i)^2 \rangle$ which enter the Debye-Waller factor in x-ray diffraction [6.59]. The latter measures atomic displacements relative to the whole lattice and is thus sampling the long-wavelength vibrational modes. The differences, $2u_i^2 - \sigma_i^2$ is twice the displacement correlation function (DCF), $\langle (u_0 \cdot r_i)(u_i \cdot r_i) \rangle$ [6.15]. The relative correlation parameter γ is defined by

$$\gamma = \langle (u_0 \cdot r_i)(u_i \cdot r_i) \rangle / 2u^2 = 1 - \sigma^2 / 2u^2. \tag{6.21}$$

For near-neighbor shells the motion is highly correlated so σ^2 is considerably less than $2u^2$ and $\gamma > 0$. For more distant shells this correlation is lost and σ^2 approaches $2u^2$, so $\gamma \to 0$. At high temperatures the correlation increases with temperature and for $T \geq \theta_D$, γ approaches a constant. For the first shell of bcc and fcc lattices, $\gamma \simeq 0.4 - 0.5$ [6.15,59]. This behavior can be seen in Table 6.2

Table 6.2. Correlation parameter, γ, calculated with the force constant (FC) and Correlated Debye Models (CDM) [6.59]

Temp. [K]	Model	Shell	Cu (fcc)	Fe (bcc)	Pt (fcc)
4	FC	1	0.178	0.196	0.197
	CDM	1	0.146	0.163	0.148
	FC	2	0.032	0.082	0.051
	CDM	2	0.001	0.068	0.002
700	FC	1	0.415	0.468	0.446
	CDM	1	0.387	0.405	0.387
	FC	2	0.167	0.276	0.218
	CDM	2	0.18	0.26	0.18

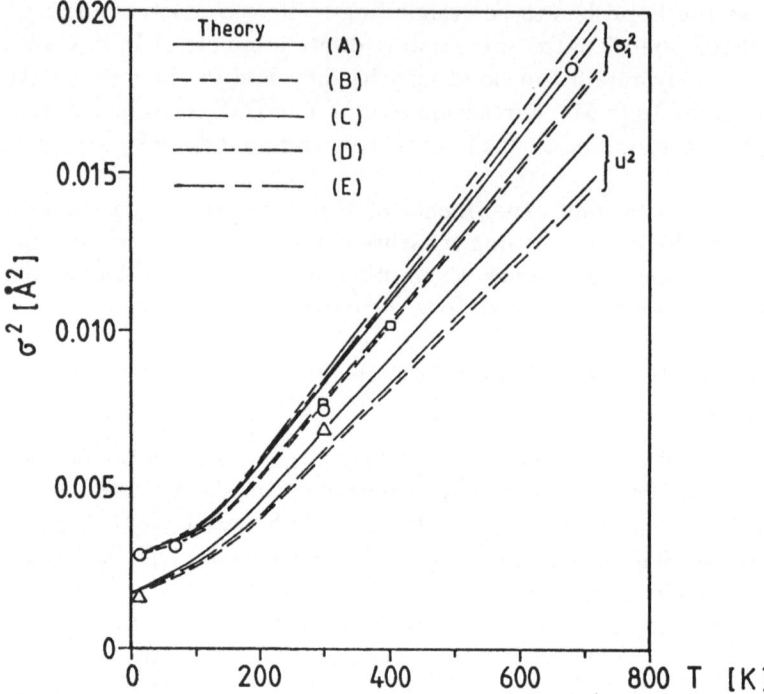

Fig. 6.5. Variation of the mean-square vibrational amplitudes u^2, and the mean-square fluctuations in distance σ_1^2, for the first neighbor shell of Cu as a function of temperature as calculated from various force constant models (A [6.60], B [6.61] and C (single parameter)), from the Debye (D) and Einstein (E) models [6.59]. The experimental data, shown by the symbols, □ [6.62], ○ [6.63] and △ [6.60] are seen to fit models B and C somewhat better than the other models below 500 K

where the results of the calculation of γ is given for two different models. This result is counter to the intuitive idea that correlations should decrease at higher temperatures due to the contributions from the higher-frequency optical modes. In reality the vibrational amplitude of the atoms becomes appreciable so that it is favorable to reduce the Coulomb energy due to the overlapping electronic wave functions by developing more correlation. Hence the results show that the lower-frequency longitudinal vibrations contribute significantly to the atomic displacements at high temperatures.

The quantities u_i^2 and σ_i^2 have been calculated for isotropic, harmonic fcc Cu as a function of temperature using several different models, as shown in Fig. 6.5 [6.59]. Models A and B are based on multiparameter harmonic force constant models which have been fit to measured phonon spectra; A [6.60] and B [6.61]. Model C is a single central force constant model. There is some ambiguity in results obtained from the Einstein and Debye approximations because of the arbitrariness in the choice of the Einstein and Debye temperatures. Model D is the correlated Debye model [6.15] using $\theta_D \simeq 315$ K. Model E is the Einstein model with the phonon mode at the frequencies

$$\omega_E(u^2) \simeq (3/5)\omega_D \quad \text{and} \quad \omega(\sigma_1^2) \simeq (3/4)\omega_D, \tag{6.22}$$

where $\omega_D = k_B \theta_D / \hbar$ [6.59].

It is seen that the various force constant models give results differing by 5–10% and that the temperature dependence is quite linear above ~200 K.

It has been found experimentally that parametric fitting to the height and widths of the Fourier transformed radial scattering functions $R(r)$ do not give very reliable σ^2 values [6.63]. Instead it is advantageous to use a method which exploits the dominant temperature dependence of the Debye-Waller factor. In this method the peak of the ith neighbor shell in $R(r)$ is backtransformed to obtain $\chi_i(k, T)$. Then to a good approximation (the change in r due to the thermal expansion is usually negligible)

$$\ln[\chi_i(T_2)/\chi_i(T_1)] = 2[\sigma_i^2(T_2) - \sigma_i^2(T_1)]k^2 + \ln[N(T_1)/N(T_2)]. \tag{6.23}$$

Thus a plot of $\ln[\chi_i(T_2)/\chi_i(T_1)]$ vs k^2 yields $2[\sigma_i^2(T_2) - \sigma_i^2(T_1)]$ from the slope and $\ln[N(T_1)/N(T_2)]$ from the intercept [6.64]. With this method many of the uncertainties in parameters such as the amplitude and phase functions cancel out and, in principle, the effects of thermal vibrations can be separated from the vacancy disorder induced by temperature. Then, by calculating or using some other parametric procedure to obtain σ^2 at the lowest temperature, $\sigma^2(T)$ can be obtained as shown for the first neighbor shell of Cu in Fig. 6.5. It is seen that the experimental data for σ_1^2 of Cu, obtained in the manner described above, fit models B and C very well. The correlated Debye model *(D)* fits more poorly, but due to its simplicity it is often used.

The results of several EXAFS measurements on metallic systems have been summarized in Table 6.3. Since $\sigma^2(T)$ is seen to be quite linear above ~200 K (Fig. 6.5), the quantity $\Delta(\sigma^2)/\Delta T$ is usually given in the literature. The quantity, $\Delta(u^2)/\Delta T$, as obtained from x-ray diffraction measurements is also listed. However, as seen from Fig. 6.5, the slope of σ^2 is temperature dependent at temperatures below ~200 K so that care must be taken in comparing $\Delta(\sigma^2)/\Delta T$ over different temperature ranges. The temperature ranges of the data are quoted in Table 6.4 when reported. The correlation as indicated by $\gamma' = 1 - \Delta(\sigma^2)/2\Delta(u^2)$, a quantity analogous to γ, is given in the last column. It is only meaningful for matched temperature ranges of $\Delta(\sigma^2)$ and $\Delta(u^2)$.

The literature contains many other values of σ obtained by parametric fits. These values are less reliable then those obtained from a plot of $\ln[(\chi(T_2)/\chi(T_1)]$ vs k^2.

b) Anisotropic, Harmonic Materials

Anisotropic materials having harmonic forces have been represented by treating σ^2 and u^2 as two-component tensors. The EXAFS spectra of superconducting Nb_3Ge, having an A15 structure, were analyzed in this manner. Using calculated phase and amplitude functions [6.12] the backtransform of the first two (unresolved) peaks in $R(r)$ for bcc Nb was used to obtain the mean-free path

Table 6.3. Temperature derivatives of the mean-squared fluctuations, $\Delta(\sigma^2)/\Delta T$, and amplitudes, $\Delta(u^2)/\Delta T$, (in units of $10^{-5}\mathring{A}^2/K$) and the correlation parameter $\gamma' = 1 - \Delta(\sigma^2)/2\Delta(u^2)$. The first element of the pair is the absorbing atom

Host	Pair	Shell	EXAFS	$\Delta(\sigma_i^2)/\Delta T$ Calc.	$\Delta(u^2)/\Delta T$ x-ray	γ'
Ti	Ti-Ti	1	5.06+0.5[a]			
Ti	Cu-Ti	1	6.1 +1.2[b]			
Fe	Fe-Fe	1		1.41[c]	1.38[c]	
		2		1.95[c]		
Co	Co-Co	1	1.1 +0.045[d]	1.14[e]	1.36[f]	0.60
		2	1.4 +0.45[d]	1.44[e]		0.48
Cu	Cu-Cu	1	2.2 ±0.2[d]	2.32[e]	1.9[f]	0.42
		1	2.1 ±0.02[g]	2.4[c]	2.0[c]	
		1	2.0[h1], 2.7[h2]			
		2	3.2 ±0.4[d]	2.96[e]		0.16
				3.4[c]		
Zn	Zn-Zn	1	5.9 ±0.6 (as deposited)[i]			
			3.4 ±0.6 (annealed)[i]			
Nb	Nb-Nb	1	1.4[j]			
Nb_3Ge	Nb-Nb	inter-chain	2.5[j]			
	Nb-Ge		1.2[j]			
	Nb-Nb	intra-chain	0.8[j]			
Nb_3Ge	Ge-Nb	1	2.1-2.5[k]			
Pt	Pt-Pt	1		1.56[c]	1.43[c]	
				2.28[c]		
$AuCu_3$	Au-Cu		2.3[l]			

[a] Ref.[6.65]; $\Delta T = 50\,K$, near 0 C. [b] Ref.[6.66]; $\Delta T = 70\,K$, near 0 C. [c] Ref.[6.59]; force constant model, B; $T_1 = 150\,K$, $T_2 = 700\,K$. [d] Ref.[6.62]; $T_1 = 80\,K$, $T_2 = 300\,K$. [e] Ref.[6.62]; correlated Debye models, $T_1 = 80\,K$, $T_2 = 300\,K$. [f] Ref.[6.67]; $T_1 = 80\,K$, $T_2 = 300\,K$. [g] Ref.[6.64]. [h] 1) Ref.[6.63]; $T_1 = 77\,K$, $T_2 = 295\,K$; 2) $T_1 = 295\,K$, $T_2 = 388\,K$. [i] Ref.[6.68]; polycrystalline films; $T_1 = 293\,K$, $T_2 = 648\,K$; parametric analysis. [j] Ref.[6.69]; $T_1 = 150\,K$, $T_2 = 300\,K$; parametric analysis. [k] Ref.[6.66]; obtained from linewidth; $T_1 = 77\,K$, $T_2 = 573\,K$. [l] Ref.[6.70]; value of $(\Delta\sigma)^2/\Delta T$.

factor (mfp) and σ^2(Nb) for the first two neighbor shells. Then using this sample mfp factor for Nb_3Ge, model calculations were fit to the backtransform of R(r) for Nb_3Ge in order to obtain σ^2 values for the first three peaks of R(r) for Nb_3Ge [6.69]. These peaks are due to intrachain and interchain Nb-Nb atoms and Nb-Ge neighbors. They all occur within 2.58 to 3.16 Å and are not well resolved. The resulting $\Delta(\sigma^2)/\Delta T$ values obtained are listed in Table 6.3. Since these were parametric fits, this type of analysis may be quite inaccurate. It was concluded that the anisotropic chain-type ordering of the Nb atoms strongly affected the vibrational correlations between the different intra- and interchain neighbor pairs.

c) Anisotropic, Anharmonic Systems

Conventional EXAFS analysis implicitly assumes that the radial distribution functions are Gaussian and that the disorder is small. If these conditions are not met, large analysis errors can occur. For example, an early conventional analysis of EXAFS data on Zn indicated an apparent room temperature contraction of the nearest-neighbor distance, relative to that at 20 K, of 0.09 Å in disagreement with the known 0.05 Å expansion [6.14]. It is now realized that Zn, which has room temperature values of $\sigma_\perp^2 = 0.04 \pm 0.005$ Å^2 and $\sigma_\parallel^2 = 0.015 \pm 0.05$ Å^2 [6.15], must be modelled with anharmonic forces and asymmetrical peaks in $R(r)$. The errors introduced by asymmetric radial distribution functions due to anharmonic potentials, inelastic scattering and large disorder are discussed and criteria for determining their importance are given in [6.4,14 and 68]. Systems not meeting these criteria must be treated with great care since the EXAFS technique has serious limitations for anharmonic systems.

By using simple analytic functions for peak shapes and comparing the calculated $\chi(k)$ obtained by backtransforming these line shapes with those obtained from the experimental EXAFS data, information has been obtained about the pair potentials and line-shapes of the distribution functions of nearest neighbor atoms in various anharmonic systems. This is discussed further in [6.4 and 68].

6.4.4 Structure of Amorphous Metals

X-ray, neutron and electron diffraction techniques have been most widely used to study the structure of amorphous metals [6.71]. These types of experiments are mainly sensitive to long range order and they provide information about the partial radial distribution functions (RDF) and short-range order only by combining many measurements. Thus it was of great interest to examine the short-range order with the EXAFS technique. However, in this case, too, the analysis is not straightforward (Sect. 6.3.2). As mentioned in Sect. 6.1.2, the particular advantage of EXAFS spectroscopy is that, being atom specific, it allows the determination of partial RDF's.

Amorphous metals have been classified into four basic types depending on their constituents.

1. T-M materials containing transition metals (T) and metalloids (M) atoms; i.e., sp elements which have a strong tendency towards covalent bonding such as B, P, Si or Ge. These form amorphous alloys over a narrow composition range near 20 at.% of M content where a deep eutectic exists in the phase diagram.
2. T-T alloys composed of one element from the beginning of a transition series having mainly itinerant d electrons such as Ti, Zr, Nb combined with another transition metal from the end of a transition series having many localized d electrons such as Fe, Co, Ni or Cu.

3. There are two types of rare-earth metal alloys (RE-metal): Those were the metal is a transition metal form amorphous metals over a wide range of compositions near the 50-50 composition. The other type contains sp metals atoms.
4. Simple sp metal alloys, such as $Be_{70}Al_{30}$.

In studies of amorphous materials the disorder represented by σ^2 as well as the asymmetry of the line shapes arise predominately from static, not thermal disorder. In the analysis it is assumed that the measurements are made at low enough temperatures that the thermal broadening is symmetric.

It has been shown that EXAFS analysis of disordered structures is particularly sensitive to the line shapes in the radial scattering function $R(r)$. Specifically it has been found that symmetric Gaussian lines yield nearest-neighbor distances that are too small by $\sim 0.1\,\text{Å}$ and coordination numbers that are too small [6.14,68,72]. In these studies it is generally assumed that the "correct" values of the bond lengths and coordination numbers are given by careful diffraction analyses; however, these techniques also suffer from similar problems [6.28,73]. All the methods of analysis of the EXAFS data on amorphous materials have the common procedure of determining the parameters by a least squares minimization of the differences between the $\chi(k)$ calculated from (6.7 or 9) and the experimental $\chi(k)$ obtained by backtransforming the proper region of $R(r)$. The inherently asymmetric line shapes in amorphous materials have been treated in three ways:

1. The first method is applicable to a single well-resolved line. This line is parameterized by the first three moments $\bar{r}$, $\overline{r^2} = \sigma^2_{total}$ and $\overline{r^3}$, the asymmetry parameter [6.14].
2. The second method is similar to the first except that a specific form of the asymmetric line shape is assumed as given by (called A1)

$$p(r_j) = (1/a)\exp(r - r_j/a) \quad r \geq r_j,$$

$$= 0 \quad r < r_j. \tag{6.24}$$

The average shell distance is $r = r_j + a$ where $2r_j$ is the distance between two touching spheres and a is the static root mean-square displacement giving the structural disorder. This method works well for T-M alloys but generally is not successful for T-T alloys.
3. The third method used is to simulate an asymmetrical radial shell distribution for each shell by two components with Gaussian line shapes when calculating $\chi^c_j(k)$. This method is widely used for T-T alloys. It unfortunately introduces an excessive number of parameters into the analysis.

A large number of EXAFS experiments have been performed on metallic amorphous systems. The experimental conditions and type of analysis are listed in Table 6.4 and the results are summarized in Table 6.5. Unless noted the EXAFS data were taken using x-rays from synchrotron radiation which generally has an energy resolution of 1–2 eV around 8 keV. Many of the experiments reported are quite sketchy and, while a summary of most results are listed in the tables, only the more complete papers are discussed.

a) T-M Metallic Glasses

Many of the structures proposed for these materials are based on the original dense-random-packed hard sphere (DRPHS) model of *Bernal* [6.74], as modified by *Polk* [6.75]. Here the voids between the hard spheres are filled by the "smaller" metalloid atoms which thus supposedly make a denser, more stable structure. Since, in the simplified model, there is one void per four hard spheres the stable compositions are near $R_{80}M_{20}$ and each M atom is coordinated to only T atoms.

An alternate model has the bonding of the covalent metalloid atom dominating, leading to a continuous random network [6.76]. It was found that this model appears to be valid for amorphous non-metals, such as Ge, but usually is not a good description of the structure of T-M metallic glasses.

Thus a structural question that has received considerable attention, and for which EXAFS is somewhat suited, is whether there is chemical ordering with M-M avoidance, as favored by the DRPHS model, in these alloys [6.77,78].

$Fe_{80}B_{20}$; $Fe_{40}Ni_{40}B_{20}$

The Fe K-edge EXAFS was analyzed with method 2 using a two shell model corresponding to nearest Fe and B neighbors. Each was assumed to have the line shape A1. Using a value of $a = 0.17$ Å, the Fe-Fe bond length was brought to within 0.02 Å of that obtained from x-ray scattering [6.79]. However, there was still a large discrepancy in the number of Fe neighbors; 8.2 from EXAFS and 2.6 from x-ray scattering. An analysis using symmetric Gaussian peaks gave bond lengths that were smaller by 0.1 Å and coordination numbers that were about a factor of two smaller than those listed in Table 6.5. This result clearly shows the sensitivity of EXAFS results to the line shapes of the RDF. It was found that replacing Fe with up to 50% Ni atoms did not alter the Fe EXAFS significantly. However, there were distinct differences seen between the Fe and Ni EXAFS. Assuming the same value for $a(= 0.17$ Å$)$ as for $Fe_{80}B_{20}$, gave an unreasonable coordination number of N = 17 for Ni. Setting N = 12 resulted in $a_{Ni} = 0.11$ Å [6.72].

In another experiment on $Fe_{80}B_{20}$ [6.80] it was found that the EXAFS signal was reduced by only 10–15% in going from 80 to 300 K. An analysis using method 2 gave different values for $a(= 0.25$ Å$)$ and a widely different Fe-B bond length than obtained in [6.72], see Table 6.5. These discrepancies indicate the sensitivity of the results obtained by EXAFS to the details of the analysis.

Table 6.4. Summary of analyses and conditions used for EXAFS experiments on the structure of amorphous metals. The notations used are:

In column 2, sync means a synchrotron source and RA( ) denotes a rotating anode source with the energy resolution in eV given in the parenthesis.

In the column marked ϕ,f,D; CA indicates calculated functions, MC indicates that model compounds and CMC that complex model compounds were used to obtain the phase ϕ, amplitude f, and attenuation factor D.

The type of analysis is noted; 2r and 2k indicate that, in the analysis done by method two, the least-squares minimization procedure was carried out in r or k-space, respectively.

The symmetry is denoted by S for a single Gaussian line shape per shell, A for the asymmetry determined by method 1, A1 for the asymmetrical line shape given by (6.24) and G2 for representing the asymmetry of the line shape of a shell by two Gaussian lines

Sample	x-ray source	Sample prep.^a	ϕ,f,D	#Osc. in x	Analysis method	Line shape	Ref.
$Fe_{80}B_{20}$	sync		Fe-MC	5-6	2k	S,A1	[6.72]
$Fe_{40}Ni_{40}B_{20}$			B,Ni-CA				
$Fe_{80}B_{20}$	sync	MS^b	CA	5-6	2k	A1	[6.80]
$Fe_{40}Ni_{40}^{-}$	sync		CMC	5-6	2r	S	[6.3]
$P_{20-x}B_x$; x=0,10,14,20							
$Fe_{80}B_{10}Ge_{10}$							
$T_{100-x}X_x$	RA(35)	MS	CMC	4-5	qualitative R(r)		[6.89]
T=Fe,Co,Ni;M=B,P							
$Co_{100-x}B_x$	sync	MS	CMC	4-5	1	A	[6.81]
$Co_{76}P_{24}$	RA(20)	ED	CA	3	3k	S	[6.82]
$Co_{80}P_{20}$	sync	CE	P-CA	4-5	2k,3k	A1,G2	[6.83]
$Ni_{80}P_{20}$			Co,Ni-CMC,CA				
Ni_xGe_{100-x}	RA(5)	CS	MC	4-5		S	[6.84]
Mo_xGe_{100-x}	sync	CS	MC	6-7		S	[6.28]
$Pd_{100-x}Ge_x$	sync	AQ,CS	MC			S	[6.90]
$Cu_{66}Ti_{33}$	sync	CS	ϕ,f-CA,D-MC	4-5 Cu,1 Ti		S	[6.91]
$Fe_{90}Zr_{10}$	RA(5)		CMC	1-2		S	[6.92]
$Ni_{66}Y_{33}$	sync	CS	ϕ,f-CA	3-4	3k	G2	[6.85]
$Cu_{60}Zr_{40}$	Y,Zr(7)	CS,MQ,RQ	D-MC				[6.85]
$Cu_{46}Zr_{54}$							[6.72,87]
$REFe_2$	sync	MS	MC		qualitative R(r)		[6.93]
$RE_{100-x}Z_x$	sync		CA	3-4	1	A	[6.88]

^a MS melt spinning, ED electrode-position, CE coevaporation, CS cosputtered, AQ arc quenched, MQ melt quenched, RQ roller quenched. ^b Allied Chemical Metglas 2605

Table 6.5. Experimentally determined amorphous alloy parameters from EXAFS studies. The first element listed in a pair is the absorbing atom. N is the number of atoms in a shell, σ is the room temperature root-mean-square thermal disorder for an asymmetric line shape plus the total disorder for a symmetric line shape, a is the asymmetry parameter in (6.24) and the attenuation factor is given in D = exp($-2r_i/\lambda$) (a constant) or the parameter $\Gamma = k/\lambda$ (a constant with units $Å^{-2}$) when noted. The symmetry notation is the same as in Table 6.4. The measurements were taken at room temperature unless noted

Alloy	r [Å]	Pair	N	D	Symm	σ [Å]	a [Å]	Ref.
		T-M						
$Fe_{80}B_{20}$	2.06	Fe-B	2.2		A1		0.17	[6.72]
	2.55	Fe-Fe	8.2				0.17	
$Fe_{40}Ni_{40}B_{20}$	2.57	Fe-Fe,Ni	12		A1		0.17[a]	
	2.51	Ni-Fe,Ni	12[a]				0.11	
$Fe_{80}B_{20}$	2.30	Fe-B			A1		0.25	[6.80]
	2.55	Fe-Fe					0.25	
$Fe_{40}Ni_{40}B_{20}$	2.51±0.05	Ni-Ni			S			[6.3]
	2.00±0.1	Ni-B						
$Fe_{40}Ni_{40}P_{20}$	2.63±0.05	Fe-Fe			S			
	2.47±0.05	Ni-Ni						
	2.26±0.1	Fe-P						
	2.30±0.1	Ni-P						
$Fe_{80}B_{10}Ge_{10}$	2.4	Ge-Fe			S			[6.3]
$Co_{81.5}B_{18.5}$	2.55±0.02[b]	Co-Co	12.7[c]		A	0.105[d]	0.12[e]	[6.81]
$Co_{78}B_{22}$	2.57±0.02[b]	Co-Co	13.0[f]		A	0.13[d]	0.14[e]	
$Co_{74}B_{26}$	2.57±0.02[b]	Co-Co	11.7[f]		A	0.13[d]	0.14[e]	
$Co_{76}P_{24}$[g]	2.28	Co-P	2.6	0.99	S	0.08		[6.82]
	2.64	Co-Co	10.0	0.15		0.12		
$Co_{80}P_{20}$	2.23±0.02	P-Co	4.5±0.5	=0.55	G2	0.08(NT)		[6.83]
	2.33±0.02	P-Co	4.5±0.5	=0.55	G2	0.1		
	2.29±0.02	P-Co	9.0±0.5	=0.55	A1	0.07	0.1	
	2.29±0.02	Co-P	1.5±0.5	=0.57	A1	0.07	0.1	
	2.47±0.02	Co-Co	10.5±0.5	=0.57	A1	0.13	0.04	

[a] Assumed values; [b] Γ from method 1; [c] Obtained from Ref.[6.94]; [d] rms of second moment from method 1; [e] cube root of third moment from method 1; [f] relative to N for $Co_{81.5}B_{18.5}$; [g] temperature varied between 10 K and room temperature

Table 6.5 (cont.)

Alloy	r [Å]	Pair	N	D	Symm	σ [Å]	a [Å]	Ref.
Ni$_{80}$P$_{20}$	2.23±0.02	P-Ni	9.5±0.5	=0.50	A1	0.05(NT)	0.12	
	2.21±0.02	Ni-P	1.5±0.5	=0.55		0.08	0.12	
	2.52±0.02	Ni-Ni	9.0±0.5	=0.55		0.08	0.14	[6.84]
Ni$_{23}$Ge$_{77}$	2.31±0.03	Ge-Ni	1.6±0.2		S			
Ni$_{48}$Ge$_{52}$	2.29±0.03	Ge-Ni	2.6±0.2					
Ni$_{55}$Ge$_{45}$	2.31±0.05	Ge-Ni	4.0±0.2					
Ni$_{23}$Ge$_{77}$	2.44±0.03	Ge-Ge	3.2±0.2					
Ni$_{48}$Ge$_{52}$	2.41±0.03	Ge-Ge	3.7±0.2					
Ni$_{55}$Ge$_{45}$	2.43±0.03	Ge-Ge	3.8±0.2					[6.28]
Mo$_{42}$Ge$_{58}$	2.59	Mo-Ge	2.6		S	0.10		
	2.48	Mo-Mo	0.11					
	2.71	Ge-Ge	1.21			0.09		
	2.56	Ge-Mo	0.75			0.06		
Mo$_{65}$Ge$_{35}$	2.63	Mo-Ge	0.8			0.09		
	2.37	Mo-Mo						
	2.75	Ge-Ge	2.8			0.09		
	2.58	Ge-Mo	1.6			0.06		
Mo$_{70}$Ge$_{30}$	2.67	Mo-Ge	1.0			0.12		
	2.52	Mo-Mo	0.03					
Pd$_{100-x}$Ge$_x$	2.49±0.01	Ge-Pd	8.6±0.5		S	0.1		[6.90]
(x=20,22)	3.38	Ge-Ge	4.0±1.0					

Alloy	r [Å]	Pair	N	Line shape	σ [Å]	a [Å]	Ref.
			T-T				
Cu$_{66}$Ti$_{33}$ ^{g}	2.51-2.53	Cu-Cu	3-4	S	0.11-0.12(RT)		[6.91]
	2.74+0.04	Cu-Ti	5-6		0.14		
	2.74+0.04	Ti-Cu	(10-12)		0.14		
	(3-3.12)	Ti-Ti	(7-8)		(0.12)		

184

Alloy	r [Å]	Pair	N	Line shape	σ [Å]	a [Å]	Ref.
$Fe_{90}Zr_{10}$	2.42	Fe-Fe	4.6	S	0.14		[6.92]
	2.75	Fe-Zr	0.8		0.00		
$Ni_{66}Y_{33}$ [g]	2.4 ±0.02	Ni-Ni	2.0±0.5	G2	0.08	=0.52±0.13	[6.85]
	2.55±0.02	Ni-Ni	4.0±0.5		0.12		
	2.71±0.02	Ni-Y	4.5±0.5				
	3.05±0.02	Ni-Y	1.5±0.5				
	2.71±0.02	Y-Ni	9.0±0.5		0.15	=0.71±0.13	
	3.05±0.02	Y-Ni	3.0±0.5				
	3.40±0.05	Y-Y	4.0±1.0		0.15		
$Cu_{60}Zr_{40}$ [g]	2.52	Cu-Cu	3.75	G2	0.10		[6.85]
	3.00	Cu-Cu	2		0.10		
	2.71	Cu-Zr	4.5				
	3.05	Cu-Zr	2		0.12		
	2.71	Zr-Cu	7				
	3.05	Zr-Cu	3		0.12		
	3.15	Zr-Zr	4				
$Cu_{46}Zr_{54}$ [g]	2.54±0.01	Cu-Cu	3.5±0.5	G2	0.10±0.01(30K)		[6.87]
	2.95±0.05	Cu-Cu	2.5±0.5		0.12±0.01		
	2.72±0.05	Cu-Zn	4.0±0.5		0.12±0.01		
	2.95±0.05	Cu-Zr	1.5±0.5		0.12±0.01		
	2.69±0.05	Zr-Cu	4.0±0.5		0.12±0.01		
	2.97±0.05	Zr-Cu	1.0±0.5		0.14±0.01		
	3.14±0.05	Zr-Zr	5.0±0.5				
$Cu_{46}Zr_{54}$	2.47±0.03	Cu-Cu	4.6±1	A1		0.12	[6.72]
	2.74±0.03	Cu-Zr	5.1±1			0	
	2.74±0.02	Zr-Cu					
	3.14±0.02	Zr-Zr					

[a] Assumed values; [b] r̄ from method 1; [c] Obtained from Ref.[6.94]; [d] rms of second moment from method 1; [e] cube root of third moment from method 1; [f] relative to N for $Co_{81.5}B_{18.5}$; [g] temperature varied between 10 K and room temperature

Table 6.5 (cont.)

Alloy	r [Å]	Pair	N	Symm	σ [Å][d]	a [Å][e]	Ref.
	(±0.03)		(±1)		(±0.03)	(±0.03)	
		RE-Metal					
$Eu_{80}Au_{20}$	3.40	Au–Eu	10[a]	A	0.2[d]	0.26[e](50 K)	[6.88]
$Tb_{80}Au_{20}$	3.12	Tb–Au	3[a]	A	0.13	0.26 (200 K)	
	3.55	Tb–Tb	9[a]		0.14	0.25	
	3.12	Au–Tb	11[a]		0.13	0.26	
$Tb_{80}Ga_{20}$	3.12	Tb–Au	3[a]	A	0.14	0.19 (200 K)	
	3.55	Tb–Tb	9[a]		0.14	0.18	
	3.12	Ga–Tb	11[a]		0.14	0.20	
$Dy_{65}Al_{35}$	3.10	Dy–Al	5[a]	A	0.15	0.14 (200 K)	[6.88]
	3.55	Dy–Dy	8[a]		0.14	0.13 (200 K)	
$Eu_5Mg_{65}Zn_{30}$	3.53	Eu–Mg	9[a]	A	0.13	-0.1 (200 K)	
	3.42	Eu–Zn	4[a]		0.13	-0.1	

[a] Assumed values; [b] $\bar{f}$ from method 1; [c] Obtained from Ref.[6.94]; [d] rms of second moment from method 1; [e] cube root of third moment from method 1; [f] relative to N for $Co_{81.5}B_{18.5}$; [g] temperature varied between 10K and room temperature

$Co_{100}B_x$, $x=18.5, 22, 26$

Using parameters determined from hcp Co and Co_3B the first three moments of the peaks in $R(r)$ were determined by method 1. From a study including polarized neutron diffraction and Mössbauer data it was concluded that the chemical short-range order prevails; the B atoms have no B nearest neighbors and are enclosed in Co structural units with characteristic features of Co_3B. The connectivity of these units depended on the concentration x [6.81].

$Co_{76}P_{24}$

An electrodeposited $Co_{76}P_{24}$ sample was measured using a rotating anode source having only 20 eV resolution full width at half maximum (FWHM). The data were analyzed using method 3. It was claimed that the asymmetry of the nn peak in the RDF, by itself, cannot be responsible for the differences in nn distances obtained from Gaussian analyzed EXAFS and x-ray scattering. It was suggested that the first shell in $R(r)$ for Co is dominated by Co-P rather than Co-Co contributions due to $\sigma_{Co-Co} > \sigma_{Co-P}$ and $D_{Co-Co} < D_{Co-P}$ [6.82]. However, the attenuation D values obtained in the analysis seem unreasonable.

$Co_{80}P_{20}$, $Ni_{80}P_{20}$

Co-evaporated amorphous films were examined at 78 K at both the P and Co or Ni K-edges [6.83]. In this work the mean-free path factor, $D_i = \exp(-2r_i/\lambda)$, is taken to have an explicit k-dependence give by $\lambda = k/\Gamma$, where Γ is a fitted constant. This gives slightly different σ values than those obtained with a model using a k-independent attenuation factor. Method 2 was used to analyze the Co and Ni EXAFS while both methods 2 and 3 were used for the P EXAFS. The two different methods gave essentially the same line shapes, showing that considering the Co shell around P as being made up of two components can, in this case, effectively simulate the asymmetric line shape of (6.24). (However, there is really no good physical basis for a two shell model and it should become increasingly invalid as the derived splitting of the two components increases.) The results obtained from EXAFS were in agreement with those of x-ray and neutron scattering. For $Co_{80}P_{20}$ it was found that the local arrangement of the atoms in the amorphous state had a strong resemblance to that of crystalline Co_2P. There was no indication of M-M first shell coordination. The generalization was made that the local order in an amorphous alloy tends to look like the closest crystalline compound when it is stable.

Ni_xGe_{110-x}

This system was studied over the range of x $=0-0.55$ [6.84]. It becomes metallic around x $=0.23-0.30$. The data were analyzed in the usual manner of backtransforming and least-squares fitting but the asymmetry of the line shapes was not taken into account. Thus the bond lengths may be in error; although the Ge-Ge and Ge-Ni bond lengths are so close in value that perhaps this two shell model adequately simulates the asymmetries of the line shape. As seen in Table 6.5 the coordination number of Ge increases with Ni content from four for amorphous Ge to 8.1 for $Ni_{55}Ge_{45}$. This is near the value corresponding to the DRPHS model.

Mo_xGe_{100-x}

This system was studied over the range of x =0–0.70 [6.28]. It becomes metallic at about 30% Mo. In the Mo-rich region the major changes with compositional variation took place around Mo, while the average environment around Ge changed little. This behavior is the reversed in the Ge-rich region. Analyses with both a one and two shell model (single Ge and Mo shells were used, not two components as in method 3) using only Gaussian line shapes were made. A careful comparison of EXAFS and differential anomalous scattering analyses were performed on this system. It was found that the average first shells are much broader than that of amorphous Ge; leading to apparent discrepancies between the EXAFS and scattering results. It was emphasized that both techniques must be applied to get an accurate description of amorphous materials.

b) T-T Amorphous Alloys

$Ni_{66}Y_{33}$, $Cu_{60}Zr_{40}$

It was found that the EXAFS data could not be fit well by using the asymmetric line shape of (6.24) for a spectra composed of lines having the bond lengths and coordination numbers determined from x-ray and neutron scattering experiments [6.85]. This was attributed to the hard core (sharp rise at r_i) of the line shape representing a strong covalent bond well, but not a metal-metal bond [6.86]. However, good agreement with the x-ray data was obtained by using a two shell Gaussian model for each pair other than Y-Y and Zr-Zr (Method 3). However, the splitting obtained between the two Gaussians used to represent the asymmetric line shape of the Cu-Cu shell is so large that it seems questionable that this is a good representation. These analyses contain a large number of parameters and were strongly guided by the crystalline structure, especially for $Ni_{66}Y_{33}$. It was concluded that the coordination numbers obtained from EXAFS and scattering measurements suggest that the local atomic arrangement in the amorphous state resembles that of the crystalline state. These numbers are different from those obtained from a statistical model. The results, in particular for unlike pairs of atoms, were taken as strong evidence for chemical ordering.

$Cu_{60}Zr_{40}$, $Cu_{46}Zr_{54}$

A large increase in intensity of the second peak of the radial scattering distribution around the Zr atoms was seen in going from 40% Zr to 54% Zr; while the environment around the Cu atoms changed very little. The similarity of the radial scattering functions around Cu was interpreted as evidence of chemical order while the increased second peak was attributed to an increase in Zr-Zr pairs [6.87].

The $Cu_{46}Zr_{54}$ alloy was measured by two different groups and analyzed using the different methods 2 and 3. The analysis using method 2 found the asymmetric line shape parameter a of (6.24) to be equal to 0.12 Å for the shells around Cu while the shells surrounding Zr were found to have symmetric line shapes (a = 0) [6.72]. The Cu-Cu distance, 2.47±0.03 Å, derived in this analysis was considerably different than that found for the average distance, 2.69 Å, from the two shell Gaussian model of method 3 [6.87]. This again illustrates how dependent the results of EXAFS analyses are on the method and details of the analysis.

c) RE-Metal Amorphous Metals

RE_{100-x}-Z_x; RE=Eu, Gd, Tb, Dy; Z=Al, Ga, Au; x=20 and 35

EXAFS at the L_{III} edge of the RE and Au and the K-edge of Ga were measured on these alloys [6.88]. The analysis was done by method 1. Data was taken at 10, 20, 50 and 200 K. All the $RE_{80}Z_{20}$ results, other than $Eu_{80}Au_{20}$, were similar except that the third moment decreased for both RE-Z and RE-RE pairs from about 0.25 to 0.19 to 0.14 in going from Au to Ga to Al. This was taken as strong evidence that short-range order is determined by Z but that the degree of structural order around both the RE and Z atoms is comparable. Typical results are listed in Table 6.5. $Tb_{65}Al_{35}$ was very similar to $Dy_{65}Al_{35}$. An analysis assuming symmetrical line shapes gave Re-Z bond lengths that corresponded to a 15% volume contraction. No obvious relationship of structural short-range order was found between the amorphous and crystalline structures.

d) Comments and Conclusions

Analysis of EXAFS experiments of amorphous metal systems is, of necessity, quite subjective and full of simplifying assumptions and compromises. As discussed in Sect. 6.3, EXAFS has severe limitations when applied to systems with broad asymmetric atomic distributions. It has been shown in many analyses that the T-M amorphous alloys can be made to agree with diffraction experiments fairly well by using the asymmetric line shape of (6.24). In contrast the T-T alloys are not fit satisfactorily by this type of skewed line shape. The two component shell model used to analyze the T-T systems provides more flexibility in shape but introduces an excessive number of parameters. It would be worthwhile investigating whether a more general line shape such as two back-to-back Gaussians or Lorentzians, for example,

$$p(r) \exp[-(r - r_i)^2/a_l^2] \quad r \leq r_i,$$

$$p(r) \exp[-(r - r_i)^2/a_h^2] \quad r > r_i$$

would provide sufficient flexibility of the line shape to give satisfactory fits or whether the T-T systems have enough structure to require more components per shell. It would clearly be desirable for different groups to analyze the same system, preferably even the same data using different models; so that some information could be obtained on the uniqueness and reliability of the derived parameters. In the case where this has been done ($Cu_{46}Zr_{54}$) the results were drastically different.

There now seems to be some accumulation of evidence for the reasonable supposition that the amorphous state tends to have some similarities to nearby stable crystalline compounds. However, much remains to be investigated on the nature, extent and evolution of such correlations.

We have discussed three applications of the EXAFS technique in metallic systems. From the results obtained, especially in cases were there exists data for comparison from other techniques, we see that caution must be exercised in applying the EXAFS technique. It is only straightforwardly applicable to radially symmetric nearest-neighbor shells. Thus it is ideal for the study of the nearest-neighbor environment around isolated solute atoms in metals where these conditions should be satisfied. It is also well suited for studying Debye-Waller factors or mean-square displacements in isotropic harmonic metals. It is less well suited when the material is anisotropic or the forces anharmonic. It is poorly suited to the study of the structure of amorphous materials; but because of the difficulty of this problem and the lack of any more applicable techniques, it has been widely used to study these materials. Experience has shown that the results obtained from EXAFS analysis can be erroneous if the assumptions and models used are not consistent with data and conclusions of as many other complementary techniques as possible, e.g., x-ray diffraction and neutron scattering. Thus, although the EXAFS technique has many unique capabilities and is a powerful technique when properly applied to appropriate problems, it should be used with caution and in many cases the quoted results should be regarded with scepticism.

References

6.1   P. Rabe, R. Haensel: Festkörperprobleme **20**, 43 (Vieweg, Braunschweig 1980)
6.2   P.A. Lee, P.H. Citron, P. Eisenberger, B.M. Kincaid: Rev. Mod. Phys. **53**, 769 (1981)
6.3   J. Wong: In *Glassy Metals I*, ed. by H.J. Güntherodt, H. Beck, Topics Appl. Phys., Vol. 46 (Springer, Berlin, Heidelberg 1981) Chap.4
6.4   S.J. Gurman: Mat. Sci. **17**, 1541 (1982)
6.5   T.M. Hayes, J.B. Boyce: Solid State Physics, **37**, 173 (1982)
6.6   A. Bianconi, L. Incoccia, St. Stipcich (eds.) *EXAFS and Near Edge Structures*, Springer Ser. Chem. Phys., Vol. 27 (Springer, Berlin, Heidelberg 1983)
        K.O. Hodgson, B. Hedman, J.E. Penner-Hahn (eds.): *EXAFS and Near Edge Structures* III, Springer Proc. Phys., Vol. 2 (Springer, Berlin, Heidelberg 1984)
6.7   D.E. Sayers, F.W. Lytle, E.A. Stern: Adv. in X-ray Analysis **13**, 248 (1970)
        D.E. Sayers, E.A. Stern, F.W. Lytle: Phys. Rev. Let. **27**, 1204 (1971)
6.8   E.A. Stern: Phys. Rev. B **10**, 3027 (1974)
6.9   C.A. Ashley, S. Doniach: Phys. Rev. B **11**, 1279 (1975)
6.10   P.A. Lee, J.B. Pendry Phys. Rev. B **11**, 2795 (1975)
6.11   P.A. Lee, G. Beni: Phys. Rev. B **15**, 2862 (1977)
6.12   B.K. Teo, P.A. Lee: J. Am. Chem. Soc. **101**, 2815 (1979)
6.13   M.B. Stearns: Phys. Rev. B **25**, 2382 (1982)
6.14   P. Eisenberger, G.S. Brown: Solid State Commun. **29**, 481 (1979)
6.15   G. Beni, P.M. Platzman: Phys. Rev. B **14**, 1514 (1976)

6.16   T.M. Hayes, A.C. Wright: In *Structure of Non-Crystalline Materials*, ed. by P.H. Gaskell, J.M. Parker, E.A. Davis (Taylor and Francis, London 1983) p.108

6.17   J.J. Rehr, E.A. Stern: Phys. Rev. B **17**, 4413 (1976)

6.18   E.A. Stern, B.A. Bunker, S.M. Heald: Phys. Rev. B **21**, 5521 (1980)

6.19   G. Beni, P.A. Lee, P.M. Platzman: Phys. Rev. B **13**, 5170 (1976)

6.20   B.K. Teo: J. Am. Chem. Soc. **103**, 3990 (1981)

6.21   N. Alberding, E.D. Crozier: Phys. Rev. B **27**, 3374 (1983)

6.22   I. Landau, W.E. Spicer: J. Electron Spec. Related Phenom. **3**, 409 (1974)

6.23   J.E. Muller, W.L. Schaich: Phys. Rev. B **27**, 6489 (1983)

6.24   J.J. Rehr, E.A. Stern, R.L. Martin, E.R. Davidson: Phys. Rev. B **17**, 560 (1978)

6.25   C. Noguera, D. Spanjaard: J. Phys. F **11**, 1133 (1981)

6.26   A. Fontaine, P. Lagarde, D. Raoux, J.M. Esteva: J. Phys. F **9**, 2143 (1979)

6.27   D.G. Stearns: Phil. Mag. B **49**, 541 (1984)

6.28   J.B. Kortright: Ph.D. Thesis, Stanford University, SSRL Report 84/05 (1984)

6.29   A. Werner, H.D. Hochheimer, B. Lengeler: Rev. Sci. Instrum. **53**, 1467 (1982)

6.30   G. Martens, P. Rabe, G. Tolkiehn, A. Werner: Phys. Stat. Sol. (a) **55**, 105 (1979)

6.31   G. Martens, P. Rabe, N. Schwentner, A. Werner: J. Phys. C **11**, 3125 (1978)

6.32   G. Martens, P. Rabe: J. Phys. C **14**, 1523 (1981)

6.33   A. Bianconi, D. Jackson, K. Monahan: Phys. Rev. B **17**, 2021 (1978)

6.34   B. Lengeler, P. Eisenberger: Phys. Rev. B **21**, 4507 (1980)

6.35   J. Goulon, C. Goulon-Ginet, R. Cortes, J.M. Dubois: J. Physique **43**, 539 (1982)

6.36   E.A. Stern, K. Kim: Phys. Rev. B **23**, 3781 (1981)

6.37   J.J. Boland, F.G. Halaka, J.D. Baldschwieler: Phys. Rev. B **28**, 2921 (1983)

6.38   B.W. Holland, J.B. Pendry, R.F. Pettifer, J. Bordas: J. Phys. C **11**, 633 (1978)

6.39   P. Rabe, G. Martens, N. Schwentner, A. Werner: Phys. Rev. B **17**, 1481 (1978)

6.40   P.H. Citrin, P. Eisenberger, B.M. Kincaid: Phys. Rev. Lett. **36**, 1346 (1976)

6.41   D.G. Stearns, M.B. Stearns: Phys. Rev. B **27**, 3842 (1983)

6.42   D. Raoux, A. Fontaine, P. Lagarde, A. Sadoc: Phys. Rev. B **24**, 5547 (1981)

6.43   A. Fontaine, P. Lagarde, A. Naudon, D. Raoux, D. Spanjaard: Phil. Mag. B **40**, 17 (1979)

6.44   C. Auvray, P. Georgopoulos, J.B. Cohen: Acta Metallurgica **29**, 1061 (1981)

6.45   B. Lengeler, P. Eisenberger: Phys. Rev. B **21**, 4507 (1980)

6.46   J. Mimault, A. Fontaine, P. Lagarde, D. Raoux, A. Sadoc, D. Spanjaard: J. Phys. F **11**, 1311 (1981)

6.47   W. Weber, H. Peisl: In *Point Defects and Defect Interactions in Metals*, Yamada Sci. Found. (University of Tokyo Press, 1982)

6.48   M. Marcus: Sol. St. Commun. **38**, 251 (1981)

6.49   R. Munch, H.D. Hochheimer, A. Werner, G. Materlik, A. Jayaraman, K.V. Rao: Phys. Rev. Lett. **50**, 1619 (1983)

6.50   M.B. Stearns, J. Appl. Phys. **52**, 1649 (1981)

6.51   J.B. Hastings, P. Eisenberger, B. Lengeler, M.L. Perlman: Phys. Rev. Lett. **43**, 1807 (1979)

6.52   R.L. Cohen, L.C. Feldman, K.W. West, B.M. Kincaid: Phys. Rev. Lett. **49**, 1416 (1982)

6.53   J.C. Mikkelsen, J.B. Boyce: 11th Annual User's Group Meeting, Stanford Synchrotron Radiation Laboratory, SSRL Report No. 84/06 (October 1984)

6.54   S. Froyen, C. Herring: J. Appl. Phys. **52**, 7165 (1981)

6.55   P.G. Tomlinson, J.P. Carbotte, G.R. Piercy: J. Phys. F **7**, 1305 (1977)

6.56   S.A. Drentje, J. Ekster: J. Appl. Phys. **45**, 3242 (1974)

6.57   J.C. Mikkelson, Jr, J.B. Boyce: Phys. Rev. Lett. **49**, 1412 (1982)

6.58   S.P. Vernon, M.B. Stearns: Phys. Rev. B **29**, 6968 (1984)

6.59   E. Sevillano, H. Meuth, J.J. Rehr: Phys. Rev. B **20**, 4908 (1979)

6.60   E.C. Svensson, B.N. Brockhouse, J.M. Rowe: Phys. Rev. **155**, 619 (1967)

6.61   R.M. Nicklow, G. Gilat, H.G. Smith, L.J. Raubenheimer, M.K. Wilkinson: Phys. Rev. **164**, 922 (1967)

6.62   W. Bohmer, P. Rabe: J. Phys. Rev. B **20**, 2465 (1979)

6.63   R.B. Greegor, F.W. Lytle: Phys. Rev. B **20**, 4902 (1979)

6.64   E.A. Stern, D.E. Sayers, F.W. Lytle: Phys. Rev. B **11**, 4836 (1975)

6.65   M. Marcus: Sol. State Commun. **38**, 251 (1981)

6.66 T. Claeson, J.B. Boyce, T.H. Geballe: Phys. Rev. B **25**, 6666 (1982)
6.67 *International Tables for X-Ray Crystallography*, Vol. III, Table 3.3.5.1, B and C (Birmingham: Kynoch)
6.68 E.D. Crozier, A.J. Seary: Can. J. Phys. **58**, 1388 (1980)
6.69 G.S. Cargill III, R.F. Boehme, W. Weber: Phys. Rev. Lett. **50**, 1391 (1983)
6.70 T. Claeson, J.B. Boyce: Phys. Rev. B **29**, 1551 (1984)
6.71 P.H. Gaskell: Nucl. Inst. Meth. **199**, 45 (1982); J. Phys. **12**, 4337 (1979)
6.72 R. Haensel, P. Rabe, G. Tolkiehn, A. Werner: Proc. of NATO Adv. St. Inst.: Liq. and Amorphous Metals, ed. by E. Luscher, H. Coufal; NATO Adv. Study Inst.Series E: Appl. Sci. **36**, 467 (Sijthoff & Noordhoff, Germantown, MA, USA 1980)
6.73 J. Chang, D.B. Dove: J. Non-Cryst. Solids **16**, 72 (1974)
6.74 J.D. Bernal: Proc. Roy. Soc. A **280**, 299 (1964)
6.75 D.E. Polk: Acta Metall. **20**, 485 (1972)
6.76 P. Chauduri, D. Turnbull: Science **199**, 11 (1978)
      N.F. Mott, E.A. Davis: *Electronic Processes in Non-Crystalline Materials*, 2nd. ed. (Clarendon, Oxford 1979) Chap. 7
6.77 J. Bletry, J.F. Sadoc: J. Phys. **F 5**, L110 (1975)
      J.F. Sandoc, J. Dixmier: Mater. Sci. Eng. **23**, 187 (1976)
6.78 Y. Weseda, S. Tamaki: Z. Phys. **B 23**, 315 (1976)
6.79 Y. Weseda, H.S. Chen: Phys. Status Solidi **A 49**,387 (1978)
6.80 M. De Crescenzi, A. Balzarotti, F. Comin, L. Incoccia, S. Mobilio, N. Motta: Solid State Commun. **37**, 921 (1981)
6.81 J.M. Dubois, G. Le Caer, P. Chieux, J. Goulon: Nucl. Inst. Meth. **199**, 315 (1982)
6.82 G.S. Cargill III: Proc. 4th Int. Conf. on RQM (Sendai 1981) p. 389
6.83 P. Lagarde, J. Rivory, G. Vlaic: J. Non-Cryst. Solids **57**, 275 (1983)
6.84 H. Oyanagi, K. Tsuji, S. Hosoya, S. Minomura, T. Fukamachi: J. Non-Cryst. Solids **35—36**, 555 (1980)
6.85 A. Sadoc, D. Raoux, P. Lagarde, A. Fontaine: J. Non-Cryst. Solids **50**, 331 (1982)
6.86 A.M. Flank, P. Lagrade, D. Raoux, J. Rivory, A. Sadoc: Proc. 4th Int. Conf. on RQM (Sendai 1981) p. 393
6.87 A. Sadoc, A.M. Flank, D. Raoux, P. Lagarde: J. de Phys. **43** (C9), 43 (1982)
6.88 M. Maurer, J.M. Friedt, G. Krill: J. Phys. **F 13**, 2389 (1983)
6.89 F. Schmucke, P. Lamparter, S. Steeb: Z. Naturforsch. **37a**, 572 (1982)
6.90 T.M. Hayes, J.W. Allen, J. Tauc., B.C. Giessen, J.J. Hauser: Phys. Rev. Lett. **40**, 1282 (1978)
6.91 D. Raoux, J.F. Sadoc, P. Lagarde, A. Sadoc, A. Fontaine: J. de Phys. **40** (C8), 207 (1980)
6.92 H. Terauchi, S. Iida, K. Tanabe, H. Maeda, M. Hida, N. Kamijo, K. Osamura, M. Takashiga, T. Nakamura: J. Phys. Soc. Jpn. **52**, 3454 (1983)
6.93 E.A. Stern, S. Rinaldi, E. Callen, S. Heald, B. Bunker: J. Magn. & Magn. Mater. **7**, 188 (1978)

7. X-Ray Photoelectron Spectroscopy

G.K. Wertheim

With 14 Figures

Photoelectron spectroscopy with characteristic x-rays was developed into a useful analytical technique by Kai Siegbahn and collaborators at Uppsala in the late sixties. The most important applications depend on the ability to measure small shifts in core-electron binding energies, which are related to the valence state and chemical bonding of the atom. In metals the core-electron binding energy is a function of the cohesive energy of the solid. Changes in the cohesive energy at the surface of a bulk metal, or in a small metal cluster, find expression in the core-electron binding energy. More detailed analysis of core-electron photoemission spectra from metals reveals the contributions of the collective excitations which accompany the ejection of an energetic electron. These include the screening response of the conduction electrons, which modifies the shape of primary line, and the excitation of plasmons and interband transitions, which manifest themselves as satellites of this line. The technique can also be used to obtain the valence band density of states of metals and alloys, and is particularly well adapted to the elemental analysis of thin (20 Å) surface layers.

7.1 Historical

X-ray photoelectron spectroscopy has its roots in the photoelectric effect which initially was concerned only with the electrons near the Fermi surface [7.1]. As photoemission spectroscopy developed, its scope widened to include the entire valence region, using photons in the vacuum ultra-violet region. Incentives for the study of core electrons were limited because they were thought to be entirely atomic in character. Moreover, since photoemission with characteristic x-radiation has inherently poor resolution and low signal levels, little work was done with this form of excitation. The core electrons remained the province of x-ray emission and absorption edge spectroscopies.

X-ray photoelectron spectroscopy owes its current prominence to the vision and pioneering efforts of Kai Siegbahn of Uppsala University in Sweden [7.2,3]. Siegbahn's accomplishments received international recognition in 1981 when he shared the Noble prize in Physics with N. Bloembergen and A.L. Schwalow. Siegbahn was motivated by the recognition that core electrons are, in fact, not entirely insensitive to their environment. They do carry information about the electronic state of the atom itself, and are sensitive also to the properties of the solid or molecule in which the atom is located. A core-electron spectroscopy with sufficient resolution could consequently make significant contributions to our understanding of the electronic structure of all types of materials.

Much of the early attention focused on valence and bonding in simple chemical systems. He consequently called his technique ESCA (Electron Spectroscopy for Chemical Analysis) to suggest applications which would probe the chemical bond. However, ESCA rapidly caught the attention of physicists and metallurgists, and became a technique with a very broad spectrum of applications. As a result the technique-oriented term, x-ray photoelectron spectroscopy, abbreviated XPS, has gained currency, and is used interchangeably with ESCA.

When Siegbahn and his group at Uppsala University began to report the results of core electron photoemission spectroscopy in the late sixties [7.2], the advantages of this technique rapidly became apparent. It makes it possible to study the lighter elements that have awkward x-ray emission energies and broad emission lines, e.g., C, O, N, and F. In ESCA, core electrons are excited to final states high above the Fermi edge so that limitations set by selection rules and the orbital character of the empty density of states disappear. One can then excite any shell and take advantage of the core levels with largest orbital angular momentum that are sharpest and tend to have the largest photoelectric cross section. More important, however, is the fact that chemically induced shifts in core-electron binding energies are much larger than shifts previously observed in x-ray emission spectroscopy. The reason for this is clear. Chemical effects shift all inner core levels by almost identical amounts. As a result, x-rays produced by transitions between two core levels show only differential, higher order shifts; ESCA shifts show the full value.

7.2 Basic Principles

7.2.1 Photoemission

When an atom absorbs a photon by photoelectric effect an energetic electron and a core hole are produced. The kinetic energy of the electron and the energy of the hole state add up to the energy of the photon. The energy of the hole state is therefore taken to define the binding energy E_B of the electron in the atom.

The natural assumption regarding the core-electron binding energy of an electron in an atom is that it will correspond to the eigenenergy of the electron in the initial state of the atom. This identification, known as *Koopmans'* Theorem [7.4], is based on the assumption that the orbitals of the other electrons remain unchanged when the photoelectron is emitted. It is consequently called the frozen-orbital approximation, and results in a final state which is not an eigenstate of the system with the photo-hole. However, even if the photoelectron is removed in a time short compared to that characterizing the motion of the other electrons in the atom, i.e., in the sudden approximation, a set of final eigenstates is actually produced. The state with lowest energy, the ground state of the hole-state atom, has the same energy as that obtained if the photoelectron were to leave the atom so slowly that an adiabatic description is appropriate. In this limit the outer orbitals relax as the electron departs from the atom, leaving a hole in an inner shell. This relaxation lowers the energy of the final state well below that of the Koopmans state. For the isolated atom, the difference between the energy of the Koopmans state and that of the fully relaxed state is called the intra-atomic relaxation energy.

If the core electron is removed more rapidly, then the relaxation process will populate not only the ground state of the hole-state atom, but also those of its excited states allowed by selection rules. A sum rule shows that, in the sudden approximation the centroid of the population of these final eigenstates will coincide with the *Koopmans* energy [7.5,6]. The final state with lowest energy is usually the strongest and is called the main line, the rest are called shake-up satellites, because they correspond to excited states of the hole-state atom.

Theoretical values for the free-atom binding energies are obtained by calculating the total energy of the atom both with and without the core hole. The difference corresponds to the binding energy measured in photoemission. For a more detailed treatment of the theory of photoemission see [7.7–9].

7.2.2 The Core-Electron Binding Energy in a Metal

The core-electron binding energy measured for an atom in a metal differs from that of a free atom for a number of reasons. Binding energies will change if there is a redistribution of charge in the *intital* state among outer orbitals with different radial expectation values. In general, this happens whenever hybridized bonds or bands are formed. This effect is particularly strong in transition metals with $s - d$ hybridization [7.10,11]. The *final*-state relaxation energy is also greatly modified, because extra-atomic relaxation channels are available in the metal. The relaxation energy contributed by the core electrons remains very much like that in the free atom, but the formation of a conduction band makes all the conduction electrons of the solid available to screen the core hole. In insulators, in the absence of conduction electron screening, the polarization of the neighboring atoms helps to lower the energy of the final state. These additional relaxation channels in solids serve to decrease the core-electron binding energy below that of the free atom.

There is also a difference in the binding-energy reference level in the two measurements. Binding energies of free atoms are measured relative to the vacuum levels; those for metals are usually given relative to the Fermi level E_F. Crudely speaking it would then suffice to add the work function to the measurement in the metal to bring both to a common vacuum reference. This approach is not rigorously correct, however, because the work function contains dipolar terms not relevant to the energy of the core hole state. A more meaningful correction could, in principle, be obtained by using the chemical potential [7.12].

The quantitative interpretation of core-electron binding energies in terms of the above parameterization is generally hampered by the fact that the contributions from screening, hybridization, and reference level are not readily separated experimentally. It does, however, provide a useful conceptual framework, and one which is closely related to theoretical methods.

An alternative approach abandons these concepts, and demonstrates that the Fermi level referenced binding energy in the metal can be related to the vacuum level referenced binding energy of the free atom through an ionization potential and two cohesive energies. It is based on the Born-Haber cycle [7.13] illustrated in Fig. 7.1. This formalism starts with the bulk metal, and goes through the following steps: (1) remove one atom, at the cost of the atomic cohesive energy E_{coh}^Z, (2) core ionize this free atom, which requires an energy $E_{B,atom}^Z$, (3) replace this core-ionized atom with a valence-ionized atom of the next-higher atomic number, relying on an equivalent cores argument for the effective identity of the two atoms (the equivalent-core approximation may be concisely stated as follows. A fully relaxed, core-ionized atom is like a valence-ionized atom with next-higher atomic number, as far as the outer electrons are concerned), (4) neutralize this atom, gaining the ionization potential I^{Z+1}, (5) condense this atom into a metal with the next-higher atomic number, gaining the cohesive energy E_{coh}^{Z+1}, (6) allow this atom to diffuse back into the original metal yielding an energy ΔE_{impl}. The net result of this procedure is a fully screened $Z + 1$ impurity atom in the original Z metal. Using the equivalent cores argument in the reverse sense, this $Z + 1$ atom corresponds to a fully screened, core-ionized Z metal atom. Consequently, the net energy of this cycle is the energy required to excite an electron from the core level to the Fermi energy, producing a screened final state, i.e. it is equal to the core-electron binding energy measured from E_F, as usually defined by

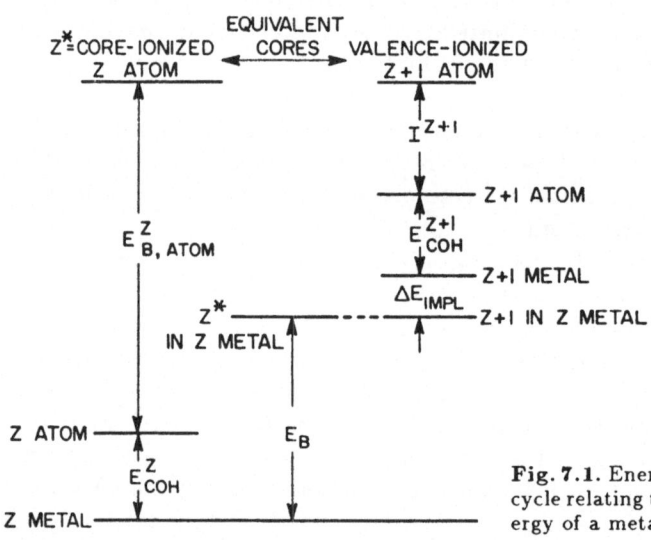

Fig. 7.1. Energy diagram for a Born-Haber cycle relating the core-electron binding energy of a metal to that of the free atom

196

$$E_B = E_{B,atom}^Z - I^{Z+1} + E_{coh}^Z - E_{coh}^{Z+1} - \Delta E_{impl}. \tag{7.1}$$

The cohesive energies of metals are generally well known, and the core-electron binding energies of many atomic species are currently being measured. Only the last term is often not well known, but has generally been found to be small. This formalism yields excellent estimates of binding energies in metals. It can be readily extended to deal with the change in binding energy at the surface, as well as to relate binding energy shifts in alloys to heats of solution, see below.

7.2.3 Core-Electron Satellites

As pointed out in Sect. 7.2.1, the main line of a core-electron photoemission spectrum is always accompanied by shake-up satellites at greater binding energy. The nature of the excited final states responsible for the satellite structure depends strongly on the environment of the photoexcited atom. In the isolated atom they are, of necessity, the discrete atomic excited states of the atom. For an atom in a simple metal they comprise the electron-hole pair and plasmon excitations of the conduction band. In other systems charge transfer or inter-band transitions may dominate. This does not, however, exhaust the possibilities, because other processes, e.g., modification of the band structure by the core hole, or final-state configuration interaction, may make additional states available. In general, it is necessary to consider the excited states of the solid with the hole-state atom, in order to assign the observed satellite structure.

The three parameters which determine the nature of the final-state effects are (1) the width of the conduction band, (2) the strength of the core-hole potential, and (3) the Coulomb correlation energy[7.14,15]. The latter is the energy required to move one electron in the solid so as to produce a unit increase above the equilibrium charge at a given lattice site. If the band width is large compared to the other two parameters, then the final state of the valence electrons closely resembles the initial state, and core-electron photoemission reflects the screening and plasmons of the initial-state band structure. This case is well approximated in the simple metals, Na, Mg, and Al. If the core-hole potential dominates, on expects to find a state split off from the bottom of the band, which may be filled or empty, resulting in two distinct final states. This behavior has been found in some transition-metal compounds. If the Coulomb correlation energy is large compared to the band width, a set of well-separated states characterizes the material. Examples of this behavior are found in the 4f levels of the rare earths.

Core-electron spectroscopy of rare-earth materials often gives unusual results because the potential of the core hole is sufficiently strong to pull the next empty 4f level below E_F. This can be understood in terms of the equivalent-cores approximation, because in the rare earths, the ion with next higher Z, in general, has an additional f electron. It means that the Coulomb correlation energy that separates successive f levels is of just the same size as the potential due to an added nuclear charge. Since the 4f wave functions are core-like in radial extent with little f-f overlap between adjacent atoms, the band width of the 4f states is extremely small. The net result is that in the final state with a core hole a previously empty 4f level appears as a sharp level below E_F. It may remain empty or become occupied, resulting in two distinct final states. Normally the 4f level remains empty, screening is accomplished within the (5d6s) conduction band. If it is filled, a lower-energy final state is obtained, because an electron in the 4f wave function produces a larger negative potential inside the atom than a conduction electron [7.16]. The resulting satellites were called "shake-down" satellites to contrast them with the shake-up satellites which lie at larger binding energy. Properly speaking, the shake-down feature should be considered the main line, since it has the lowest energy. However, since it is usually relatively weak it is treated as a satellite. In this case the main line and satellite simply represent alternate screening channels. Recently, related behavior has been identified in the spectra of metallic transition metal oxides with d conduction bands [7.17]. A metal core hole may be screened either by the conduction band itself, or by an excitation from the valence to the conduction band [7.18].

Satellites due to final-state configuration interaction require a further generalization, because there are cases in which the state corresponding to the satellite does not include the hole of the main line. A satellite of the 3s state of K is of this character, corresponding to a final state with two 3p holes and an electron promoted into a normally empty 3d level [7.19]. The spin and orbital angular momenta of the holes and the excited electron in the final state are coupled to give a spin of 1/2 and a vanishing orbital angular momentum, producing a ^{2}S state. This is the same as that of the final state of the primary line which a single 3s hole, see also [7.20].

7.2.4 Plasmons, Electron Mean-Free Path, and Surface Aspects of XPS

The motion of the electron through the solid, after it has been liberated in the photoelectric process, is perturbed by the interaction with the valence electrons, exciting interband transitions as well as collective oscillations known as plasmons. If the plasmon dispersion is weak and the lifetime long, then the ESCA spectrum may exhibit satellite lines due to the excitation of one or more plasmons (Fig. 7.2) [7.21a]. The contributions to the plasmon satellites from the energy loss of the propagating photoelectron are called extrinsic plasmons. In addition, there is a contribution due to the excitation of plasmons as part of the primary photoemission event. These are called intrinsic plasmons, and are part of the shake-up structure discussed above. They are the analogs in the solid state of the shake-up satellites produced by the photoionization of a free atom.

At the surface of the sample the plasmon energy is modified by the vacuum interface. A simple calculation shows that the energy is reduced by $1/\sqrt{2}$ at the surface. Both bulk and surface plasmons can be identified in Fig. 7.2. Intrinsic surface plasmons are excited during the photoexcitation of surface atoms.

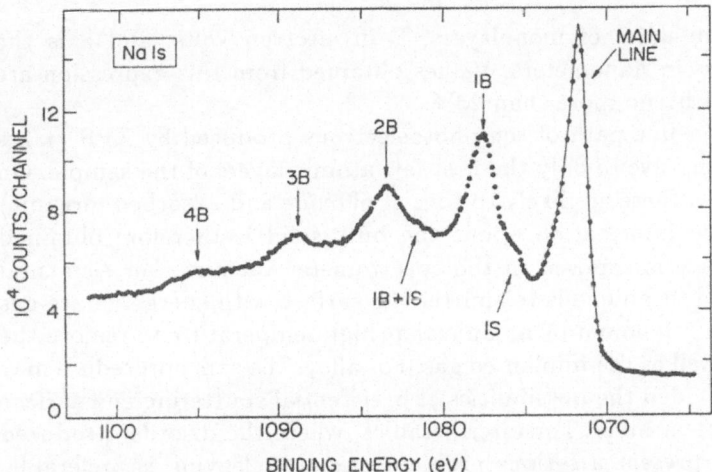

Fig. 7.2. Plasmon satellites of a core-electron emission line in metallic sodium. Note that both bulk, B, and surface, S, plasmons are readily identified [7.21b]

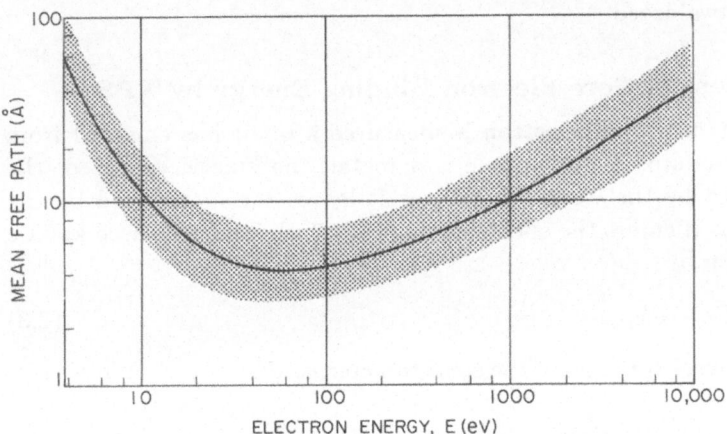

Fig. 7.3. The mean-free path of electrons in solids. Measurements in a wide range of materials lie well within the shaded band

The plasmon coupling is so strong that the mean-free path λ of ESCA electrons in a solid is generally no greater than ten lattice spacings. It is energy dependent with a minimum of 4 Å for kinetic energies in the vicinity of 50 eV. For larger kinetic energies it increases approximately as $\sqrt{E}$. It has been found empirically that the mean-free path of electrons of a given energy is of comparable magnitude in a wide range of materials [7.22,23]. This observation gave rise to the term "universal curve" to describe the energy dependence of λ shown in Fig. 7.3. In the absence of information about a particular material an empirical expression [7.24] can be used to estimate λ

$$\lambda = 538/E^2 + 0.41(aE)^{1/2} \tag{7.2}$$

where λ is given in units of monolayers, E in electron volts, and a is the monolayer thickness in nanometers. Values obtained from this expression are likely to be in error by no more than 25%.

The small mean-free path of the photoelectrons produced by XPS makes the spectroscopy sensitive to only the first few atomic layers of the sample. On an air-exposed metal these are likely to consist of oxide and adsorbed (organic) contaminants. Little information about the bulk solid is therefore obtained unless a fresh surface is exposed in the spectrometer vacuum. For elemental metals the preferred technique is to sputter the surface with energetic rare-gas ions (argon at 1 keV), followed by an anneal at high temperature to remove the lattice damage, as well as the implanted gas. For alloys the same procedure may be appropriate, provided the possibilities of preferential sputtering and surface segregation are kept in mind. For intermetallics, where the disorder produced by sputtering may present a serious problem, vacuum cleaving is preferable. Other common techniques are abrasion with a diamond or corundum file, and scraping with a tungsten carbide cutter. These serve to expose a fresh, but damaged surface. Whether they are adequate for a particular investigation must be carefully evaluated.

7.2.5 Measurement of Core-Electron Binding Energy by XPS

The kinetic energy of a photoelectron is measured after it has emerged from the sample into vacuum. If the sample is a metal, the kinetic energy of the electron is modified by the work function as it leaves the metal, and by the contact potential as it enters the spectrometer (Fig. 7.4). The measured kinetic energy is then given by

$$E'_{kin} = h\nu - E_B - e\Phi_{spec}, \tag{7.3}$$

where Φ_{spec} is the work function of the spectrometer, and $h\nu$ the photon energy.

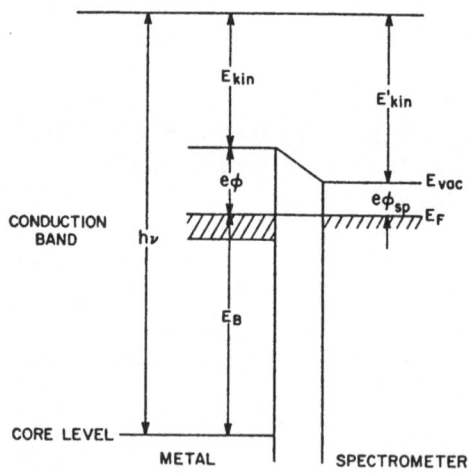

Fig. 7.4. Measurement of binding and kinetic energies of an electron photoexcited from a metal which is electrically connected to a spectrometer. Since the kinetic energy is measured in the spectrometer, its work function appears in the equation connecting the binding energy to the kinetic energy

The first ESCA instruments utilized Mg or Al K_α radiation with photon energies of 1253.6 or 1486.6 eV, respectively, to excite photoelectrons. These low-energy radiations were chosen because the photoelectric cross sections of core electrons decreases rapidly with increasing photon energy. Elements with lower Z, which would yield photons with even lower energy, either present technical problems, e.g., Na is not readily fashioned into a stable anode, or do not offer a narrow K_α emission line, e.g. in C the 2p level is part of the broad valence band.

Even Mg and Al radiations are far from ideal, however, because the 2p spin-orbit splittings are 0.3 and 0.4 eV, respectively. This splitting, combined with the lifetime width of the 1s hole state gives a complex line shape and limits the resolution to about 0.9 and 1.1 eV for Mg and Al radiations, respectively. In addition, the x-ray spectrum contains other line radiations, e.g., the $K_{\alpha3,4}$ and K_β emission, as well as the broad bremsstrahlung background. Each of the line radiations will produce its own photoemission line for each core electron in the sample. Various subtraction techniques have been developed to remove the resulting x-ray satellite lines from the spectrum. They produce qualitatively reasonable results, but are inadequate when a satellite overlaps a much weaker structure.

In order to overcome these problems instruments were built employing spherically bent quartz crystals to monochromatize and focus Al K_α radiation onto the sample (Fig. 7.5). Multielement detection is then often used to increase the counting rate to an acceptable level. The best resolution which has been obtained [7.25,26] corresponds to a Gaussian with 0.2 eV FWHM, (full width at half maximum) close to the theoretical limit set by the properties of quartz.

The measurement of the kinetic energy of the photoelectron is generally carried out with an electrostatic deflection analyzer with cylindrical or

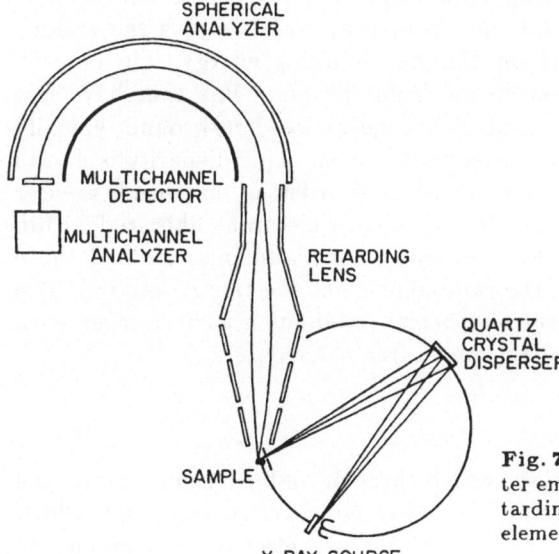

Fig. 7.5. Diagram of an ESCA spectrometer employing an x-ray monochromator, a retarding lens, a spherical analyzer, and multielement detection

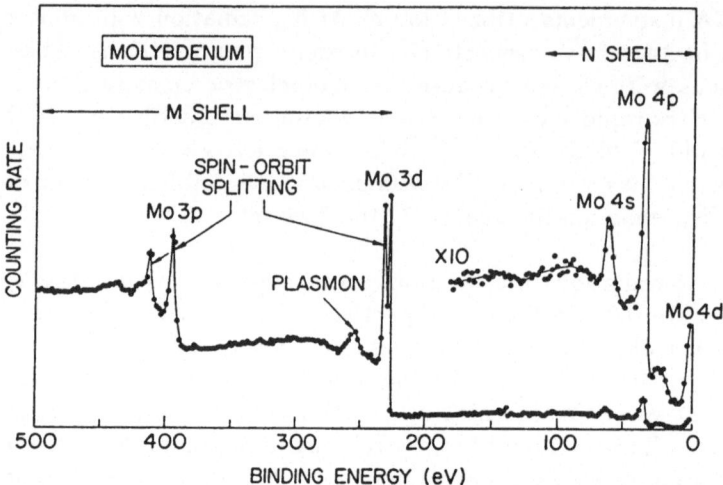

Fig. 7.6. A wide scan of the core-electron spectrum of Mo. The M shell cross sections are about an order of magnitude greater than those of the N shell. The energy loss tail exhibits some resolved plasmon excitations

hemispherical geometry. With a cylindrical analyzer detection is usually limited to electrons of a single energy. Spherical analyzers have an image plane in which kinetic energy is linearly related to a space coordinate. This makes them suitable for multielement detection, which can increase the rate of data acquisition by one or two orders of magnitude. For more details about these analyzers see *Barrie* [7.27].

A wide-scan photoemission spectrum produced by a spectrometer employing monochromatized Al K_α radiation and multielement detection is illustrated in Fig. 7.6. The emission from the localized core levels appears as sharp spikes. The spin-orbit splitting helps to identify the orbital character of the levels. The other feature of the spectrum that stands out is the sudden increase in the background level on the high-binding energy side of each emission line. It is due to photoelectrons from the main line that have lost energy before emerging from the sample. This energy-loss background typically has an area much greater than the emission line itself. This disparity is due to the fact that the absorption length for the x-rays is orders of magnitude greater than the mean-free path of the photoelectrons. Only electrons liberated within a few mean-free paths will contribute to the main photoemission line; those generated deep within the bulk of the sample produce the energy loss tail. The narrow main line, therefore, provides information about a surface layer with thickness defined by the photoelectron mean-free path.

7.3 Related Methods

There are many other techniques which provide information about the electronic states of metals. The oldest is x-ray emission spectroscopy which gave the first experimental information about core electrons. Most closely related to XPS is ultraviolet photoemission spectroscopy, including the recently developed angle-resolved version. Inverse photoemission and x-ray absorption edge spectroscopy make the empty band structure accessible. X-ray emission and Auger electron spectroscopies look at the deexcitation of a deep core hole by core and valence electrons. Electron energy loss spectroscopy is a very versatile technique which can be used to study core-electron states as well as vibrational spectra of adsorbates on metals. The essential features of these techniques are briefly outlined below. Others, including low-energy electron diffraction and extended x-ray absorption fine-structure spectroscopy [7.28], which provide spatial rather than electronic information, have been discussed in Chap. 6.

7.3.1 Angle-Resolved Photoemission Spectroscopy (ARPES)

This technique [7.29] represents the most sophisticated development of the well known discipline of ultra-violet photoelectron spectroscopy. It has produced the most detailed experimental information about the band structure of solids within a few eV of the Fermi surface. With a tunable photon source, an oriented single-crystal surface, and a rotatable electron spectrometer with small angular aperture one can determine the k-conserving transitions between occupied and empty bands. This makes it possible to map out the band structure and construct an E vs k diagram which can be directly compared with the results of band-structure calculations. In general, it is necessary to postulate that the empty states are free-electron-like in order to map the occupied states. The results which have been obtained by this technique are impressive, but it is of necessity restricted to materials which are available in the form of single crystals.

7.3.2 Inverse Photoemission Spectroscopy (IPES)

The unoccupied density of states can be studied by inverse photoemission [7.30–32], also known as bremsstrahlung isochromate spectroscopy. In this spectroscopy one detects the photon emitted when an energetic electron is captured into a normally empty state. In the usual implementation, photons of a fixed energy are detected as the energy of the incident electron is scanned. This technique has been used to detect the empty d-band states of the transition metals as well as the 4f multiplets produced when an electron is added to the 4f shell of a rare earth metal atom. It has quite general utility, but signal levels are low, especially in the angle-resolved mode.

7.3.3 X-Ray Absorption Edge Spectroscopy (XAS)

In XAS a tunable photon source is used to probe the empty band structure by exciting transitions from the core levels to states above E_F [7.33]. The threshold excitation energy in a metal provides a measure of the core-electron binding energy. Excitations to higher lying states map the empty band structure, very much like IPES. However, the final state in XAS, unlike that in IPES, contains a core hole. The data consequently do not correspond directly to the properties of the initial unoccupied density of states. They may, in fact, be quite strongly perturbed. For example, sharp lines may be obtained below the absorption edge due to the formation of a core exciton, i.e. a bound state involving a normally empty state which is pulled down below the Fermi energy by the Coulomb potential of the core hole. Even in the absence of this phenomenon the response at the absorption edge is modified by the screening response of the conduction electrons, which will change the shape of the edge in a way that depends on the orbital angular momentum of the core hole. The advantage of XAS lies in the possibility of discriminating between the various orbital components of the density of states (DOS) by using the dipole selection rule to populate selected components from s, p, d or f core levels.

7.3.4 X-Ray Emission Spectroscopy (XES)

X-ray emission is excited by bombarding the sample with energetic electrons which create core holes. Radiative transitions in which a core electron from a higher-lying state makes an allowed dipole transition into the initial state hole give rise to line radiation. It carries relatively little information about the nature of the solid because all core states tend to be shifted by similar amounts by chemical effects. The measured shifts are typically very small, less than 0.1 eV, because they depend on differences in the overlap between core and valence orbitals. Of greater interest are transitions from the conduction band, which give an image of that part of the occupied density of states which has dipole-allowed transitions to the core orbital. This spectroscopy is particularly useful in the study of alloys, because the contributions to the density of states of the various elemental constituents are separately obtained by looking at the emission from the core levels of each element [7.34]. A drawback of this technique is that self-absorption seriously distorts the valence band emission spectra. Moreover, the lifetime width and spin-orbit splitting of the initial state core hole is always folded into the spectrum. Significant problems also arise from many-body effects which change the shape of the emission edge. One of the advantages of the technique is that it gives information about the bulk solid, because the mean-free path of soft x-rays is large compared to a typical lattice spacing.

7.3.5 Auger Electron Spectroscopy (AES)

This technique (Chap. 8) uses the same method of excitation as XES, but focuses on the products of a non-radiative, two-electron decay process [7.35]. It usually dominates the radiative one, and is not limited by dipole selection rules. Like XPS, it depends on the detection of energetic electrons and is consequently surface sensitive. Its major advantages over XPS are speed and higher sensitivity for low Z elements. Auger signals are usually plotted out directly on a chart recorded. The disadvantage of this technique lies in the inherent complexity of the Auger final state, which has two interacting holes in outer orbitals. The nature of this interaction when both holes lie in the valence band has been elucidated only recently, as has the chemical shift of Auger electrons. Auger spectroscopy is widely used to characterize surfaces and to determine chemical composition.

7.3.6 Electron Energy Loss Spectroscopy (EELS)

This term is applied to two quite different types of experiment. When practiced with high energy electron (100 keV) it is closely related to x-ray absorption spectroscopy, except that one has the option of turning off the dipole selection rule by looking at transitions in which the incident electron is scattered through a finite angle. A great deal of useful information about the empty density of states and final state effects in core-electron excitation has been obtained by this method [7.36]. The general approach is to accelerate electrons to high kinetic energy, let them interact with a thin specimen, and then retard them back to low kinetic energy. Small changes in kinetic energy which took place during the interaction with the sample can then be measured with high accuracy.

The second embodiment of EELS is in the low-energy range, using electrons with kinetic energies of only a few electron volts. These are typically reflected from the surface of a metal in order to study the vibrational properties of an adsorbate [7.37]. Resolution of a few meV has been achieved.

7.4 Applications

7.4.1 Chemical Analysis

Core-electron spectra provide an immediate identification of the chemical constituents of a sample. Although Auger spectroscopy is more commonly used for elemental analysis because of its greater sensitivity for the common low-Z elements, this most primitive use of XPS is often of considerable value. Qualitative analysis is based on the fact that the core-electron binding energies of the elements are unique (Fig. 7.7). Tables of core-electron binding energies needed for this purpose, are available from various sources [7.2,7,8,38]. As a result an XPS scan including the outer shells, typically encompassing 0–500 eV in binding energy (Fig. 7.6) should allow unambiguous identification of the constituent elements. Any remaining doubt can be removed by verifying the spin-orbit splitting of the dominant p, d, or f photoemission line. Line widths and relative intensities are also helpful in making these identifications.

Quantitative analysis, based on the areas of XPS lines and known photoelectric cross sections, is also possible. Properly speaking, for this purpose the area of the line should include the intrinsic satellite structure. However, since the intrinsic and extrinsic satellite structures cannot be separated without a great deal of effort, it has been customary to use the area of only the main line in both calibration and measurement. (Calibration with a known standard is always necessary, because neither the x-ray flux nor the spectrometer efficiency are usually known on an absolute scale.) There is always a large margin for error, since there is no assurance that the ratio of the area in the main line to that of the intrinsic satellites is the same for all the lines used in both cases. Other factors which affect the intensities of photoemission lines are (1) the escape depth, (2) the energy-dependent spectrometer sensitivity, and (3) the

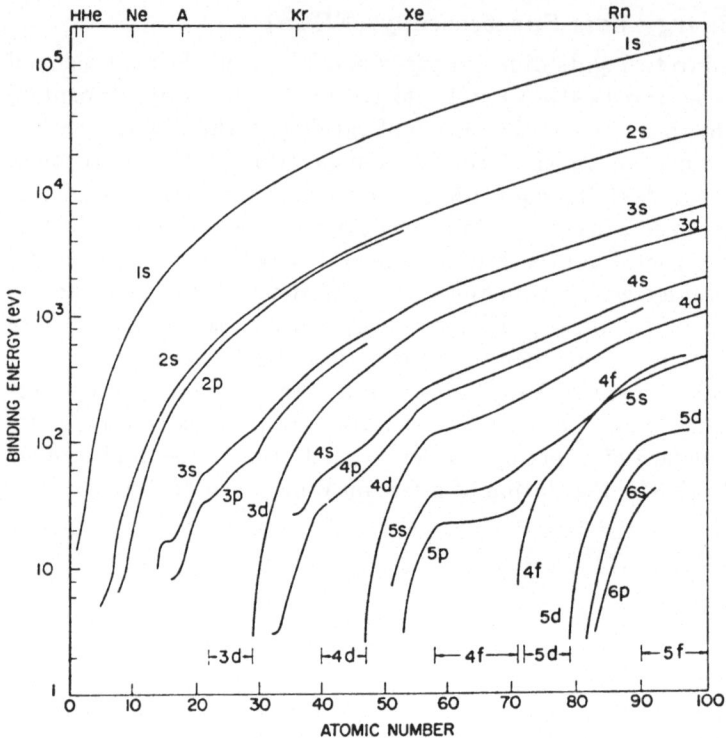

Fig. 7.7. The core-electron binding energies of the elements, plotted from [Ref. 7.2, Appendix I]. Note the logarithmic energy scale

angular-momentum-dependent photoelectron emission pattern. Corrections for all these should be made to obtain a meaningful quantitative analysis.

In practice, one is often content to measure only relative concentrations and make the assumption that all elements in a given material will have the same fractional intrinsic shake-up intensity. The area of the photoemission lines is then taken as that lying above an arbitrary sloping base line, excluding all of the shake-up structure. The photoelectric cross sections are usually taken from the calculations of *Scofield* [7.39], but tables of empirically adjusted values [7.40] are also widely used. These sources may differ by as much as a factor of two, which provides a worst-case estimate of the reliability of the tabular values. As a check on the reliability of such a determination, it should be repeated with a second core level. For more details on chemical analysis by XPS, see *Briggs* [7.41].

Chemical analysis by XPS samples a surface layer with thickness determined by the mean-free path of the photoelectrons. The surface sensitivity can be further enhanced by collecting only those electrons that leave the sample at a glancing angle. This technique can be used to good advantage in the studies of surface segregation in alloys and surface chemical reactions.

7.4.2 Binding Energy Shifts

Binding energy shifts are often considered to be the most useful parameters to emerge from ESCA. The early studies of molecular systems [7.2,3] suggest that one should attempt to relate these shifts to bonding and valence. When applied to metals, this approach encounters the problem that shifts between Fermi level referenced binding energies include the effect of the contact potential between the two substances. Since correction based on the work function are not reliable for the reason mentioned above, we will assume for the moment that the difference in chemical potential can be determined. The basic concept then is that increasing the valence charge on an atom in the *initial* state will decrease the binding energies of the inner core electron by the Coulomb potential (e^2/r), r being the mean radius of the added valence charge. An additional initial-state Coulomb term arises from the charge on neighboring atoms, or, more accurately, from the Madelung potential of all the other charges in the solid. This approach, which has been used with considerable success in ionic compounds in which charges are localized on lattice sites, is more difficult to implement in covalent solids or metals.

The measured binding energy shifts, however, also contain a contribution from the change in the *final*-state relaxation energy. In metals the dominant screening mechanism is the collective response of the conduction electrons. This relaxation energy can be separated from the initial-state Coulomb shift only with the help of some additional measurements. One approach is to measure the shift in both the XPS binding energy and in the Auger kinetic energy, and combine them in such a way that initial-state shifts cancel, leaving a measure of the relaxation energy [7.42]. This technique is based on the recognition that initial-state shifts are of Coulombic origin and affect all hole states in the same way, while final state relaxation shifts depend on the square of the final-state screening charge.

The shift in XPS **core-electron** binding energy between two chemical states may be written

$$\delta E_B = \delta E_{init} - \delta E_{relax}. \tag{7.4}$$

The shift in the Auger kinetic energy can analogously be written

$$\delta E_A = \delta E_{init} - \delta E_{relax} - (\delta E'_{init} - \delta E'_{relax}), \tag{7.5}$$

where the prime refers to the two-hole final state of the Auger process. For this two-hole state

$$\delta E'_{init} = 2\delta E_{init} \quad \text{and} \tag{7.6}$$
$$\delta E'_{relax} = 4\delta E_{relax}. \tag{7.7}$$

The factor of four arises from the interaction of the two holes of the final state with the two electron screening charge. It is valid provided the screening response is linear, a condition met in simple metals, but not in transition metals with strongly modulated densities of state. Combining these four equations, one obtains the following expression for the relaxation energy,

$$\delta E_{relax} = 0.5(\delta E_A + \delta E_B). \tag{7.8}$$

The quantity in parentheses is called the Auger parameter.

With the help of the Auger shift one is consequently able to determine the change in relaxation energy. The remaining shift contains all the initial-state effects discussed above. These will not be separable on the basis of experiment alone. This difficulty traces back to the fact that the analysis is based on a hypothetical valence charge, an ill-defined concept in a metal. Even when the spatial charge distribution in the solid has been calculated, atomic charges can be assigned only in terms of an arbitrarily chosen (Wigner-Seitz) cell. It is therefore not surprising that core-electron binding energy shifts also refuse to yield a value for the charge transfer.

A more useful approach to the interpretation of core-electron binding energy shifts in alloys is that of *Steiner* et al. [7.43], based on a Born-Haber cycle. It is closely related to the cycle discussed above which connects free-atom binding energies to those in metals through the cohesive energies. For alloys the analogous result shows that the shift depends on the heat of solution. It has been proposed that heats of solution which are difficult to obtain by thermodynamic means, could well be determined from core-electron binding energy shifts [7.44].

7.4.3 Valence Electron Density of States

XPS has enjoyed considerable success in delineating the densities of states of metal, alloys and compounds, see for example the valence band data of the gold intermetallics shown in Fig. 7.8. Here XPS competes directly with ultra-violet photoemission spectroscopy, but has the significant advantage that the spectrum is insensitive to the photon energy, because the photoelectron is excited into the high-energy continuum of empty states. As a result the data reflect the properties of only the *occupied* density of states (DOS) but with distortions which we will consider below.

An example, taken from early XPS studies of the Ag-Pd alloy system [7.45], illustrates the utility of XPS valence band structures. Data covering the entire range of composition show that the Ag and Pd d-band remain identifiable at all compositions (Fig. 7.9). The rigid-band model, based on a common d-band, is therefore inappropriate, even though it is capable of explaining many of the properties of the system. The high density of states at E_F is seen to disappear because the Pd d-band narrows, eventually to become a virtual bound-state resonance 2 eV below the Fermi level. A good theoretical description of this alloy system is obtained from the coherent potential approximation [7.46]. Core-electron binding energy shifts indicate that the charge transfer between the two components remains small.

Studies in the virtual bound-state regime are particularly interesting because they show not only the states contributed by the dilute Pd impurity, but also the loss of states in the Ag host [7.47]. By taking a difference spectrum between the dilute alloy and the pure host, normalized to identical areas, the

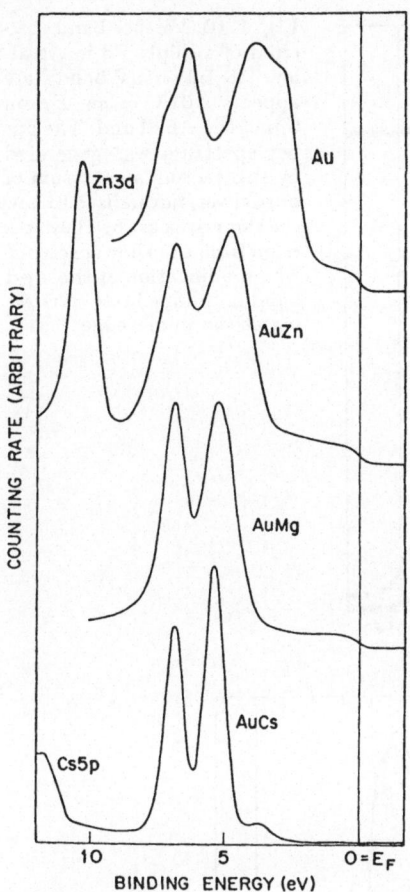

Fig. 7.8. Valence band spectra of gold and some of its alloys. Note the narrowing of the *d* band as the overlap between the *d* wavefunctions of adjacent Au atoms is reduced, and as the binding becomes more ionic. AuCs is a CsCl structure semiconductor, the others are metals

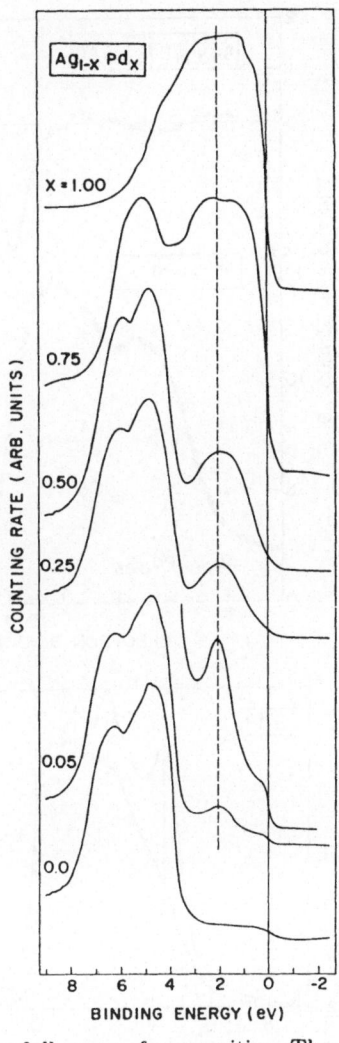

Fig. 7.9. Valence band spectra of the Ag-Pd alloys over the full range of composition. The system exhibits complete solid solubility

subtle changes produced by alloying become evident (Fig. 7.10). The dominant effect on the d-band of the Ag host is a loss of states from the upper part of the band, resulting in some slight band narrowing as the long-range order is disturbed by the impurity Pd.

Detailed comparisons between XPS valence band structures and theoretical DOS have generally yielded encouraging results. The cases of Pd, Ag, Pt, and Au (Fig. 7.11) show that most of the observed features agree with those in the band structure calculation [7.48]. However, if the agreement were measured in terms of a least-squares criterion, it would be judged to be poor. It is, of course, possible that the DOS are not reliable, but we shall see that most of the discrepancies arise from the photoemission process.

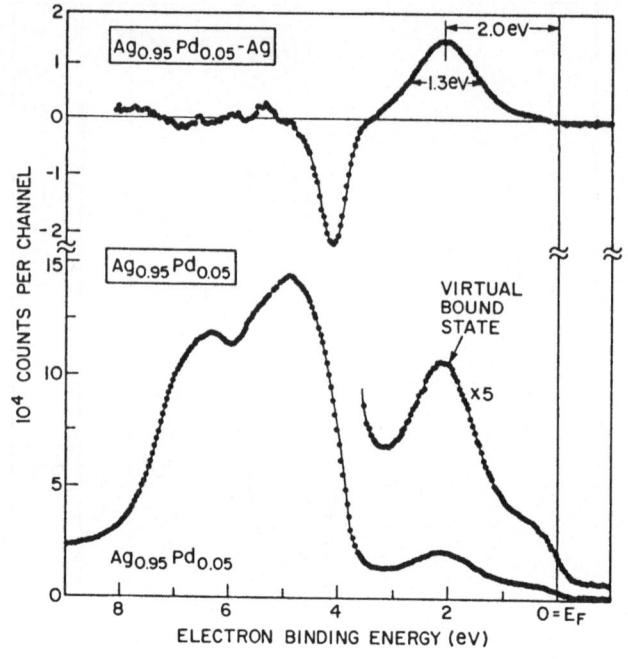

Fig. 7.10. Valence band spectrum of a dilute Pd in Ag alloy. The Pd virtual bond state appears 2.0 eV below E_F on top of the Ag s band. The upper spectrum was generated by subtracting a spectrum of pure silver, normalized to have the same area as the alloy spectrum from the alloy spectrum. The modification of the Ag d band is confined to a narrowing at the upper edge

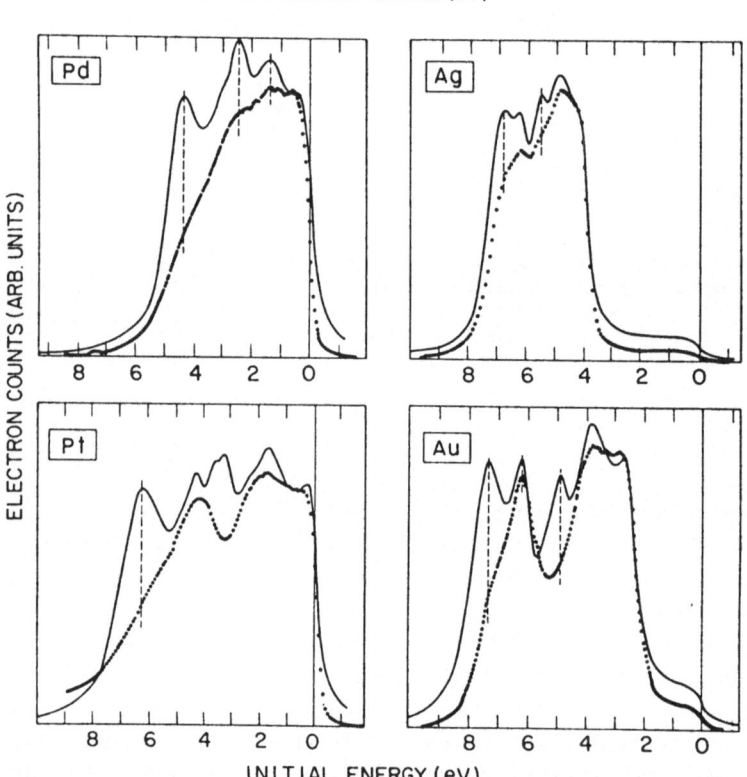

Fig. 7.11. Comparison of valence band XPS data for Pd, Ag, Pt, and Au with total densities of states [7.48]

The XPS response differs from the DOS because: (1) Different orbital states will be excited with distinct cross sections. The effects produced by this selectivity may be quite dramatic, generally favoring d and f states over s and p states. The cross sections of Scofield are of considerable help in assessing the extent of this effect. For example in the compound ReO_3 the cross section of the Re 5d states which contribute to both the valence and conduction bands is about 30 times larger than that of the O 2p states which one would expect to dominate in the valence band. As a result the XPS valence band spectrum corresponds much more closely to the Re 5d component of the DOS than to the total DOS [7.49]. Similarly in Au the cross section of a 5d electron is 10 times that of a 6s electron. As a result most of the signal from the 6s conduction band is due to the 5d admixture into that band. (2) The lifetime of the hole states in the valence band produced by photoemission will depend on their energy below E_F [7.50]. At the Fermi level only the broadening due to the Fermi function itself is seen [7.51]. With increasing energies below E_F the life time broadening becomes more and more important, resulting in a reduction of peak heights and a tailing out toward greater binding energy. (3) The data contain a background which rises toward greater binding energy. It is due in large part to electrons generated by photoemission deep within the sample which emerge only after having been inelastically scattered. However, it also contains the intrinsic shake-up and plasmon parts of the primary photoemission process. The background is therefore not necessarily featureless as is often assumed, but will contain structure due to the quantized excitations of the solid. (4) The data are broadened by the instrumental resolution function.

The agreement between band-structure theory and photoemission data can be greatly improved by incorporating these phenomena into the theoretical description. One proceeds in the following manner: Starting with the projected partial densities of states, weight them according to the transition probabilities appropriate for the photon energy employed, broaden the spectrum with a binding-energy-dependent lifetime Lorentzian, as well as with the known instrumental resolution function, and finally add a representation of the plasmons and the inelastic background to approximate the collective effects and the spectrum of degraded electrons.

Successful though it may be, this methodology does not properly introduce the collective and final-state aspects of the photoemission process. These effects are particularly dramatic for the localized f and strongly correlated d states.

The satellites in the valence band of Ni have been shown to be part of the intrinsic photoexcitation response [7.52,53]. Even the valence band spectra of insulators containing transition metal ions have now been shown to require a full configuration interaction treatment to elucidate the origin of the complex structure which is observed [7.54]. Finally, the 4f spectra of the rare-earth metals are well known to exhibit the multiplets of the $4f^{n-1}$ final state populated by the removal of an f electron, and consequently bear no resemblance to the $4f^n$ initial state [7.55,56]. With increasing localization, final-state effects thus become more and more significant. It is therefore clear that valence-band photoemission spectra do not, in general, correspond in detail to the density of states or even to the orbitals of the initial state.

For the delocalized states the nature of the band itself is generally reproduced with reasonable accuracy, even if many-body features distort and add structure to the high binding energy side. A rigorous treatment demands a major theoretical effort. A full many-body calculation has so far been implemented only for a simple free-electron metal, where satisfactory agreement with data for Na was found [7.57]. The shape of the conduction band remains similar to the DOS, but merges into the intrinsic plasmon structure at the bottom of the band. Consequently, it is usually assumed that XPS data for delocalized bands can be profitably compared with densities of states derived from band-structure calculations. Historically this procedure has enjoyed considerable success, and the failures have ultimately served to point to cases where collective effects cannot be ignored.

7.4.4 Conduction-Electron Screening

The collective many-body response of the conduction electrons to the creation of a fixed charge in the lattice, namely the core hole, manifests itself in all core-electron spectroscopies. When its effects were first observed in the shape of x-ray absorption edges of metals, it was termed the x-ray absorption edge anomaly. A theoretical explanation was given by *Mahan* [7.58] 25 years after the original observation. For XPS the effect of screening on core-electron line shapes was predicted theoretically by *Doniach* and *Šunjić* [7.59] some years before it was recognized in experimental data [7.60].

Screening serves to lower the energy of the hole state, even though it requires the excitation of electron-hole pairs at the Fermi surface, a process which consumes energy. The spectrum of these excitations appears as a tail on the high-energy side of the core electron photoemission line. The energy of the optimally screened state defines the core-electron binding energy in a metal. According to the work of *Mahan* [7.58], and *Doniach* and *Šunjić* [7.59] the many-body tail has the form of a one-sided singularity:

$$I(\omega) = \hbar(\omega - \omega_o)^{\alpha - 1}; \quad \omega > \omega_o \tag{7.9}$$

where ω_o is the energy of the optimally screened state, and α is the singularity index. For a free-electron metal α is simply related to the Friedel phase shifts, δ_l, which are normalized so that the screening charge, Z, is given by

$$Z = 2 \sum_{l=0}^{\infty} (2l + 1)(\delta_l/\pi) \quad \text{and } \alpha \text{ by} \tag{7.10}$$

$$\alpha = 2 \sum_{l=0}^{\infty} (2l + 1)(\delta_l/\pi)^2. \tag{7.11}$$

It can be readily shown that for a free-electron metal α can be no greater than 0.5.

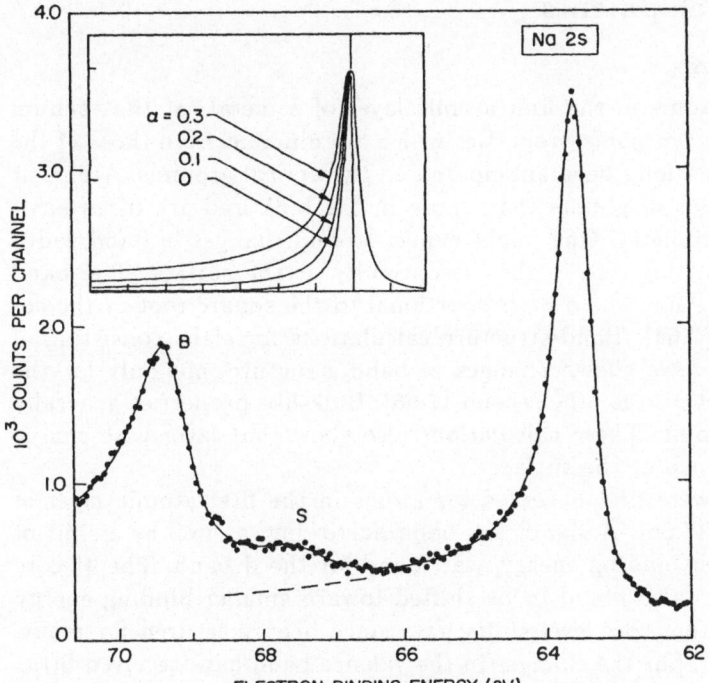

Fig. 7.12. The asymmetric 2s core-electron spectrum of metallic Na, fitted with the many-body line shape. The inset shows the α dependence of the line shape

In order to facilitate comparison with experimental spectra the effect of the finite lifetime of the core hole must be included in the experimental description. By folding a Lorentzian into the singularity, Doniach and Šunjić obtained

$$I(\omega) = \frac{\cos[\pi\alpha/2 + (1 - \alpha)\,\arctan(\varepsilon/\gamma)]}{(\varepsilon^2 + \gamma^2)^{(1-\alpha)/2}} \qquad (7.12)$$

where $\varepsilon = \hbar(\omega - \omega_0)$, and γ is the half width at half height of the Lorentzian. A plot of this function for various values of α is shown as an inset in Fig. 7.12.

This formula has proved to be successful in representing experimental photoemission lines from metals [7.61,62], and the singularity index α has been found to be in accord with calculated scattering phase shifts for the simple metals [7.63]. Figure 7.12 shows the Na 2s XPS line, least-squares fitted with the above equation. The bulk and surface plasmons were represented by the same equation, but with larger Lorentzian widths. The singularity index was found to be 0.20, in good agreement with theoretical evaluations [7.63].

7.5 Recent Developments

7.5.1 Surface Atoms

The fact that the atoms in the first atomic layer of a metal (at the vacuum interface) must have electronic properties which are different from those of the atoms in the bulk had long been anticipated on theoretical grounds. Atoms at the surface have fewer neighbors than those in the bulk and are in an environment of lower symmetry. One might expect to find changes in band width and crystal field splitting due to the discontinuity at the surface. The band width W has been estimated to be proportional to the square root of the co-ordination number [7.64]. Band-structure calculations for slabs consisting of a few atomic layers have shown changes in band structure, not only for the first atomic layer but also for the second [7.65]. Bulk-like properties generally exist beyond that point. These calculations also show that layer-wise charge neutrality is maintained at the surface.

Surface effects were first observed for atoms in the first atomic layer of polycrystalline gold [7.66]. A significant band narrowing, as well as a shift of 0.5 eV toward smaller binding energy, was found for the d band. The 4f core levels of the Au were also found to be shifted toward smaller binding energy by 0.4 eV. Surface-atom core-level shifts have since been measured for many other elements [7.67], but the changes in the valence band have received little attention from XPS, because of the overlap of bulk and surface features.

The changes in band structure at the surface manifest themselves in the core electron binding energies, because, as noted earlier, changes in the radial distribution of valence charge shift core electrons by a Coulomb mechanism. In the noble metals changes in band width and hybridization at the surface cause a transfer of charge from extended s bands to more localized d bands,

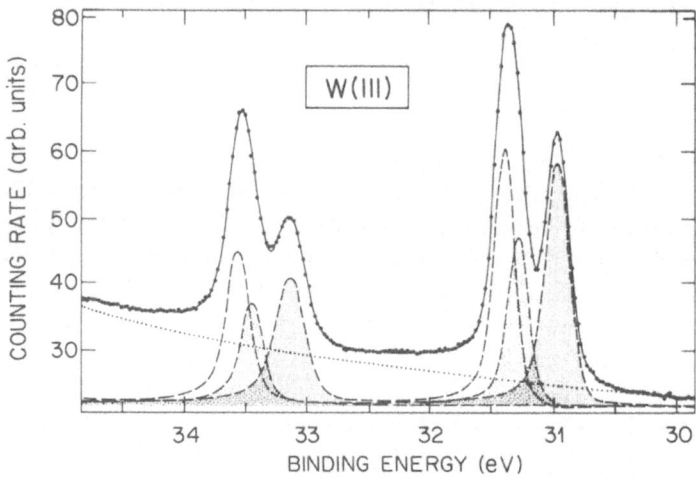

Fig. 7.13. The W 4f core-electron spectrum taken with synchrotron radiation at high resolution. The data have been fitted with three components in each spin-orbit line, all having the same many-body line shape. The shaded component at smallest binding energy is due to the atoms in the surface layer. The weakest component is due to subsurface atoms

accounting for the decrease in core-electron binding energy. Shifts for both the first and second atomic layers have been resolved in surfaces with relatively open structure (Fig. 7.13) [7.68], while on close-packed surfaces only a single shift is usually resolved. It has also been shown that these surface-atom shifts have distinct values for the different crystallographic surfaces [7.69]. The ability to resolve the surface atoms has made possible detailed studies of the effects of adsorbed gases and chemical reactions at the surface [7.70].

Surface-atom core-level shifts have been successfully interpreted using a Born-Haber cycle [7.71] in which surface effects arise from a reduction of the cohesive energy due to the lower coordination number. A reduction to 80% of the bulk value predicts surface atom core level shift of the transition metals with reasonable accuracy. For the transition metals, surface atom core level shifts are positive in the first half of the d-group and negative for the second half.

7.5.2 Metal Clusters

A natural extension of surface atoms studies leads to small metal clusters. In these systems the average coordination number can be reduced much more drastically than at the surface. The study of random-sized, supported clusters of noble and transition metals has made considerable progress in the past decade. The anticipated valence band narrowing has been confirmed for a number of different metals, including Pd (Fig. 7.14) [7.71], Ag [7.73], and Au [7.74]. One

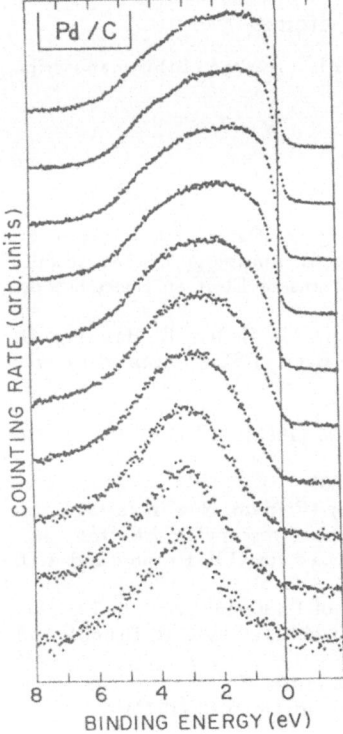

Fig. 7.14. Valence-band spectra of Pd clusters supported on amorphous carbon. Starting with the smallest clusters at the bottom, the cluster size (mass) doubles between successive spectra. The top spectrum corresponds closely to that of bulk Pd

215

surprising result is that the **core-electron** binding energy shifts were all found to be positive for clusters supported on insulating or semimetallic substrates, even for metals like Au and Pt where the surface atom shifts are negative. It has now been shown that these positive shifts are due to an entirely different mechanism, one not operative in surface layers of bulk metals [7.75]. It arises from the charge left on the cluster by the ejection of the photoelectron. This produces a macroscopic Coulomb shift of magnitude inversely proportional to the radius of the cluster. In smaller clusters there is a cross-over to a regime where conduction electron screening is replaced by local screening [7.76], resulting in a positive, size-independent core-electron binding energy shift. This occurs even before the loss of metallic properties in the smallest clusters [7.77]. Band-structure and rehybridization effects, similar to those observed for surface atoms, must also contribute to the shifts in clusters, but are apparently overshadowed by the charging and screening phenomena.

The transition from the isolated atom to the few-atom cluster is of special interest. Recent experimental developments utilizing a nozzle expansion technique, have made possible studies of metal atom clusters of uniform size, ranging from a few atoms up to a few hundred atoms [7.78,79]. So far these studies have been restricted to clusters in flight, and photoemission measurements have been limited to the measurement of the threshold ionization potential. The preparation of samples containing clusters of a single size on a substrate or in a matrix is currently under consideration. Such samples will make possible much more detailed studies of the development of band structure and metallic properties in small aggregates of atoms.

Acknowledgements. I am indebted to S.B. DiCenzo for a critical reading of this manuscript, as well as for many helpful suggestions.

References

7.1   A. Einstein: Ann. Physik **17**, 132 (1905)
7.2   K. Siegbahn, C. Nordling, A. Fahlman, R. Nordberg, K. Hamrin, J. Hedman, G. Johansson. T. Bergmark, S.E. Karlsson, I. Lindgren, B. Lindberg: ESCA; Atomic, Molecular, and solid State Structure Studies by Means of Electron Spectroscopy, Nova Regia Soc. Sci. Upsaliensis, Sev IV, **20** (1967)
7.3   K. Siegbahn, C. Nordling, G. Johansson, J. Hedman, P.F. Heden, K. Hamring, U. Gelius, T. Bergmark, L.O. Werme, R. Manne, Y. Baer: ESCA, *Applied to Free Molecules* (North Holland. Amsterdam 1969)
7.4   T. Koopmans: Physica **1**, 104 (1934)
7.5   B. Lundqvist: Phys. Kondens. Mater. **6**, 193 and 206 (1967); **7**, 117 (1968); **9**, 236 (1969)
7.6   R. Manne, T. Aberg: Chem. Phys. Lett. **7**, 282 (1970)
7.7   T.A Carlson: *Photoelectron and Auger Spectroscopy* (Plenum, New York 1975)
7.8   C.S. Fadley: "Basic Concepts of X-ray Photoelectron Spectroscopy" in *Electron Spectroscopy, Theory. Techniques and Applications*, Vol. 2, ed. by C.R. Brundle and A.D. Baker (Academic, New York 1978) Chap.1, pp.1-145
7.9   R.L. Martin, D.A. Shirley. "Many Electron Theory of Photoemission", in *Electron Spectroscopy, Theory, Techniques and Applications*, Vol. 1, ed by C.R. Brundle and A.D. Baker (Academic, New York 1977)
7.10   A.R. Williams, N.D. Lang: Phys. Rev. Lett. **40**, 954 (1978)
7.11   D.D. Gelatt, M. Ehrenreich, R.E. Watson: Phys. Rev. B **15**, 1613 (1977)

7.12   C. Herring. M.H. Nichols: Rev. Mod. Phys. **21**, 185 (1949)
7.13   B. Johansson, N. Mårtensson: Phys. Rev. B **21**, 4427 (1980)
7.14   J. Friedel: Comm. Solid State Phys. **2**, 21 (1969)
7.15   M. Combescott, P. Nozieres: J. Phys. (Paris) **32**, 913 (1971)
7.16   G. Crecelius, G.K. Werthein, D.N.E. Buchanan: Phys. Rev. B **18**, 6519 (1978)
7.17   G.K Wertheim: (unpublished)
7.18   Since the valence band is made up largely of anion states while the conduction band
        has metal character, the satellite can be described as a charge-transfer state.
7.19   G.K. Wertheim, A. Rosencwaig: Phys. Rev. Lett. **26**, 1179 (1971)
7.20   S. Svensson, N. Mårtenson, E. Basillier, P.A. Malmqvist, U. Gelius, K. Siegbahn:
        Phys. Scr. **14**, 141 (1976)
        G. Wendin, M. Ohno: ibid **14**, 148 (1976)
7.21   P. Steiner, H. Höchst, S. Hüfner: Simple Metals in *Photoemission in Solids II*, ed. by
        L. Ley and M. Cardona, Topics Appl. Phys., Vol. 27 (Springer, Berlin, Heidelberg
        1979), Chap.7, pp. 357-464
        P.H. Citrin: Phys. Rev. B **8**, 5545 (1973)
7.22   C.J. Powell: Surf. Sci. **44**, 29 (1974)
7.23   I. Lindau, W.E. Spicer: J. Electron Spectrosc. **3**, 417 (1974)
7.24   M.P. Seah, W.A. Dench: Surf. Interface Anal. **1**, 2 (1979)
7.25   Y. Baer, G. Bush, P. Cohn: Rev. Sci. Instr. **46**, 466 (1975)
7.26   K. Siegbahn: Science **217**, 111 (1982)
7.27   A. Barrie: "Instrumentation for Electron Spectroscopy" in Handbook of X-ray and
        Ultraviolet Photoelectron Spectroscopy, ed. by. D. Briggs, (Heyden, London 1977)
        p.79
7.28   A Bianconi, L. Incoccia, S. Stipcich (eds): *EXAFS and Near Edge Structures*,
        Springer Ser. Chem., Vol. 27 (Springer, Berlin, Heidelberg 1983)
        K.O Hodgson, B. Hedman, J.E. Penner-Hahn (eds): *EXAFS and Near Edge Struc-
        tureIII*, Springer Proc. Phys. Vol. 2 (Springer, Berlin, Heidelberg 1984)
7.29   N.V. Smith: "Angular Dependent Photoemission" in *Photoemission in Solids I*, ed.
        by M. Cardona and L. Ley, Topics Appl. Phys., Vol. 26 (Springer, Berlin, Heidelberg
        1978) Chap.6, pp.237-263
7.30   V. Dose: Prog. Surf. Sci. **13**, 225 (1983)
7.31   J.K. Lang, Y. Baer: Rev. Sci. Instru. **50**, 221 (1979); Solid State Commun. **31**, 945
        (1979)
7.32   N.V. Smith: Vacuum **33**, 803 (1983)
7.33   D.J Fabian, L.M. Watson, C.A.W. Marshall: "Soft X-ray Spectroscopy and the
        Electronic Structure of Solids", Rept. Progr. Phys. **34**, 601 (1972)
7.34   C. Bonnelle, C.Mande (eds.): *Advances in X-ray Spectroscopy* (Pergamon, Oxford
        1982)
7.35   H. Ibach (ed.): *Electron Spectroscopy for Surface Analysis*, Topics Current Phys.,
        Vol. 4 (Springer, Berlin Heidelberg 1977)
7.36   S.E. Schnatterly: "Inelestic Electron Scattering Spectroscopy" in *Solid State Physics*
        (Academic, New York 1979) p. 275-358
7.37   H. Froitzheim: "Electron Energy Loss Spectroscopy", in [Ref.7.35, Chap.6]
7.38   J.C. Fuggle, N. Mårtensson: J. Electron Spectr. **21**, 275 (1980)
7.39   J.H. Scofield: J. Electron Spectrosc. **8**, 129 (1976)
7.40   R.C.G. Leckey: Phys. Rev. A **13**, 1043 (1976)
        V.I. Nefedov, N.P. Sergushin, I.M. Band, M.B. Trzhaskovskaya; J. Electron Spectrosc.
        2, 383 (1973); **7**, 175 (1975)
7.41   D. Briggs: "X-Ray Photoelectron Spectroscopy as an Analytical Technique" in *Hand-
        book of X-ray and Ultra-violet Photoelectron Spectroscopy*, ed. by D. Briggs (Hey-
        den, London 1977) p.153
7.42   C.D. Wagner: "The Role of Auger Lines in Photoelectron Spectroscopy", in *Hand-
        book of X-Ray and Ultra-violet Photoelectron Spectroscopy*, ed. by D. Briggs (Hey-
        den, London 1977) p.249
7.43   P. Steiner, S. Hüfner, N. Mårtensson, B. Johansson: Solid State Commun. **37**, 73
        (1981)
7.44   P. Steiner, S. Hüfner: Solid State Commun. **37**, 79 (1981);
        N. Mårtensson, R. Nyholm, H. Calen, J. Hedman, B. Johansson: Phys.Rev. B **24**,
        1725 (1981)

7.45 S. Hüfner, G.K. Wertheim, J.H. Wernick, A. Melera: Solid State Commun. **4**, 259 (1972)
      S. Hüfner, G.K. Wertheim, J.H. Wernick: Phys. Rev. B **8**,4511 (1973)
7.46 P. Soven: Phys. Rev. **156**, 809 (1967); **178**, 1136 (1969);
      B. Velicky, S. Kirkpatrick, H. Ehrenreich: Phys. Rev. **175**, 747 (1968)
7.47 S. Hüfner, G.K. Wertheim, J.H. Wernick: Solid State Commun. **17**, 1585 (1975)
7.48 N.V. Smith G.K. Wertheim. S. Hüfner, M.M. Traum: Phys. Rev B **10**, 3197 (1974)
7.49 G.K. Wertheim, L.F. Mattheiss, M. Campagna, T.P. Pearsall: Phys. Rev. Lett. **32**, 997 (1974)
7.50 H. Höchst, S. Hüfner, A. Goldmann: Phys. Lett. **57A**, 265 (1976)
      H. Höchst, A. Goldmann, S. Hüfner: Z.Physik B **24**, 245 (1976)
7.51 G.K.L. Cranstoun, R.G. Egdell, M.D. Hill, R. Samson: J. Electron Spectrosc. **33**, 23 (1984)
7.52 G.C. Tibbetts, W.F. Egelhoff, Jr: Phys. Rev. Lett. **41**, 188 (1978)
7.53 D.R. Penn: Phys. Rev. Lett. **42**, 921 (1979)
7.54 Y. Kakehashi, A. Kotani: Phys. Rev. B **29**, 4292 (1984);
      A. Fujimori, F. Minami: Phys. Rev. B **30**, 957 (1984)
7.55 M. Campagna, G.K. Wertheim, Y. Baer: "Unfilled Inner Shells: Rare Earths and their Compounds", in *Photoemission in Solids II*, ed. by L. Ley and M. Cardona, Topics Appl. Phys., Vol. 27 (Springer, Berlin, Heidelberg 1979) Chap.4, pp.217-360
7.56 K. Lang, Y. Baer, P.A. Cox: J. Phys. F **11**, 121 (1981)
7.57 S. Prutzer, S.M. Bose: Phys. Rev. **B31**, 762 (1985)
7.58 G.D. Mahan: Phys. Rev. **169**, 612 (1967)
7.59 S. Doniach, M. Šunjić: J.Phys. C **3**, 285 (1970)
7.60 S. Hüfner, G.K. Wertheim, D.N.E. Buchanan, K.W. West: Phys. Lett. **46**A, 420 (1974)
7.61 G.K. Wertheim, S. Hüfner: Phys. Rev. Lett. **35**, 53 (1978)
7.62 G.K. Wertheim, P.H. Citrin: "Fermi Surface Excitations in XPS Line Shapes from Metals", in *Photoemission in Solids I*, ed. by M. Cardona and L. Ley, Topics Appl. Phys., Vol. 26 (Springer, Berlin, Heidelberg 1978) Chap.5, pp.197-234
7.63 P.H. Citrin, G.K. Wertheim, Y. Baer: Phys. Rev. Lett. **35**,885 (1975); Phys.Rev. B **16**, 4256 (1977)
7.64 F. Cyrot-Lackmann: Adv.Phys. **16**, 393 (1967); J. Phys. Chem. Solids **29**, 1235 (1968)
7.65 J.A. Appelbaum, D.R. Hamann: Solid State Commun. **27**, 882 (1978)
7.66 P.H. Citrin, G.K. Wertheim, Y. Baer: Phys. Rev. Lett. **41**, 1425 (1978), Phys. Rev. B **27**, 3160 (1983)
7.67 P.H. Citrin, G.K. Wertheim: Phys. Rev. B **27**, 3176 (1983)
7.68 J.F. Van der Veen, P. Heinmann, F.J. Himpsel, D.E. Eastman: Solid State Commun. **37**, 555 (1981)
7.69 J.F. van der Veen, F.J. Himpsel, D.E. Eastman: Phys. Rev. Lett. **44**, 189 (1980)
      P. Heinmann, J.F. van der Veen, D.E. Eastman: Solid State Commun. **38**, 595 (1981)
7.70 T.M. Duc, G. Guillot, Y. Lasailly, J. Lecante, G. Jugnet. J.C. Vedrine: Phys.Rev.Lett. **43**, 789 (1980)
7.71 A. Rosengren, B. Johansson: Phys. Rev. B **22**, 3706 (1980)
      A. Rosengren: Phys.Rev. B **24**, 7393 (1981)
7.72 R. Unwin, A.M. Bradshaw: Chem. Phys. Lett. **58**, 58 (1978)
7.73 G.Apai, S-T. Lee, M.G. Mason: Solid State Commun. **37**, 213 (1981)
7.74 M.G. Mason: Phys.Rev. B **27**, 748 (1983)
7.75 G.K. Wertheim, S.B. DiCenzo, S.E. Youngquist: Phys. Rev. Lett. **51**, 2310 (1983)
7.76 G.K. Wertheim, S.B. DiCenzo, D.N.E. Buchanan, P.A. Bennett: Solid State Commun. **53**, 377 (1985)
7.77 M. Cini: Surf. Sci. **62**, 148 (1977)
7.78 A. Herrmann, E. Schumacher, L. Woste: J. Chem Phys. **68**, 2327 (1978)
7.79 A. Schmidt-Ott, P. Schurtenberger, H.C. Siegmann: Phys. Rev. Lett. **45**, 1284 (1980)

8. Auger Electron Spectroscopy

M.P. Seah
With 22 Figures

Auger electron spectroscopy is the most general and popular of the current surface analysis methods. What do we mean by surface analysis? Typically we mean the chemical analysis of the outermost atom layers at a free surface and, sometimes, in addition, the analysis of the atom layers below the surface to a depth of $1\,\mu m$ or so. The free surface is simply the surface that we see and is concerned, here, mainly with solid metals. Internal surfaces or interfaces may also be studied if they can be converted to free surfaces for the analysis. We shall discuss this later in Sect. 8.5.1.

The generality of surface analysis may be judged by the statistics of there being considerably more than 1000 instruments currently operating in over 37 countries, with commercial instruments typically retailing at £100 000 to £300 000. The areas of industrial application are shown by the boxed lists in Fig. 8.1. Each list is associated with a sector of the pie diagram which, in turn,

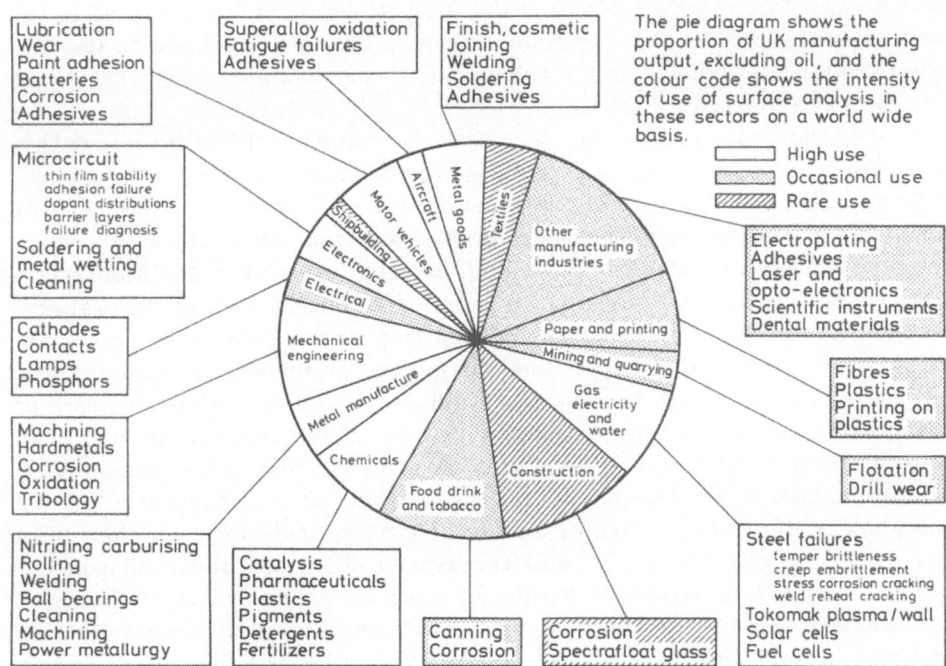

Fig. 8.1. Topics of applications of surface analysis and their relation to the manufacturing sectors of the UK [8.1]

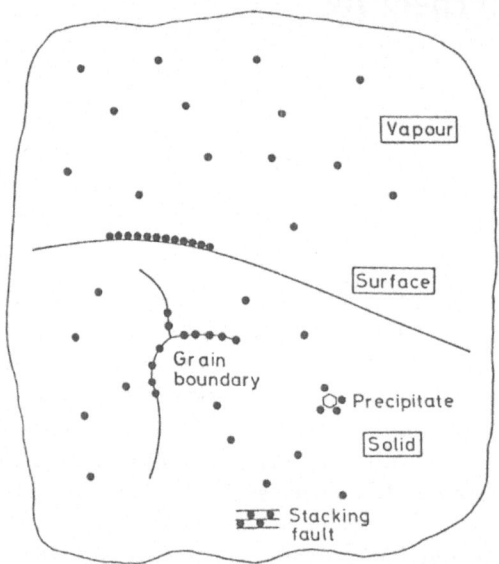

Fig. 8.2. Concentrations of solute atoms in a solid containing defects and in its environment [8.2]

Labels in figure: Vapour, Surface, Grain boundary, Precipitate, Solid, Stacking fault

shows the manufacturing contribution to UK output. The analysis shown in Fig. 8.1 relates to the three main surface analysis techniques, Auger electron spectroscopy (AES), x-ray photoelectron spectroscopy (XPS) and secondary ion mass spectroscopy (SIMS). The major effort, 53%, is in AES, with 33% in XPS and 14% in SIMS. Many of the AES studies have been focused in the areas of metallurgical and electronic research. This is not related to the way the technique developed, that development, as we shall see in Sect. 8.2, being somewhat haphazard.

Why should the precise chemistry at surfaces and interfaces or through thin films differ from that of the bulk and therefore be interesting? Figure 8.2 shows schematically the situation of a polycrystalline solid A, with certain defects and with a free surface. When held in an isothermal enclosure with a small quantity of a second element B we find, at equilibrium after the atoms have been allowed to redistribute, that B atoms are concentrated at dislocations, stacking faults, grain boundaries and the free surface. The presence of the B atoms lowers the dislocation line energy, the grain boundary energy, the surface energy, the grain boundary self-diffusivity and may either increase or decrease the grain boundary cohesion [8.2–4]. These changes, in turn, can have a strong detrimental or beneficial effect on the brittle fracture of steels and other metals, on densification in the sintering of metal powders, on the morphology of surfaces, cavities and second phase inclusions, on the life of barrier layers, etc. Figure 8.2 and the aspects discussed above all concern atomic composition variations within an atom layer or so of the defect site. Variations over longer distances occur due to interdiffusion at phase interfaces and reactions at surfaces, the most common of the latter being the oxidation and corrosion processes.

With modern instrumentation, AES has developed into a very powerful analytical technique to study problems associated with both the surface/interface monolayer and the surface/interface zone. These will be illustrated in Sect. 8.5. Let us first go back to the roots of the technique and see why the technique has evolved the way it has.

8.1 History

The start of Auger electron spectroscopy must be taken as the observation of fixed-length tracks in a cloud-chamber experiment by *Auger* in 1925 [8.7]. The interpretation of this phenomenon concerns the electron core level energies in atoms as shown in Fig. 8.3. The alternative notations for describing these levels are also shown as readers from different disciplines may be used to different schemes [8.8,9]. The energies of the core levels differ for each element but are well known from x-ray measurements and are thoroughly tabled [8.10]. First a core level electron is ejected from a level of binding energy E_A by some ionizing radiation (exampled by the K shell in Fig. 8.3). The excited atom then relaxes when an electron from a higher level E_B (e.g., L_1) falls into the unoccupied state. The released quantum of energy $E_A - E_B$ is then either emitted as an x-ray or is given to a third electron of binding energy E_C (e.g., L_3) so that electron is emitted from the atom with an energy E_{ABC}, given approximately by

$$E_{ABC} = E_A - E_B - E_C. \tag{8.1}$$

The characteristic x-ray emitted above forms the basis of the well known electron probe x-ray microanalysis technique described by *Reed* [8.11] and *Zaluzec* [8.12]. The electron emitted with the characteristic energy E_{ABC} is the electron first observed by Pierre Auger and is now termed an Auger electron. In

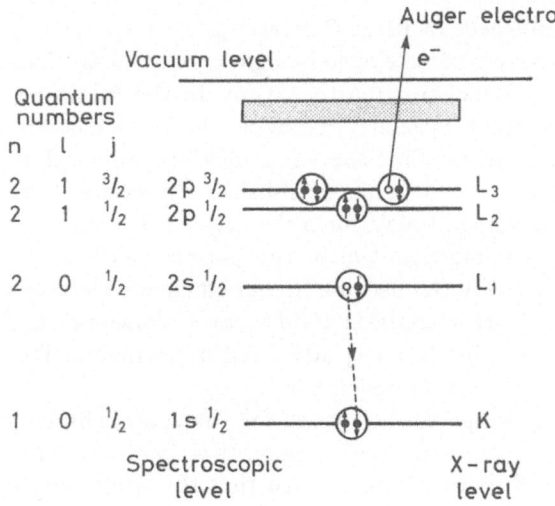

Fig. 8.3. Core-level notation in atoms and the principal and other quantum numbers in relation to a typical Auger electron process

Auger electron spectroscopy the ionizing radiation is usually a beam of electrons in the energy range 2 to 20 keV and the emitted Auger electrons have energies in the range 20 to 2000 eV. The eV unit of energy used here is the electron-volt and is the energy gained by an electron on being accelerated across a potential of one volt $(1 eV = 1.602 \times 10^{-19} J)$.

Although Auger observed these electrons from atoms in the gas phase at a very early date, no real advance in Auger electron spectroscopy for surface studies came until 1967. Some work on higher energy Auger electrons occurred in the 1920's and 1930's but it was not until 1953 that *Lander* [8.13] proposed their use for surface analysis. It was important to recognise that, for surface studies, the lower-energy Auger electrons should be used since these suffer intense inelastic scattering in solids and only survive to be ejected and measured if originating in the outermost atom layers. Also, in the emitted electron energy spectra the Auger electron peaks are small in comparison with the overall secondary electron background. In 1967 these problems were recognised by *Harris* [8.14a] who built and operated the first practical AES system incorporating an electron beam, a sample manipulator and a low-energy electron spectrometer. Two further important aspects were also incorporated. To ensure that surfaces prepared in the system remained unaffected by the vacuum environment, the apparatus was mounted in a full bakeable ultra-high vacuum system. These stainless steel systems, after baking to 200°C or so for 24 hours, achieve pressures of the order of 10 nPa in which a fully day's working is generally possible before significant contamination occurs. The second aspect introduced by Harris was the use of potential modulation on the spectrometer to obtain the differential of the energy spectrum. This differentiated form enabled the small Auger electron peaks to be more clearly observed and measured.

Thus, in one step, Harris had established the main aspects of the analytical technique as we know it today. In a consecutive publication *Harris* [8.14b] went on to demonstrate the use of the technique in monitoring the surface segregation of impurities in iron and nickel based alloys. A number of important phenomena were now rapidly established. In 1968, *Palmberg* and *Rhodin* [8.15] showed, by depositing single-atom layers on to cleaned single-crystal faces, that AES characterized a surface zone of thickness 0.5 to 1.0 nm. In the following year *Marcus* and *Palmberg* [8.16] showed, by slowly removing the surface layers in situ with a low-energy argon ion beam, that the variations in composition through the surface layers could be traced. This established the concept of sputter-depth profile measurements which, today, form the major use of surface analysis instruments [8.17]. A continuing problem of the period was that of poor signal quality. This was overcome by the use of a higher efficiency electron spectrometer called the cylindrical mirror analyser (CMA) by *Palmberg* et al. [8.18] which survives to this day, even in the most advanced instruments. We shall consider electron spectrometers briefly in Sect. 8.3.

A number of important further steps set the scene for the establishment of the methods, data bases and instrument development that comprise the modern technique. In 1969 *Haas* and *Grant* [8.19] showed that the Auger electron peak shape and energy could give chemical state information as well as elemental identification. The changes involved were easily observed and quite significant. In the following year, high spatial resolution measurements were demonstrated and in 1971 Auger electron images with submicron resolution [8.20] were presented.

Thus, many of the important steps were compressed into the first four years. Since that time improvements have been made in understanding the theory, in quantification, in the specification of instruments, in the provision of ultra-high vacuum airlocks for rapid sample changing, in the growth of multitechnique instruments, in the data base and in understanding how to tackle different samples. By the end of the last decade many industrial firms used the technique not only for research but also for product development and quality control. The technique had become of age.

8.2 Principles

A range of principles is involved in AES. We shall not describe all of them but merely those from which we may learn more about the sample being studied or those that limit what we may learn.

8.2.1 The Auger Energies

Equation (8.1) allows us to deduce, approximately, the energies of the Auger electron transitions of each element. The equation is not quite correct, however, since the presence of the initial core hole in the level E_A causes the levels at E_B and E_C to become more tightly bound, and, in addition, the screening effect of the conduction or valence-band electrons and the interaction of the two final hole states must all be determined. An approximation to this analysis was proposed by *Chung* and *Jenkins* [8.21] who argued that part of the relaxation in an atom of nuclear charge, Z, was similar to the behaviour of an atom with a nuclear charge of $Z + 1$. Thus empirically they found that

$$E_{ABC} = E_A(Z) - \frac{1}{2}[E_B(Z) + E_B(Z + 1)] - \frac{1}{2}[E_C(Z) + E_C(Z + 1)]. \quad (8.2)$$

Calculations based on this equation together with an indication of their relative intensities are given by *Coghlan* and *Clausing* [8.22]. Fuller calculations of the relaxation and hole-hole interaction terms were given by *Larkins* [8.23] with an estimated accuracy of 1 to 2 eV and with a full tabulation of final states. The simpler relation of (8.2) is accurate to 10 eV or so and provides the energy for a transition which is commonly labelled ABC where A, B and C denote the x-ray levels shown in Fig. 8.3 and may, for example, be KL_1L_3. In the more thorough analysis the final state of the atom would have two holes, one in L_1 and one in L_3 so that the spin and orbital angular momentum quantum numbers may be combined to give two final states separated by several eV. These effects are important in the study of the Auger process itself and in the study of the effects

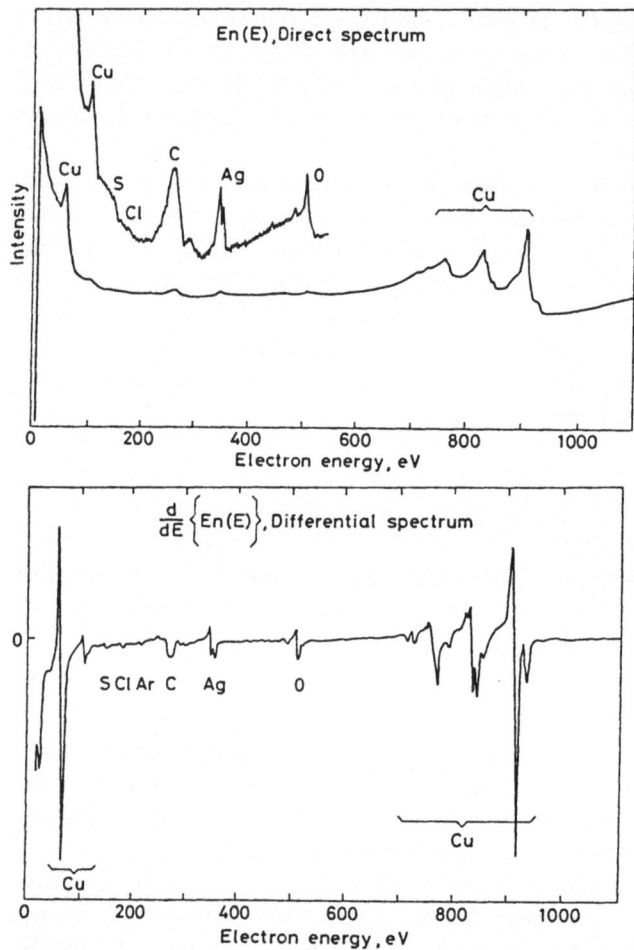

Fig. 8.4. Auger electron spectra from contaminated copper; top, direct spectrum, and bottom, the differential spectrum popularly used

of the environment on the spectra of a species, however, in practice they are rarely used for microscopic analysis and fingerprint methods are used.

A typical Auger electron spectrum is shown in the upper portion of Fig. 8.4 with some of the elements identified. In the lower portion of the figure is the differential of this spectrum and it is evident that, here, the peaks have greater visibility, e.g., the small S and Cl peaks are easier to measure. Traditionally, therefore, the differential spectrum has been used for microanalysis and the direct spectrum reserved for more basic studies. With the advent of cheap microcomputers the situation is now changing and data acquired in one mode can be transformed to the other mode at the touch of a button. In the differential mode the peak energy is quoted as the maximum negative excursion and gives a value a few eV greater than that for the peak in the direct energy spectrum.

For microanalysis it is customary to refer to experimentally determined reference spectra and peak positions which are available for most elements in both the differential [8.24,25] and direct [8.25] modes in published handbooks. The spectra are reliable, show the relative intensities of all peaks from a given element and allow the intensities from one element to be related to those

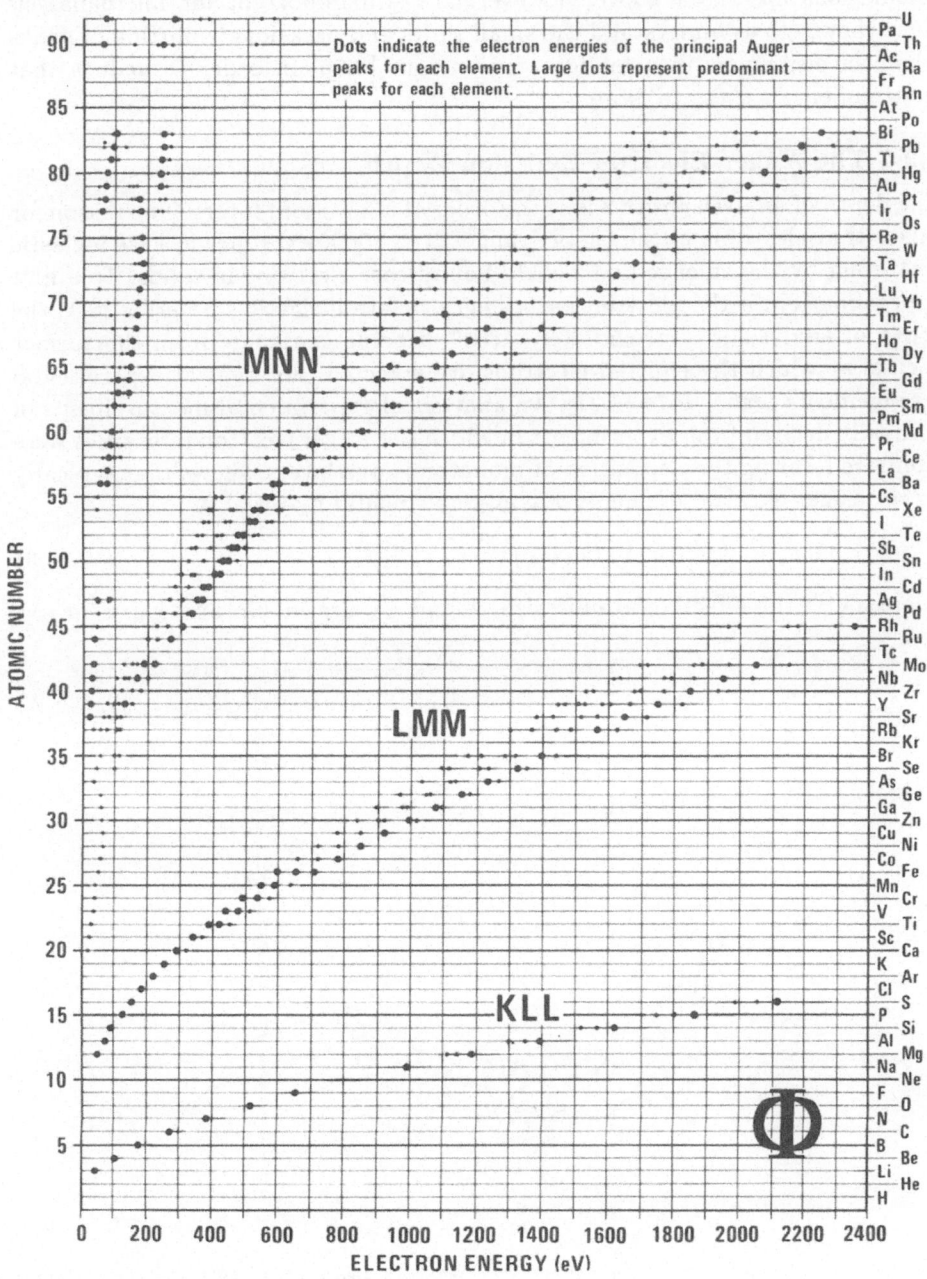

Fig. 8.5. Positions of the principal Auger electron peaks, after *Davis* et al. [8.24]

of another. The positions of the peaks and their approximate intensities are shown in Fig. 8.5. Note that even though peak overlaps may occur, as, for instance, between Na, Zn and the rare-earth metals at energies around 1000 eV, no confusion occurs since each element has its own group of peaks, each of which has a well-defined intensity with respect to the others in the group. The calculations described above, however, are still important for microanalysis since they help us understand the small shifts and occasional splitting of peaks when an element is in a foreign environment [8.26] in order to deduce that environment or chemical state.

8.2.2 The Auger Electron Emission Depth

The depth of analysis in AES is of the order of a few atom layers. The reason for this is that electrons in the energy range 10 to 10000 eV suffer intense inelastic scattering by the valence and weakly bound core electrons in solids. To a first approximation, if the electron path length before inelastic scattering is λ, the depth of analysis is approximately $\lambda \cos\theta$, θ being the angle from the surface normal at which the emitted electrons are detected. For typical experimental arrangements $\cos\theta = 0.74$ [8.27]. An analysis of experimental measurements of λ, mostly derived from experiments in which the Auger electron intensities were monitored during the deposition of monatomic overlayers, shows that typically for an Auger electron of energy E_A in an elemental material A

$$\lambda = 538 a_A E^{-2} + 0.41 a_A (a_A E_A)^{1/2} \tag{8.3}$$

where a_A is the effective atomic size, λ and a_A are in nm and E_A is in eV.

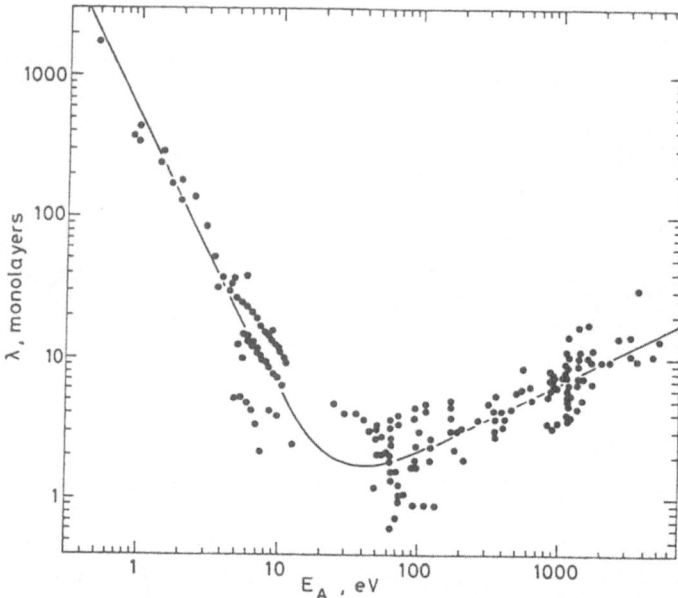

Fig. 8.6. Correlation of the universal curve of *Seah* and *Dench* [8.28] with the experimental λ data for elements

In SI units, a_A is deduced from the relation $1000 \varrho N a_A^3 = M$ where ϱ is the density, N is Avogadro's number, and M is the atomic weight. The experimental data and the fit of (8.3) with λ expressed in atom layers, is shown in Fig. 8.6. The experimental data are not very accurate and both theory and selected experimental data appear to show an energy dependence nearer $E^{0.7}$ than $E^{0.5}$ at the higher energies [8.29].

8.2.3 Quantitative Analysis by AES

The simplest approach to quantitative AES is that often adopted in the literature [8.24] and makes use of both the measured peak heights I_i for i equal to the elements A, B, C ... in the spectrum, and the relative peak heights of each of those elements in the pure solid form, I_i^∞. Thus, the molar fractional composition of A, X_A, is given by

$$X_A = \frac{I_A/I_A^\infty}{\sum\limits_{i=A,B,C} I_i/I_i^\infty}. \tag{8.4}$$

This relation generally gives a result accurate to within a factor of two for solids that are homogeneous over the analysis depth. The relative sensitivities I_i^∞ are tabulated for the differential mode by *Davis* et al. [8.24] for several electron beam energies. An example, with additional data, is given in Fig. 8.7.

A full analysis of the quantification may be found elsewhere [8.30] but a few of the aspects should be considered here. With some approximation we may write the detected Auger electron current as

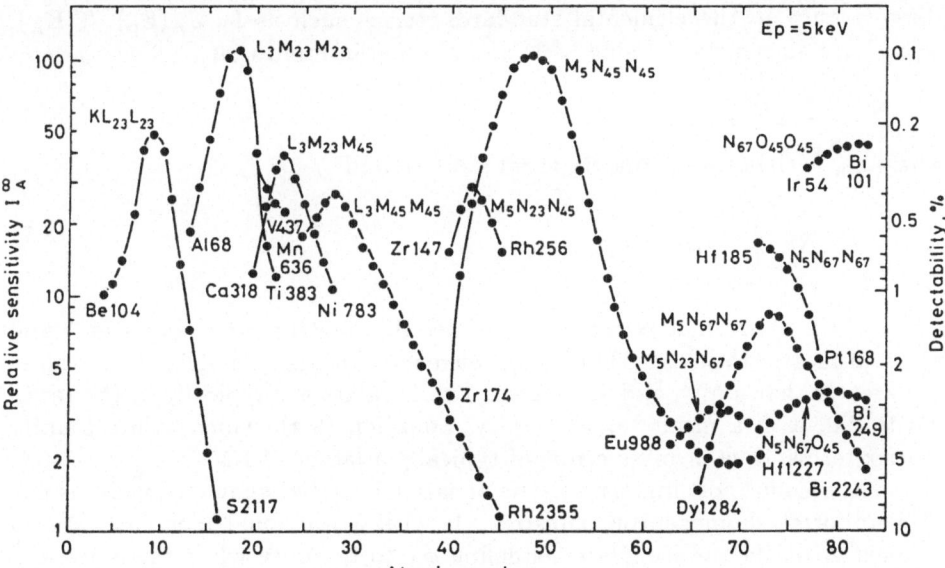

Fig. 8.7. Relative sensitivity factors, I_i^∞, for derivative spectra at 5 keV, based on the data of *Davis* et al. [8.24]. The numbers at the ends of the curves indicate the energies of the peaks and the elements at these points

$$I_A = I_0\sigma_A(E_p)[1 + r_m(E_A)]$$

$$T(E_A)D(E_A) \int_0^\infty N_A(z) \exp[-z/\lambda_m(E_A)\cos\theta]dz \qquad (8.5)$$

where I_0 is the primary electron beam current, and $\sigma_A(E_p)$ is the cross section for excitation of the core level and decay via the particular level scheme being measured for an electron beam energy E_p to produce an Auger electron of energy E_A. The product of the first two terms gives the total Auger electron intensity produced by the incident electron beam. The electron beam penetrates through the surface zone and is scattered in the solid. Some energetic electrons are backscattered to pass through the surface zone and produce further Auger electrons. The ratio of this additional contribution to that produced by the incident beam is given by the term $r_m(E_A)$ so that the whole function $[1 + r_m(E_A)]$ allows for the backscattering. For Auger electrons in the energy range 500 to 1000 eV, r_m increases from 0.4 in aluminium to 0.7 in iron and 0.9 in gold. The last function integrates the number of A atoms, $N_A(z)$, as a function of depth z, times the probability of escape of the Auger electron through the overlying atom layers. Typically, this integral will be over the surface zone to a depth of 2 nm, beyond which depth little signal survives. The remaining terms $T(E_A)D(E_A)$ relate to the efficiency of the electron spectrometer at the energy E_A for detecting those electrons.

At the present time no analyst uses (8.5) directly but, instead, uses one of a number of simpler equations for particular cases. For the case of a material homogeneous over the analysis depth (2 nm), the final integral becomes $N_A\lambda_m(E_A)\cos\theta$. If we measure ratios of intensities in a given solid and ratio these in turn to the elemental standards, terms such as I_0, $\sigma_A(E_p)$, $T(E_A)$, $D(E_A)$ are all removed. Using (8.3) and bearing in mind that

$$N_A = X_A/a_m^3 \qquad (8.6)$$

where a_m is the average matrix atom size, we find

$$X_A = \frac{I_A/I_A^\infty}{\sum_i F_{iA}I_i/I_i^\infty} \qquad (8.7)$$

where the F_{iA} are matrix correction terms allowing for the changes in λ and r_m from matrix to matrix. For a light element in a heavy matrix the F values are greater than unity, and vice versa. The F values are typically in the range 0.5 to 2.0, with a mean value of unity. Equation (8.4), which assumes unity matrix terms, thus involves errors of typically a factor of 1.5.

The second most important type of sample that we should consider is one of a uniformly homogeneous substrate B with a fractional monolayer ϕ_A of element A on its surface. Here we define ϕ_A to be unity when there are a_A^{-2} atoms of A per unit area. In this case the substrate intensity is that of pure B from a fraction $(1-\phi_A)$ of the surface but is reduced by $\exp[-a_A/\lambda_A(E_B)\cos\theta]$

from the remaining area, ϕ_A. The complementary signal from the A atoms is $\{1 - \exp[-a_A/\lambda_A(E_A)\cos\theta]\}$ from the fractional area ϕ_A which, with some modification for the backscattering, gives us

$$\phi_A = Q_{AB}\frac{I_A/I_A^\infty}{I_B/I_B^\infty} \quad \text{where} \tag{8.8}$$

$$Q_{AB} = \frac{\lambda_A(E_A)\cos\theta}{a_A}\left(\frac{1 + r_A(E_A)}{1 + r_B(E_A)}\right). \tag{8.9}$$

Equation (8.8) is similar to (8.7) but in the sense that F_{iA} is of the order of unity, Q_{AB} is of the order of the overlayer Auger electron escape depth in monolayers [8.30].

8.2.4 Composition Depth Profiling

The above principles describe how the spectral energies and intensities relate to the outermost atom layers of a solid, however, much of the use of the AES technique is for determining the composition through thin layers such as those deposited in the manufacture of electronic devices or those formed during oxidation and corrosion processes. To obtain a composition-depth profile it is customary to measure the Auger electron signal from the point to be profiled and then record the necessary spectral regions as the surface is removed by ion sputtering. Figure 8.8 shows the essence of this in the inset and below is the profile through a tantalum pentoxide layer grown anodically on tantalum.

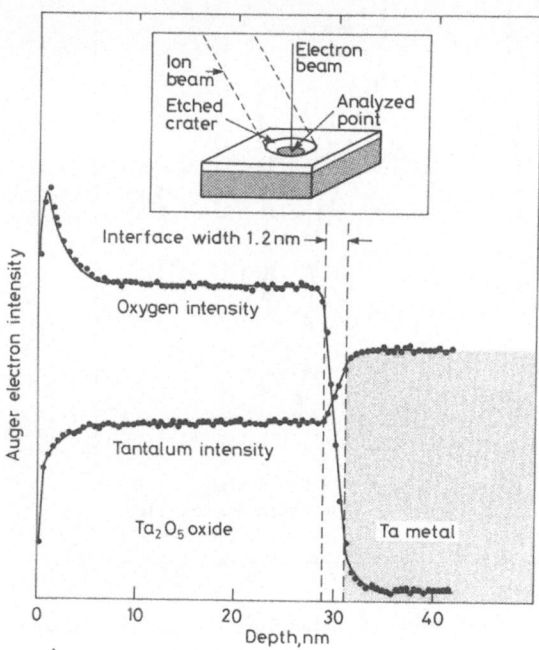

Fig. 8.8. Composition depth profile by AES through a reference sample of anodically grown tantalum pentoxide on tantalum with, inset, a schematic of the experimental arrangement

The profile shows the oxygen and tantalum Auger electron peak intensities as a function of sputtering time and this time axis has subsequently been converted to a depth scale using the known value of the oxide thickness.

Three aspects are of importance when interpreting the depth profiles; (i) quantification of the intensities, (ii) calibration of the depth scale, and (iii) the resolution of interfaces. Quantification of the intensities is not accurate at the present time since (8.7) must be modified to take into account the preferential sputtering that occurs for different elements in any given mixture or compound. Equation (8.7) is still correct but it only provides the composition of the sputter modified surface. The correction for the sputter modification is still the subject of some debate and at the present time we do not have general predictive rules to use [8.31,32]. Often the problem does not arise since we know what layers have been deposited; they are often elemental and we are more concerned with the way they redistribute under ageing conditions. The calibration of the depth scale is also less important but may be achieved, for pure elements, to an accuracy of 10% or so using the tabulated sputtering yields S in atoms per ion. The conversion from time t to depth z is then simply

$$z = SJa^3t/e \qquad (8.10)$$

where J is the ion beam flux density and e is the electron charge. Figure 8.9

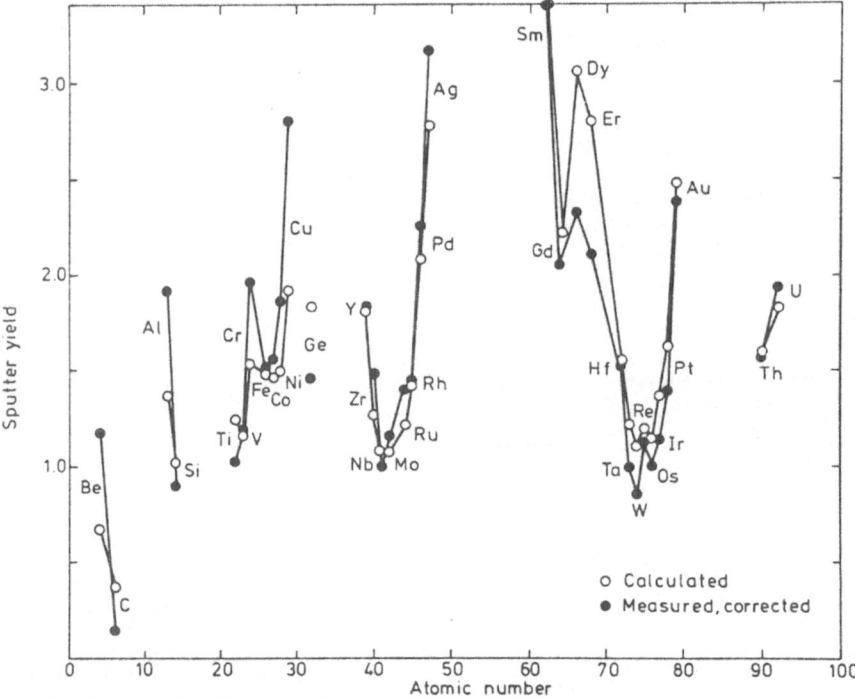

Fig. 8.9. A comparison between the sputtering yields calculated for pure elements (o) and the corrected experimental data (•) [8.33]

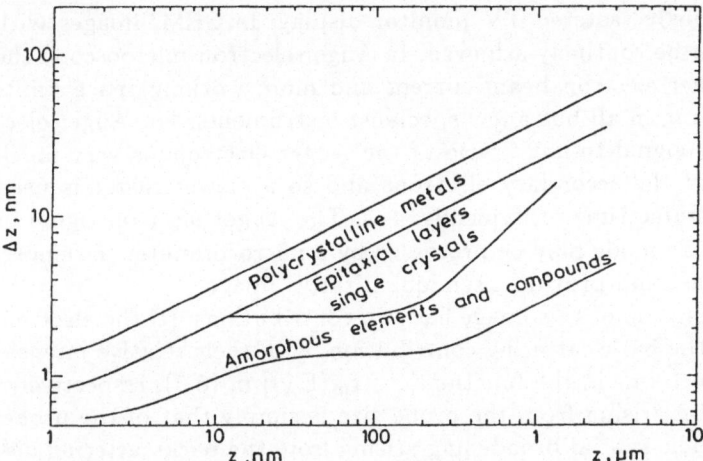

Fig. 8.10. The depth resolutions for different systems as a function of the depth sputtered using argon ions [8.34]

shows the comparison of experimental and theoretical pure element sputtering yields using 500 eV argon ions. For ions of energy E_i up to 3 keV the sputtering yields may be determined from Fig. 8.9 and the approximate relation

$$S \propto E_i^{1/2}. \tag{8.11}$$

Although, above, we have only mentioned argon-ion sputtering, other gases may be used. For improving the depth resolution both xenon and nitrogen have been cited although a full analysis has yet to be established. This depth resolution Δz is taken to be the depth over which the signal changes between 16% and 84% of the values either side of the interface. The optimum depth resolutions in different systems are shown in Fig. 8.10 as a function of the depth z sputtered. The relatively poor result for polycrystalline metals is thought to be due to roughening of the surface by the ion sputtering [8.35]. This roughening, in turn, is thought to be caused by channeling of the ions in the crystal lattice. In amorphous layers this effect disappears and the resolution is then limited by atomic mixing in the surface layers [8.36]. Single-crystal surfaces and epitaxial layers appear to fall in the middle range with some workers reporting results equivalent to the amorphous films and some to the polycrystalline films.

8.2.5 Spatial Resolution in Auger Microscopy

Above we have considered the information depth and the depth resolution in sputter-depth profiling, however, we should also consider the resolution laterally across the surface, i.e. the spatial resolution.

Images of the surface may be obtained in most AES systems in the manner of the scanning electron microscope (SEM). The focused electron beam is rastered at TV rates over the surface and the secondary electron emission is detected by a scintillator/photomultiplier combination to modulate the brightness of a synchronously rastered TV monitor display. In SEM, images with 10 nm resolution can be routinely achieved. In Auger electron microoscopy the need to have a higher electron beam current and more working space limits the resolution to 50 nm in all but a few specialist instruments. For Auger electron microscopy the signal-to-noise ratio of the Auger electrons is very much poorer than those of the secondary electrons and so a slower raster is used for imaging with a frame time of, typically, 40 s. The Auger electron signal in the direct or derivative mode may be processed by a microcomputer to remove topographical contrast and provide a semiquantitative image.

The spatial resolution of this image has two contributions: (i) the electron probe size, and (ii) the backscattering contribution, with their relative intensities given by the two terms in the function $[1 + r_m(E_A)]$ in (8.5), respectively. The spatial resolution arising from the probe size is simply that of the probe and may be 50 nm. The spatial broadening arising from the backscattering depends on the atomic number of the substrate. For light elements the primary electron beam penetrates the sample to some depths and is finally scattered out of the sample producing Auger electrons over a radius of around 1 μm. For heavy elements this figure is reduced by an order of magnitude. Thus, the backscattered electrons form a halo around the incident point of the electron beam but with a very weak contrast. The half width at half maximum, H, of this halo may be estimated from scattering rules developed for the electron probe microanalyser and, as a function of atomic number, is illustrated in Fig. 8.11. The three elemental symbols in Fig. 8.11 show the results of full calculations by *Gomati* and *Prutton* [8.38]. Thus, for instance, the backscattered halo for Al has a full width at half maximum intensity of 2 μm but with a total intensity of only half that produced by the incident beam. If the beam is of 50 nm diameter the contrast ratio between the backscattered and direct beam contributions to the image is around 1:3200. On the other hand, for Au the halo has a full width

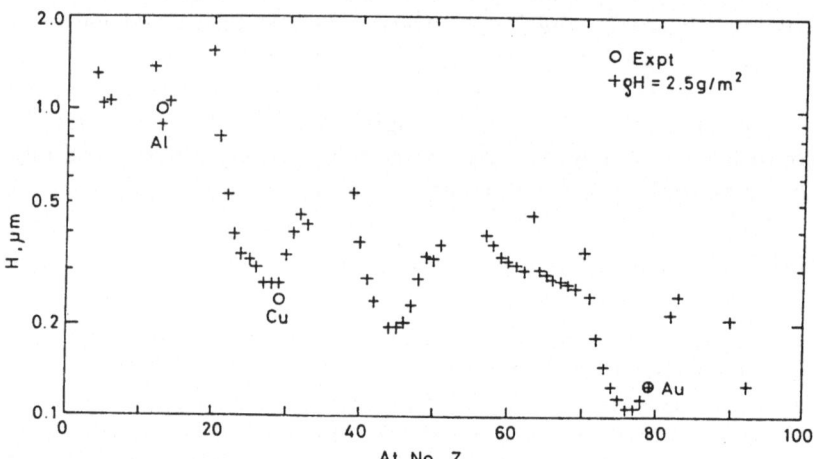

Fig. 8.11. The spatial resolution factor, H, for AES using 20 keV electrons [8.37]. Here ϱ is the specimen density

232

at half maximum reducing to 250 nm so that the contrast ratio is still as weak as 1:25. Thus the spatial resolution for Auger electron images is considerably better than that in the electron probe microanalyser and may be taken as the probe size. Of course, for quantitative analysis, the backscattering contribution remains important.

8.3 The Instrument

A discussion of the instruments and how they work is one of the most important considerations of any analytical method. For practical studies the commercially available instruments define the extent of the experiments that are possible in a reasonable timescale.

As a minimum we require an electron gun to provide a beam of electrons, a specimen stage to hold the sample and an electron spectrometer to detect the emitted Auger electrons. This is all contained in a vacuum system with a working vacuum of the order of 10 n Pa. Figure 8.12 shows one of the modern commercial arrangements. In this arrangement a relatively compact electrostatically focused electron gun with 200 nm resolution, and providing beam currents of up to 1 μA at a coarser resolution, can irradiate selected points on the sam-

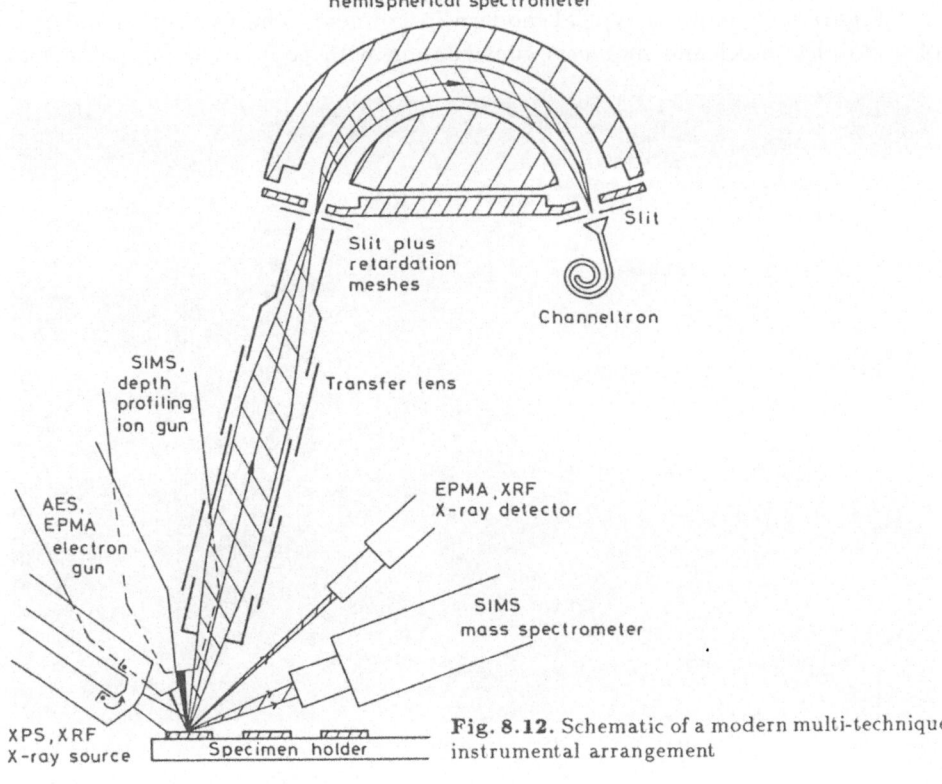

Fig. 8.12. Schematic of a modern multi-technique instrumental arrangement

ple. The high efficiency electron spectrometer can then be used to measure the emitted electron energy spectrum. With modern electron spectrometers the energy resolution may be selected to provide energy resolutions between 0.25 and 10 eV, the coarser resolutions being used to maximise the sensitivity. For those interested, details of electron spectrometers may be found in recent reviews [8.39,40]. The electron spectrometer shown in Fig. 8.12 is exactly the same as that used for x-ray photoelectron spectroscopy (XPS), as described in Chap. 7. Because the specimen chamber, sample handling and electronics are the same for XPS and AES, the same instrument can be used for both if an x-ray source is added. Also shown in Fig. 8.12 is an x-ray detector which, with the electron source, allows x-ray microanalysis (EPMA) and, again, with the x-ray source allows x-ray fluorescence analysis (XRF). In order to perform depth-profiling experiments an ion gun has been added and, with the further addition of a mass spectrometer with an ion energy filter, secondary ion mass spectroscopy (SIMS) [8.41,42] is possible.

Figure 8.12 serves to show that AES and other surface techniques are not exclusive to one another but can all be used to study different aspects of the same sample. The only constraint that does exist is that the systems using an electromagnetically focused electron gun, and which are much more efficient for high spatial resolution in AES (50 nm), are generally not recommended for XPS since some stray magnetic fields always occur and the XPS is difficult to optimize.

Figure 8.13 shows a typical modern instrument. The vacuum chamber is of a stainless steel and mumetal construction with copper sealing gaskets to

Fig. 8.13. A typical modern AES/XPS instrument

provide a working vacuum of 10 n Pa. Samples are either mounted on a small stub or inserted into a loading cartridge for fracture. An airlock system allows one or a set of samples to be translated into the ultra-high vacuum region in about 10 to 15 minutes. Surface analysis or depth profiling may then be done, either under manual or computer control. The software in typical commercial systems permits the smoothing and averaging of data, addition, subtraction and comparison of spectra, quantification, depth-profile evaluation, mapping, the removal of topography contrast in mapping, etc.

8.4 Related Methods

In order to compare related methods it is best to start by summarising the points we have established for AES. AES is a technique which can detect all elements except H and He to a sensitivity of one part in 1000 in a layer between 2 and 10 atoms thick at a solid surface. The method is easy to use, easy to quantify with a modest accuracy, and can be used to resolve spatial information at about 50 nm laterally across a surface. Combined with argon ion sputtering, composition – depth profiles to depths of $1\,\mu$m may be measured with good-depth resolution.

There are many other surface analysis techniques [8.43] and it would be fruitless to consider them all here. We merely consider the most popular ones. After AES, the most popular is XPS. As described in Chap 7, XPS has all of the attributes of AES except as follows. Until recently the spatial resolution in XPS was of the order of 2 mm, however, recent developments have reduced this to $100\,\mu$m. It is unlikely that this will be reduced further in the near future. The poor lateral resolution leads to poor resolution in sputter depth profiles and because of the need to sputter a large area this approach is not very popular. The main advantages of XPS are the ability to define chemical states with reasonably large data banks available and the clearly resolved peak shifts and structure associated with particular compounds. XPS is also favoured for studying insulators where, in AES, strong charging may destroy the spectrum. Finally, for a given analysis the sample degradation in organic and other sensitive compounds is far less in XPS.

The next most important technique is SIMS which traditionally has been used in two different modes with different types of instrument. In SIMS secondary ions sputtered from the surface are detected by a mass analyser. SIMS can therefore be used to study both H and He which are not accessible to AES and XPS. A second important advantage is that SIMS has detection limits below 1 ppm for many elements. Quantification of the spectra can, however, by very difficult. If the primary ion beam has a very low current density the surface is relatively undisturbed throughout the analysis. In this mode, called static SIMS, very large complete molecules may be sputtered from the surface giving detailed information about the whole range of chemical groups present. The spatial resolution in this mode is rarely better than $50\,\mu$m. In the older mode of dynamic SIMS the surface is eroded during the measurement and the instrument is used to study bulk properties or the properties of thin films. Using liquid metal ion sources spatial resolutions of 50 nm are now possible although most work is still completed on instruments with argon or oxygen ion beams with a resolution around 1 μm. Because of the exceptional sensitivity, dynamic SIMS finds considerable use in assessing the dopant distributions in semiconductor devices. Over five orders of magnitude dynamic range are possible with final detection limits as low as 2 parts in 100 million for B in Si [8.44].

Although it is not a surface-analytical technique the electron probe microanalyser (EPMA) should be mentioned. The EPMA is a very popular and powerful technique for general microanalysis of metallurgical sections. As discussed earlier, it makes use of the x-ray emitted when atoms de-excite after irradiation with an electron beam of 10–30 keV. This x-ray de-excitation actually competes with the Auger process. For heavy elements the competition is fairly even but for light elements the Auger process dominates. This factor, combined with the lack of sensitivity of the popular solid state detectors for x-rays of less than 1 keV energy, makes the detection of the elements below Ne very difficult in the EPMA. For this reason the Auger electron microprobe is beginning to take over some work on bulk materials analysis where identification of B, C, N or O is required.

8.5 Applications

The applications of a technique serve to illustrate its full potential and give a guide to the prospective user of the way it may be applied to his own problems. In the following sections we consider the problems of sample handling, the method of measurement, the interpretation of results and the general perspective in selected applied topic areas involving metals.

8.5.1 Grain Boundary Segregation Studies

A very well known problem that occurs in steels, iron, copper, nickel, tungsten and a number of other structural metals is that of brittle intergranular fracture. These metals, when containing certain impurities, lose their ductility and fail catastrophically in a brittle intergranular manner.

If such samples are fractured in situ in the UHV AES system, the spectra show peaks due to impurities which have concentrated prior to fracture at the grain boundary sites and which have then promoted grain boundary decohesion. It should be noted that samples fractured outside the UHV system, or exposed to air after the UHV fracture, are valueless and generally show only C, O and the metal base but no segregants. Figure 8.14 shows the spectrum taken for an in situ fracture specimen from the rotor that initiated the failure at Hinkley Point B Power Station [8.46]. The actual rotor and the analysed intergranular surface are shown in Fig. 8.15.

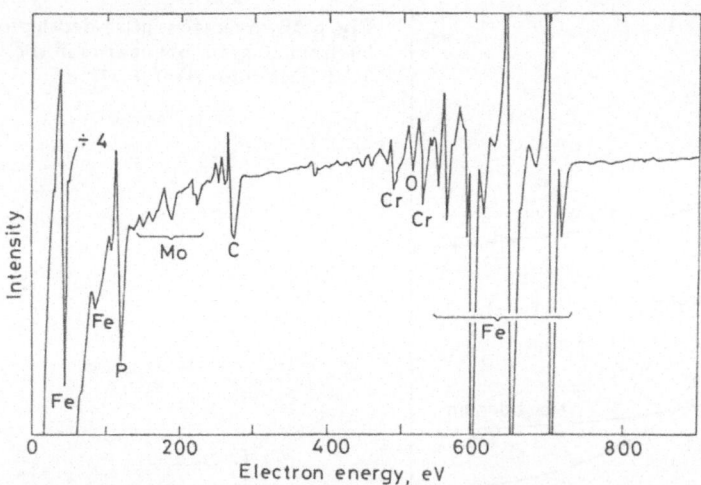

Fig. 8.14. Auger electron spectrum from a grain boundary surface in the 3Cr1/2Mo rotor initiating the failure at Hinkley Point B Power Station [8.45]

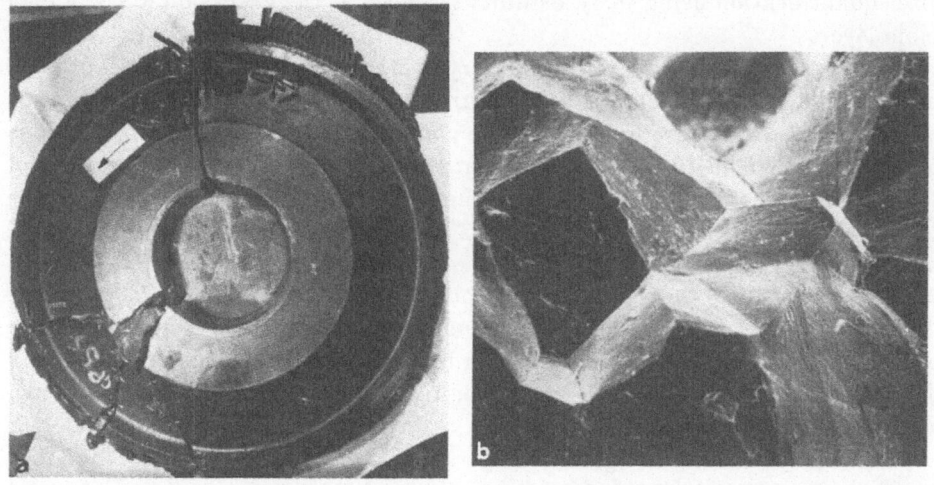

Fig. 8.15. (a) The rotor that initiated the failure at Hinkley Point B Power Station [8.46] and **(b)** a fracture surface from close to the site of fracture initiation (field of view 200 μm)

Figure 8.14 shows peaks due to the Fe matrix, Cr and Mo carbides and segregated P. If we sputter-erode the surface to determine the localisation of these elements at the grain boundary we get the results shown in Fig. 8.16. These show that the Cr and Mo are largely present as grain boundary carbides but that the P is present precisely at the grain boundary atom layer. Indeed this result, also found for a whole host of impurity elements [8.5], shows that the replacement of the metal atoms by impurities along the grain boundary atom layer has caused the weakness. After fracture, on average half of the impurity atoms are left on each fracture surface and so the spectrum shown in Fig. 8.14,

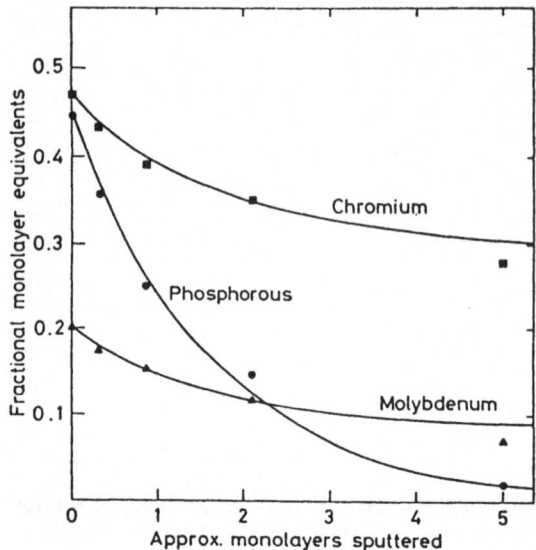

Fig. 8.16. Average sputter-depth profile from 10 grain boundaries in the 3Cr1/2Mo rotor steel [8.47]

after quantification using (8.8), exhibits a P level at the grain boundary of 0.45 monolayers.

Here, P is termed a grain boundary segregant and its level is expressed in fractions of a monolayer at the grain boundary. This presentation is necessary if the measurements are to be compared with theoretical predictions of segregation [8.48,49] or to be correlated with other data. Measurements of the level of segregation in many grain boundaries in a given sample show in some cases a very small scatter and in others a wider scatter [8.5]. However, detailed measurements by *Suzuki* et al. [8.50] show two very significant points. Firstly, that during fracture the segregant does not always partition evenly between the two fracture surfaces but adheres more strongly to the higher index fracture plane, and secondly the amount on each fracture surface depends on the index of that surface. This effect, shown in the polar diagram of Fig. 8.17, is known as the anisotropy of segregation.

In many situations good image quality is important or the true intergranular areas in service failures may not be properly identified. A good example of the way imaging may be used is shown in fracture studies by *Johnson* [8.51]. These studies involve grey (brittle) cast iron and cast iron with Mg additions to render it ductile. In the Mg treated iron the graphite inclusions are spherical whereas without treatment the graphite is in a flake form. AES studies of the fracture surfaces showed pure carbon for the graphite side of the graphite/iron interface in both cases, however different results occurred for the iron side of this interface in the two cast irons. In the ductile iron very small levels of Mg and P are seen whereas in the brittle iron considerable S and O occur. It is thought that the S and O poison the growth sites for the graphite basal plane so that flakes are produced. The addition of Mg precipitates the S and O so that the basal plane sites are active and graphite

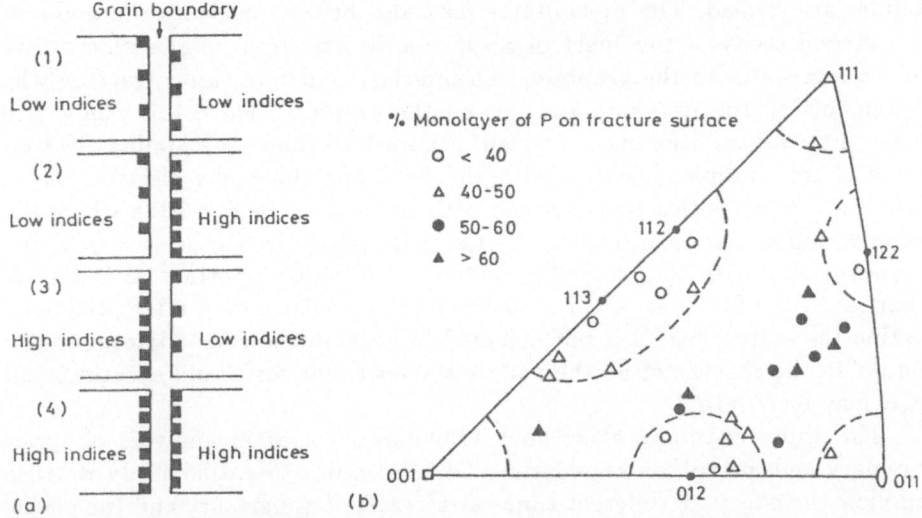

Fig. 8.17a,b. The segregation of phosphorus in iron – 1%P, (a) showing the segregating atoms (dark squares) in relation to the fracture process, and (b) the level of segregation in relation to the relative orientation of each grain to the grain boundary [8.50]

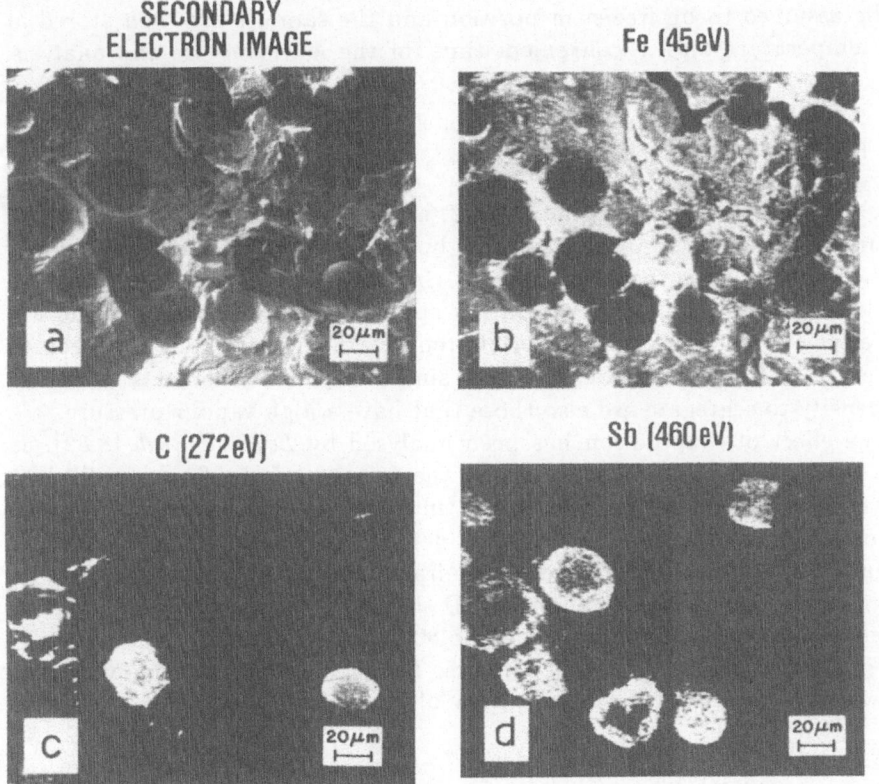

Fig. 8.18a–d. Scanning electron images of ductile iron doped with 0.07%Sb (a) secondary electron image, (b) Fe, (c) C and (d) Sb Auger electron maps [8.51]

nodules are formed. The precipitates may also help to nucleate the nodules. In a second study of the effect of Sb in ductile cast iron he also shows that the Sb segregates to the graphite/iron interface and that, after fracture, the Sb remains on the iron side and not on the graphite. Thus, this study also shows that the partitioning is uneven. Figure 8.18 shows a scanning electron image of this sample, together with the Fe, Cand Sb Auger electron maps. Note that the Sb is clearly associated with the hollows in Fig. 8.18a where the graphite nodule has been removed by the mating fracture surface, whereas the C is associated with the remaining nodules. The Sb is important as it acts as a barrier to the free flow of C across the interface. Thus, even after prolonged heating the matrix region is not depleted in Fe_3C in the zone adjacent to the nodule. In the absence of Sb the nodule acts as a sink for C and eventually all Fe_3C may be removed.

The above examples show how AES may be used in analyses of grain boundary and interphase segregation. Of course, in a scientific study we wish to follow the effects of different times at a certain temperature and the effects of other impurities or alloying elements in the metal. These experiments are all carried out by treatments of the solid material in the laboratory with a rapid quench at the end of the heat treatment. Substitutional solutes, on the whole, may be assumed to be frozen in position and the samples are then stored at room temperature until a convenient time for the AES fracture and analysis. The results of many experiments are discussed by *Seah* [8.5].

8.5.2 Surface Segregation

Surface segregation, as shown in Fig. 8.2, is the segregation of solutes to the free surface of the metal. Unlike the grain boundary case, the whole experiment must be done in situ in the AES appartus. Specimens are mounted on a heating stage and are first cleaned by argon ion sputtering. Then measurements are made after certain times at appropriate temperatures. Care must be taken to avoid excessive evaporation of segregants since many of the elements that have a propensity to segregate are also those that have a high vapour pressure.

The effect of evaporation has been analysed by *Lea* and *Seah* [8.52], as shown in Fig. 8.19. The ordinate shows the fraction of the final equilibrium segregation achieved and the abscissa the time on a log scale in units of $\alpha^2 d^2/D$. Here α is the surface enrichment ratio defined as the fractional monolayer segregation at equilibrium divided by the fractional molar content in the bulk, d is the solute monolayer thickness and D the solute diffusivity in the matrix. In steels in the temperature range 550° to 800°C the interesting events usually occur in the timescale 10s to a few hours. The evaporation rate parameter E may be evaluated from the kinetic theory of gases [8.53] and is given by

$$E = \frac{3.513 \times 10^{26} \alpha^2 a^3 dkp}{(MT)^{1/2}D} \tag{8.12}$$

where M is the atomic weight of the solute atoms, a is the size of the matrix

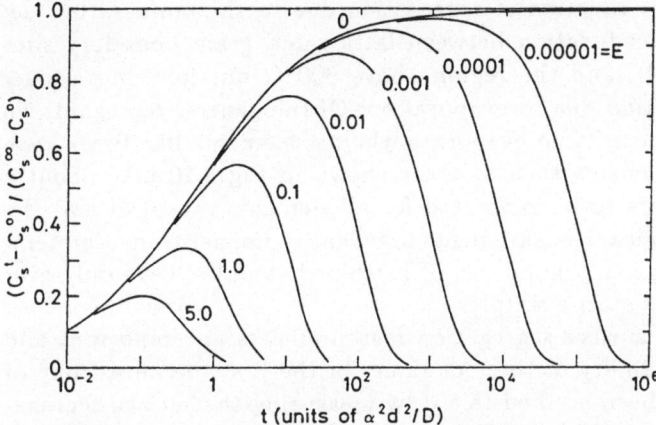

Fig. 8.19. The time dependence of the segregation level at a free surface with various solute evaporation rates [8.52]

atoms defined as in (8.3), k is a parameter describing the solute-solvent interaction and p is the vapour pressure of the evaporating species at temperature T. As is clear from Fig. 8.19, if the solute evaporates easily the segregation reaches a maximum at an early time and then, because of the rapid loss of solute, the material below the surface becomes solute depleted and the segregation falls, eventually to zero. With this kind of experiment we may generate C-curves analogous to those describing temper brittleness [8.54], as shown in Fig. 8.20. The data is generated by measuring the Sn segregation as a function of time at temperatures in the range 450° to 800°C at the surface of a piece of 5NiCrMoV rotor steel containing 0.022 wt.% Sn. The low temperature data shows simple bulk diffusion limitation, whereas the region above 600°C at times shorter than

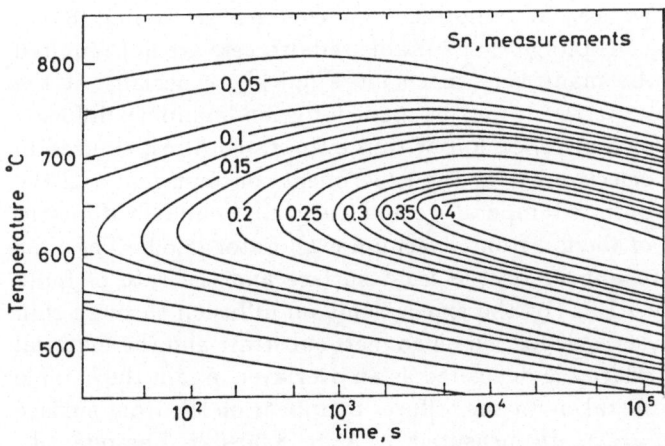

Fig. 8.20. The time dependence of the Sn surface segregation at different holding temperatures for a 5NiCrMoV rotor steel [8.53]. The numbers against the curves denote the fraction of full segregation measured by AES

1000 s shows the simple reduction of segregation due to the temperature dependence of the partition function between lattice and grain boundary sites (thermodynamic control), and the region above 600°C but for times longer than 1000 s shows depletion due to evaporation. Of the general segregants Sn has a relatively low propensity to evaporate whereas elements like P are soon depleted [8.53]. Measurements such as those shown in Fig. 8.20 take about a week and permit contours to be generated for all elements except H and He. The equivalent metallurgical measurements use Charpy-impact transition temperatures and, requiring many hundreds of machined samples, have only ever been completed for a few steel examples.

Two other types of surface segregation measurements are important and require the imaging capability of the technique. In the first the anisotropy of surface segregation has been studied [8.55] by measuring the surface segregation, in situ, for each crystal in a polycrystalline foil, the orientation of each subsequently being determined by selected area electron channeling patterns in the SEM. The segregations varied over an order of magnitude between different grains but considerably more data need to be established before a full explanation of the results can be given.

Surface segregation is also important in mechanical failures involving creep cavitation. Failures are common where these cavities occur profusely on grain boundaries. Both grain boundary and surface segregation can be important [8.56]. An excellent example of the microanalytical capability is given by *Franzoni* et al. [8.57] in which creep embrittled 2 1/4 Cr1Mo steel was fractured in situ in an AES apparatus. The segregation in the 1 μm diameter cavities was then shown to be of up to a third of a monolayer of each of Sn, Sb and Cu.

8.5.3 Grain Boundary Diffusion

Grain boundary diffusivity (D_{gb}) measurements may be made by AES although the traditional method has been by radiotracer and chemical sectioning [8.58]. The AES method has the advantage that suitable radiotracers are not required and measurements may be made very much more quickly, especially at the lower temperatures. The experiment may be made for grain boundary diffusion in bulk material in the form of rolled foil or thin evaporated films, as used in the manufacture of microelectronics. In the experiments on bulk material the foil is heated to the appropriate temperature and the grain boundary diffusant is deposited on the back of the foil from a vacuum evaporator [8.59]. The time taken for the diffusant to appear on the front surface and the rate of build up are then measured by AES. For the experiments on diffusion through thin evaporated films the diffuser is deposited on an inert substrate and the material containing the grain boundaries is deposited as an overlayer. Again the sample is heated and both the time taken for the diffuser to appear on the front surface and the rate of build up are again measured by AES [8.60–62]. The quantity measured on the surface, ϕ_A monolayers, increases with time t approximately as

$$\phi_A = \frac{2D_{gb}\delta}{agl}\left(t - \frac{l^2}{6D_{gb}}\right) \tag{8.13}$$

where ϕ_A and a are as previously defined, δ is the grain boundary width, g is the average grain size and l is the layer or foil thickness. At long times ϕ_A increases linearly with t and if this line is projected back to the time axis the intercept occurs at

$$t = l^2/6D_{gb}. \tag{8.14}$$

A full analysis is given by *Holloway* et al. [8.61] but the above describes the general behaviour. Most studies of this sort are for evaporated overlayers, the precise structure of which is unclear. Measurements for the better defined metal foils are not common but the recent data for tin in iron by *Bernardini* et al. [8.59], and shown in Fig. 8.21, illustrates two points, (i) the agreement with radiotracer measurements is exceptionally good, and (ii) studies are possible in the lower temperature range.

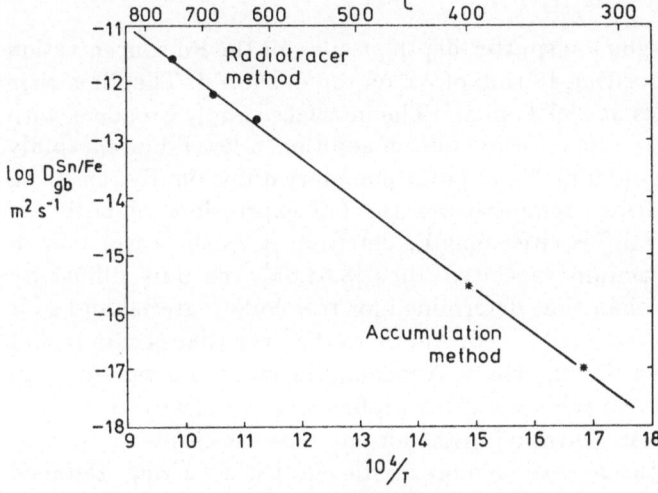

Fig. 8.21. The grain boundary diffusivity of Sn in Fe in the temperature range 300° to 750°C using radiotracer measurements (•) and AES (∗) [8.59]

8.5.4 Defect-Enhanced Diffusion

Above we have considered a purely surface method of determining grain boundary diffusivity. Alternatively, both the grain boundary and bulk diffusivities may be measured by using the composition-depth profiling approach. This method is of particular importance in thin film technology relevant to the coating and microelectronics sectors. Data gathered on bulk samples is generally not relevant since the high defect level in thin films leads to much faster diffusion there.

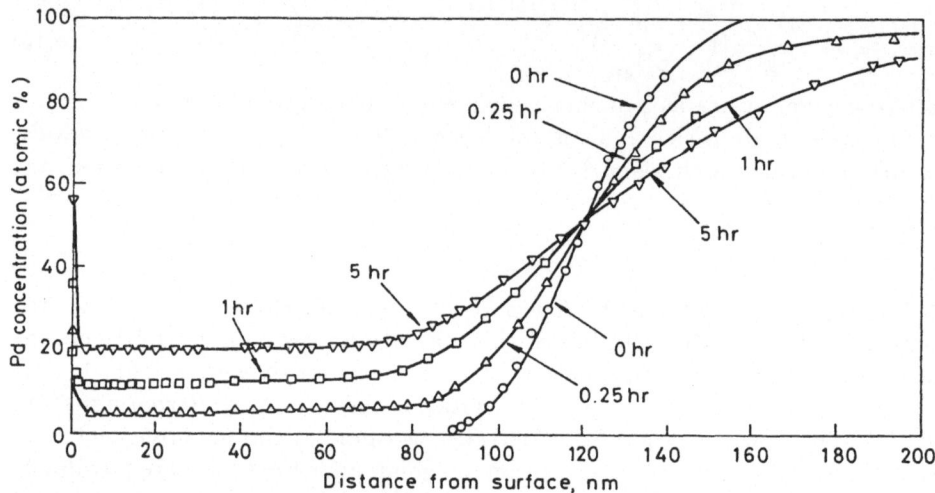

Fig. 8.22. Pd composition profiles in the Au/Pd system at 250°C as a function of the ageing treatment [8.63]

Figure 8.22 shows argon ion sputter-depth profiles for the Pd concentration in a sample made by depositing 120 nm of Au on top of the Pd. The films were annealed for various times at 250°C in air. The interface simply broadens with the usual error function of bulk diffusion but, in addition, a low Pd level rapidly extends throughout the gold film due to grain boundary diffusion. By repeating the measurements at various temperatures the full expressions of both the bulk and grain boundary diffusivities may be determined. In this case, as with most evaluations with vacuum deposited films [8.64,65], the bulk diffusivity measured is much faster than that determined for true bulk material and so is generally thought to be defect-enhanced. Another difference that occurs is that the annealing environment is important. Annealing in vacuum gives times an order of magnitude longer to achieve a given profile. It is not clear if this result arises because impurities are burnt off in the air and that the cleaner gold layer then has higher diffusivities or if some more subtle effect is occurring. Detailed analyses of this sort are very easy with AES but not by alternative techniques.

8.5.5 Other Studies

The examples of applications of AES given above are fairly commonplace. Equally important but less popular are the high-resolution studies of the initial forms and mechanisms of growth of metal films on various substrates. The imaging capability in Auger electron microscope systems allows the detection of islands at submonolayer levels which are important in understanding the early stages of film nucleation [8.66,67]. With specially designed instruments spatial resolution of 50 nm have been achieved at this level.

Although not strictly studies of metals, we should mention here the considerable volume of composition-depth profiling work that has been done in studies of the oxidation and corrosion of steels [8.6], aluminium alloys [8.68], etc. Again, the greatest single sector using AES is the electronics sector with analyses of devices, contacts, bonds, etc. A classic example of interfacial phase formation is shown in the work of *Singer* [8.69] in the formation of nickel silicide Schottky contacts between nickel and silicon. Other metallurgical studies cover the sintering of powdered nickel based superalloys [8.70], the study of fracture in WC-Co hardmetals [8.71] and of the integrity of heavy metals [8.72]. Most of these studies relate to joining and fracture. Other studies exist in machining [8.73], wear [8.74] and a host of other sectors [8.75]. For those seeking a source of references to the surface analysis literature the review of *Turner* et al. [8.76] is recommended.

8.6 Future Developments

The last few years have seen very great changes in AES as an applied technique not because the science has changed but because the instruments are much more user-friendly. The introduction of sample airlocks on the one hand has made it far easier to gain access to instruments, enabling a far wider range of problems to be investigated, and the introduction of dedicated microcomputers has led to the more sophisticated concepts of analysis to become routine. For instance, we may now envisage the sputter-depth profiles for diffusion analysis being stored directly in the computer during the experiment and being processed completely through the theoretical analysis with a least-squares fitting routine to print out the diffusivities – without the use of graph paper. The addition of multidetector systems [8.40] to the output of the electron spectrometer shown in Fig. 8.12 will increase the signal strengths some 10 to 15 times. This will lead to improved detection levels or, alternatively, faster rates of working. Combined with new microfocus ion guns, composition-depth profiling may then be carried out at 200 nm in 5 mins instead of the 50 mins required for Fig. 8.22. This means that, compared with a few years ago when one had to plan quite long experiments simply to determine grain boundary diffusivities in pure materials, one can now envisage experiments covering the matrix of variables which may affect that diffusivity. Thus, simply because the technology of surface analysis is improving rapidly, driven by the research requirements of the microelectronics industry, the complexity of the experiments we can envisage as practicable is expanding rapidly.

The developments of the next decade will give us better spatial resolution, higher sensitivities, and the full facilities of SEM with x-ray analysis including selected area channeling patterns for crystal orientation. The high energy resolution currently available will lead to a greater awareness of the potential of chemical state information and hence to the use of AES as a probe of chemical state with high spatial resolution.

References

8.1   M.P. Seah: Surf. Interface Anal. **2**, 222-239 (1980)

8.2   E.D. Hondros, M.P. Seah: Int.Met.Revs. **22**, 262-301 (1977)

8.3   M.P. Seah: J.Phys. F: **10**, 1043-1064 (1980)

8.4   E.D. Hondros, M.P. Seah: "Interfacial and Surface Microchemistry" in *Physical Metallurgy*, 3rd ed., ed. by R.W. Cahn and P. Haasen (Elsevier, Amsterdam 1983) Chap. 13, pp. 855-931

8.5   M.P. Seah: "AES in Metallurgy" in *Practical Surface Analysis by Auger and X-Ray Photoelectron Spectroscopy*, ed. by D. Briggs and M.P. Seah (Wiley, Chichester 1983) Chap. 5, pp. 247-282

8.6   N.S. McIntyre: "Uses of Auger Electron and Photoelectron Spectroscopies in Corrosion Science" in *Practical Surface Analysis by Auger and X-Ray Photoelectron Spectroscopy*, ed. by D. Briggs and M.P. Seah (Wiley, Chichester 1983) Chap. 10, pp. 397-427

8.7   P. Auger: J. Phys. Radium **6**, 205-208 (1925)

8.8   R.M. Eisberg: *Fundamentals of Modern Physics* (Wiley, New York 1961) pp. 391-474

8.9   D. Briggs, J.C. Riviere: "Spectral Interpretation" in *Practical Surface Analysis by Auger and X-Ray Photoelectron Spectroscopy*, ed. by Dr. Briggs and M.P. Seah (Wiley, Chichester 1983) Chap. 3, pp. 87-139

8.10   J.A. Bearden, A.F. Burr: Rev. Mod. Phys. **31**, 49-66 (1967)

8.11   S.J.B. Reed: *Electron Microprobe Analysis* (University Press, Cambridge 1975)

8.12   N.J. Zaluzec: "Quantitative X-ray Microanalysis. Instrumental Considerations and Applications to Materials Science" in *Introduction to Analytical Electron Microscopy*, ed. by J.H. Hren et al. (Plenum, New York 1979) pp. 121-195

8.13   J.J. Lander: Phys. Rev. **91**, 1382-1387 (1953)

8.14   L.A. Harris: GE R&D Report 67-C-201 May 1967, later as J. Appl. Phys. **39**, 1419-1427 (1968)
       L.A. Harris: GE R&D Report 67-C-199 May 1967, later as J. Appl. Phys. **39**, 1428-1431 (1968)

8.15   P.W. Palmberg, T.N. Rhodin: J. Appl. Phys. **39**, 2425-2432 (1968)

8.16   H.L. Marcus, P.W. Palmberg: T. Met. Soc. AIME **245**, 1666 (1969)

8.17   H. Oechsner (ed.): *Thin Film and Depth Profile Analysis*, Topics Current Phys., Vol. 37 (Springer, Berlin, Heidelberg 1984)

8.18   P.W. Palmberg, G.K. Bohn, J.C. Tracy: Appl. Phys. Lett. **15**, 254-255, (1969)

8.19   T.W. Haas, J.T. Grant: Appl. Phys. Lett. **16**, 172-173 (1970)

8.20   N.C. MacDonald, J.R. Waldrop: Appl. Phys. Lett. **19**, 315-318 (1971)

8.21   M.F. Chung, L.H. Jenkins: Surf. Sci. **22**, 479-489 (1970)

8.22   W.A. Coghlan, R.E. Clausing: Atomic Data **5**, 317-469 (1973)

8.23   F.P. Larkins: Atomic Data and Nuclear Data Tables **20**, 311-387 (1977)

8.24   L.E. Davis, N.C. MacDonald, P.W. Palmberg, G.E. Riach, R.E. Weber: In *Handbook of Auger Electron Spectroscopy*, 2nd ed. (Physical Electronics Ind. Inc, Eden Prairie, MN 1976)

8.25   T. Sekine, Y. Nagasawa, M. Kudoh, Y. Sakai, A.S. Parkes, J.D. Geller, A. Mogami, K. Hirata: Handbook of Auger Electron Spectroscopy, (JEOL, Tokyo 1982)

8.26   P. Weightman: Repts. Prog. Phys. **45**, 753-814 (1982)

8.27   M.P. Seah: Surf. Sci. **32**, 703-728 (1972)

8.28   M.P. Seah, W.A. Dench: Surf. Interface Anal. **1**, 2-11 (1979)

8.29   C.J. Powell: Scanning Electron Microscopy P**4**, (1984) 1649-1664

8.30   M.P. Seah: "Quantification of AES and XPS" in *Practical Surface Analysis by Auger and X-Ray Photoelectron Spectroscopy*, ed. by D. Briggs and M.P. Seah (Wiley, Chichester 1983) Chap. 5, pp. 181-216

8.31   R. Kelly: Surf. Sci. **100**, 85-107 (1980)

8.32   R. Kelly: Surface Interface Anal. **7**, 1-7 (1985)

8.33   M.P. Seah: Thin Solid Films **81**, 279-287 (1981)

8.34   M.P. Seah, C.P. Hunt: Surf. Interface Anal. **5**, 33-37 (1983)

8.35   M.P. Seah, M.E. Jones: Thin Solid Films **115**, 203-216 (1984)

8.36   H.H. Andersen: Appl. Phys. **18**, 425-426 (1979)

8.37   M.P. Seah: Proc. of 9th Intern. Vac. Congr. and 5th Intern. Conf. on Solid Surfaces (Invited Speakers' Volume) ed. by J.L. Segovia (ASEVA, Madrid 1983) pp. 63-75

8.38   M.M. El Gomati, M. Prutton: Surf. Sci. **72**, 485-494 (1978)

8.39   D. Roy, J.D. Carette: "Design of Electron Spectrometers for Surface Analysis" in *Electron Spectroscopy for Surface Analysis*, ed. by H. Ibach, Topics Current Phys., Vol. 4 (Springer, Berlin, Heidelberg 1977) Chap.2, pp. 13-58

8.40   M.P. Seah: "Electron and Ion Energy Analysis", in *Methods of Surface Analysis: Techniques and Applications*, ed. by J.M. Walls (University Press, Cambridge 1986) Chap.3

8.41   A. Benninghoven: Appl. Phys. **1**, 3 (1973)
        A. Benninghoven et al (eds.): *Secondary Ion Mass Spetrometry SIMS* II-IV, Springer Ser. Chem. Phys., Vol. 9, 19 and 36 (Springer, Berlin, Heidelberg 1979, 1982 and 1984)
        A. Benninghoven et al (eds.): *Secondary Ion Mass Spectrometry SIMS* V, Springer Proc. Phys. Vol. 6 (Springer, Berlin, Heidelberg 1986)

8.42   A. Brown, J.C. Vickerman: Surf. Interface Anal. **6**, 1-14 (1984)

8.43   J.A.D. Matthew, M. Prutton: *Handbook of Surface Science* (Longman, London, to be published)

8.44   J.B. Clegg, A.E. Morgan, H.A.M. de Grefte, F. Simondet, A. Huber, G. Blackmore, M.G. Dowsett, D.E. Sykes, C.W. Magee, V.R. Deline: Surf. Interface Anal. **6**, 162-166 (1984)

8.45   M.P. Seah: Surf. Sci. **53**, 168-212 (1975)

8.46   D. Kalderon: Proc. Inst. Mech. Eng. **186**, 341-377 (1972)

8.47   M.P. Seah: J. Vac. Sci. Technolog. **17**, 16-24 (1980)

8.48   E.D. Hondros, M.P. Seah: Met. Trans **8A**, 1363-1371 (1977)

8.49   M. Guttmann, D. McLean: "Grain Boundary Segregation in Multicomponent Systems" in *Interfacial Segregation*, ed. by W.C. Johnson and J.M. Blakely (ASM Metals Park, Ohio 1979) Chap.9, pp. 261-348

8.50   S. Suzuki, K. Abiko, H. Kimura: Scripta Metall. **15**, 1139-1143 (1981)

8.51   W.C. Johnson: "Interphase Boundary Segregation and Materials Properties", in *Interfacial Segregation*, ed. by W.C. Johnson and J.M. Blakely (ASM Metals Park, Ohio 1979) Chap. 10, pp. 351-379

8.52   C. Lea, M.P. Seah: Phil. Mag. **35**, 213-228 (1977)

8.53   C. Lea, M.P. Seah: Met. Sci. **18**

8.54   M.P. Seah: Acta Metall. **25**, 345-357 (1977)

8.55   W.C. Johnson, N.G. Charka, R. Ku, J.L. Bomback, P.P. Wynblatt: J. Vac. Sci. Technolog. **15**, 467-469 (1978)

8.56   M.P. Seah: Phil. Trans. Roy. Soc. A **295**, 265-278 (1980)

8.57   U. Franzoni, H. Goretski, S. Sturlese: Scripta Met. **15**, 743-748 (1981)

8.58   J. Bernardini, P. Gas, E.D. Hondros, M.P. Seah: Proc. Roy. Soc. A **379**, 159-178 (1982)

8.59   J. Bernardini, C. Lea, E.D. Hondros: Scripta Met. **15**, 649-652 (1981)

8.60   S. Danyluk, G.E. McGuire, K.M. Koliwad, M.G. Yang: Thin Solid Films **25**, 483-489 (1975)

8.61   P.H. Holloway, D.E. Amos, G.C. Nelson: J. Appl. Phys. **47**, 3769-3775 (1976)

8.62   G.C. Nelson, P.H. Holloway: "Determination of the Low Temperature Diffusion of Chromium through Gold Films by Ion Scattering Spectroscopy and Auger Electron Spectroscopy" in *Surface Analysis Techniques for Metallurgical Applications, ASTM STP 596* (ASTM Pa, 1976) pp. 68-78

8.63   P.M. Hall, J.M. Morabito, J.M. Poate: Thin Solid Films **33**, 107-134 (1976)

8.64   P.M. Hall, J.M. Morabito, N.T. Panousis: Thin Solid Films **41**, 341-361 (1977)

8.65   T.J. Chuang, K. Wandelt: Surf. Sci. **81**, 355-369 (1979)

8.66   J.A. Venables, J. Derrien, A.P. Janssen: Surf. Sci. **95**, 411-430 (1980)

8.67   A.P. Janssen, J.A. Venables: Surf. Sci. **77**, 351-364 (1978)

8.68   C. Lea, C. Molinari: J. Mater. Sci. **19**, 2336-2352 (1984)

8.69   K. Singer: Thin Solid Films **57**, 115-126 (1979)

8.70   R.E. Waters, J.A. Charles, C. Lea: Metals Technolog. **8**, 194-200 (1981)

8.71   C. Lea, B. Roebuck: Metal Sci. **15**, 262-266 (1981)

8.72   C. Lea, B.C. Muddle, D.V. Edmonds: Met. Trans. **14A**, 667-677 (1983)

8.73   C.T.H. Stoddart, C. Lea, W.A. Dench, P. Green, H.R. Pettit: Metals Technolog. **6**, 176-184 (1979)

8.74   D.H. Buckley: J. Vac. Sci. Technolog. **13**, 88-95 (1976)

8.75   G.E. McGuire, P.H. Holloway: "Applications of Auger Spectroscopy in *Electron Spectroscopy, Theory Techniques and Applications*", Vol. 4, ed. by C.R. Brundle and A.D. Baker (Academic Press, New York 1981) Chap. 1, pp. 1-84

8.76   N.H. Turner, B.I. Dunlap, R.J. Colton: Anal. Chem. **56**, 373R-416R (1984)

9. Positron Annihilation

W. Triftshäuser

With 28 Figures

It is intended to describe the use of positron annihilation spectroscopy in metals and alloys. In the sections following the introduction of the basic principles, applications of positron annihilation are presented to study defect and defect-free materials. The investigation of Fermi-surface properties in single crystals of metals and alloys is one of the very interesting topics in positron annihilation. Even more exciting and in some way unique is the positron method in the field of defects in metallic materials, such as vacancies in thermal equilibrium, quenched-in defects, dislocations, grain boundaries associated with phase transitions, amorphous structures, and radiation-induced defects and their agglomerates, together with implanted gases.

9.1 Background

Experiments in physics, involving the interaction of matter with antimatter, are fascinating. The first antiparticle discovered was the positive electron, the positron. Later more and more antimatter particles were found, and as of today each elementary particle has its counterpart in antimatter. But from all the antimatter particles only the positron has succeeded so far to become a valuable probe in solid state physics investigations. The work carried out in this field, from the collection of the basic facts to the verification of the existence of the positron itself, and then further to the wide perspectives as a potential tool for applied purposes, can serve as a nice example for exhibiting the development from pure basic science to applied research and to a standard method in metal physics and material science.

In his relativistic theory *Dirac* predicted in 1930 the existence of a particle which could either be interpreted as a hole in the continuum of the negative electron states or as a positively charged particle, the positron, with the rest mass of an electron [9.1]. *Anderson* found in 1932 the first experimental evidence of the existence of a positively charged electron in his studies of cosmic radiation [9.2]. Later it was confirmed that this positively charged electron, the positron, is the antiparticle to the electron and that it was the particle predicted by *Dirac*. The rest mass of the positron is that of the electron $(m = 9.10956 \times 10^{-31} \text{kg})$ and the charge is the positive elementary charge $(e = 1.60219 \times 10^{-19} \text{As})$.

From the discovery of the positron to the first application of positron annihilation in solid-state physics it took about ten years. The first attempt to measure the angular correlation of annihilating photons in metals was made by *Beringer* and *Montgomery* in 1942 [9.3]. In 1949 *De Benedetti* et al performed more accurate angular correlation measurements and also gave a theoretical estimate of the mean momentum of the annihilating electron-positron pair [9.4]. It was recognized that the annihilation of a positron with the electrons in a metal is a many-body effect. Some years later several groups published data on the momentum distribution of electrons in metals by positron annihilation [9.5,6]. The application of this method to metal physics has been rapidly developed, especially in the last two decades. The utility of positron annihilation studies of metals relies on the fact that the characteristics of the annihilation process which, in principle, involve sophisticated considerations of quantum electrodynamics, nevertheless depend almost entirely on the initial state of the positron-many-electron system. The study of the lifetime of positrons in matter was initiated by *De Benedetti* and *Riching* [9.7]. It has become a standard technique especially for the investigation of defects in metals and alloys.

A second break-through of the positron annihilation method occurred in the late 1960s when the high sensitivity of positrons to vacancies, vacancy-like defects and vacancy agglomerates was detected [9.8–13]. In defects, where atoms are missing or their density is locally reduced, the repulsion between the positron and the ion cores is decreased, and the redistribution mainly of the conduction electrons in metals causes a negative electrostatic potential at this site. Thus positrons see vacancies, dislocations and voids as strongly attractive centers in the metal, and they therefore represent a very sensitive microscopic probe for these kinds of defects. Since changes in the positron annihilation characteristics can already be observed for vacancy concentrations as low as 10^{-7}, this method is especially useful for low defect concentrations. Because of this high sensitivity, positrons can also be used to study the dynamical processes of vacancy agglomeration to voids in metals, which is of great technological interest with respect to materials used in high-flux fission and fusion reactors.

The positron annihilation method and its application to material science has gained further momentum through monoenergetic positron beams which have come into operation in recent years [9.14–18]. The development of positrons for use as a probe of surface phenomena and near-surface defects ($\leq 1\,\mu m$) awaited the discovery to produce an intensive flux of monoenergetic positrons with variable energy [9.19–22].

Since this chapter is only a part of a book on *Microscopic Methods in Metals* it cannot (because of space limitations) be a complete review on positron annihilation and its applications. More extended reviews on this subject are given in [9.23–26].

9.2 Basic Principles

9.2.1 Positron Thermalization

When an energetic positron (energy distribution of a beta spectrum) from a radioactive source enters a metal, it interacts with electrons and nuclei, and in this way loses rapidly its kinetic energy until it becomes thermalized. Inelastic collisions with electrons will result in ionization or electronic excitations depending on the energy transferred. Because the positron is a light particle, it will undergo appreciable angular deflections during the inelastic collisions with the electrons. Especially in the last stages of the slowing-down process the angular deflection by collision becomes so large that the correlation between the positron path in the metal and its initial direction is completely lost. Electronic excitations can be in the nature of electron-hole pair production or of collective plasmon excitations of the conduction electrons, and finally of positron-phonon scattering [9.27–29]. This last step of thermalization requires a longer time than the first stages of electronic slowing-down. The total thermalization time of about 10^{-11}s, however, is still short compared to the mean lifetime $\bar{\tau} \geq 10^{-10}$s of positrons in metals. The first theoretical calculations yielded times of 10^{-12}s for the positron to reach thermal energy [9.30]. At low positron energies positron-phonon scattering is more effective in dissipating the positron energy than electron-hole pair creation, and hence considerably shortens the time required for thermalization [9.27]. Theoretically a positron in a solid should reach thermal equilibrium down to temperatures close to absolute zero before being annihilated [9.31]. Precise measurements have revealed evidence for thermalization down to about $10\,\mathrm{K}$ [9.32].

Because of the short mean lifetime ($\bar{\tau} \geq 10^{-10}$s) of positrons in metals together with conventional positron-source intensities (up to several 10^{-1} Curies (Ci)), there is on the average only one positron present in the specimen at a time, and therefore the positron can be in the lowest possible energy state. Therefore thermalization justifies the assumption that the positron momentum is small compared to the momenta of the electrons with which it annihilates: positron annihilation characteristics convey information about electronic states sampled by the positron wave function. The thermalized positron, still scattered by phonons, diffuses in the material until it finally annihilates with an electron. The mean diffusion length depends strongly on the status of the specimen [9.33,34]. In perfect crystals this length is of the order of $1000\,\text{Å}$ or less. In a single crystal the lattice structure can influence the motion of positrons [9.35]. Channelling along crystallographic planes or directions is expected to favour deeper penetration in the oriented crystal than in random directions. But the acceptance criterion for positron channeling limits the fraction of channelled positrons from a isotropic source to a few percent [9.36,37].

9.2.2 Annihilation Process

The transformation of mass into energy, as in the case for the interaction of matter with antimatter, is called annihilation. The process involves certain general laws of conservation: the conservation of total energy, the conservation of total linear momentum, the conservation of total angular momentum, the conservation of total electric charge and of parity. For the annihilation of the electron-positron pair the total rest mass m_0 of the two particles is converted into energy ($E = 2m_0c^2$).

Therefore the study of the nature of the annihilation photons will give information about the state of the annihilating electron-positron pair immediately before the transition. From quantum electrodynamics there are several possibilities in which the annihilation of a positron with an electron will occur. The non-photon and the single-photon annihilation will only occur in the presence of a third body, an electron or a nucleus, which can absorb the recoil momentum. These processes are very rare and are only interesting from a theoretical point of view, but they are of no practical importance. From the requirements of conservation of angular momentum and parity it follows that the electron-positron pair must decay under emission of three photons if the spins of the two particles are parallel, and under emission of two photons if the spins are antiparallel. The two-photon annihilation process requires that the photons are emitted in opposite directions in the center of a mass system (conservation of linear momentum), each of the photon carrying half of the energy of the process. In the three-photon annihilation process the photons can be emitted in various directions relative to each other, and the energy of the photons can vary from zero to the maximum energy of $m_0c^2 = 511$ keV.

The cross section σ_{2-ph} for the two-photon annihilation of a free positron and a stationary electron, summed over all spin directions, was calculated by *Dirac* [9.1] as

$$\sigma_{2-ph} = \frac{\pi r_0^2}{\gamma + 1} \left\{ \frac{\gamma^2 + 4\gamma + 1)}{\gamma^2 - 1} \ln[\gamma + (\gamma^2 - 1)^{1/2}] - \frac{\gamma + 3}{(\gamma^2 - 1)^{1/2}} \right\} \tag{9.1}$$

where $\gamma = (1 - v^2/c^2)^{-1/2}$, v is the velocity of the positron, and $r_0 = e^2/4\pi\varepsilon_0 m_0c^2$ is the classical electron radius. In the limit of low kinetic energies of the electron and the positron, (9.1) reduces to

$$\sigma_{2-ph} = \frac{\pi r_0^2 c}{v}. \tag{9.2}$$

This cross section for two-photon annihilation is about 371 times larger than the one for three-photon annihilation, and therefore for almost all investigations this decay mode is used in positron annihilation experiments. The annihilation probability λ (inverse lifetime τ) for this process becomes independent of the positron velocity and proportional to the electron density at the site of the positron $n_e = |\psi(0)|^2$

$$\lambda = \sigma_{2-ph} v |\psi(0)|^2 = \pi r_0^2 c n_e. \tag{9.3}$$

In this expression neither the electron-positron nor the electron-electron Coulomb interaction is included. The changes required to take into account the complete many-body interaction (one positron-many electrons) encountered in real systems, such as a positron annihilating in an electron gas or in condensed matter, have been the subject of many theoretical investigations [9.38–45].

A positron in an electron gas attracts a cloud of electrons around it that effectively screens its positive charge. Hence, the electron density at the position of the positron is higher than obtained by the independent-particle model and consequently the annihilation rate will be enhanced. This effect is clearly seen in positron lifetime experiments where the measured lifetimes are about an order of magnitude shorter than predicted by the independent-particle theory. Inspite of this enhancement, the annihilation rate depends only weakly on the momentum of the electron with which the positron annihilates [9.46–49]. For electron momenta less than the Fermi momentum the electron-positron interactions cancel to a great extent the contributions resulting from the electron-electron correlations [9.39]. The quantitative explanation of these facts is one of the successes of the many-body theory.

In order to interpret the characteristics of the annihilation process, a relation is required for the probability $\Gamma(\boldsymbol{p})$ that a positron in an arbitrary system of electrons will annihilate under emission of two photons having total momentum $\boldsymbol{p}$. The first, most general, treatment of this problem was supplied by *Ferrel* [9.38] and *Chang Lee* [9.50]. This is, in particular, a many-body problem and cannot be solved exactly. In the case that the initial state of the electron-positron system can be represented by the product of a positron wave function and one-electron wave functions (independent-particle model) $\Gamma(\boldsymbol{p})$ can be calculated [9.4]; i.e.,

$$\Gamma(\boldsymbol{p})\mathrm{d}\boldsymbol{p} = \frac{\pi r_0^2 c}{(2\pi)^3} \varrho(\boldsymbol{p})\mathrm{d}\boldsymbol{p} \quad \text{with} \tag{9.4}$$

$$\varrho(\boldsymbol{p}) = \sum_{k=1}^{n} \left| \int \exp\left(-\mathrm{i}\frac{\boldsymbol{p}\cdot\boldsymbol{r}}{\hbar}\right) \psi_+(\boldsymbol{r})\psi_-^k(\boldsymbol{r})\mathrm{d}\boldsymbol{r} \right|^2, \tag{9.5}$$

where $\varrho(\boldsymbol{p})$ is the photon-pair momentum density, $\psi_+(\boldsymbol{r})$ and $\psi_-^k(\boldsymbol{r})$ represent the positron and k^{th} electron wave function, respectively. The summation extends over all occupied n electron states.

At a temperature T the thermalized positron is near the bottom of the positron conduction band. Therefore its momentum is determined by a Boltzmann distribution

$$f_+(p_+) = (2\pi m^* k_B T)^{-3/2} \exp\left(-\frac{p_+^2}{2m^* k_B T}\right) \tag{9.6}$$

where m^* is the positron effective mass, and k_B is the Boltzmann constant. The momentum distribution of the electrons over occupied states labelled by the wave vector $\boldsymbol{k}$ is given by the Fermi-Dirac distribution

$$f_-[E(\boldsymbol{k}_-)] = \frac{1}{\exp\left(\frac{E(\boldsymbol{k}_-) - E_F}{k_B T}\right) + 1}, \tag{9.7}$$

where E_F is the Fermi energy. Since the two annihilation photons carry away the combined momenta of the electron and the positron, the momentum distribution at a finite temperature T is given by the convolution of (9.6) with (9.4). At T = 0, the positron is in its ground state $k_+ = 0$, and $f_+(p_+)$ is a delta function at $p_+ = 0$, while $f_-[E(k_-)]$ is unity for all occupied states k_- and zero for all unoccupied states.

In a crystalline solid (periodic crystal) the positrons move in an array of repulsive ion potentials, and $\psi_+(r)$ and $\psi^k_-(r)$ become Bloch waves

$$\psi_{k_+}(r) = u_{k_+}(r)\,\exp\,(ik_+\cdot r) \quad \text{and} \tag{9.8}$$

$$\psi_{k_-}(r) = u_{k_{-,j}}(r)\,\exp\,(ik_-\cdot r) \tag{9.9}$$

where $u_{k_{-,j}}(r)$ has the lattice periodicity, and j labels the energy band. At finite temperatures (9.5) can be written as

$$\varrho(p, T) = \sum_{k_+, k_{-,j}} f_-(E_{k_{-,j}}, T) f_+(E_{k_+}, T)|A_j(p, k_-, k_+)|^2 \tag{9.10}$$

where $A_j(p, k_-, k_+)$ represents the Fourier transform of the positron and electron wave function

$$A_j(p, k_-, k_+) = \int_V \exp\left(-\frac{i}{\hbar}p\cdot r\right)\psi_{k_+}(r)\psi_{k_-}(r)dr \quad \text{or} \tag{9.11}$$

$$A_j(p, k_-, k_+) = \frac{1}{V}\int_V \exp\left[-i\left(\frac{p}{\hbar} - k_- - k_+\right)\cdot r\right]u_{k_{-,j}}(r)u_{k_+}(r)dr. \tag{9.12}$$

The integration in (9.11 and 12), which extends over the total volume V of the crystal, may be reduced to an integration only over the volume v_u of the unit cell by making use of the periodicity of $u_{k_{-,j}}(r)$ and $u_{k_+}(r)$. This results in

$$A_j(p, k_-, k_+) = \frac{1}{v_u}\delta\left(\frac{p}{\hbar} - k_- - k_+ - G\right)$$
$$\int_{v_u} \exp\,(-G\cdot r)u_{k_{-,j}}(r)u_{k_+}(r)dr \tag{9.13}$$

where G is a reciprocal lattice vector. The delta function regulates conservation of crystal momentum. The momentum carried away by the annihilation photons equals that of the electron-positron pair up to a reciprocal lattice vector G. At zero temperature where the positron is in its ground state ($k_+ = 0$), an electron state k does not only contribute to $\varrho(p)$ at $p = \hbar\cdot k$ but also at $p = \hbar(k+G)$ ("Umklapp" process) with each contribution having a weighting factor $|A_j(p, k_-)|^2$ [9.51].

The total annihilation rate λ is obtained by integrating $\Gamma(p)$ over all photon momenta, i.e.

$$\lambda = \int\limits_{-\infty}^{+\infty} \Gamma(\boldsymbol{p})\mathrm{d}\boldsymbol{p} = \frac{1}{\tau}.$$ (9.14)

The inverse of λ is the positron lifetime τ.

9.3 Experimental Methods

9.3.1 Positron Sources

Isotopes emitting positrons in a nuclear transition can be used as positron sources. Although there are many isotopes which decay via the emission of positrons, there are only a few suitable as positron sources in experimental studies, because of their half-lives and their positron-decay efficiencies. The most commonly applied sources are listed in Table 9.1. The first four isotopes (^{11}C to ^{18}F) are mainly used in nuclear medicine applications (positron-emission tomography), because many organic compounds can be labelled relatively easily with these isotopes. ^{22}Na produced by a charged-particle reaction, e.g. ^{24}Mg(d, α)^{22}Na, is a very convenient positron source, because of its long half-life, its high positron-decay efficiency, and the fact that within a time interval of a few picoseconds a positron and a high-energy γ-ray (1.28 MeV) is emitted. This γ-ray can be used as indication signal for the "birth" of a positron, particularly in positron lifetime studies. In ^{44}Ti the emission of a positron is also accompanied by a γ-ray. However, the achievable specific activity of this isotope is much lower than that for ^{22}Na. ^{58}Co can be produced from nickel irradiated with fast neutrons in a nuclear reactor through the reaction ^{58}Ni(n, p)^{58}Co. It is separated from nickel by chemical methods (ion exchange); relative high specific activities ($\sim$500 mCi) are obtainable. By the absorption of thermal neutrons in a nuclear reactor the isotope ^{64}Cu is produced: ^{63}Cu(n, γ)^{64}Cu. The maximum obtainable saturation activity per unit mass is determined by the neutron flux. For a flux of 10^{15} n/s cm^2 and an irradiation time of 24 hours this activity is 580 Ci per gram copper. Thus positron sources of high specific activity can be obtained by this process.

Table 9.1. Positron sources

Nuclide	Half-life	Positron-decay efficiency [%]	Maximum emitted energy [MeV]
^{11}C	20 min	99	0.96
^{13}N	9.96 min	100	1.20
^{15}O	2.05 min	100	1.74
^{18}F	1.83 h	97	0.635
^{22}Na	2.62 y	90.6	0.545
^{44}Ti	47 y	94	1.47
^{58}Co	71.3 d	15.0	0.475
^{64}Cu	12.8 h	19.0	0.656
^{68}Ge	275 d	88	1.88

In recent years positrons made by pair-production came into use [9.21,22]. Through the Bremsstrahlung from the stopping of high-energy electrons of an electron accelerator (e.g., LINAC), electron-positron pairs are produced in a converter target. This source of positrons is primarily used in positron-beam systems and can, in principle, yield high positron intensities, because of the large average current in electron accelerators [9.52,53]. However, the real useful power of such sources has yet to be demonstrated.

The principles of the positron annihilation methods are indicated schematically in Fig. 9.1: angular correlation, lifetime and Doppler broadening.

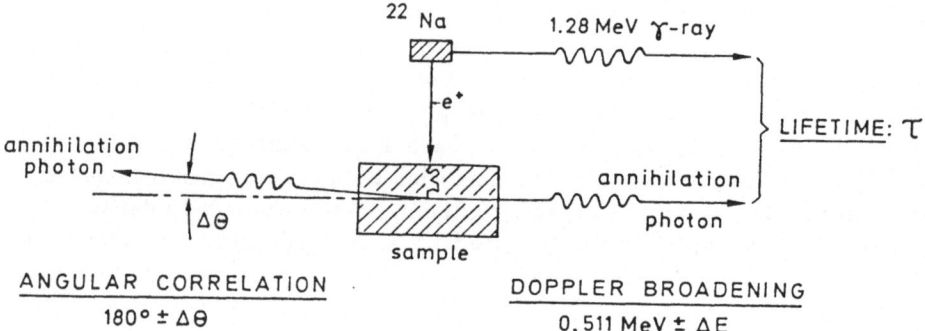

Fig. 9.1. The schematic of positron annihilation and the experimental methods: angular correlation, lifetime and Doppler broadening

9.3.2 Angular Correlation of Annihilation Photons

The annihilation of an electron-positron pair produces two photons with energy of nearly 511 keV, as illustrated in Fig. 9.1. Conservation of momentum requires that in the center-of-mass system they are emitted collinearly. In the laboratory system the momentum components p_x and p_z perpendicular to the direction of the annihilation photons cause a deviation from 180 degrees. This transverse momentum can be determined by a two-photon angular correlation system via measuring the pair-coincidence rate of the annihilation photons as a function of the angle between the two quanta. A typical setup of such a system is shown schematically in Fig. 9.2. The precise measurement of the angle between the photons requires a high angular resolution which is achieved by a small detector opening in the z direction and a large distance (specimen-detector) in the y direction. By the use of detectors, which subtend a large angle (100–250 milliradians (mrad)) in the x direction, the opening angle of the detectors and hence the coincidence-count rate is increased while a high angular resolution (0.2–0.8 mrad) along the z direction is maintained. Since the Doppler energy shift $\Delta E \leq 5$ keV is small compared to the energy resolution of the NaI detectors, and the opening angle in the horizontal plane is now much wider than the angular distribution $\Delta\theta \leq 20$ mrad, all information concerning p_x and p_y is lost and the coincidence-count rate $N(p_z)$ is given from (9.4) by

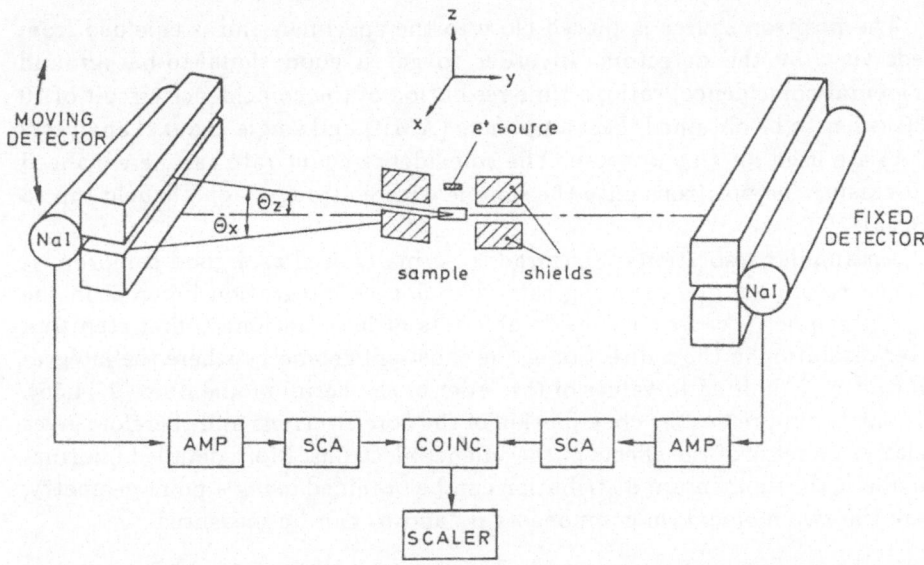

Fig. 9.2. The schematic experimental setup of a long-slit angular-correlation system

$$N(p_z)\Delta p_z = \Delta p_z \int_{-\infty}^{+\infty} dp_x \int_{-\infty}^{+\infty} \Gamma(\boldsymbol{p})dp_y \qquad (9.15)$$

where Δp_z corresponds to the detector opening in the z direction, $p_z = m_0 c\theta_z$ and $p_x = m_0 c\theta_x$. This is the so-called long-slit geometry employed in angular correlation experiments [9.5,6]. $N(p_z)$ is proportional to $\Gamma(\boldsymbol{p})$ integrated over a slice of thickness Δp_z in momentum space perpendicular to z (Fig. 9.3, left section).

LONG–SLIT GEOMETRY   POINT GEOMETRY

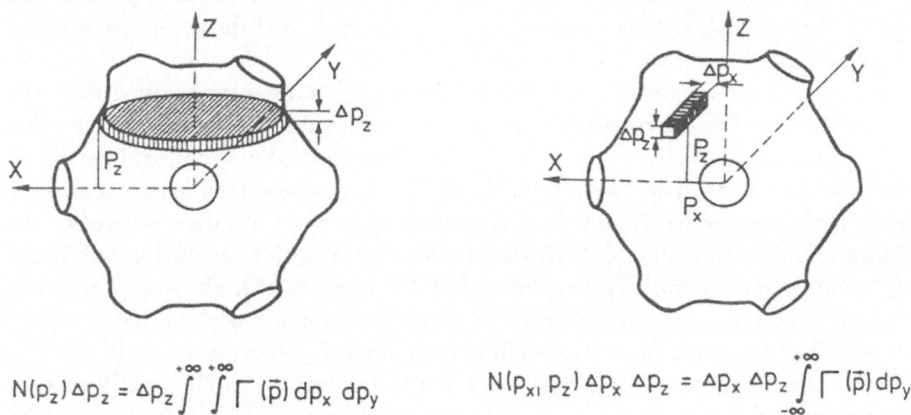

$$N(p_z)\Delta p_z = \Delta p_z \int_{-\infty}^{+\infty}\int_{-\infty}^{+\infty} \Gamma(\bar{p}) \, dp_x \, dp_y \qquad\qquad N(p_{x},p_z)\Delta p_x \Delta p_z = \Delta p_x \Delta p_z \int_{-\infty}^{+\infty} \Gamma(\bar{p}) \, dp_y$$

Fig. 9.3. Comparison of positron angular correlation using long-slit or point geometry

257

The positron source is placed close to the specimen and is shielded from direct view by the detectors. In order to get a good signal-to-background (accidental-coincidence) ratio, a time resolution of the coincidence circuit of 10 to 30 ns has to be obtained. Fast amplifiers (AMP) and single-channel analyzers (SCA) are used for this purpose. The coincidence-count rate can be enhanced by focussing the positrons onto the sample with a strong magnetic field (up to 6 Tesla).

An angular resolution of 0.5 mrad is normally used as a good compromise between resolution and counting rate. The double integration inherent in the long-slit geometry causes a considerable loss of information. A first step to a better resolution in the x direction is the short-slit geometry where the integration over p_x is limited to values of the order of the Fermi momentum [9.54,55]. This partly suppresses the contribution of the core electrons and therefore gives a relative increase of the effects of the valence electrons. More detailed information about the momentum distribution can be obtained using a point geometry, where the two momentum components p_x and p_z can be measured

$$
N(p_x, p_z)\Delta p_x \Delta p_z = \Delta p_x \Delta p_z \int\limits_{-\infty}^{+\infty} \Gamma(\boldsymbol{p})dp_y. \tag{9.16}
$$

Here $N(p_x, p_z)$ is again proportional to $\Gamma(\boldsymbol{p})$ but integrated only over a channel of width $\Delta p_x \times \Delta p_z$ along the y direction in momentum space (Fig. 9.3, right section). In the first point-detector arrangement the counting rate was very low, but the data exhibited a considerable amount of detail not obtainable with a long-slit geometry [9.56]. The considerable loss in counting rate inherent in the use of point detectors or crossed slits can be overcome by the use of multiple-detector systems that allow truly two-dimensional measurements. The first position-sensitive detector system for positron annihilation measurements was proposed by *Triftshäuser* in 1971 [9.57] and a prototype using ion-implanted germanium was developed [9.58]. A more conventional approach was taken by *Berko* with the use first of 11 pairs of small NaI-detectors [9.59] and then of 32 pairs as a multicounter system [9.60].

Since then other position-sensitive detector systems were applied for two-dimensional angular correlation devices. The technique of using a large slice of a NaI crystal optically coupled to many photomultipliers for localization of γ-rays was introduced by *Anger* [9.61,62]. A pair of modern Anger cameras is used for a very effective system resulting in a final angular resolution of 0.65 mrad × 0.65 mrad [9.63,64]. However, the best angular resolution will most likely be obtained by multi-wire proportional chambers of high γ-ray detection efficiency. A combination of high-resolution lead converters with multi-wire proportional chambers has been successfully developed by *Jeavons* et al. [9.65-67]. A 10 cm × 10 cm prototype was first used at the University of Geneva [9.66,68]. New cameras of this type are currently installed in Geneva, in Petten and in Munich. The efficiency of these new systems is up to 17% for a 511 keV γ-

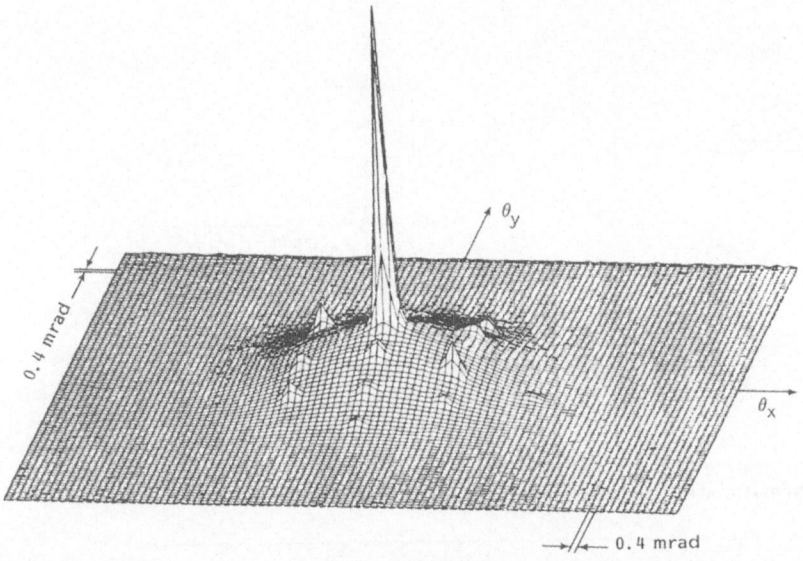

Fig. 9.4. Two-dimensional angular correlation distribution in quartz at 77 K with a (0.7×0.7) mrad^2 resolution as obtained by *Manuel* [9.67]

ray at an area of 31 cm × 31 cm and with a geometrical resolution between 1.0 and 1.5 mm. Therefore, an angular resolution of 0.2 mrad × 0.2 mrad with high counting efficiency can be achieved. In the near future possibly more systems of this kind will come into operation. The performance and potential power of such a system is clearly demonstrated in Fig. 9.4 in the case of a quartz single crystal [9.67]. The sharp peak at $\theta_y = \theta_y = 0$ and the "Umklapp" processes at momenta corresponding to the reciprocal lattice, caused by the annihilation of para-positronium from a Bloch state, are well resolved [9.69–71].

9.3.3 Doppler Broadening of Annihilation Radiation

In angular correlation experiments only two momentum components (the transverse components p_x and p_z) of the total momentum can be determined. The third component p_y, parallel to the main direction of the annihilation photons, can be measured by detecting the energy of the two photons E_1 or E_2, i.e.,

$$E_{1,2} = E_0 \pm \Delta E = m_0 c^2 - E_B/2 \pm c p_y/2. \tag{9.17}$$

Typical atomic momenta lead to Doppler shifts of a few keV (for a 10 eV electron ΔE is 1.598 keV). Thus the smaller contribution of the binding energy E_B of the electron and the positron in the system can be neglected (the positron annihilates mainly with the outer electrons). The subsequent broadening of the annihilation-radiation line can be measured with a solid-state detector (e.g., high-purity germanium diode). A typical setup of such a system is shown in Fig. 9.5. The energy resolution of modern detector systems is 1.15 keV at

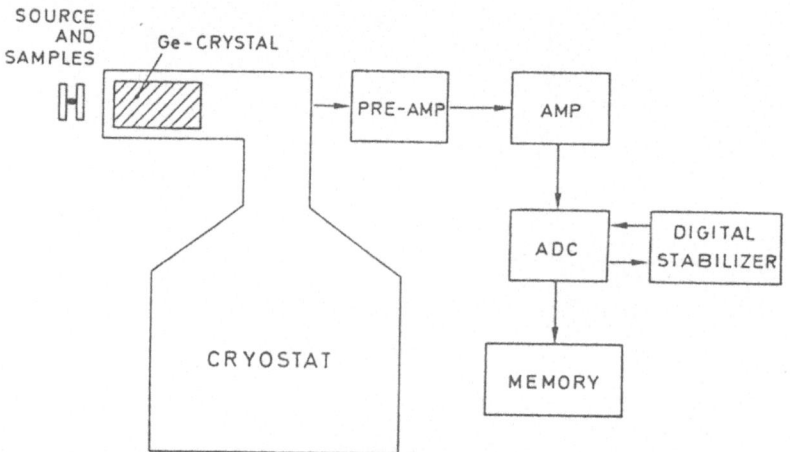

Fig. 9.5. Schematic arrangement of a system for Doppler broadening measurements

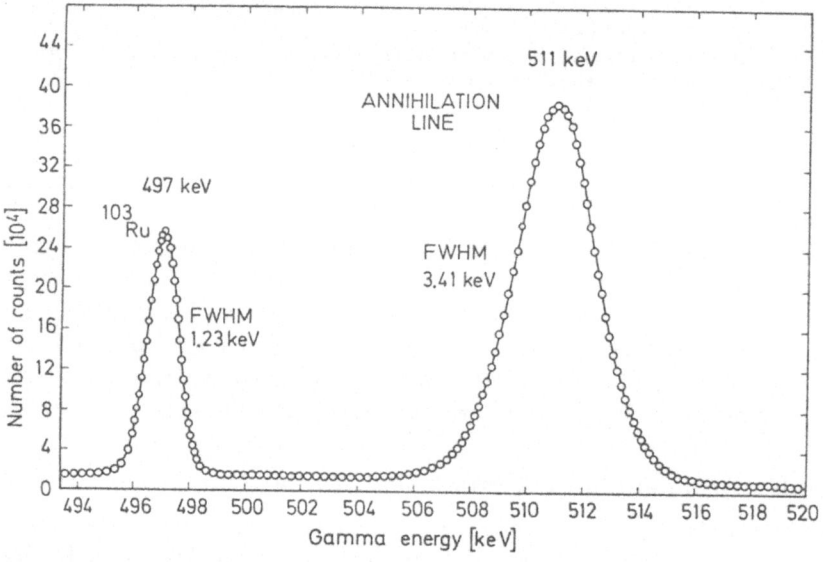

Fig. 9.6. Measured energy spectrum of the Doppler broadened annihilation line compared with the γ-line of ^{103}Ru at 497 keV

497 keV which corresponds to about 4 mrad. The information obtained is similar to that in the long-slit geometry. However, the resolution is more than one order of magnitude lower than in angular correlation measurements. Therefore the Doppler broadening technique is not so well suited for Fermi-surface studies, but it is quite powerful for investigating material properties (e.g., lattice defects) where high resolution is not of primary importance. The experiments are relatively easy to perform with modern electronic equipment. Since coincidence counts need not to be taken, the data-acquisition rate is very high and the intensity of the positron source can be relatively low (10 to 30 μCi). With a

multichannel analyzer (analog-digital converter and memory) the whole energy spectrum is accumulated. A typical Doppler broadening spectrum is shown in Fig. 9.6 compared with the natural line width of the 497 keV γ-ray of ^{103}Ru.

9.3.4 Lifetime Measurements of Positrons

As illustrated in Fig. 9.1, the lifetime of a positron can be determined by measuring the time elapsed between the appearance of the γ-ray from the source and one of the annihilation photons. The measuring system consists of a so-called fast-slow coincidence system with a fast branch for the time measurement and a slow branch for the energy selection of the photons. Figure 9.7 gives a schematic representation of a typical lifetime setup. The slow-coincidence circuit is used to focus the detectors on the annihilation photons and on the 1.28 MeV γ-ray (if a ^{22}Na source is used), respectively. The coincidence signal gates the timing signal from the fast coincidence circuit into the analog-digital converter (ADC). The anode outputs of the two photo-multipliers are fed into constant-fraction discriminators (CFD). The time difference between the start pulse from the 1.28 MeV γ-ray and the stop pulse from one of the annihilation photons is transformed by the time-to-amplitude converter (TAC) into an electronic signal with an amplitude proportional to the time difference. The output signal from the TAC is fed into the ADC and stored in the memory. For precise measurements a digital stabilizer for the ADC is necessary in order to compensate possible electronic shifts. For small time differences (short positron lifetimes) the lifetime spectrum (the relationship between counts and

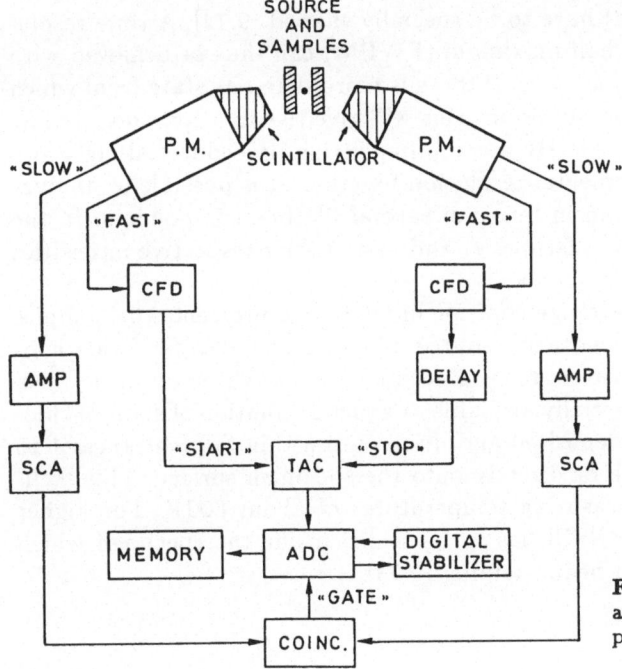

Fig. 9.7. Schematic diagram of a fast-slow coincidence system for positron lifetime measurements

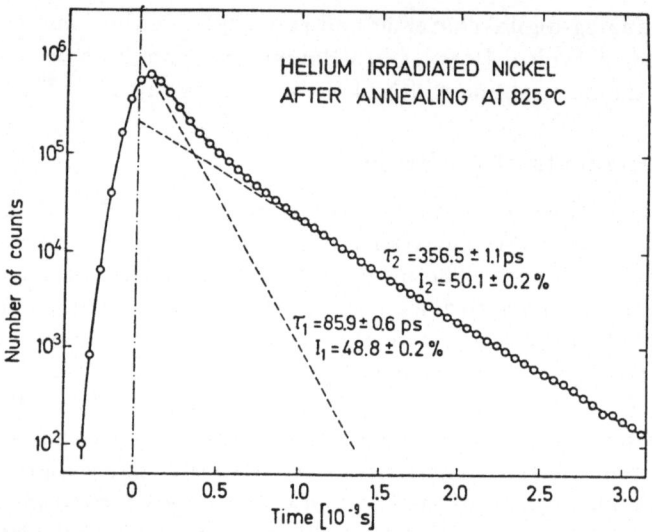

Fig. 9.8. Positron lifetime spectrum for helium-irradiated nickel. Two lifetimes and their intensities are derived from the measured spectrum

the time interval) is smeared by the time-resolution function of the apparatus. The instrumental-resolution function (prompt curve) can be obtained by replacing the positron source by a ^{60}Co source which simultaneously emits two γ-rays. In most cases this resolution function can be best approximated by a sum of several Gaussian functions. For the best instrumental resolution, the scintillator crystals (plastic) have to be specially shaped [9.72]. A time resolution of 165 ps full width at half maximum (FWHM) can thus be achieved with modern electronic instrumentations. If there is more than one state from which the annihilation of positrons can occur, this will give rise to a complex lifetime spectrum, as shown in Fig. 9.8. By a computer-fitting procedure taking properly into account the instrumental resolution function, it is possible to analyze a measured lifetime spectrum in terms of several lifetime components. In the example given in Fig. 9.8 two lifetimes τ_1 and τ_2 and their respective intensities I_1 and I_2 are resolved.

In the most common arrangement for lifetime measurements the samples to be investigated and the positron source are in a so-called sandwich configuration, i.e. the radioactive source is placed in-between the specimens. ^{22}Na positron sources are commercially available as aqueous solution of sodium chloride. This solution can be deposited and dried onto a thin foil (thickness 1 to 5 μm) of titanium or nickel, or directly onto the specimen surface. This technique is sufficient up to measuring temperatures of about 600 K. For higher sample temperatures the ^{22}NaCl has to be sealed inside the specimen which can be achieved by electron beam welding [9.73].

9.3.5 Monoenergetic Positron Beams

The development of monoenergetic positrons of low energy for the use as a probe of near-surface phenomena started after it was found that thermalized positrons are re-emitted from metal surfaces [9.14–18]. As early as 1950 *Madansky* and *Rasetti* [9.74] made a first attempt of forming a beam of thermal-energy positrons by moderating the energetic positrons emitted from ^{64}Cu. Many years later *Groce* et al. [9.75,76] succeeded in producing a beam of a few positrons per second and measured the total scattering cross section of positrons in helium.

Positrons from a radioactive source are thermalized in the metal and diffuse back to the surface. Under certain conditions a reasonable amount of these positrons are re-emitted with energies between 1 and 2 eV. Using specially prepared surfaces and single crystals in ultra-high vacuum conditions, the conversion yield of moderated to unmoderated positrons was improved from 10^{-6} to 2×10^{-3} [9.16,77–80].

The maximum energy of the re-emitted positrons is interpreted as the positron negative work function of the surface [9.79,81,82]. These positrons with an energy spread of about 1 eV can then be accelerated to form a more or less monoenergetic positron beam of almost any energy desired. By varying the energy of the positrons, the range of the positrons into the specimen can be adjusted and different regions in various depths from the surface can be sampled. The first higher-energy beam system which allowed positron energies up to 30 keV was operated by *Triftshäuser* and *Kögel* [9.83]. This system, using a magnetic guidance field, is shown schematically in Fig. 9.9. It has specially been designed as a compact system under ultra-high-vacuum conditions and is appropriate to operate at positron energies up to 60 keV or even higher.

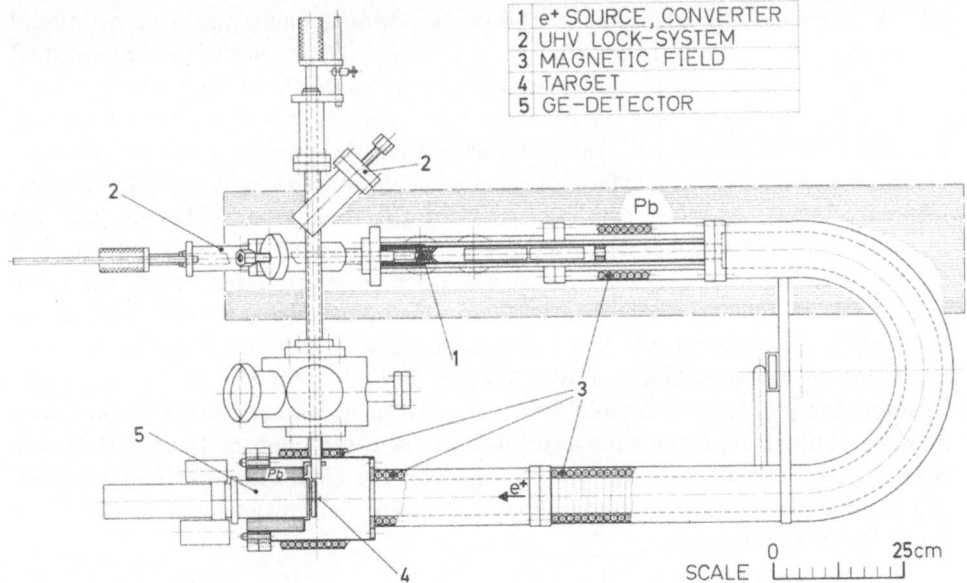

1 e⁺ SOURCE, CONVERTER
2 UHV LOCK–SYSTEM
3 MAGNETIC FIELD
4 TARGET
5 GE–DETECTOR

Fig. 9.9. Experimental arrangement of a positron-beam system for positron energies up to 30 keV using a magnetic guidance field

Positrons from a 30 mCi ^{22}Na source are thermalized in a tungsten moderator (1 mm wide and 25 μm thick tungsten foils) [9.77]. By electrical lenses the re-emitted positrons are focused and accelerated to 150 eV in order to form a positron beam of about 4 mm diameter, and are then deflected 180 degrees in a longitudinal magnetic field of about 10^{-2} Tesla. The final positron energy is achieved by applying a high voltage to the target [9.83]. In later designs of higher-energy beams (e.g., Brockhaven), the target is at ground potential and the source is floating at high voltage.

For the detection of the annihilating positrons inside the target or at the surface mainly three methods have been used so far. One is the conventional Doppler broadening method (Sect. 9.3.3) where the line shape is measured as a function of the incident positron energy [9.20,34]. The second one, applied only recently to positron beams, is the angular correlation method using two-dimensional detector systems. In the third method the fraction of the positronium (Ps) is measured, which is formed at the surface [9.84–88]. The para-Ps state decays into two photons with the energy of each approximately equal to 511 keV. The ortho-Ps decays mainly into three photons producing a continuous energy distribution ranging from about zero to 511 keV for each of the quanta. The energy sum of these three photons equals 2×511 keV. If ortho-Ps is formed at the surface, the probability of decaying into three photons can be reduced by a pick-off or spin-exchange process resulting in a two-photon annihilation. The measured photon-energy spectrum of those positrons annihilating from the ortho-Ps state is markedly different from the two-photon process when positrons decay either in metal or from the para-Ps state. Figure 9.10 shows a typical Doppler broadening spectra for annealed nickel at two different positron energies. The Doppler broadened spectrum at the incident positron energy of 24 keV is practically identical to the corresponding one using a conventional positron source. At the lower energy of 1.65 keV, where positrons are stopped close to the surface and therefore are able to diffuse to the surface, the influence of the specimen surface is clearly visible. This results in a considerable narrowing of the annihilation peak, very similar to that observed for positron trapping at defects (Sect. 9.4.3). Surface properties such as defects, lattice distortions, impurity layers are responsible factors for this behaviour. The annihilation from para-Ps and pick-off from ortho-Ps will also contribute to this effect. The effects of helium irradiation will be discussed in more detail later. All three measured energy spectra in Fig. 9.10 are normalized to the same area under the annihilation spectrum. With this procedure it is found that the number of counts at energies below about 504 keV is higher for the positrons at low incident energy (1.65 keV) than for the positrons at higher energy (10 keV and 24 keV). This is evidence that annihilation via three photons from ortho-Ps is present. By simply measuring the change over the total annihilation spectrum, the Ps fraction F can be determined if certain assumptions are made. After *Lynn* [9.88] this fraction is

$$F = \left(1 + \frac{P_1}{P_0} \frac{R_1 - R_F}{R_F - R_0}\right)^{-1} \tag{9.18}$$

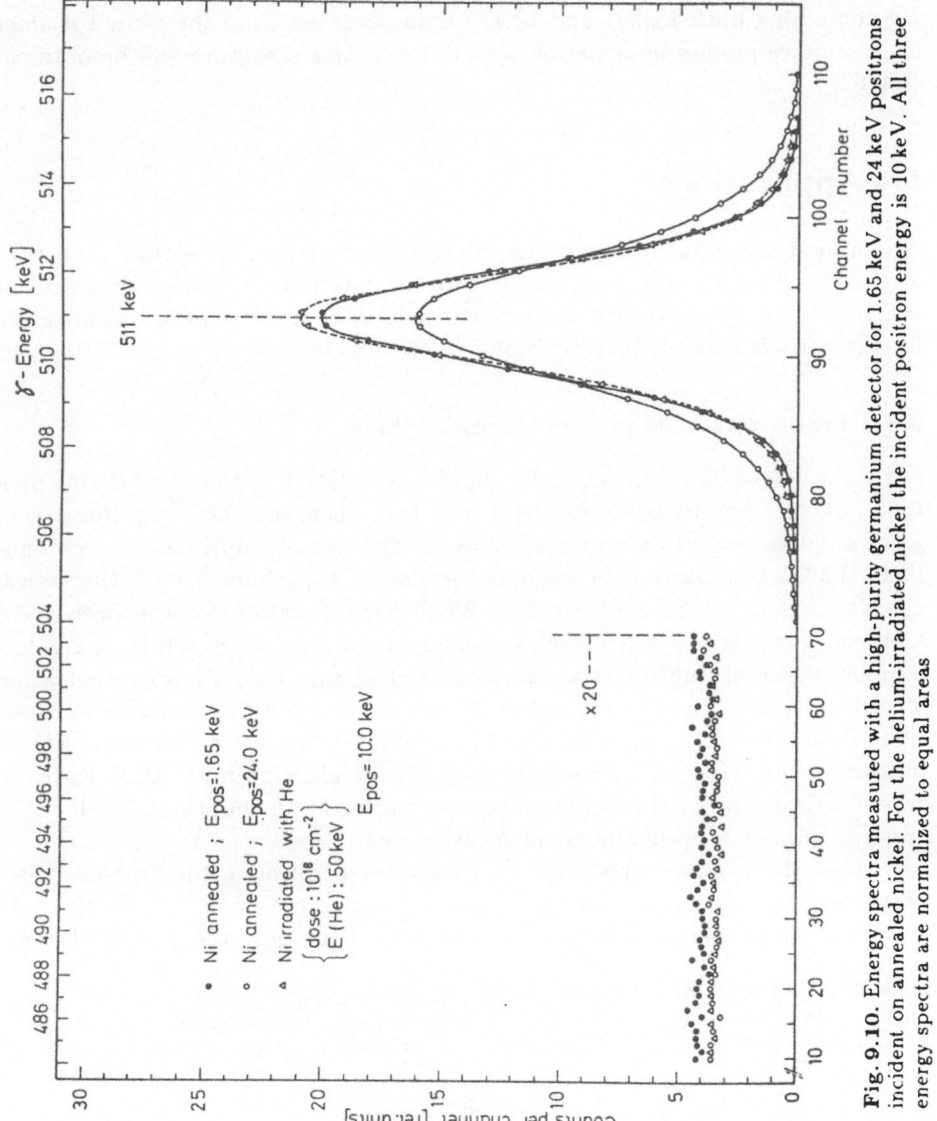

Fig. 9.10. Energy spectra measured with a high-purity germanium detector for 1.65 keV and 24 keV positrons incident on annealed nickel. For the helium-irradiated nickel the incident positron energy is 10 keV. All three energy spectra are normalized to equal areas

with $R_F = (T_F - P_F)/P_F$, where T_F is the total and P_F is the peak region of the measured energy spectrum. The subscripts 0 and 1 refer to 0% and 100% Ps formation, respectively. However, there are still uncertainties remaining for the determination of the experimental values for R_0, R_1 and P_1/P_0.

For low-energy positron-diffraction (LEPD) experiments an electrostatic guidance field for the positron beam probably is the best solution [9.89]. Analogous to LEED, monoenergetic positrons are diffracted at the metal surface.

By combining both LEPD and LEED data obtained from the same specimen surface more precise information about the surface structure will be obtained [9.90].

9.4 Applications

All the experimental methods described in the preceding section have been applied to the investigation of metals. Most of the experimental results available are from the work performed in the last two decades. Typical examples are given for the various fields of investigation.

9.4.1 Fermi Surfaces in Metals and Alloys

For the study of Fermi surfaces the angular correlation of the annihilating photons (ACAR) has to be determined to a high accuracy. This requires a very good angular resolution and preferentially the use of multi-detector systems (Sect. 9.3.2). Comparison of angular correlation experiments with theoretical calculations, in which a delocalized Bloch wave function for the positron is assumed, requires the use of well-annealed, metallurgically carefully characterized and oriented single-crystal samples which should have a low concentration of dislocations, vacancies and other defects to ensure no appreciable localization of the positron at such defects. Figure 9.11 shows two-dimensional ACAR measurements $N(p_y, p_x)$ according to (9.16) for aluminum at 100 K with the crystal orientation of the Brillouin zone as indicated in the insert [9.60]. Each crossing of lines is an independent measurement.

After the angular correlation data have been collected, the problem arises of extracting information about the underlying momentum distribution and the shape of the Fermi surface, i.e., to evaluate $\Gamma(\mathbf{p})$ from the measured $N(p_z)$

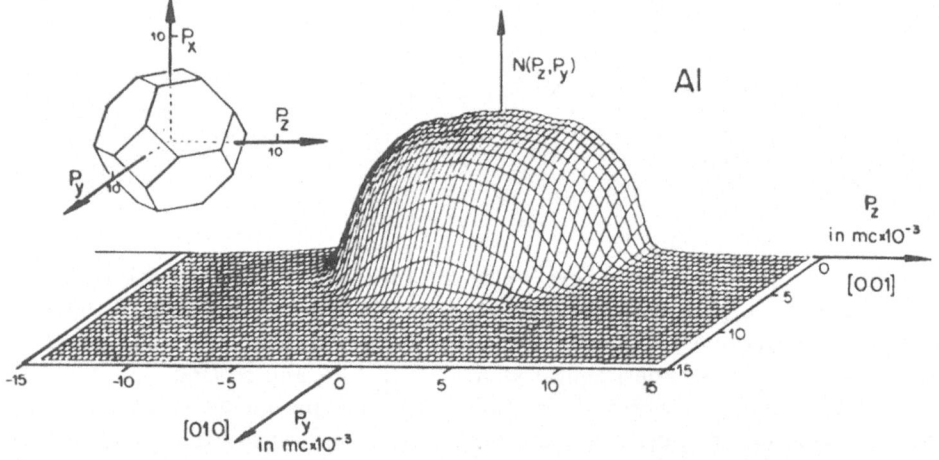

Fig. 9.11. Two-dimensional angular correlation distribution of aluminum at 100 K by *Berko* et al. [9.60]. The orientation of the crystal is illustrated by the Brillouin zone in the insert

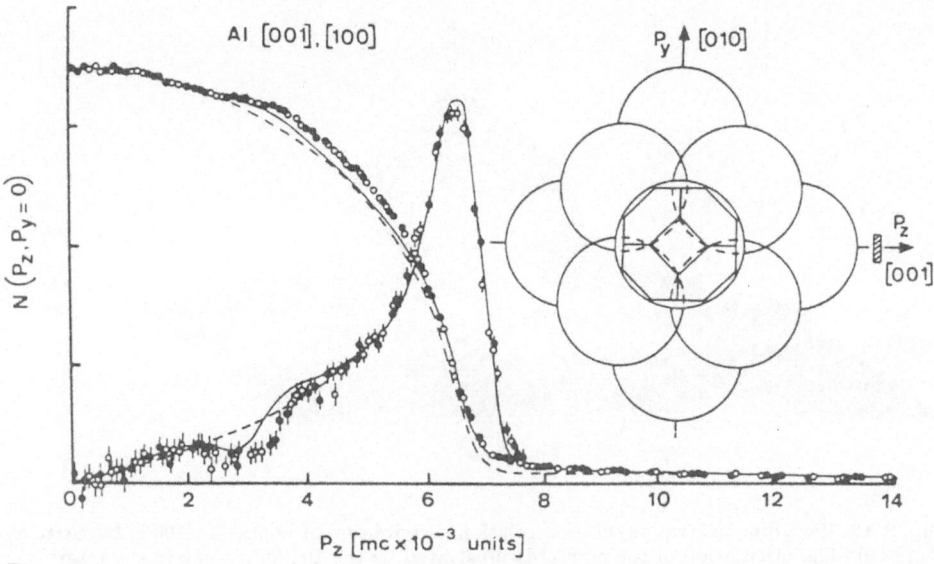

Fig. 9.12. Angular correlation in aluminum single crystal together with its first derivative by *Berko* et al. [9.60]. The full curve is the theoretical prediction from an OPW calculation

or $N(p_x, p_z)$. Various methods have been proposed and are being used for the reconstruction of $\Gamma(\boldsymbol{p})$ of (9.4,15 and 16) [9.91–96]. Thereby the propagation of statistical errors in the reconstruction methods deserves special attention. Experimental data used for reconstruction have to be of high statistical accuracy and should be taken in several crystal directions. For several metals and alloys band-structure calculations (e.g., using orthogonalized plane waves "OPW") are available, and the electron-momentum distribution predicted by theory can be directly compared with experimental results. Thereby a momentum-dependent enhancement for the conduction and core electrons has to be taken into account [9.39–45,97,98]. In Fig. 9.12 such an OPW calculation including enhancement [9.47] is compared with experimental data on aluminum [9.60]. The data are taken from the results in Fig. 9.11 at $p_y = 0$. A reasonably good agreement between theory and experiment is obtained.

Figure 9.13 shows two-dimensional ACAR measurements for copper at 100 K [9.60]. This metal has been used extensively as a test material for positron annihilation experiments [9.54,91,99,100]. Copper exhibits the typical band structure of metals having d electrons. It has an experimentally well-known yet sufficiently complex Fermi surface [9.101] and is theoretically well understood. The structure due to the Fermi surface necks around $p_y = p_z = 0$ is clearly exhibited. The agreement with the Fermi surface from de Haas-van Alphen measurements [9.101] is excellent as well as with theory [9.102]. Fermi surface studies on other pure metals, including ferromagnetic metals, can be found in [9.23,26].

The main advantage of the positron angular correlation method, however, is in the field of disordered systems such as random substitutional alloys. For

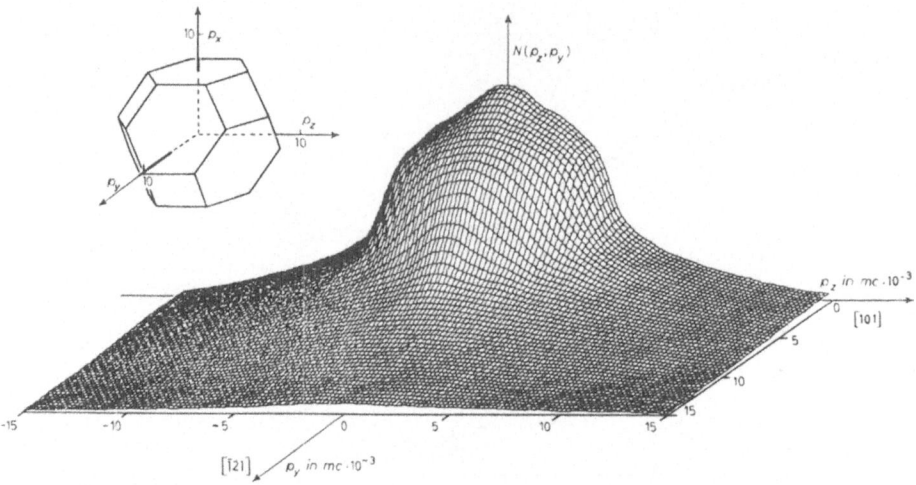

Fig. 9.13. Two-dimensional angular correlation distribution of copper at 100 K by *Berko* et al. [9.60]. The orientation of the crystal is illustrated by the Brillouin zone in the insert

these systems the ACAR technique provides an almost unique method for measuring the Fermi surface. The standard solid-state Fermi surface technique, such as the de Haas-van Alphen effect, cannot be applied to high-concentration random alloys, because of the restriction of a long mean-free path of the electrons. Positron annihilation, not requiring coherent electron wave packets, is free from such limitations, although the presence of the positron itself causes some problems.

Theories of the electronic structure of concentrated random-substitutional alloys enable realistic calculations of energy bands and density of states [9.103, 104]. An essential result of these calculations is the fact that the Fermi surface even in disordered alloys seems to be well defined since the smearing at the Fermi surface is still a small fraction of the total dimension, although larger than the thermal smearing. Two-dimensional ACAR measurements in α-brass single crystals were performed by *Haghgooie* et al. [9.105] and compared with pure copper. The α-brass momentum density is found to be quite similar to that of copper, except for the growth of the Fermi surface. This similarity is shown in Fig. 9.14 where the contour maps have been plotted for Cu and Cu-30at.% Zn. The Fermi surface is found to expand essentially uniformly with zinc concentration. The (111)-neck radius of (1.71 ± 0.07) mrad fits the prediction of 1.7 mrad from the rigid-band model. *Haghgooie* et al. [9.105] also concluded that the smearing at the Fermi surface, if present, is not observed with the available angular resolution. In the near future higher-resolution two-dimensional ACAR data may be possible and more alloy systems will be investigated.

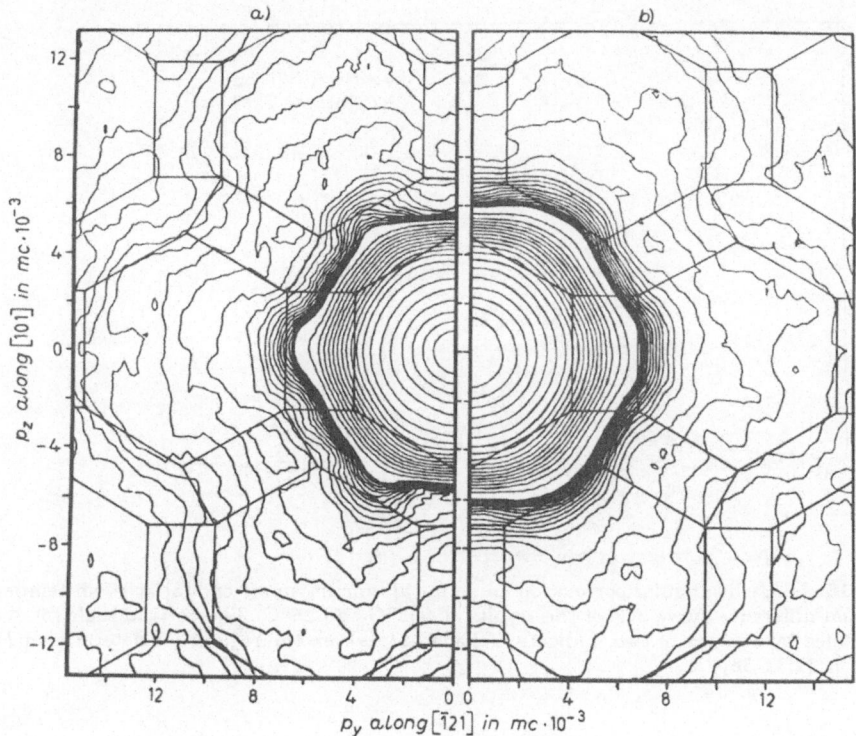

Fig. 9.14a,b. Contour map of two-dimensional angular correlation data for copper (a) and copper-30at.% zinc (b) single crystal by *Haghgooie* et al. [9.105]

9.4.2 Metals at Various Temperatures

In metals without lattice defects small changes in the positron annihilation parameters with temperature are expected due to the thermal expansion of the lattice. If a specimen contains defects, the positron can form a localized state at these sites and the annihilation characteristics are altered.

a) Vacancies in Thermal Equilibrium

The response of a metal to a finite temperature is, besides lattice vibrations, the formation of a finite concentration of atomic defects, predominantly vacancies. In 1964 changes of the angular distribution of annihilation photons in cadmium, indium and zinc have been reported when the temperature of these metals was raised from room temperature to close to their melting points [9.106]. But yet some years later the reversible temperature dependence of the mean positron lifetime in cadmium and zinc suggested equilibrium vacancies to be responsible for this behaviour [9.8,107]. This was the beginning of extensive studies of vacancies in thermal equilibrium with positrons. All the experimental techniques described (Sect. 9.3) have been applied to these investigations. Effects due to vacancies can already be observed at concentrations of about 10^{-7}. Figure 9.15 shows an angular correlation curve of an aluminum single

269

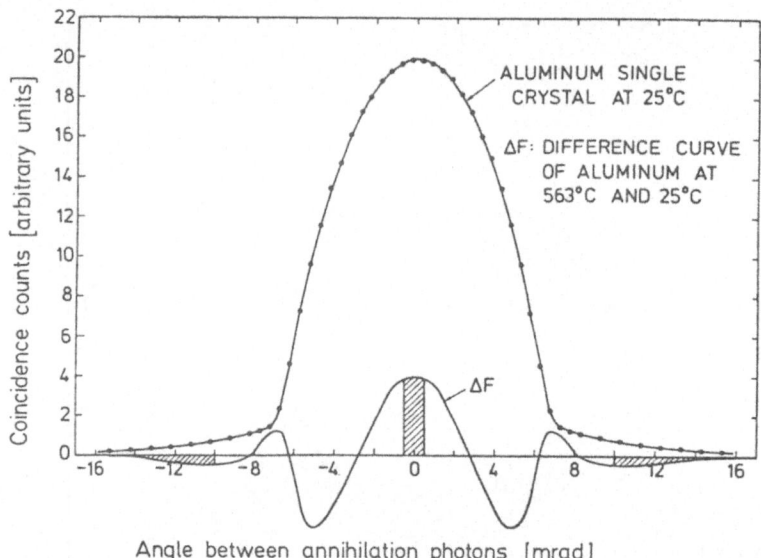

Fig. 9.15. Long-slit angular correlation curve for aluminum single crystal at room temperature and difference curve ΔF of the results at 563°C and 25°C. The vertical scale for ΔF is expanded by a factor of two. Indicated (shaded areas) are the regions for determining F^v and F^c in (9.32–36)

crystal at room temperture and a difference curve obtained at high temperature [9.108]. At the vicinity of a vacancy the electron distribution deviates from that of the perfect lattice. For a positron localized at the vacancy the overlap with the core electrons is diminished as an ion is missing from the defect. This leads to a reduced probability of annihilations with core electrons (8–16 mrad). On the other hand, the conduction electron density is also depleted at the vacancy region with a concomitant narrowing in momentum distribution. Thus the angular correlation curve for a trapped positron is always more peaked at small momenta (± 3 mrad) than for a Bloch-like positron. The effect due to the thermal energy of the positron is clearly visible in the interval from about 6 to 8 mrad.

The change of the peak-count rate (shaded area around zero angle in Fig. 9.15) with temperature in aluminum is shown in Fig. 9.16 [9.108]. The peak-count rate increases proportionally to the thermal expansion up to about 200°C. Above this temperature when vacancy trapping becomes effective the peak-count rate increases more than linearly with temperature and tends to saturate just below the melting point. This kind of behaviour has been observed also in positron mean lifetime and Doppler broadening line shape measurements and is typical for most metals and alloys [9.109].

The equilibrium concentration of single vacancies

$$c_v(T) = \exp\left(S_{1v}^F/k_B\right)\,\exp\left(-H_{1v}^F/k_BT\right) \tag{9.19}$$

is determined by the formation entropy S_{1v}^F and the formation enthalpy H_{1v}^F.

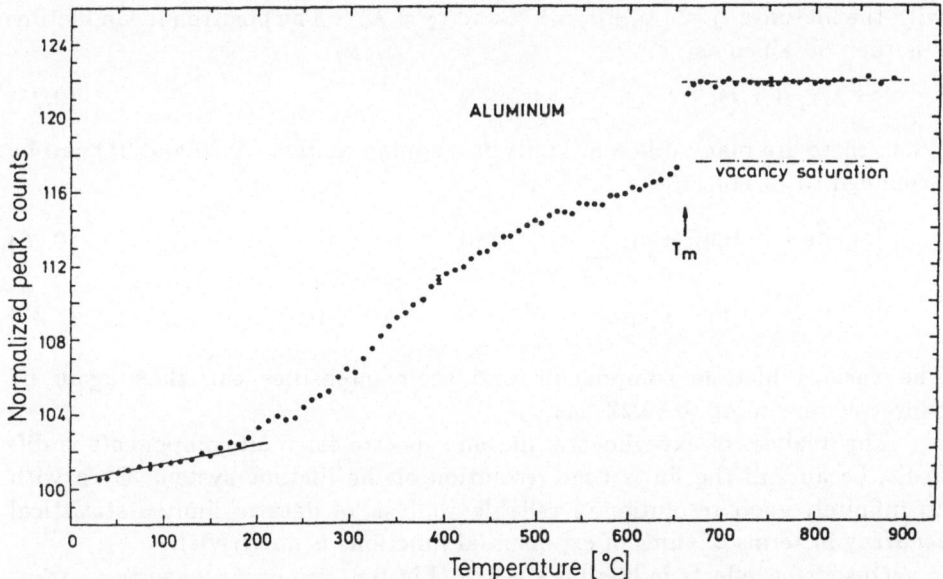

Fig. 9.16. Normalized peak counts for aluminum as a function of temperature. The full curve indicates the thermal expansion of the lattice. At the melting point T_m (660.2°C) the counting rate rises abruptly and remains then independent of temperature

If the temperature in a metal is raised, vacancies are created, and positrons are localized at these point defects. At the melting point of aluminum the vacancy concentration will be 9.4×10^{-4} [9.110]. The two-state trapping model [9.111–113] distinguishes between annihilation of positrons in the perfect lattice and annihilation of positrons trapped in a defect. For a positron annihilating in the perfect lattice (bulk) the lifetime is τ_b and the annihilation rate (9.14) is λ_b with a relative intensity I_b. For annihilation at a vacancy the corresponding values are τ_v, λ_v and I_v. The number of free and trapped positrons, n_b and n_v, respectively, are determined by the rate equations

$$dn_b/dt = -\lambda_b n_b - \mu_v c_v n_b, \tag{9.20}$$

$$dn_v/dt = -\lambda_v n_v + \mu_v c_v n_b \tag{9.21}$$

where μ_v is the trapping probability of a vacancy and $\mu_v c_v$ is the total trapping rate. Using the initial condition that no positrons are trapped at time $t = 0$, the probability for the positron to annihilate at time t is

$$n(t) = I_1 \exp(-\lambda_b - \mu_v c_v)t + I_2 \exp(-\lambda_v t). \tag{9.22}$$

The corresponding intensities are

$$I_1 = \frac{\lambda_b - \lambda_v}{\lambda_b - \lambda_v + \mu_v c_v}, \tag{9.23}$$

$$I_2 = \frac{\mu_v c_v}{\lambda_b - \lambda_v + \mu_v c_v} \tag{9.24}$$

271

with the lifetimes $\tau_1 = (\lambda_b + \mu_v c_v)^{-1}$ and $\tau_2 = \lambda_v^{-1}$. The positron mean lifetime $\bar{\tau}$ is then obtained as

$$\bar{\tau} = I_1 \tau_1 + I_2 \tau_2. \tag{9.25}$$

When there are many different kinds of trapping centers, (9.20 and 21) can be expanded to [9.114,115]

$$dn_b/dt = -\lambda_b n_b - n_b \sum_i \mu_i c_i \quad \text{and} \tag{9.26}$$

$$dn_i/dt = -\lambda_i n_i + \mu_i c_i n_b. \tag{9.27}$$

The various lifetime components and their intensities can then again be expressed according to (9.22–24).

The analysis of experimental lifetime spectra for multicomponents is difficult, because of the finite time resolution of the lifetime system. Even with an infinitely good resolution, a reliable analysis of data of limited statistical accuracy in terms of sums of exponential functions is not trivial.

Observable effects in lifetime spectra (Fig. 9.8) can occur whenever a trapping rate $\mu_i c_i$ is of the same order as λ_b. The trapping probability depends on several parameters such as the crystal structure, the lattice constant, the radius of the ion core, the valence of the atom, the size of the defect and the relaxation around the defect. The removal of an atom in a metal from its lattice site leads to an excess negative charge due to the conduction electrons. This results in a negative potential and will therefore attract a positron. The energy difference between the free positron state and the bond state in the vacancy is the binding energy E_b of the positron which increases with the number of the valence electrons of the metal. For several metals *Hodges* [9.116] has calculated the binding energy and the value of the trapping potential in a vacancy, as well as the trapping probability. The binding energies obtained are between 0.8 eV for cadmium and 2.0 eV for aluminum. For the alkali metals E_b is about 1.0 eV. Subsequent calculations yielded E_b values for aluminum between 0.8 and 6 eV [9.117–121]. If the spatial extent of the trap state is small compared to the positron thermal wavelength then the trapping probability is temperature independent [9.122]. The positron mobility has little effect on the overall trapping probability for localized trap states [9.123].

The change of the peak-counting rate (Fig. 9.16) with temperature can be evaluated on the basis of the two-state trapping model. In principle, any annihilation characteristic which is a linear function of the positron states can provide quantitative information about state populations or trap concentrations [9.124]. It is essential that this characteristic parameter exists for all positron states and that it can be determined with good accuracy. Such measurable parameters are the mean lifetime $\bar{\tau}$, (9.25), the line shape parameter in Doppler broadening, and the normalized peak counts in angular correlation. If F(T) is the coincidence-count rate as a function of temperature, F_b is the counting rate if all positrons annihilate in the perfect lattice, and F_v is the one expected if all positrons annihilate from the state trapped at vacancies [9.108], i.e.,

$$F(T) = p_b F_b + p_v F_v \tag{9.28}$$

where p_b and p_v are relative probabilities, corresponding to n_b and n_v of (9.20 and 21)

$$p_b = \frac{\lambda_b}{\lambda_b + \mu_v c_v}, \tag{9.29}$$

$$p_v = \frac{\mu_v c_v}{\lambda_b + \mu_v c_v}. \tag{9.30}$$

With $F_b = F_0(1 + \beta \Delta l/l)$, F_0 being the count rate at low temperature (e.g., 273 K) where trapping at vacancies is negligible and $\Delta l/l$ is the linear thermal expansion, and combining (9.19,28–30), the following relation results

$$F(\overset{.}{T}) = \frac{F_0(1 + \beta \Delta l/l)\lambda_b + F_v \mu_v \, \exp{(S_{1v}^F/k_B)} \, \exp{(-H_{1v}^F/k_B T)}}{\lambda_b + \mu_v \, \exp{(S_{1v}^F/k_B)} \, \exp{(-H_{1v}^F/k_b^T)}}. \tag{9.31}$$

The fitting of (9.31) to the experimental data (Fig. 9.14) with β, F_v and $\mu_v \exp{(S_{1v}^F(k_B)}$ as adjustable parameters yields very accurate values for H_{1v}^F. This method was first applied by *McKee* et al. [9.9].

For a number of metals the monovacancy-formation enthalpy obtained from various measurements is listed in Table 9.2. Attempts to extract from the data values for the divacancies have not been very successful. In diluted alloys values for the vacancy-impurity binding energy have been obtained [9.155]. For most of the concentrated alloys only an effective formation enthalpy has been given [9.156–158]. In order to extract values for the vacancy-binding energy, different vacancy-neighbour complexes have to be taken into account [9.159,160].

b) Liquid Metals

Investigations of liquid metals and studies of the effect of melting involved mainly angular correlation and lifetime measurements [9.161–166]. Before it was recognized that positrons can be trapped by temperature-induced defects below melting, the observed changes in the angular distributions of the annihilation photons were, to a large extend, attributed to electronic structure changes. The changes at the melting point and in the liquid phase were considered to be due to the effect of disorder on the conduction electron states [9.167].

Significant changes in angular correlation experiments between the solid and the liquid phase were found in several metals. The very pronounced changes in mercury are qualitatively similar to those arising from vacancy trapping in solid metals. This is supported by accurate measurements of the coincidence-counting rate at the peak of the angular correlation curve in solid and liquid

Table 9.2. Vacancy formation enthalpies

Metal	Formation enthalpy H_{1v}^F [eV]	Experimental method[a]	Ref.
Al	0.66±0.04	PC	[9.9]
	0.68±0.03	LT	[9.124]
	0.69±0.03	LT	[9.125]
	0.66±0.01	PC	[9.108]
	0.66±0.02	LT,LS	[9.126]
	0.66±0.01	Q	[9.127]
	0.68±0.05	PC	[9.128]
Au	0.97±0.01	PC	[9.129]
	0.98±0.03	LT	[9.125]
	0.92±0.04	LS	[9.130]
	0.96±0.02	PC	[9.131]
	0.94±0.02	Q	[9.132]
Ag	1.16±0.02	PC	[9.129]
	1.20±0.05	LS	[9.133]
	1.11±0.05	LS	[9.134]
	1.10±0.04	Q	[9.135]
Cu	1.28	PC	[9.136]
	1.29±0.02	PC	[9.129]
	1.28±0.04	PC	[9.137]
	1.26±0.07	LS	[9.138]
	1.31±0.05	LS	[9.139]
	1.27±0.05	Q	[9.140]
	1.30±0.05	Q	[9.141]
In	0.55±0.02	PC	[9.9]
	0.46±0.03	LT	[9.124]
	0.48±0.01	PC	[9.108]
	0.39±0.04	LS,LT	[9.142]
	0.59±0.03	LS	[9.143]
Pb	0.50±0.03	PC	[9.9]
	0.49±0.03	LT	[9.124]
	0.54±0.02	PC	[9.108]
	0.58±0.02	LT	[9.144]
α-Fe (bcc)	1.60±0.15	LS	[9.145]
	1.60±0.10	PC	[9.146]
	1.40±0.10	PC	[9.147]
γ-Fe (fcc)	1.54±0.15	LS	[9.145]
	1.40±0.15	PC	[9.146]
	1.7	PC	[9.147]
Co	1.34±0.07	PC	[9.146]
Ni	1.72±0.10	PC	[9.128]
	1.74±0.06	PC	[9.137]
	1.58±0.05	Q	[9.148]
	1.54±0.02	LT	[9.149]
	1.55±0.05	PC	[9.146]
	1.8±0.1	LS	[9.150]
Cr	3.51±0.13	LS	[9.151]
V	2.1±0.2	LS	[9.152]
Nb	2.6±0.3	LS	[9.152]
Ta	2.8±0.6	LS	[9.152]
Mo	3.0±0.2	LS	[9.152]
	2.9	Q	[9.153]
W	4.0±0.3	LS	[9.152]
	3.6±0.2	Q	[9.154]
Zn	0.54±0.02	PC	[9.9]
	0.51±0.025	LT	[9.124]

[a] PC: positron angular correlation peak counts; LT: positron lifetime; LS: Doppler broadening line shape; Q: quenching

aluminum, indium and lead [9.108]. At the melting point the coincidence rate arises abruptly (Fig. 9.16). The transition from the solid to the liquid phase is also accompanied by a volume increase [9.167]. Because of this volume increase (6% for aluminum) and the nonregular arrangements of the atoms in the liquid, the structure of the positron traps in the liquid is different compared to the vacancies in the solid. The Coulomb repulsion between the positron and the positively charged ions in the liquid metal causes an increase in the effective volume of the trapping site and this leads to a stronger localization of the positron wave function, and hence to a change in the annihilation characteristic [9.108]. Similar results are obtained from positron lifetime measurements in gallium [9.165]. In sodium only a small increase of the lifetime on melting takes place which is typical for all alkali metals [9.165]. Large "melting effects" occur in metals where vacancy trapping is either missing or very weak in the solid phase. The smearing and narrowing of the conduction-electron part of the angular distribution is almost the same as observed in vacancy trapping and can be similarly interpreted in terms of positron localization and local density effects. The nature of the effective trapping sites in the liquid phase is of considerable interest. In spite of a possible size distribution of potential positron traps and their transient nature because of density fluctuations in the liquid, it is most likely that a positron creates its own hole and trapping site [9.108,168,169]. Because of this hole-creating effect in the liquid phase and the resulting localized positron state, the measured angular distribution of the annihilation photons is not representative for the properties of the electrons in the liquid itself, but rather those in ion gaps. Therefore, particular care has to be taken for the interpretation of positron annihilation data in liquid metals, especially if the observed changes are attributed to electronic structure effects due to the disorder in liquids [9.170]. Most likely the desired information about the disordered and amorphous structure of liquid metals cannot be obtained from positron annihilation. This applies to positron studies in liquid alloys as well.

c) Phase Transitions

In many metals and alloys the crystallographic structure changes with temperature. *Dekhtyar* et al. [9.171] investigated order-disorder transitions in copper-gold and nickel-manganese alloys. In CuAu and Ni_3Mn small differences in the angular correlation curves for the ordered and disordered phase were observed. More accurate studies of Cu_3Au, CuZn and alloys showed anomalous effects in the temperature dependence of the angular correlation curve attributable to order-disorder transitions [9.172,173]. Alloys with a first-order transition exhibit a continuous change of the peak-count rate in the neighbourhood of the transition, whereas for a second-order transition a discontinuous change at the transition temperature is observed [9.174]. Experiments on metals (Co,Fe [9.146], U [9.175]) and various iron alloys [9.160] indicate that positron annihilation is also very sensitive to structural changes. Most recent positron lifetime and Doppler broadening measurements on uranium up to the melting point ex-

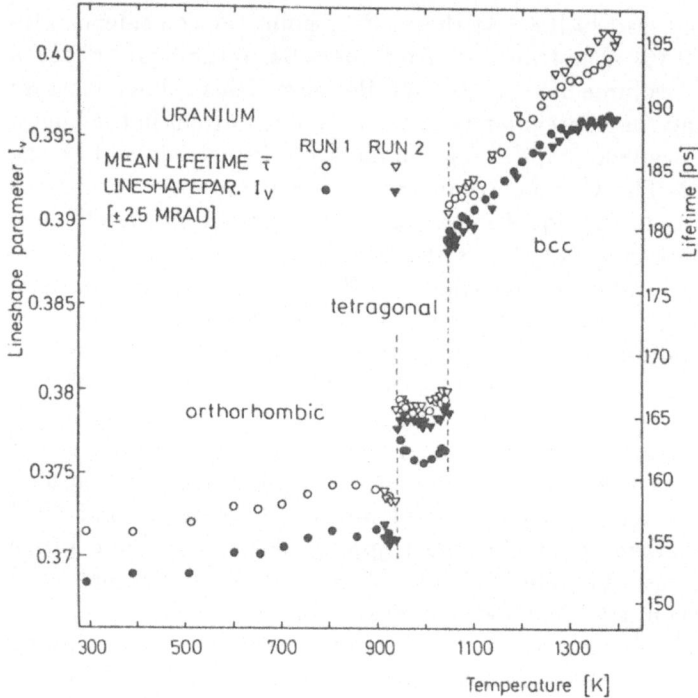

Fig. 9.17. The line shape parameter l_V and the mean lifetime $\bar{\tau}$ calculated from the two resolved lifetimes τ_1 and τ_2 according to (9.25), for uranium as a function of temperature

hibit very clearly the several phase transformations [9.176]. These results are shown in Fig. 9.17. A significant dependence of the lifetime $\bar{\tau}$ and the line shape parameter on crystal structure and temperature is observed. Trapping at thermally produced vacancies is only detected in the bcc-phase with a surprisingly low total trapping rate. It is concluded that the vacancy-formation parameters are temperature dependent with an average value of 0.3 eV for H_{1V}^F [9.176]. In this bcc-phase, uranium shows an anomalous self-diffusion behaviour which is also observed in other metals [9.177,178].

9.4.3 Radiation Induced Defects

When an energetic particle such as an electron, a proton, a neutron, a deuteron or an α-particle penetrates into a metal crystal, an atom at a lattice site is displaced and becomes an interstitial atom. The radiation damage consists mainly of single interstitials and single vacancies when the energy transferred to the atom is just above the displacement energy, and if the irradition takes place at very low temperatures both types of defects are retained in the crystal.

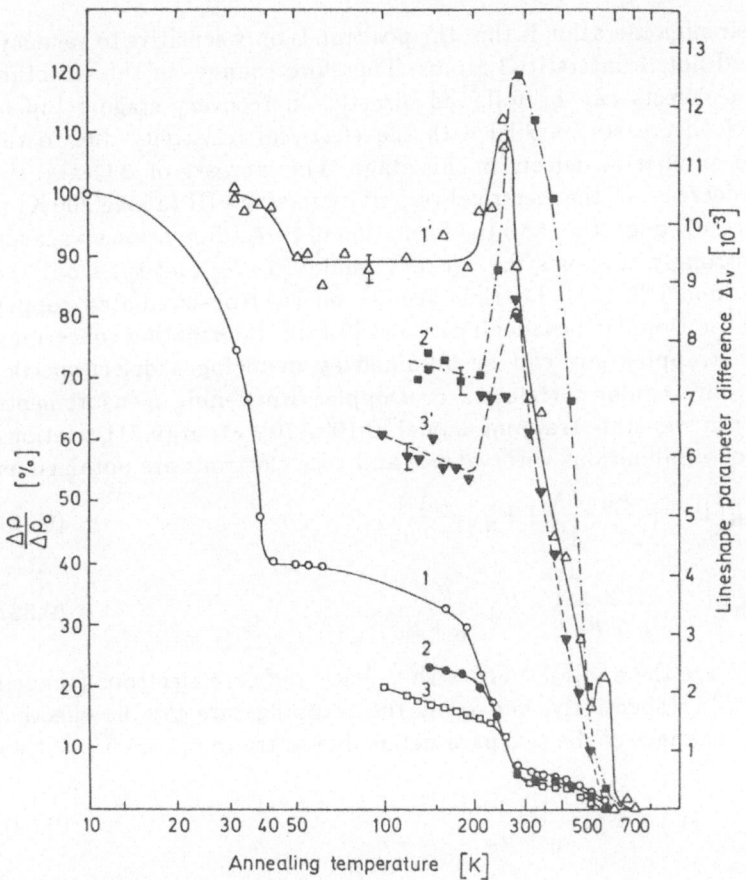

Fig. 9.18. Line shape parameter difference ΔI_v and electrical resistivity $\Delta\varrho/\Delta\varrho_0$ for copper after irradiation at 4.2 K with 3 MeV electrons as a function of the annealing temperature. The curves marked from 1 to 3 correspond to an initial Frenkel-defect concentration of 5×10^{-4}, 2.6×10^{-4} and 1.2×10^{-4}, respectively

a) Electron and Neutron Irradiation

Irradiation at 4.2 K with electrons possessing energies of the order of 3 MeV results predominantly in single vacancies and interstitials, both immobile at this temperature. The maximally transferred energy by a 3 MeV electron to a copper atom bound in the crystal lattice is about 100 eV. Electrical resistivity measurements as a function of the annealing temperature have revealed various recovery stages [9.179]. The first investigations combining electrical resistivity and positron annihilation measurements in annealing studies after low-temperature electron irradiation have been reported by *Mantl* and *Triftshäuser* [9.13]. Figure 9.18 shows the Doppler broadening line shape parameter difference ΔI_v and the electrical resistivity as a function of the annealing temperature for electron-irradiated copper at three different doses [9.180]. The advantage of using positrons as a probe in the investigations of

277

defects and their agglomeration is that the positron is only sensitive to vacancy-type defects and not to interstitial atoms. Therefore changes in the structure of vacancy-type defects can be followed directly. In recovery stage I (up to about 40 K) ΔI_v decreases parallel with the electrical resistivity due to the decrease in the number of defects in this stage. The increase of ΔI_v and the concommitant decrease of the electrical resistivity in stage III (above 200 K) is evidence for vacancy migration and the formation of three-dimensional vacancy clusters. This strongly supports the vacancy model [9.178] and rules out the two-interstitial model [9.181]. Lifetime studies on electron-irradiated copper have confirmed the Doppler broadening results [9.182]. Information concerning the type of the trapping site can be obtained by deducing a defect-specific parameter R from angular correlation or Doppler broadening measurements on the basis of the two-state trapping model [9.108,179]. From (9.31) relations for the change of annihilations with valence and core electrons are obtained as

$$F^v(T) = F_b^v \left(1 - \frac{\mu_d}{\lambda_b + \mu_d}\right) + F_d^v \frac{\mu_d}{\lambda_b + \mu_d}, \tag{9.32}$$

$$F^c(T) = F_b^c \left(1 - \frac{\mu_d}{\lambda_b + \mu_d}\right) + F_d^c \frac{\mu_d}{\lambda_b + \mu_d} \tag{9.33}$$

where F^v and F^c are the annihilations with valence and core electrons (shaded areas in Fig. 9.15), respectively, and μ_d is the trapping rate for the effective defect. Then the changes of the two parameters due to trapping can be written as

$$F^v(T) - F_b^v = (F_d^v - F_b^v) \frac{\mu_d}{\lambda_b + \mu_d}, \tag{9.34}$$

$$F^c(T) - F_b^c = (F_d^c - F_b^c) \frac{\mu_d}{\lambda_b + \mu_d}. \tag{9.35}$$

These differences still depend on the concentration of the defects. The defect-specific parameter R which is independent of the defect concentration is obtained by taking the ratio of (9.34 and 35)

$$R = \left| \frac{F^v(T) - F_b^v}{F^c(T) - F_b^c} \right| = \left| \frac{F_d^v - F_b^v}{F_d^c - F_b^c} \right|. \tag{9.36}$$

In annealing studies, T is the corresponding annealing temperature. Equations (9.32–36) can be applied to Doppler broadening measurements accordingly [9.180]. This R parameter characterizes the type of the trapping defects and shows a clear correlation with the size of the defects [9.108,180], as observed for the positron lifetime, too. Annealing studies in electron-irradiated molybdenum using the lifetime method are shown in Fig. 9.19 [9.183]. Up to 200°C positrons are trapped at single vacancies exhibiting a lifetime of about 200 ps. The increase of τ_2 between 200°C and about 300°C is interpreted as the formation of vacancy clusters due to vacancy migration. At higher annealing temperature voids are formed for which the corresponding lifetime is 500 ps [9.183]. A similar

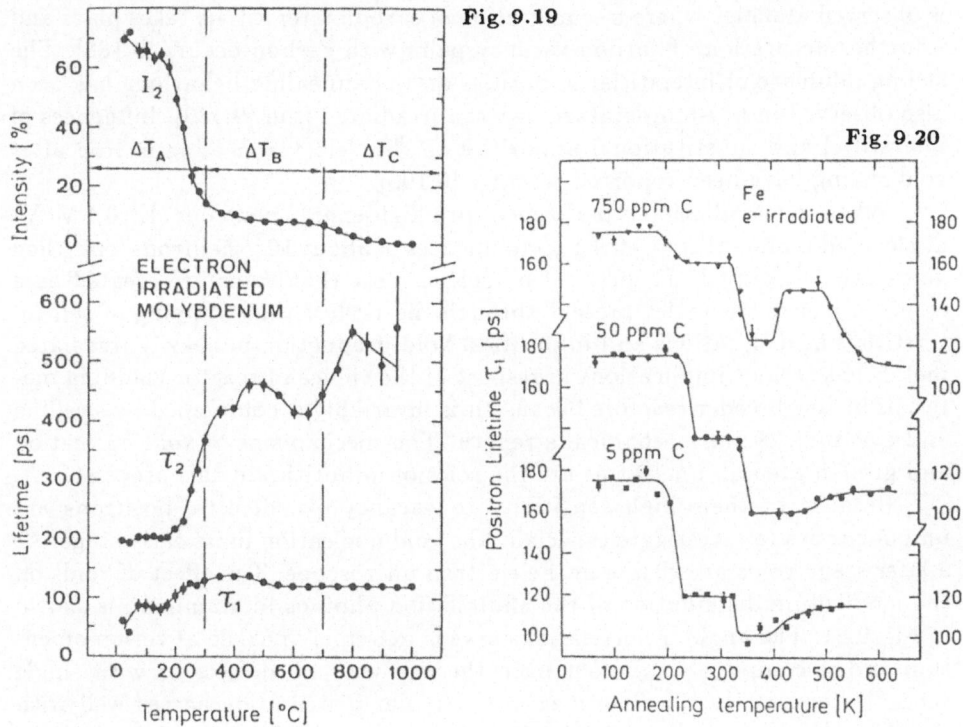

Fig. 9.19

Fig. 9.20

Fig. 9.19. Positron lifetime parameters for molybdenum irradiated at room temperature with 10 MeV electrons, as a function of the annealing temperature as determined by *Eldrup* et al. [9.183]. The temperature intervals ΔT_A, ΔT_B and ΔT_C indicate different annealing states

Fig. 9.20. The short positron lifetime τ_1 in iron after irradiation at 20 K with 3 MeV electrons as a function of the annealing temperature as determined by *Vehanen* et al. [9.188]. The specimens contained carbon impurities ranging from 5 ppm to 750 ppm

behaviour has been reported for molybdenum after neutron irradiation at 60°C [9.184] and after plastic deformation [9.185].

The influence of interstitial impurities on the annealing behaviour of vacancies has clearly been demonstrated in the case of iron containing various amounts of carbon [9.186–188]. After low-temperature electron irradiation, the migration of vacancies is observed in an annealing stage at 220 K, verified by the increase of lifetime τ_2 through the formation of vacancy clusters. The lifetime τ_1 for specimens containing different concentrations of carbon impurities is shown in Fig. 9.20 as a function of the annealing temperature. In addition to vacancy agglomerates, carbon-vacancy pairs are also formed during the free migration of vacancies in stage III. Carbon interacts strongly with vacancies. Trapping into a vacancy-carbon pair still exists, indicating an asymmetric position for the carbon impurity inside the vacancy which is also supported by internal-friction results. The tendency of vacancies to form clusters decreases as the concentration of carbon impurities is increased. A further annealing stage is observed at 350 K where free migration of carbon interstitials takes place and a further decoration of carbon-vacancy pairs with carbon occurs [9.188]. The strong influence of interstitial impurities on the annealing behaviour has been also observed in low-temperature neutron-irradiated iron [9.189]. Influences of interstitial and substitutional impurities on the defect annealing in iron after cold rolling have been reported recently [9.190].

When the irradiation is performed with high-energy neutrons (E>0.1 MeV) at elevated temperature ($\sim$0.4 T_m) to fluences of about 10^{21} neutrons/cm^2 then voids can be formed. Theoretical models suggest that voids are created as a result of excess vacancies present through the preferential trapping of self interstitials at dislocations and impurities. Void production in heavily irradiated metals has serious implications in respect of the choice of reactor-cladding materials in fast-breeder reactors because it is invariably accompanied by swelling and eventual loss of mechanical strength. The mechanisms of void nucleation and growth are still unclear as are the roles of impurities in this process.

Because of their high sensitivity to vacancy-like defects, positrons are unique probes to investigate especially the void nucleation in its early stage. At a later stage voids are visible in the electron microscope. The effect of voids on the momentum distribution of the annihilation photons in aluminum is shown in Fig. 9.21. The angular correlation curve is extremely peaked at zero momentum and is completely different from the one with no defects or with single vacancies present [9.12]. The measured angular distribution agrees well with the theoretical prediction by *Hodges* and *Stott* [9.191] that the positron annihilates from a localized state at the inner surface of the void. A very similar behaviour has been reported for voids in molybdenum [9.10] and nickel [9.192].

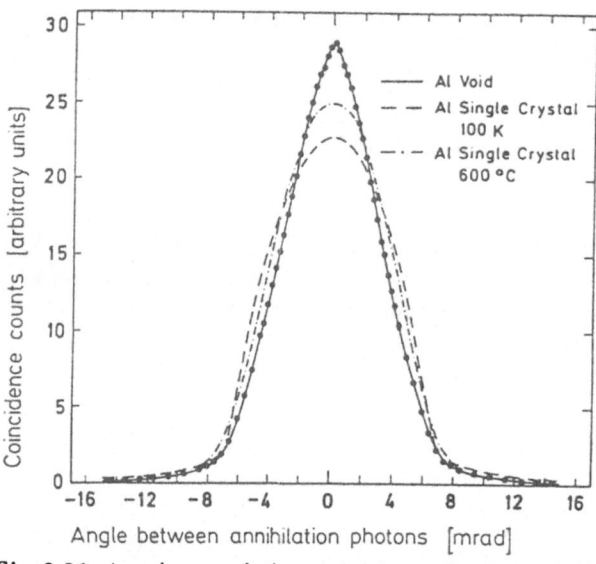

Fig. 9.21. Angular correlation curves in an aluminum single crystal at 100 K (no defects), at 600°C (vacancies in thermal equilibrium) and after neutron irradiation (voids) [9.12]. The curves are normalized to equal areas

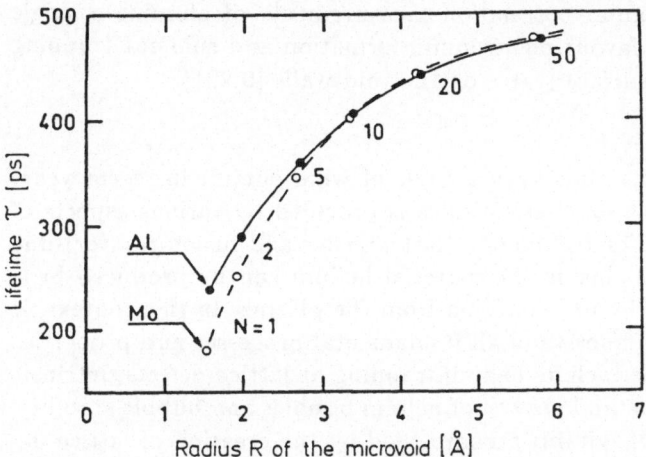

Fig. 9.22. Calculated positron lifetimes in vacancy clusters of aluminum and molybdenum by *Hautojärvi* et al. [9.196]

Positron lifetime measurements of specimens containing voids exhibit lifetime components around 500 ps [9.11,193,194]. Annealing studies show that voids in aluminum disappear at 590 K [9.193–195]. Results of theoretical calculations of positron lifetimes as a function of vacancy-cluster size (spherical shape) in aluminum and molybdenum are shown in Fig. 9.22 [9.196]. The rapid increase of τ with the number of vacancies in the microvoid is noticeable. The effect of divacancies on lifetime has also been investigated using a ellipsoidal model [9.197]. The trapping probability calculated as a function of void size increases with the number of vacancies in the void [9.198]. For large void sizes ($\sim$500 Å diameter) the trapping probability is independent of the void size but it might become temperature dependent since the positron thermal wavelength is no longer large compared to the dimension of the defect. *Nieminen* et al. [9.199] have found a change of the positron density at the void boundary as a function of temperature and deduced a temperature-dependent trapping probability suggesting both transition and diffusion limited regions.

It is most likely that a positronium is formed in voids, but only under certain conditions this can be experimentally verified. A positronium is the analogon to a hydrogen atom where the proton is replaced by a positron. In the studies of the irradiation-induced voids in aluminum, indications for positronium decay have been observed [9.12,195]. Recent angular correlations studies show evidence for positronium formation in niobium containing voids which are covered with a very thin oxide layer [9.200].

The observed very long lifetimes are associated with impurities (e.g., silicon, oxygen) segregated onto the void surface. These impurities act as a buffer inhibiting or reducing surface-state formation, and positronium pick-off processes are responsible for the long lifetime.

Two-dimensional angular correlation measurements of aluminum single crystals containing voids favour positronium formation and rule out trapping of positrons in extended surface states on the void walls [9.201].

b) Ion Implantation

Helium in metals and alloys has been a topic of wide pursuit in recent years [9.202]. The main impetus for these studies is provided by various aspects of material problems in reactor technology (fast breeder and fusion reactor) due to the presence of helium. Inside the material helium can be produced by a nuclear reaction (n, α) or by implantation from the plasma. In this context, it has become necessary to understand all fundamental processes and properties of helium in metals. These include helium trapping at lattice defects, intrinsic helium mobility, nucleation and growth of helium bubbles and bubble stability.

Ion implantation is inevitably accompanied by the creation of lattice defects. While the irradiation with heavy ions (e.g., metal ions) of several MeV energy leads to strongly damaged regions around and up to the range of the primary ion (few 10^{-6}m), the irradiation with light ions (e.g., hydrogen, deuterium, helium) results in lattice defects and interstitial or trapped

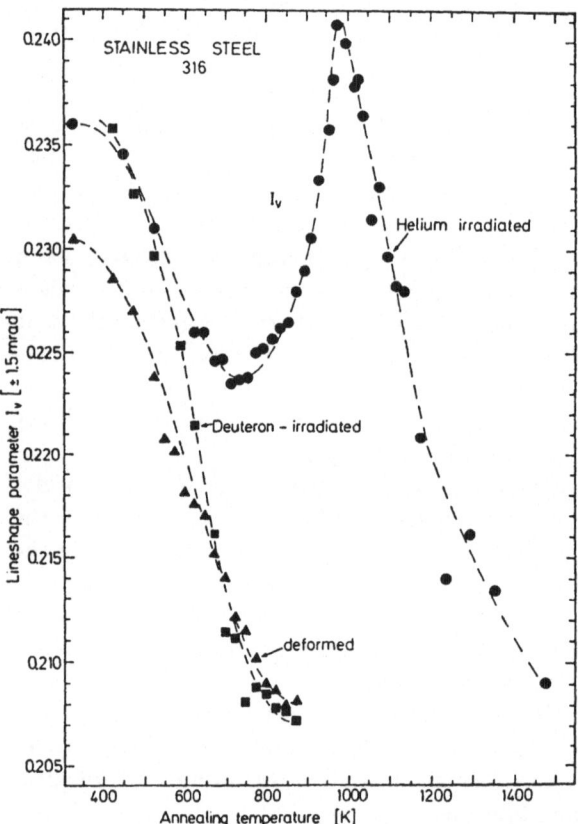

Fig. 9.23. Line shape parameter I_v for room-temperature helium-irradiated, deuterium-irradiated and deformed stainless steel 316 as a function of the annealing temperature [9.203]

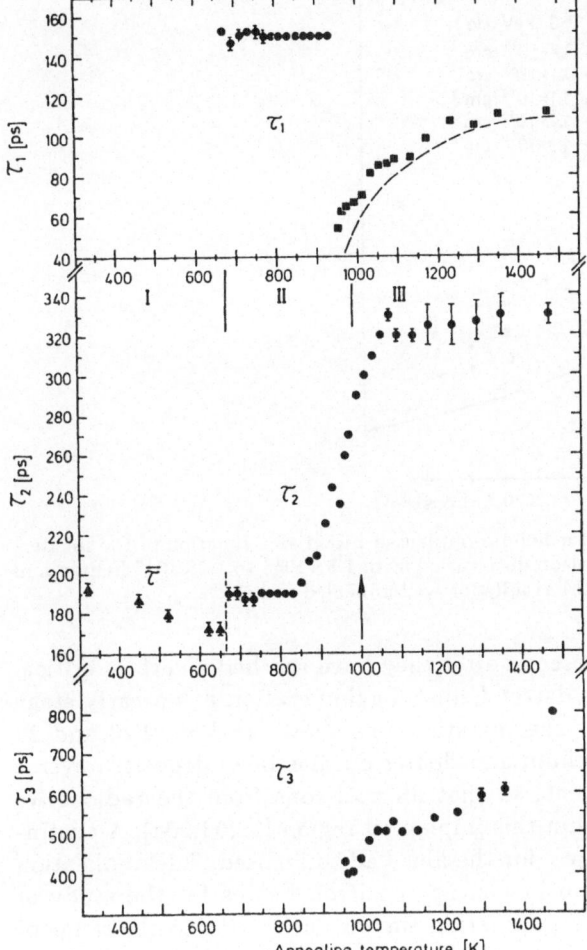

Fig. 9.24. Lifetime components in helium-irradiated stainless steel 316 as a function of the annealing temperature [9.203]. The arrow marks the temperature, where helium bubbles are visible in the electron microscope

gaseous impurities. The positron annihilation techniques have been applied in recent years to these ion implantation problems. The influence of ion implantation on the annealing behaviour in stainless steel is shown in Fig. 9.23. The Doppler broadening line shape-parameter I_v is plotted for helium- and deuteron-irradiated specimens [9.203]. For comparison the behaviour for cold-deformed stainless steel is indicated as well. The defects introduced by ion implantation (500 atomic ppm) and by rolling at room temperature disappear after annealing at about 800 K. At higher temperatures only in the helium-irradiated specimens new defects are observed (increase of I_v and R parameter) which are identified as helium bubbles. The positron lifetime results of Fig. 9.24 are indicative for the growth of the helium bubbles (τ_2) above 800 K. The observed lifetime component τ_3 is due to positronium formation inside the helium bubbles. Marked differences in the relation τ_2 versus I_v are found between voids and helium bubbles [9.204]. Helium bubbles can only be seen in

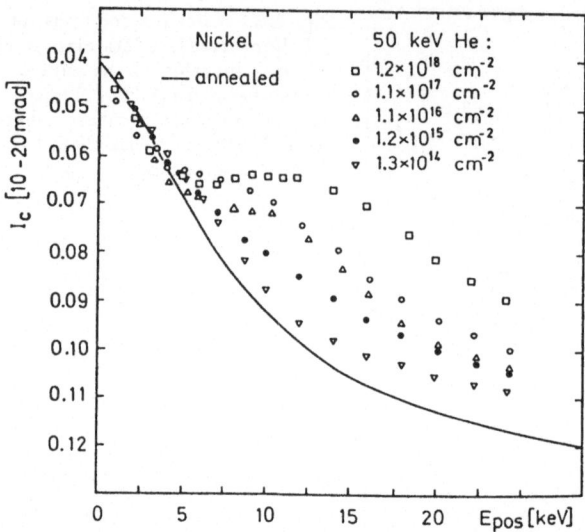

Fig. 9.25. Line shape parameter I_c for helium-implanted nickel as a function of the incident positron energy [9.34]. The implantation dose varied from 1.3×10^{14} to 1.2×10^{18} helium ions per cm^2. The full curve represents the results for well-annealed nickel

the transmission electron microscope after they have reached a certain critical size [9.205], whereas positrons detect helium agglomeration at an early stage or at the nucleation itself. For the investigations shown in Figs. 9.23 and 24 an uniform concentration of helium and deuterium has been deposited over a range of 10^{-4}m in the specimens, so that all positrons from the radioactive isotope (^{22}Na) are stopped within this implanted region [9.203,204]. A continuous energy ranging up to 28 MeV for the ions had to be used. The application of monoenergetic positrons from low-energy positron beams for the study of ion implantation requires only implantation energies according to the range of the positrons, e.g. a few keV up to about 100 keV. By varying the incident positron energy, various depths from the surface can be sampled. Figure 9.25 shows first typical results for helium-implanted nickel at different implantation doses [9.34,83]. The Doppler broadening line shape parameter I_c (annihilations with core electrons) is plotted as a function of the incident positron energy. The effect due to the helium implantation and due to the implantation dose is clearly exhibited. The defects associated with the helium implantation extend more than three times deeper into the material than the calculated mean deposition depth of the helium ions [9.205]. In order to investigate the full extent of the damaged region, the incident positron energy has to be increased to at least 50 keV. The determination of the nature of the defects together with their concentration and depth distribution will be the topic for further studies. From transmission electron microscopy it is known that for doses of about 10^{17}He/cm^2 both helium bubbles and dislocations are produced [9.206,207]. By a combined ion-implantation and sputtering method krypton has been deposited into copper [9.208]. Positron lifetime and angular correlation studies suggest positron trapping in submicroscopic krypton-vacancy clusters [9.209]. Most recently hydrogen defect interactions in proton-irradiated molybdenum have been reported [9.210]. The interaction of hydrogen with vacancies has also been observed in copper by Doppler broadening measurements [9.211].

9.4.4 Amorphous Alloys

Amorphous metallic alloys are metastable solids with atomic arrangements which are not spatially periodic. Metallic glasses can be produced in a variety of ways: evaporation, sputtering, or fast quenching from the liquid state. Most investigations have been performed on amorphous alloys which are stable at room temperature. The metallic glasses may be classified conventionally into two groups according to their constituents: metal-metalloid (M1 tpye) and metal-metal (M2 type) systems. Several structural models have been proposed ranging from local strain, packing faults, chemical short-range order and dislocations to the dense random packing of hard spheres [9.212]. Experimental evidence for chemical and configurational short-range order [9.213–215] and for local density fluctuations [9.216] has been found in many amorphous alloys. For positrons, amorphous alloys look like irregular arrays of potential wells with different strength. There are several possible states for a thermalized positron: (i) The positron is in a free-particle state, if the scattering of the positron by the irregular array of potentials is weak. Then the annihilation characteristics reflect the bulk properties. (ii) The scattering is strong and the mean-free path of the positron is of the order of the average atomic spacing. The annihilation characteristics are statistically averaged quantities over the annihilation sites. (iii) The positron is localized with its wave function confined to a small region of space. Then the annihilation characteristics reflect the local environment of the annihilation sites.

The first indication that the positron might be in a localized state has been the detection of a relatively long lifetime in amorphous alloys [9.217]. In the M1-type alloys only a single lifetime is observed which is remarkably insensitive to the alloy composition [9.218,219]. Lifetime results for various metallic glasses are presented in Fig. 9.26. In spite of the wide variation of the alloy constituents, the lifetimes do not vary more than a few picoseconds, indicating that positrons annihilate from one definite state. This is supported by the Doppler broadening results shown in Fig. 9.27 for the same specimens [9.170]. The difference curve in respect to well annealed pure iron shows only a very small variation with alloy composition and agrees very well with the curve for 50% deformed iron [9.220] where positrons are localized in the core of the dislocation. The difference to vacancies in iron is quite obvious indicating that the volume where the positron is localized is definitely smaller than that of a vacancy. The open space is very similar to that around the core of a dislocation in the crystalline matrix. This picture is also consistent with the observed temperature dependence in the amorphous state [9.221–223], which shows the same behaviour as for dislocation loops [9.224]. An explanation for the temperature dependence of M1 alloys has been given by *Kögel* et al. [9.225].

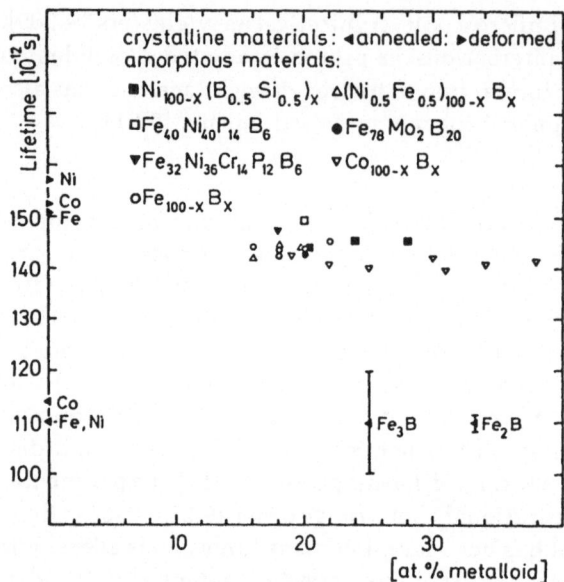

Fig. 9.26. Positron lifetimes for various amorphous and crystalline alloys [9.218]. The lifetime values for well-annealed and deformed pure iron, cobalt and nickel are included for comparison

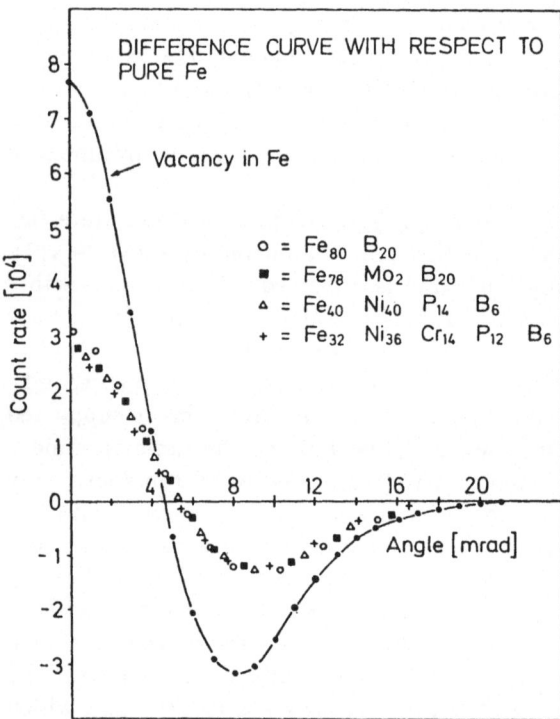

Fig. 9.27. Doppler broadening difference curves with respect to pure iron for various amorphous alloys and for iron containing vacancies [9.218]

The effect of low-temperature electron irradiation on the mean lifetime has been studied in amorphous alloys [9.220,226,227]. In addition to the intrinsic defects, vacancies are detected and the recovery occurs through annealing of close Frenkel pairs [9.226].

In the M1-type alloys positrons are localized at intrinsic empty spaces of about 0.7 atomic volumes with a concentration of about 10^{-3} to 10^{-2} [9.226]. This is confirmed by the results from positron beam experiments and the derived short positron diffusion lengths (10 – 200 Å) [9.220,228]. The situation for M2-type alloys is slightly different. From the two observed lifetimes and their intensities it can be concluded that annihilations take place from Bloch states in the amorphous matrix as well as from localized states at defects [9.218]. The fraction of positrons trapped at these vacancy-like defects is between about 3% to 10%.

9.4.5 Surfaces

The availability of monoenergetic positrons is intimately connected with solid surfaces. The condition of the metal surface is critically responsible for the fraction of re-emitted positrons (Sect. 9.3.5.). The study of surface phenomena with positrons awaited the development of positron beams of high intensity. There are several processes that have to be considered when slow positrons collide with a surface. Some of the incident positrons will scatter elastically from the target and are diffracted. The rest of the incident positrons will penetrate into the material, become thermalized, diffuse through the bulk, annihilate or are ejected from the surface. Positrons reaching the surface may become bound in a potential well at the surface or leave the surface as a free particle, or as a positronium atom in the ground state or in an excited state. The application of positron beams to surface science is still in an early stage, although a large number of experiments have been proposed and already reported [9.79]. Positronium formation at the metal surface [9.229] and thermionic positronium emission [9.88] as a source of thermal positronium in vacuum may become a valuable tool for surface investigations [9.230]. Low-energy electron diffraction (LEED) is used as a method to study the periodic arrangement of atoms on a surface. Low-energy positron diffraction (LEPD) could become a useful complement to LEED because positrons interact with a crystal without exchange effects and the positrons are repelled from the ion cores which attract and strongly influence low-energy electrons. The first LEPD results have been reported for a copper single-crystal surface [9.231,232]. A reasonable agreement with theory is obtained [9.233]. Positron and electron energy-loss spectra for the same single-crystal surface (tungsten and silicon) have been reported [9.234]. Significant differences between the positron and the electron spectra are detected, although with low statistical accuracy.

Positron energy losses due to the excitation of adsorbed carbon monoxide on the surface of nickel have been measured [9.235]. Well-defined peaks in the energy spectrum of the re-emitted positrons are detected which are identified as the vibrational states of C-O and Ni-C bonds. It is suggested that the re-emitted positron energy loss spectroscopy might become a valuable tool for the analysis of adsorbed molecules and possibly superior to electron energy loss spectroscopy [9.235].

Low-energy positrons can be applied in the momentum spectroscopy of electron surface states. First preliminary results [9.236] of two-dimensional angular correlation measurements on aluminum [9.237] and copper single-crystals [9.238] with positron beams have been reported recently. An asymmetric angular distribution with respect to the surface is found for low-energy incident positrons which is mainly due to positronium moving away from the surface. Whether there is an asymmetry in the electron-momentum distribution as theoretically predicted by *Rozenfeld* et al. [9.239] remains an open question, until results of better statistical accuracy (i.e., high-intensity positron beams) are available.

An alternative method for the investigation of the electron density of states at the surface has been proposed by *Mills* et al. [9.240]. Feasibility has also been demonstrated. By measuring the velocity distribution of positronium emitted from the surface into the vacuum, information about the electron-momentum distribution is obtained. The kinetic energy E_k of the positronium atom is

$$E_k = E_b - (\Phi_+ + \Phi_- + \Delta E) \tag{9.37}$$

where $E_b = 6.8 \, \text{eV}$ is the positronium binding energy, and Φ_+ and Φ_- are the positron and electron work functions, respectively. ΔE is the energy difference between the Fermi energy and the initial energy state. Since the positronium formation occurs on a time scale of 10^{-18}s, the electron plasma has not enough time to re-adjust to minimum energy, and therefore the measured E_k reflects the single electron density of states.

But not only the metal surface itself and the impurities that cover the surface are of interest. Also the near-surface region is accessible to positrons of variable energy. Defects or other crystalline imperfections in near-surface layers have been investigated with monoenergetic positrons whose penetration depth into the solid can be varied from a few Angstroms to thousands of Angstroms [9.19,20]. The final aim of these investigations will be to obtain evidence about the type of the defects, their concentration and their depth distribution. The possibility to reveal this information from the measured data has been proposed by *Kögel* [9.241].

9.5 Conclusions and Outlook

Considerable progress has been made in the understanding of the electronic structure in metals using positrons and the prospects for interesting new applications look most promising.

For the high-precision studies of complex Fermi surfaces, especially in alloy systems, high angular resolution at a good counting efficiency will be important. The use of two-dimensional multi-detector devices, such as large-size multi-wire-proportional chambers with several lead converters for increased efficiency, will not only enable the observation of small details of the Fermi surface geometry, but also the amount of smearing at the Fermi momentum in disordered alloys. The quality of the single-crystal specimens, especially in respect to structural lattice defects, will be of primary importance. More precise band-structure calculations will become available to provide accurate comparison between theory and experiment. With the development of producing slow-positron beams of increasing intensity, two-dimensional angular correlation studies can be made as a function of the positron energy. Besides the electron states at clean surfaces, the study of adsorbed atoms and molecules on solid surfaces can be anticipated.

The isotopes ^{64}Cu and ^{58}Co produced in high-flux nuclear reactors by thermal and fast neutrons, respectively, will become available as very intense primary positron sources.

However, not only the total intensity of the positron beam must be increased, but also the brightness of the beam has to be enhanced. The basic limitations on how narrow a particle beam can be compressed is set by Liou-

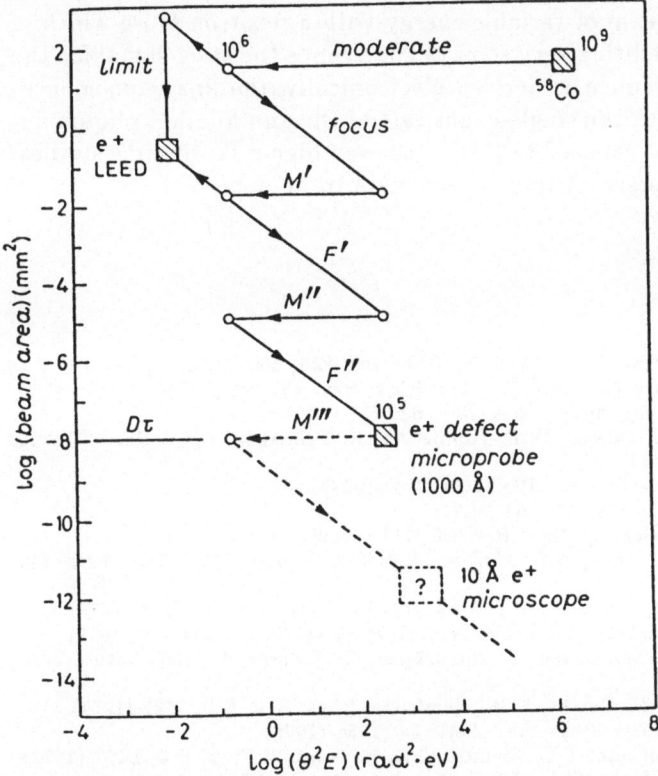

Fig. 9.28. Brightness per unit-energy phase space for slow-positron moderation as proposed by *Mills* [9.242]. The starting intensity of 10^9 positrons/second is reduced by a factor of 10^3 through the first moderation process. M′, M″ and M‴ indicate further moderation stages, whereas F′ and F″ represent focussing of the moderated positrons under constant brightness. Dτ is the product of the diffusion constant and the lifetime of the positron

ville's theorem. If the beam is modified by conservative forces (e.g., electric and/or magnetic fields), then under this theorem the phase-space volume of the beam is constant. The brightness B per unit energy of a beam of energy E and intensity I having a cross section d^2 with an angular divergence θ is

$$B = \frac{I}{E\,d^2\,\sin^2\theta} \tag{9.38}$$

and constant. If, however, the beam encounters non-conservative forces, as occurs during thermalization in a metal moderator, then B need not be a constant. A scheme for brightness enhancement was proposed some years ago and is shown in Fig. 9.28 [9.242]. In this diagram a horizontal path from right to left indicates moderation, a sloping path represents focussing under constant brightness, and a vertical path means passing the beam through an aperture. Through several moderating stages it is feasable to have a slow-positron source with a diameter determined and limited only by the diffusion length of the thermalized positrons ($\sim 1000\,\text{Å}$). Efforts are under way to produce thin (1500 – 2000 Å) tungsten foils or even single crystals for these moderation purposes.

A pulsed positron beam of variable energy with a positron pulse-width of 10^{-10}s or less will be operating very soon [9.243]. Since the start signal for the positron lifetime measurement is derived electronically, the final coincidence-count rate is identical with the single-count rate of the annihilation photon detector. This enables count rates of $2 \times 10^4 \text{s}^{-1}$ or even higher for lifetime studies. Many exciting experiments with positrons remain to be done.

References

9.1   P.A.M. Dirac: Proc. Roy. Soc. **117**, 610 (1928); and **126**, 360 (1930)
9.2   C.D. Anderson: Science **76**, 238 (1932) and Phys. Rev. **43**, 491 (1933)
9.3   R. Beringer, C.G. Montgomery: Phys. Rev. **61**, 222 (1942)
9.4   S. De Benedetti, C.E. Cowan, W.R. Konnecker, H. Primakoff: Phys. Rev. **77**, 205 (1950)
9.5   G. Lang, S. De Benedetti: Phys. Rev. **108**, 914 (1957)
9.6   A.T. Stewart: Can. J. Phys. **35**, 168 (1957)
9.7   S. De Benedetti, H. Riching: Phys. Rev. **85**, 377 (1952)
9.8   I.K. MacKenzie, T.L. Khoo, A.B. McDonald, B.T.A. McKee: Phys. Rev. Lett. **19**, 946 (1967)
9.9   B.T.A. McKee, W. Triftshäuser, A.T. Stewart: Phys. Rev. Lett. **28**, 258 (1972)
9.10 O. Mogenson, K. Petersen, R.M.J. Cotterill, B. Hudson: Nature **239**, 98 (1972)
9.11 R.M.J. Cotterill, I.K. MacKenzie, L. Smedskjaer, G. Trumpy, J. Träff: Nature **239**, 99 (1972)
9.12 W. Triftshäuser, J.D. McGervey, R.W. Hendricks: Phys. Rev. B **9**, 3321 (1974)
9.13 S. Mantl, W. Triftshäuser: Phys. Rev. Lett. **34**, 1554 (1975)
9.14 K.F. Canter, P.G. Coleman, T.C. Griffith, G.R. Heyland: J. Phys. B **5**, L167 (1972)
9.15 K.F. Canter, A.P. Mills, S. Berko: Phys. Rev. Lett. **33**, 7 (1974)
9.16 T.S. Stein, W.E. Kauppila, L.O. Roellig: Rev. Sci. Instr. **45**, 951 (1974)
9.17 P.W. Zitzewitz, J.C. van House, A. Rich, D.W. Gidley: Phys. Rev. Lett. **43**, 1281 (1979)
9.18 P.G. Coleman, J.D. McNutt, J.T. Hutton, L.M. Diana, J.L. Fry: Rev. Sci. Instr. **51**, 935 (1980)

9.19   K.G. Lynn: Phys. Rev. Lett. **44**, 1330 (1980)
9.20   W. Triftshäuser, G. Kögel: Phys. Rev. Lett. **48**, 1741 (1982)
9.21   R.H. Howell, R.A. Alvarez, M. Stanek: Appl. Phys. Lett. **40**, 751 (1982)
9.22   M. Begemann, G. Gräff, H. Heminghaus, H. Kalinowsky, R. Ley: Nucl. Instr. Meth. **201**, 287 (1982)
9.23   P. Hautojärvi (ed.): *Positrons in Solids*, Topics Current Phys., Vol. 12 (Springer, Berlin, Heidelberg 1979)
9.24   R.R. Hasiguti, K. Fujiwara (eds.): *Positron Annihilation*, Proc. 5th Intern. Conf., Lake Yamanaka (The Japan Institute of Metals, Aoba Aramaki, Sendai 1979) p. 980
9.25   P.G. Coleman, S.C. Sharma, L.M. Diana (eds.): *Positron Annihilation*, Proc. 6th Intern. Conf., Arlington, USA (North Holland, Amsterdam 1982)
9.26   W. Brandt, A. Dupasquier (eds.): *Positron Solid-State Physics*, Proc. Intern. School of Phys. "Enrico Fermi", Varenna, Italy (1981) (North Holland, Amsterdam 1983)
9.27   A. Perkins, J.P. Carbotte: Phys. Rev. B **1**, 101 (1970)
9.28   J.P. Carbotte, H.L. Aurora: Can. J. Phys. **45**, 387 (1967)
9.29   R.M. Nieminen, J. Oliva: Phys. Rev. **22**, 2226 (1980)
9.30   G.E. Lee-Whiting: Phys. Rev. **97**, 1557 (1955)
9.31   B. Bergersen, E. Pajanne: Appl. Phys. **4**, 25 (1974)
9.32   P. Kubica, A.T. Stewart: Phys. Rev. Lett. **34**, 852 (1975)
9.33   W. Brandt, R. Paulin: Phys. Rev. B **8**, 4125 (1973)
9.34   W. Triftshäuser, G. Kögel, J. Bohdansky: J. Nucl. Mat. **111/112**, 687 (1982)
9.35   E. Uggerhoj, J.U. Anderson: Can. J. Phys. **46**, 543 (1968)
9.36   W. Brandt, R. Paulin: Phys. Rev. B **15**, 2511 (1977)
9.37   R. Behnisch, F. Bell, R. Sizmann: Phys. Status Solidi **33**, 375 (1969)
9.38   R.A. Ferrell: Rev. Mod. Phys. **28**, 308 (1956)
9.39   S. Kahana: Phys. Rev. **129**, 1622 (1963)
9.40   J.P. Carbotte, S. Kahana: Phys. Rev. **139**, A213 (1965)
9.41   J.J. Zuchelli, T.G. Hickmann: Phys. Rev. **136**, A1728 (1964)
9.42   J. Arponen, P. Jauho: Phys. Rev. **167**, 239 (1968)
9.43   J. Arponen, E. Pajanne: J. Phys. F**9**, 2359 (1979)
9.44   E. Boronski, Z. Szotek, H. Stachowiak: Phys. Rev. B **23**, 1785 (1981)
9.45   A. Rubaszek, H. Stachowiak: Phys. Status Solidi **124**, 159 (1984)
9.46   J.J. Donaghy, A.T. Stewart: Phys. Rev. **164**, 396 (1967)
9.47   J.J. Mader, S. Berko, H. Krakauer, A. Bansil: Phys. Rev. Lett. **37**, 1232 (1976)
9.48   S. Wakoh, S. Berko, M. Haghgooie, J.J. Mader: J.Phys. F **9**, L231 (1979)
9.49   N. Shiotani, T. Okada, H. Sekizawa, S. Wakoh: J. Phys. Soc. Jpn. **50**, 498 (1981)
9.50   Chang Lee: Zh. Eksp. Teor. Fiz. **33**, 365 (1975) [English transl.: Sov. Phys. JETP **6**, 281 (1958)]
9.51   P.E. Mijnarends: In [Ref. 9.26, p. 146]
9.52   R.E. Sund, R.B. Walton, N.J. Norris, M.H. Mac Gregor: Nucl. Instr. Meth. **27**, 109 (1964)
9.53   G. Gräff, R. Ley, A. Osipowicz, G. Werth, J. Ahrens: Appl. Phys. A **33**, 59 (1984)
9.54   K. Fujiwara, O. Sueoka: J. Phys. Soc. Jpn. **21**, 1947 (1966)
9.55   M. Hasegawa, T. Suzuki, M. Hirabayashi: J. Phys. Soc. Jpn. **43**, 89 (1977)
9.56   P. Colombino, B. Fiscella, L. Trossi: Nuovo Cimento **27**, 589 (1963)
9.57   W. Triftshäuser: *Proc. 2nd Intern. Conference on Positron Annihilation*, Kingston, Canada (1971)
9.58   G. Riepe, D. Protic, R. Kurz, W. Triftshäuser, Zs. Kajcsos, J. Winter: In [Ref. 9.24, p. 371]
9.59   S. Berko, J.J. Mader: Appl. Phys. **5**, 287 (1975)
9.60   S. Berko, M. Haghgooie, J.J. Mader: Phys. Lett. A **63**, 335 (1977)
9.61   H.O. Anger: Rev. Sci. Instr. **29**, 27 (1958)
9.62   H.O. Anger: *Instrumentation in Nuclear Medicine*, Vol. 2, ed. by G.H. Hine, J.A. Sorenson (New York, N.Y., 1974) p. 61
9.63   J. Mayers, J.D. McGervey, P.A. Waters, R.N. West: In [Ref. 9.24, p. 417]
9.64   R.N. West: Inst. Phys. Conf. Ser. **55**, 35 (1980)
9.65   A.P. Jeavons, D.W. Townsend, N.L. Ford, K. Kull, A. Manuel, O. Fischer, M. Peter: IEEE Trans. NS-**25**, 164 (1978)
9.66   A.A. Manuel, S. Samoilov, O. Fischer, M. Peter, A.P. Jeavons: Helv. Phys. Acta **52**, 255 (1979)

9.67   A.A. Manuel: In [Ref. 9.26, p. 581]
9.68   A.A. Manuel, O. Fischer, M. Peter, A.P. Jeavons: Nucl. Instr. Meth. **156**, 67 (1978)
9.69   W. Brandt, G. Coussot, R. Paulin: Phys. Rev. Lett. **23**, 522 (1969)
9.70   A. Greenberger, A.P. Mills, A. Thompson, S. Berko: Phys. Lett. A **32**, 72 (1970)
9.71   C.H. Hodges, B.T.A. McKee, W. Triftshäuser, A.T. Stewart: Can. J. Phys. **50**, 103 (1972)
9.72   G. Kögel: In [Ref. 9.24, p. 383]
9.73   D. Herlach, K. Maier: Appl. Phys. **11**, 197 (1976)
9.74   L. Madansky, F. Rasetti: Phys. Rev. **79**, 397 (1950)
9.75   D.E. Groce, D.G. Costello, J.W. McGowan, D.F. Herring: Bull. Am. Phys. Soc. **13**, 1397 (1968)
9.76   D.G. Costello, D.E. Groce, D.F. Herring, J.W. McGowan: Phys. Rev. B **5**, 1433 (1972)
9.77   J.M. Dale, L.D. Hulett, S. Pendyala: Surf. Interface Anal. **2**, 199 (1980)
9.78   A.P. Mills, P.M. Platzman, B.L. Brown: Phys. Rev. Lett. **41**, 1076 (1978)
9.79   A.P. Mills: Appl. Phys. Lett. **35**, 427 (1979); **37**, 667 (1980)
9.80   A.P. Mills: In [Ref. 9.26, p. 432]
9.81   C.H. Hodges, M.J. Stott: Phys. Rev. B **7**, 73 (1973)
9.82   R.M. Nieminen, C.H. Hodges: Solid State Commun. **18**, 1115 (1976)
9.83   W. Triftshäuser, G. Kögel: In [Ref. 9.25, p. 142]
9.84   A.P. Mills: Phys. Rev. Lett. **41**, 1828 (1978)
9.85   K.G. Lynn: Phys. Rev. Lett. **43**, 391 (1979)
9.86   A.P. Mills: Solid State Commun. **31**, 623 (1979)
9.87   K.G. Lynn, H. Lutz: Phys. Rev. B **22**, 4143 (1980)
9.88   K.G. Lynn: In [Ref. 9.26, p. 609]
9.89   I.J. Rosenberg, A.H. Weiss, K.F. Canter: Phys. Rev. Lett. **44**, 1139 (1980)
9.90   J.M. Dale, L.D. Hulett, S. Pendyala: Appl. Spectrosc. **19**, 105 (1983)
9.91   P.E. Mijnarends: Phys. Rev. **160**, 512 (1967)
9.92   P.E. Mijnarends: In *Compton Scattering*, ed. by B. Williams (London 1977) p.323
9.93   D.G. Lock, V.H.C. Crisp, R.N. West: J. Phys. F **3**, 561 (1973)
9.94   F.M. Müller: Phys. Rev. B **15**, 3039 (1977)
9.95   C.K. Majumdar: Phys. Rev. B **4**, 2111 (1971)
9.96   D.A. Chesler, S.J. Riederer: Phys. Med. Biol. **20**, 632 (1975)
9.97   B.B. Hede, J.P. Carbotte: J. Phys. Chem. Solids **33**, 727 (1972)
9.98   J.P. Carbotte: In [Ref. 9.26, p. 32]
9.99   S. Cushner, J.C. Erskine, S. Berko: Phys. Rev. B **1**, 2852 (1970)
9.100   B. Rozenfeld, S. Chabik, J. Pajak, K. Jerie, W. Wierzchowski: Acta Phys. Pol. A **44**, 21 (1973)
9.101   M.R. Halse: Phil. Trans. Roy. Soc. London (A) **27**, 668 (1968)
9.102   P.E. Mijnarends, R.M. Singru: Phys. Rev. B **9**, 6038 (1979)
9.103   G.M. Stocks, W.M. Temmerman, B.L. Györfy: Phys. Rev. Lett. **41**, 339 (1978)
9.104   R. Prasad, S.C. Papadopoulos, A. Bansil: Phys. Rev. B **23**, 2607 (1981)
9.105   M. Haghgooie, S. Berko, U. Mizutani: In [Ref. 9.24, p. 291]
9.106   I.K. MacKenzie, G.F.O. Langstroth, B.T.A. McKee, C.G. White: Can. J. Phys. **42**, 1836 (1964)
9.107   J.H. Kusmiss, J.W. Swanson: Phys. Lett. A **27**, 517 (1968)
9.108   W. Triftshäuser: Phys. Rev. B **12**, 4634 (1975)
9.109   R.N. West: In [Ref. 9.23, p. 89]
9.110   R.W. Siegel: J. Nucl. Mat. **69/70**, 117 (1978)
9.111   W. Brandt: In *Positron Annihilation*, ed. by A.T. Stewart, L.O. Roellig (Academic, New York 1967) p. 155
9.112   B. Bergersen, M.J. Stott: Solid State Commun. **7**, 1203 (1969)
9.113   D.C. Connors, R.N. West: Phys. Lett. A **30**, 24 (1969)
9.114   M. Doyama: J. Phys. Soc. Jpn. **33**, 1495 (1972)
9.115   A. Seeger: Appl. Phys. **4**, 183 (1974)
9.116   C.H. Hodges: Phys. Rev. Lett. **25**, 284 (1970)
9.117   J. Arponen, P. Hautojärvi, R.M. Nieminen, E. Pajanne: J. Phys. F **3**, 2092 (1973)
9.118   M. Manninen, R.M. Nieminen, P. Hautojärvi, J. Arponen: Phys. Rev. B **12**, 4012 (1975)

9.119   W. Brandt: Appl. Phys. **5**, 1 (1974)
9.120   G. Mori: J. Phys. F **7**, L89 (1977)
9.121   J.P. Gupta, R.W. Siegel: Phys. Rev. Lett. **39**, 1212 (1977)
9.122   R.M. Nieminen: In [Ref. 9.26, p. 359]
9.123   T. McMullen: J. Phys. F **7**, 3041 (1977); **8**, 87 (1978)
9.124   R.N. West: Adv. Phys. **22**, 263 (1973)
9.125   T.M. Hall, A.N. Goland, C.L. Snead: Phys. Rev. B **10**, 3062 (1974)
9.126   M.J. Fluss, L.C. Smedskjaer, M.K. Chason, D.G. Legnini, R.W. Siegel: Phys. Rev. B **17**, 3444 (1978)
9.127   A.S. Berger, S.T: Ockers, M.K. Chason, R.W. Siegel: J. Nucl. Mat. **69/70**, 734 (1978)
9.128   G. Dlubek, O. Brümmer, N. Meyendorf: Phys. Status Solidi (a) **39**, K 95 (1977)
9.129   W. Triftshäuser, J.D. McGervey: Appl. Phys. **6**, 177 (1975)
9.130   D. Herlach, H. Stoll, W. Trost, H. Metz, T.E. Jackman, K. Maier, H.E. Schaefer, A. Seeger: Appl. Phys. **12**, 59 (1977)
9.131   G. Dlubek, O. Brümmer, N. Meyendorf: Appl. Phys. **13**, 67 (1977)
9.132   R.P. Sahn, K.C. Jain, R.W. Siegel: J. Nucl. Mat. **69/70**, 264 (1978)
9.133   J.L. Campbell, C.W. Schulte, J.A. Jackman: J. Phys. F **7**, 1985 (1977)
9.134   Y.C. Jean, K.G. Lynn, J.E. Dickman: Phys. Rev. B **21**, 2655 (1980)
9.135   M. Doyama, J.S. Koehler: Phys. Rev. **127**, 21 (1962)
9.136   M. Doyama, K. Kuribayashi, S. Nanao, S. Tanigawa: Appl. Phys. **4**, 153 (1974)
9.137   S. Nanao, K. Kuribayashi, S. Tanigawa, M. Doyama: J. Phys. F **7**, 1403 (1977)
9.138   P. Rice-Evans, Tin Hlaing, D.B. Rees: J. Phys. F **6**, 1079 (1976)
9.139   M.J. Fluss, L.C. Smedskjaer, R.W. Siegel, D.G. Legnini, M.K. Chason: J. Phys. F **10**, 1763 (1980)
9.140   R.R. Bourassa, B. Lengeler: J. Phys. F **6**, 1405 (1976)
9.141   A.S. Berger, S.T. Ockers, R.W. Siegel: J. Phys. F **9**, 1023 (1979)
9.142   K.P. Singh, G.S. Goodbody, R.N. West: Phys. Lett. A **55**, 237 (1975)
9.143   P. Rice-Evans, Tin Hlaing, I. Chaglar: Phys. Lett. A **60**, 368 (1977)
9.144   S.C. Sharma, S. Berko, W.K. Warburton: Phys. Lett. A **58**, 405 (1976)
9.145   H.E. Schaefer, K. Maier, M. Weller, D. Herlach, A. Seeger, J. Diehl: Scripta Metall. **11**, 803 (1977)
9.146   H. Matter, J. Winter, W. Triftshäuser: Appl. Phys. **20**, 135 (1979)
9.147   S.M. Kim, W.J.L. Buyers: J. Phys. F **8**, L103 (1978)
9.148   W. Wycisk, M. Feller-Kniepmeier: J. Nucl. Mat. **69/70**, 616 (1978)
9.149   K.G. Lynn, C.L. Snead, J.J. Hurst: J. Phys. F **10**, 1753 (1980)
9.150   L.C. Smedskjaer, M.J. Fluss, D.G. Legnini, M.K. Chason, R.W. Siegel: In [Ref. 9.25, p. 526]
9.151   J.L. Campbell, J.H. Jackman, C.W. Schulte: Appl. Phys. **16**, 29 (1978)
9.152   K. Maier, M. Peo, B. Saile, H.E. Schaefer, A. Seeger: Phil. Mag. A **40**, 707 (1979)
9.153   I.A. Schwiertlich, H. Schultz: Philos. Mag. A **42**, 601 (1980)
9.154   M.Suezawa, H. Kimura: Philos. Mag. **28**, 901 (1973)
9.155   W. Triftshäuser: In *Festkörperprobleme*, Advances in Solid State Physics, **15**, 381 (Pergamon/Vieweg, Braunschweig 1975)
9.156   K. Kuribayashi, S. Tanigawa, S. Nanao, M. Doyama: Solid State Commun. **12**, 1179 (1973)
9.157   O. Sueoka: J. Phys. Soc. Jpn. **39**, 969 (1975)
9.158   H. Fukushima, M. Doyama: J. Phys. F **6**, 677 (1976)
9.159   Th. Hehenkamp, L. Sander: Z. Metallkunde **70**, 202 (1979)
9.160   W. Triftshäuser, H. Matter, J. Winter: Appl. Phys. A **28**, 179 (1982)
9.161   J.H. Kusmiss, A.T. Stewart: Adv. Phys. **16**, 63 (1967)
9.162   D.R. Gustafson, A.R. Mackintosh, D.J. Zaffarano: Phys. Rev. **130**, 1455 (1963)
9.163   R.N. West, R.E. Borland, J.R.A. Cooper, N.E. Cusak: Proc. Phys. Soc. **92**, 195 (1967)
9.164   O.E. Mogensen, G. Trumpy: Phys. Rev. **188**, 639 (1969)
9.165   W. Brandt, H.F. Waung: Phys. Lett. A **27**, 700 (1968)
9.166   M.V. Chu, C.J. Jan, P.K. Tseng, W.F. Huang: Phys. Lett. A **43**, 423 (1973)
9.167   L.E. Ballentine: Can. J. Phys. **44**, 2533 (1966)
9.168   F. Itoh: J. Phys. Soc. Jpn. **41**, 824 (1976)
9.169   M. Doyama, S. Tanigawa, K. Kuribayashi, H. Fukushima, K. Hiude, F. Saito: J. Phys. F **5**, L230 (1975)

9.170   W. Triftshäuser: In *Liquid and Amorphous Metals*, ed. by E. Lüscher, H. Coufal (Sijthoff and Noordhoff, Alphen aan den Rijn 1980) p. 479

9.171   I. Ya. Dekhtyar, S.G. Litovchenko, V.S. Mihalenkov: Sov. Phys. Doklady **7**, 1135 (1963)

9.172   H. Morinaga: Phys. Lett. A **34**, 384 (1971)

9.173   M. Doyama, K. Kuribayashi, S. Tanigawa, S. Nanao: J. Phys. Soc. Jpn. **36**, 1706 (1974)

9.174   K. Kuribayashi, S. Tanigawa, S. Nanao, M. Doyama: Solid State Commun. **17**, 143 (1975)

9.175   H. Matter, J. Winter, W. Triftshäuser: J. Nucl. Mat. **88**, 273 (1980)

9.176   G. Kögel, P. Sperr, W. Triftshäuser, S.J. Rothman: J. Nucl. Mat. **131** (1985)

9.177   N.L. Peterson: Comments Solid State Phys. **8**, 93 (1978)

9.178   J.M. Sanchez, D. de Fontaine: Phys. Rev. Lett. **35**, 227 (1975)

9.179   W. Schilling, K. Sonnenberg: J. Phys. F **3**, 322 (1973)

9.180   S. Mantl, W. Triftshäuser: Phys. Rev. B **17**, 1645 (1978)

9.181   W. Frank, A. Seeger: Rad. Eff. **1**, 117 (1969)

9.182   K. Hinode, S. Tanigawa, M. Doyama: J. Nucl. Mat. **69/79**, 678 (1978)

9.183   M. Eldrup, O. Mogenson, J.H. Evans: J. Phys. F **6**, 499 (1976)

9.184   K. Petersen, J.H. Evans, R.M.J. Cotterill: Phil. Mag. **32**, 427 (1975)

9.185   K. Petersen, B. Nielsen, N. Thrane: Phil. Mag. **34**, 693 (1976)

9.186   P. Hautojärvi, T. Judin, A. Vehanen, J. Yli-Kauppila, J. Johansson, J. Verdone, P. Moser: Solid State Commun. **29**, 855 (1979)

9.187   P. Hautojärvi, J. Johansson, A. Vehanen, J. Yli-Kauppila, P. Moser: Phys. Rev. Lett. **44**, 1326 (1980)

9.188   A. Vehanen, P. Hautojärvi, J. Johansson, J. Yli-Kauppila, P. Moser: Phys. Rev. B **25**, 762 (1982)

9.189   M. Weller, J. Diehl, W. Triftshäuser: Solid State Commun. **17**, 1223 (1975)

9.190   G. Dlubek, O. Brümmer, V.S. Mikhalenkov, J. Yli-Kauppila, A. Vehanen, P. Hautojärvi: Crystal Res. Techn. **19**, 627 (1984)

9.191   C.H. Hodges, M.J. Stott: Solid State Commun. **21**, 293 (1973)

9.192   M. Hasegawa, T. Suzuki: Rad. Eff. **21**, 201 (1974)

9.193   K. Petersen, N. Thrane, G. Trumpy, R.W. Hendricks: Appl. Phys. **10**, 85 (1976)

9.194   V.W. Lindberg, J.D. McGervey, R.W. Hendricks, W. Triftshäuser: Phil. Mag. **36**, 117 (1977)

9.195   G. Kögel, J. Winter, W. Triftshäuser: In [Ref. 9.24, p. 707]

9.196   P. Hautojärvi, J. Heiniö, M. Manninen, R.M. Nieminen: Phil. Mag. **35**, 973 (1977)

9.197   T. McMullen, R.J. Douglas, N. Etherington, B.T.A. McKee, A.T. Stewart, E. Zaremba: J. Phys. F **11**, 1435 (1981)

9.198   R.M. Nieminen, J. Laakkonen: Appl. Phys. **20**, 181 (1979)

9.199   R.M. Nieminen, J. Laakkonen, P. Hautojärvi, A. Vehanen: Phys. Rev. B **19**, 1397 (1979)

9.200   M. Hasegawa, Y.J. He, K.R. Hoffmann, R.R. Lee, S. Berko: In [Ref. 9.236, p. 260]

9.201   A. Alam, P.A. Walters, R.N. West, J.D. McGervey: J. Phys. F **14**, 761 (1984)

9.202   D.J. Reed: Rad. Eff. **31**, 125 (1977)

9.203   B. Viswanathan, W. Triftshäuser, G. Kögel: Rad. Eff. **78**, 231 (1983)

9.204   G. Kögel, Qin-min Fan, P. Sperr, W. Triftshäuser, B. Viswanathan: J. Nucl. Mat. **131**, 148 (1985)

9.205   D.K. Brice: In *Ion Implantation Range and Energy Deposition Distribution*, Vol. I (Plenum, New York 1975)

9.206   W. Jäger, J. Roth: J. Nucl. Mat. **93/94**, 756 (1980)

9.207   W. Jäger, J. Roth: J. Nucl. Instr. Meth. **182/183**, 975 (1981)

9.208   D.S. Whitmell: Rad. Eff. **53**, 209 (1981)

9.209   M. Eldrup, J.H. Evans: J. Phys. F **12**, 1265 (1982)

9.210   H.E. Hansen, R. Talja, H. Rajainmäki, H.K. Nielsen, B. Nielsen, R.M. Nieminen: Appl. Phys. A **36**, 81 (1985)

9.211   B. Lengeler, S. Mantl, W. Triftshäuser: J. Phys. F **8**, 1691 (1978)

9.212   J.L. Finney: J.Phys. Paris Coll. **36**, C-2, 1 (1975)

9.213   G. LeCaer, J.M. Dubois, H. Fischer, U. Gonser, H.G. Wagner: Nucl. Instr. Meth. in Phys. Research B **5**, 25 (1984)

9.214 Y. Waseda, H.S. Chen: Phys. Status Solidi (a) **49**, 387 (1978)
9.215 P. Panissod, D. Aliaga-Guerra, A. Amamou, J. Durand, W.L. Johnson, W.L. Carter, S.J. Poon: Phys. Rev. Lett. **44**, 1465 (1980)
9.216 E. Nold, S. Steeb, P. Lamparter, G. Rainer-Harbach: J. Physique **41**, C8-186 (1980)
9.217 N. Shiotani: In [Ref. 9.25, p. 561]
9.218 W. Triftshäuser, G. Kögel: In *Metallic Glasses: Science and Technology*, ed. by C. Hargitai, I. Bakonyi, T. Kemeny (Kultura, Budapest 1981) p. 347
9.219 F. Itoh, T. Honda, M. Hasegawa, K. Suzuki: Nucl. Instr. Meth. in Phys. Research **199**, 323 (1982)
9.220 G. Kögel, W. Triftshäuser: In [Ref. 9.25, p. 595]
9.221 S. Tanigawa, K. Hinode, R. Nagai, M. Doyama, N. Shiotani: Phys. Status Solidi (a) **51**, 249 (1979)
9.222 Zs. Kajcsos, J. Winter, S. Mantl, W. Triftshäuser: Phys. Status Solidi (a) **58**, 77 (1980)
9.223 E. Cartier, F. Heinrich: Helv. Phys. Acta **53**, 226 (1981)
9.224 S. Mantl, W. Kesternich, W. Triftshäuser: J. Nucl. Mat. **69/70**, 593 (1978)
9.225 G. Kögel, J. Winter, W. Triftshäuser: In *Metallic Glasses: Science and Technology*, ed. by C. Hargitai, I. Bakonyi and T. Kemeny (Kultura, Budapest 1981) p.311
9.226 P. Moser, P. Hautojärvi, J. Yli-Kauppila, C. Corbel: In [Ref. 9.25, p. 592]
9.227 J. Yli-Kauppila, P. Moser, H. Künzi, P. Hautojärvi: Appl. Phys. A **27**, 31 (1982)
9.228 A. Vehanen, K.G. Lynn, P.J. Schultz, A.N. Goland, C.L. Snead, H.J. Güntherodt, E. Cartier: In [Ref. 9.25, p. 587]
9.229 K.F. Canter, A.P. Mills, S. Berko: Phys. Rev. Lett. **33**, 7 (1974)
9.230 A.P. Mills, L.N. Pfeiffer: Phys. Rev. Lett. **43**, 1961 (1979)
9.231 I.J. Rosenberg, A.H. Weiss, K.F. Canter: Phys. Rev. Lett. **44**, 1139 (1980)
9.232 A.P. Mills, P.M. Platzman: Solid State Commun. **35**, 321 (1980)
9.233 M.N. Read, D.N. Lowy: Surf. Sci. **107**, L313 (1981)
9.234 J.M. Dale, L.D. Hulett, S. Pendyala: Appl. of Surf. Sci. **8**, 472 (1981)
9.235 D. Fisher, K.G. Lynn, W. Frieze: Phys. Rev. Lett. **50**, 1149 (1983)
9.236 P.C. Jain, R.M. Singru, K.P. Gopinathan (eds.): *Positron Annihilation*, Proc. 7th Intern. Conf., New Delhi (World Scientific Publ. Co., Singapore 1985)
9.237 S. Berko, K.F. Canter, A.P. Mills, K.G. Lynn, L.O. Roellig, R.N. West: In [Ref. 9.236, p. 951]
9.238 R.H. Howell, P. Meyer, I.J. Rosenberg, M.J. Fluss: In [Ref. 9.236, p. 801]
9.239 B. Rozenfeld, K. Jerie, W. Swiatkowski: Acta Phys. Pol. A **64**, 93 (1983)
9.240 A.P. Mills, L. Pfeiffer, P.M. Platzmann: Phys. Rev. Lett. **51**, 1085 (1983)
9.241 G. Kögel: In [Ref. 9.236, p. 965]
9.242 A.P. Mills: Appl. Phys. **23**, 189 (1980)
9.243 D. Schödlbauer, P. Sperr, G. Kögel, W. Triftshäuser: In [Ref. 9.236, p. 957]

10. Muon Spectroscopy

J. Chappert and A. Yaouanc
With 13 Figures

In this chapter we present a microscopic method for the study of metals which, like many modern techniques for the investigation of condensed matter, was initiated by nuclear physicists: the *Muon Spin Rotation* (or *Relaxation*) technique, often called by the acronym μSR.

As early as 1957 *Garwin* et al. [10.1] suggested that *"muons will become a powerful tool for exploring magnetic fields in ... interatomic regions"*. The first paper in the field of metal physics came fifteen years later [10.2]. After an exploratory period in the use of the technique, significant results began to be obtained in 1976–77. It was then shown that the positive muon μ^+ could be used to study diffusion mechanisms and the localization of a light isotope of the proton [10.3–5]. Over the next five years, the problem of the effect of impurities and defects present in metals dominated the subject. Only recently has the μSR technique been used to study problems of current interest in metal physics. In many of these applications, several specific characteristics of the technique have allowed physicists to obtain new and original results. Basic diffusion processes [10.6,7], hydrides [10.8], fluctuations in superconductors and magnetic metals [10.9–11], phase transitions [10.12] are typical examples of the kind of extensive research at present conducted with the μSR method.

Two types of muons exist: the positive muon μ^+ and its anti-particle the negative muon μ^-. Here we shall deal only with the μ^+ because many more experimental results have been obtained in solid state physics for the μ^+ than for the μ^-. In a metal the formation of muonium (bound state of μ^+ and an electron) has never been detected. Therefore in the following we take the free μ^+ as the basic probe.

Several review articles or books have already been devoted to the μ^+SR technique [10.13–18]. Thus we limit ourselves here to the basic physical principles and the unique features of μ^+SR applied to metal physics. Due to the lack of space, we mention only a few specific experimental results typical of possible applications in the field.

10.1 Basic Principles of the Experimental Techniques

Muon beams are produced near large accelerators or at the so-called meson factories [10.19]. A common means of production is to bombard a light target like beryllium with protons. Several nuclear reactions take place leading to μ^+

production. The muons are implanted into the sample under study. Depending essentially on their energy, a sample of 10 to 0.1 g/cm^2 is needed to stop the muons.

There are three fundamental points concerning the μ^+SR technique. First, the μ^+ beam is naturally *polarized*, i.e. the 1/2 muonic spin is collinear with the μ^+ momentum. Muon thermalization in a metal has a negligible effect on the polarization [10.13]. A second essential point is that the μ^+ is an *unstable* particle: its lifetime is $\tau_\mu = 2.2 \times 10^{-6}$s. It decays by a weak process to a positron e^+ and two associated neutrinos. The μ^+ mass is $m_\mu = 105.6$ MeV $\simeq 0.11$ mass of the proton. Third, the e^+ is produced with an average energy of $\simeq 35$ MeV and, due to the parity violation of weak processes, the e^+ *emission probability is not isotropic* but is given by $W(\theta) = 1 + a_0\cos\theta$, θ being the angle between the μ^+ spin direction and the e^+ trajectory. In practice $a_0 \simeq 0.25$.

Basically the experimental technique consists of building a histogram by recording the number of e^+ entering a detector at a fixed position. The most common method is the *time-differential technique* which requires a μ^+ beam continuously distributed in time. Before being implanted into the sample, each μ^+ passes through a telescope which starts a clock. If the e^+ emitted after a time t enters a positron telescope, it will stop the clock. The number of e^+ detected at a fixed angle versus the elapsed time t between the start and stop signals is recorded. Here the main limitation is set by the requirement that only one μ^+ at a time is present in the sample. This condition is needed to be sure that the detected e^+ originates from the μ^+ which started the clock. Depending on the accelerator, the e^+ counting rate (good events) varies from some hundreds to two thousand per second, and the time to record a histogram ranges between a few hours and fifteen minutes. Usually a telescope consists of plastic scintillators. In standard experiments the overall time resolution is about one nanosecond and the dead time of the order of ten nanoseconds.

Two types of experimental arrangements are generally used. They differ in the direction of the magnetic field with respect to the μ^+ beam polarization.

In the first arrangement (Fig. 10.1), the applied field is *perpendicular* to the μ^+ beam. The number of e^+ detected at time t and angle θ in a plane perpendicular to the field is given by

$$N_{e^+}(t, \theta) = N_0(\theta) \exp(-t/\tau_\mu)[1 + a_0 P_\perp(t) \cos(\omega_\mu t + \theta)]. \qquad (10.1)$$

$N_0(\theta)$ is a normalization constant and the exponential term accounts for the trivial μ^+ decay. The e^+ counting rate oscillates because the μ^+ spin precesses with an angular velocity ω_μ and, as a consequence, the anisotropic e^+ decay pattern rotates in a plane perpendicular to the field. Two measured quantities are of interest. First, the angular velocity $\omega_\mu = \gamma_\mu B_\mu$, γ_μ being the μ^+ gyromagnetic ratio ($\gamma_\mu/2\pi = 13.554$ kHz/G) and B_μ the field seen by the μ^+, which can be different from the applied field as will be discussed in Sect. 10.4.1. Second, $P_\perp(t)$ is the transverse depolarization function which accounts for the damping of the oscillations (Fig. 10.1). The shape of $P_\perp(t)$ depends on whether the distribution of local fields at the μ^+ site is static or dynamic.

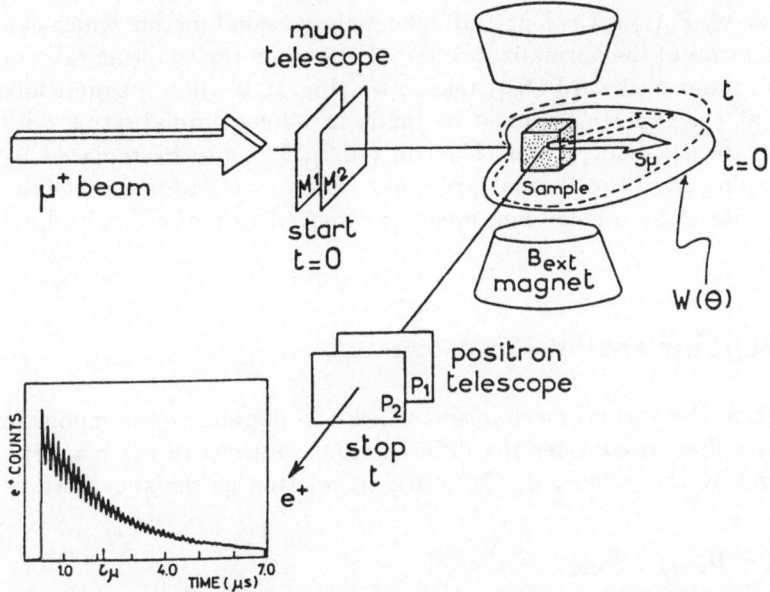

Fig. 10.1. Schematic experimental setup in the transverse field geometry. The positron count rate in a given direction, $N_{e^+}(t, \theta)$, shows a periodic modulation, superimposed on the trivial exponential decay. This modulation results from the precession of the μ^+ spin in the magnetic field applied in a direction perpendicular to the incident μ^+ beam. In the insert, the sample is the diamagnetic alloy CuNi. The temperature is 5 K and the applied magnetic field 500 G

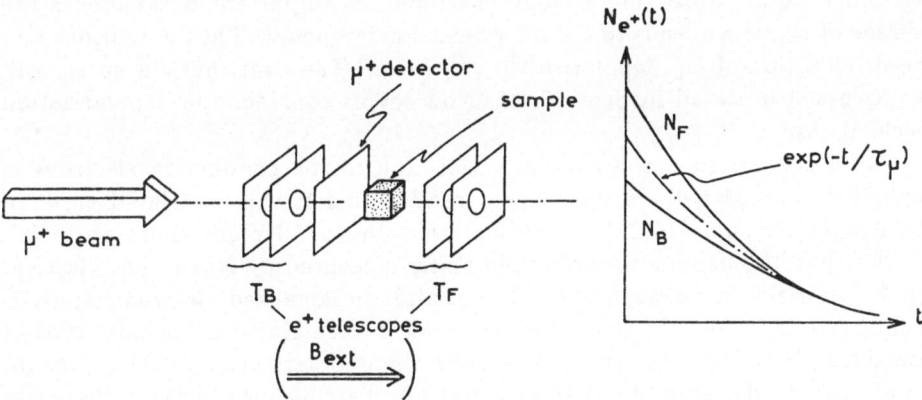

Fig. 10.2. Schematic experimental setup in the longitudinal field geometry. One observes a difference in count rates, N_F and N_B, recorded in the forward (T_F) and backward (T_B) positron telescopes respectively. A magnetic field may be present

In the second type of arrangement (Fig. 10.2) called longitudinal, the field is *parallel* to the μ^+ beam polarization. For muons stopped in a nonmagnetically ordered metal, $N_{e^+}(t, \theta)$ is given by

$$N_{e^+}(t, \theta) = N_0(\theta) \exp(-t/\tau_\mu)[1 + a_0 P_{||}(t)\cos\theta]. \tag{10.2}$$

The telescopes are placed along the μ^+ beam direction on each side of the sample ($\theta = 0$ or π). $P_{\parallel}(t)$ is the longitudinal depolarization function which can be expressed in terms of the normalized difference between the counting rates of the forward (N_F) and backward (N_B) telescopes (Fig. 10.2). In a magnetically ordered material, the experiment must be made in a longitudinal setup, with or without an external field [10.20]. Then in (10.2), θ has to be replaced by $\theta + \omega_\mu t$, ω_μ coming from the Larmor precession due to the internal magnetic field at the μ^+ site. A zero-field experiment is a special case of a longitudinal experiment.

10.2 The Depolarization Functions

In order to discuss the various mechanisms which can depolarize the muons in a metal, we have first to consider the different contributions to the magnetic field experienced by the muons, B_μ. It is usually written as the sum of three terms

$$B_\mu = B_{\text{ext}} + B_{\text{cond}} + B_{\text{dip}} \tag{10.3}$$

where B_{ext} is the externally applied magnetic field (if any), B_{cond} is the magnetic field due to the conduction electrons around the muons, and B_{dip} is the dipolar field felt by the muons. This field is due to the localized nuclear and electronic moments.

Since we are observing a muon ensemble, an important parameter is the average of B_μ which leads to the μ^+ precession frequency. The fluctuations and the distribution of B_μ depolarize the μ^+ beam. The contributions to ω_μ will be discussed in detail in Sect. 10.4.1. Now let us consider the depolarization mechanisms.

As a general rule, the depolarization due to the conduction electrons is negligible. As it will be mentioned in Sect. 10.4.1, this is related through the Knight formula [10.21] to the fact that the μ^+ Knight shifts are much smaller than the usual nuclear Knight shifts measured by NMR. The classical dipolar interaction between the μ^+ spin and the localized electronic spins is usually stronger than the RKKY (Ruderman, Kittel, Kasuya, Yoshida) type of interaction between these spins. Therefore the depolarization of the μ^+ beam is only due to dynamic fluctuations or static distributions of B_{dip}. We write $\omega_{\text{dip}} = \gamma_\mu B_{\text{dip}}$. For convenience we first discuss slow fluctuations and then consider the fluctuations from a more general point of view.

10.2.1 Slow Dipole Fluctuations

Let us first consider the effect of *slowly fluctuating* dipoles on the μ^+ beam polarization. We take slowly fluctuating to mean that the Larmor precession frequency of the dipole, ω_I, is such that $\omega_I \tau_c > 1$, τ_c being the correlation time of B_{dip} as seen by μ^+. In addition, we suppose that $\omega_\mu \tau_c > 1$. This hypothesis implies that we neglect the μ^+ spin-lattice relaxation [10.21].

We first discuss the *transverse* geometry. Muons precess in fields which vary slightly from site to site. For instance, in a diamagnetic metal such as Al, the nuclear dipoles produce a field with a Gaussian distribution at the μ^+ because $k_B T > \hbar \omega_I$ in usual conditions. Therefore we can write [10.22]

$$P_\perp(t) = \left\langle \cos\left[\int_0^t dt' \omega_{dip}(t') \right] \right\rangle$$

where $\langle \ldots \rangle$ represents the average over the field distribution collinear to B_μ. Taking the average with the hypothesis of an exponential decay of the field correlation gives the so-called Abragam formula

$$P_\perp(t) = \exp\{ - 2\sigma^2 \tau_c [\exp(-t/\tau_c) - 1 + t/\tau_c] \} \qquad (10.4)$$

where σ is the Van Vleck static line width well-known in NMR

$$\sigma^2 = \frac{1}{2} \gamma_\mu^2 \langle (B_{dip}^z)^2 \rangle = \frac{1}{6} I(I+1) \hbar^2 \gamma_\mu^2 \gamma_I^2 \sum_i (1 - 3\cos^2\theta_i)^2 r_i^{-6}. \qquad (10.5)$$

Here the sum extends over all nuclei of polar angle θ_i, located at a distance r_i from μ^+. $\langle (B_{dip}^z)^2 \rangle$ is the second moment of the distribution collinear to B_μ. The value of σ allows the μ^+ site localization to be determined, especially by examining its field dependence [10.5,23]. Usually $\tau_c \simeq \tau_s$, τ_s being the μ^+ mean stay time at a given site. We see therefore that information on the μ^+ diffusion in metals can be obtained. If the muons are immobile $(t \ll \tau_c)$, $P_\perp(t)$ has a *Gaussian* form and (10.4) becomes

$$P_\perp(t) \simeq \exp(-\sigma^2 t^2). \qquad (10.6)$$

At the other limit $(t \gg \tau_c)$, motional narrowing occurs as in NMR and $P_\perp(t)$ becomes *exponential*

$$P_\perp(t) \simeq \exp(-2\sigma^2 \tau_c t) = \exp(-\lambda t) \qquad (10.7)$$

with $\lambda = 2\sigma^2 \tau_c = \gamma_\mu^2 \langle (B_{dip}^z)^2 \rangle \tau_c$.

Let us now turn to the case of the *longitudinal* geometry. Since the magnetic field is collinear to the beam polarization, the μ^+ depolarization described by $P_\parallel(t)$ can only be due to magnetic fluctuations transverse to the μ^+ beam. If we consider that the magnetic fluctuations come from an isotropic static field distribution such as that experienced by static muons surrounded by nuclear dipoles we can write

$$P_\parallel(t) = \frac{1}{3} + \frac{2}{3} \langle \cos(\omega_{dip} t) \rangle$$

because only 2/3 of the polarization can be influenced. For a Gaussian distribution we obtain the Kubo-Toyabe formula [10.24]

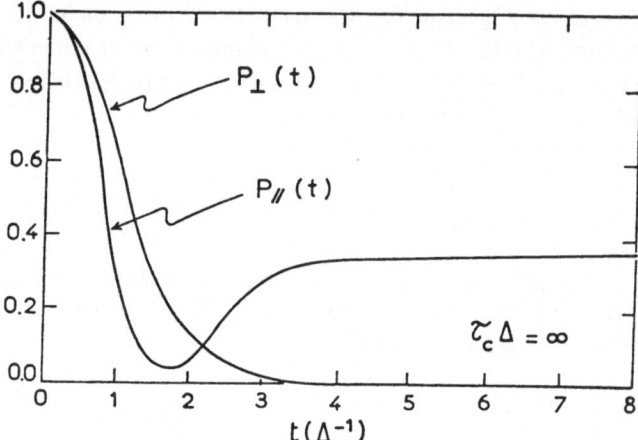

Fig. 10.3. Comparison of the static ($\tau_c \Delta = \infty$) relaxation functions, $P_\parallel(t)$ and $P_\perp(t)$, showing important differences both in the initial and final parts of the curves. Note in particular the characteristic 1/3 recovery of $P_\parallel(t)$ at large t

$$P_\parallel(t) = \frac{1}{2} + \frac{2}{3}(1 - \Delta^2 t^2) \exp(-\Delta^2 t^2/2). \tag{10.8}$$

For the isotropic distribution $\Delta^2 = 2\sigma^2 = \gamma_\mu^2 \langle (B_{dip}^z)^2 \rangle$. For a static case a comparison of $P_\parallel(t)$ and $P_\perp(t)$, given by (10.8 and 4), respectively, is shown in Fig. 10.3. A remarkable fact is that after a time $\sim 3\Delta^{-1}$, $P_\parallel(t)$ recovers a value 1/3. Extension of (10.8) to the case of dynamic fluctuations is easily obtained [10.24]. It shows that the long tail given by (10.8) is very sensitive to very low fluctuation rates. These slow rates cannot be reached by the transverse method.

Let us emphasize that the fluctuations of B_{dip} can be due to the μ^+ diffusion and/or to the dynamical behaviour of the dipoles as, for example, in magnetic metals.

10.2.2 Dipole Fluctuations and Correlation Functions

To study fluctuations, a second approach to the depolarization mechanisms is very useful. It is based on a quantum mechanical treatment of fluctuations. We write

$$P(t) = \exp[-\Lambda(t)t] \tag{10.9}$$

and assume $\Lambda(t)\tau_c < 1$.

For a *longitudinal* setup the depolarization parameter is written [10.21]

$$\Lambda_\parallel(t)t = \gamma_\mu^2 \int_0^t d\tau (t - \tau)[\phi_{xx}(\tau) + \phi_{yy}(\tau)]\cos \omega_\mu t \tag{10.10}$$

where x and y are the directions perpendicular to the μ^+ beam. $\phi_{\alpha\alpha}(\tau)$, with $\alpha = x, y$, are the symmetrized field correlation functions

302

$$\phi_{\alpha\alpha}(\tau) = \frac{1}{2}[\langle B^{\alpha}_{dip}(\tau)B^{\alpha}_{dip}(0)\rangle + \langle B^{\alpha}_{dip}(0)B^{\alpha}_{dip}(\tau)\rangle],$$

$\langle \ldots \rangle$ means thermal average. We notice that the depolarization function is exponential if the fluctuations are fast enough, i.e. $\tau_c < t \simeq \tau_\mu$. $\Lambda_{||}(t)$ depends on the applied field through ω_μ. This dependence gives information on the shape of the correlation functions, as seen in Sect. 10.4. If the fluctuations are slow compared to $1/\omega_\mu$ ($\omega_\mu\tau_c > 1$), $\Lambda_{||} = 0$, i.e., $P_{||}(t) = 1$. On the other hand, in the case of fast fluctuations ($\omega_\mu\tau_c < 1$), information can be obtained from the value of the longitudinal damping rate.

For a *transverse* setup we have [10.21]

$$\Lambda_\perp(t)t = \gamma_\mu^2 \int_0^t d\tau(t-\tau)\phi_{zz}(\tau) + \Lambda_{||}(t)t/2. \qquad (10.11)$$

This formula clearly shows that the transverse depolarization rate is due to two mechanisms. The first one, given by the term $\phi_{zz}(\tau)$, describes the depolarization due to the field distribution along the z axis which is the main field direction. The second term takes into account the possible transitions between the two μ^+ energy levels split by the Zeeman effect.

In conclusion, when the fluctuations are fast enough, the depolarization functions are exponential [10.25] but the damping rate has a meaning which depends on the experimental conditions.

10.3 Diffusion Studies by μ^+SR

The ratio of the proton mass to the μ^+ mass is about 9. It is known that depending on the metal and the temperature, hydrogen diffuses more quickly or more slowly than deuterium and tritium. Therefore there is interest in studying the μ^+ diffusion and comparing it with its heavier isotopes. Up to now the muon diffusion studies and lattice site determinations are based on the observation of the *depolarization* rate due to the interaction of the μ^+ spin with the neighboring magnetic dipoles. In diamagnetic metals, such as niobium, the depolarization is due to the nuclear magnetic moments. In magnetically ordered metals such as α-iron the depolarization comes mainly from the interaction of the localized electronic magnetic moments with the μ^+ spin. Because the electronic moments are larger than the nuclear moments by three orders of magnitude, much smaller τ_c are measured in α-iron ($5\times10^{-13}\text{s} \leq \tau_c \leq 2\times10^{-10}\text{s}$) than, for example, in niobium ($10^{-7}\text{s} \leq \tau_c \leq 10^{-5}\text{s}$). As explained in the previous section, the value of the damping rate for static muons permits the determination of the μ^+ site. In addition, from the same data, it is possible to extract the electric field gradient created by a μ^+ on its neighboring nuclei and the local lattice expansion [10.5,23].

Before discussing the experimental results, we first give some elements of the standard theory of the diffusion of a light interstitial in a metal.

10.3.1 Standard Theory of the Diffusion of a Light Interstitial in a Metal

The study of the diffusion of a particle in a metal presents two aspects: the elementary step and the space and time evolution of the diffusion process. The μ^+SR spectroscopy provides information mainly on the elementary step. Therefore we concentrate here on this aspect of the theory. A recent review can be found in [10.6,26].

It is believed that the μ^+ forms a *small polaron* in metals (at least, for those having the fcc structure): the μ^+ has the time to interact with the surrounding metal atoms and to displace them so as to create a deep potential well. This self-trapping is intrinsic and therefore should occur at random in the lattice. It has been argued that a delay to self-trapping may sometimes occur and, if so, impurities have a catalytic effect [10.27]. A priori a μ^+ can always transfer between two sites by *quantum tunneling*. But if the transparency of the barrier is too small, the only effective process is the *classical over-barrier* jump. It is described by an Arrhenius law

$$\tau_s = \tau_{s_0} \exp\left(E_a/k_B T\right) \tag{10.12}$$

where the activation energy E_a represents the barrier height and $1/\tau_{s_0}$ is the oscillation frequency of the μ^+ in its energy well. τ_{s_0} is of the order of $10^{-12} - 10^{-13}$s.

If the barrier to μ^+ diffusion has a non-negligible transparency, μ^+ can transfer between sites by coherent or incoherent quantum tunneling. In the *coherent* case, the interstitial μ^+ tunnels along with its deformation cloud which is not strongly perturbed by phonons: the cloud lifetime τ_{cl} has a meaning. There is a distribution of interstitial energies due to the strain field created by the metal impurities. The diffusion has a band character like electrons in a metal only if the energy difference between the sites is small compared to $\hbar/\tau_{cl}$. In that case, the μ^+ mobility increases when the temperature is decreased. On the other hand, when μ^+ diffuses *incoherently*, the phonon bath is strongly modified during the transfer process. In a fcc crystal at least two phonons are needed for this transfer [10.28]. Starting at about half of the Debye temperature of the metal, the diffusion is thermally activated and occurs via a multi-phonon process. But there are two differences compared to the classical diffusion mechanism: (a) The activation energy is now the energy required to bring the energy of the sites between which the transfer occurs into coincidence. (b) Unless the transfer matrix element J is large (easy transfer) [10.29], the pre-factor is very small because it is proportional to J^2. Because of the crystal structure, J is expected to be small for fcc metals and large for bcc metals. In any case τ_s now decreases when the temperature is increased.

The information on the μ^+ diffusion is mainly deduced from the effect of impurities and defects on the μ^+ damping rate, as it will be discussed now.

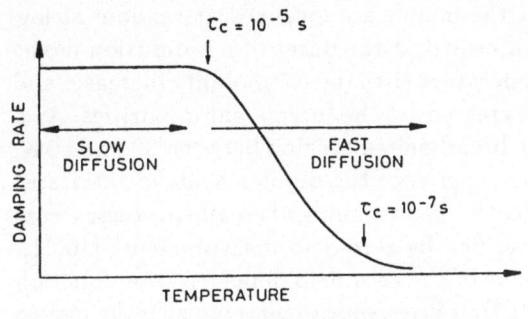

Fig. 10.4. Expected variation of the damping rate with temperature in a pure diamagnetic metal. The μ^+ diffusion rate increases with temperature

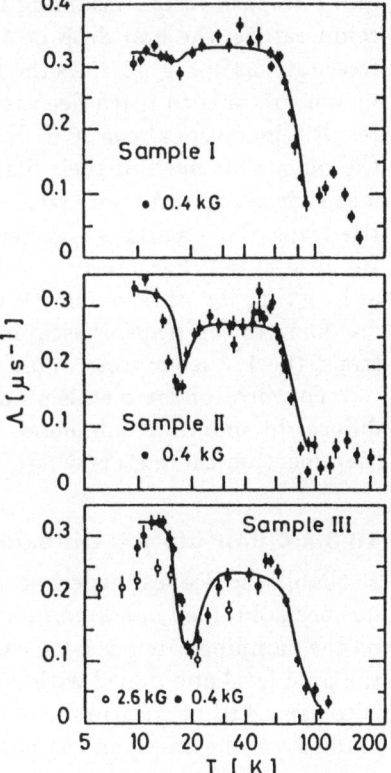

Fig. 10.5. Depolarization rate parameter Λ as a function of temperature for muons in Nb with different controlled amounts of impurities [10.30]; see the main text for details

10.3.2 Effects of Impurities and Defects on the μ^+ Damping Rate

Using the model just described for the μ^+ diffusion, we conclude that, if the elementary step of the muon diffusion in a pure metal is not a band motion, the temperature dependence of the μ^+ damping rate should behave, as shown in Fig. 10.4. At low temperature, the muons are static during their lifetime. Therefore the damping rate is both temperature independent and high because of the inhomogenous broadening. Upon increasing the temperature, the damping rate decreases because of motional narrowing.

Historically, it was soon shown experimentally that the textbook behavior of Fig. 10.4 was not really observed and that even minute amounts of impurities could drastically change the temperature dependence of the damping rate. In Fig. 10.5, we present results obtained for muons diffusing in niobium containing different controlled amounts of interstitial impurities [10.30]. Sample I contained 3700 ppm (parts per million) of N; sample II, 60 ppm of N and O; sample III, 15 ppm of N and O. In addition, all samples contained about 100 ppm of substitutional Ta impurities. While the data from the most impure sample (I) almost show the temperature behavior of Fig. 10.4, a characteristic dip appears in the damping rate $\Lambda(T)$ at about 18 K for the two other samples. The data have been analysed quantitatively with a two-state model: a *free* and a *trapped state*. Assuming that the muons are localized at random at low temperature, the first drop in Λ is ascribed to the onset of μ^+ diffusion in the free state, as in Fig. 10.4. As the temperature rises the μ^+ mobility increases and muons are able to reach deep traps created by the interstitial impurities. As a result Λ increases above 20 K. In the broad plateau region between 30 and 60 K, the muons for most of their lifetime experience the dipolar fields in the traps. The sharp drop of Λ above 60 K indicates the beginning of escape processes from the traps. This analysis has been verified by zero-field measurements [10.31]. Because of the characteristic 1/3 value of the zero-field depolarization function at long time for static muons (Fig. 10.3), it is possible to distinguish if the muons are going to the traps or escaping from them. If the muons were escaping from traps, the 1/3 asymptotic limit should not be seen.

For most of the metals studied up to now, it has been shown that the μ^+ diffuses by quantum tunneling. Therefore we first discuss this set of data and then mention the metals where the μ^+ diffuses by classical jumps.

10.3.3 Quantum μ^+ Diffusion in Metals

Probably the deepest experimental insight into μ^+ diffusion processes in metals has been obtained in *aluminium* where the influence of vacancies and impurities on the damping rate has been extensively studied. This metal can be purified to the ppm level and doped with selected impurities. In pure Al the damping rate is too small to be measured, at least down to 100 mK. The μ^+ diffusion can be studied via the effect on the damping rate of the trapping at impurities [10.6] or vacancies [10.32]. The study of μ^+ trapping above 50 K in vacancies created by electron irradiation shows that the energy transfer matrix element is very small (J $\simeq$ 2.3 meV). Via the effect of impurities, it is possible to study the low-temperature region down to the lowest accessible temperature (experimentally 25 mK).

In Fig. 10.6 we present results obtained with ultrapure 99.9999% Al and the same metal doped with various amounts of Mn. The fact that the damping rate is temperature dependent shows that the muons are diffusing. At these temperatures the only possibility is quantum diffusion. The temperature dependence of the damping rate was analysed by dividing the data into two regions, the border temperature being about 2 K. In the high-T region a bump in depolarization is observed. It is characteristic of the nature of the impurity. This bump is due to the μ^+ trapping at the impurity. A quantitative analysis shows that in the free state $\tau_s \sim T^\alpha$ with $\alpha \simeq -1$. In the low-T region the main finding is that the correlation time has a universal functional form (independent of the impurity). For the free state, the analysis gives $\tau_s \sim T^\beta c^{-\gamma}$ with $\beta \simeq 0.6$ and $\gamma \simeq 0.8$, c being the impurity concentration. In Fig. 10.7, we represent the temperature dependence of $1/\tau_s$ derived from the Al data. In addition, one may deduce from μ^+SR that the μ^+ localizes in an octahedral site at 50 mK and in a tetrahedral site at 15 K. It is worth mentioning that the same localization sites are observed for hydrogen by a channeling technique [10.33]. Depending on the impurity the self-trapping is catalyzed [10.34].

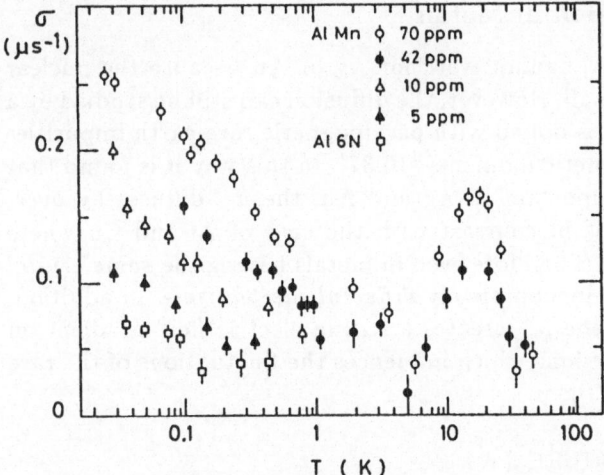

Fig. 10.6. Gaussian depolarization rate for Al and AlMn polycrystalline samples. The external field is $B_{ext} = 520\,G$ for AlMn$_{42\,ppm}$ and 150 G for the others [10.6]

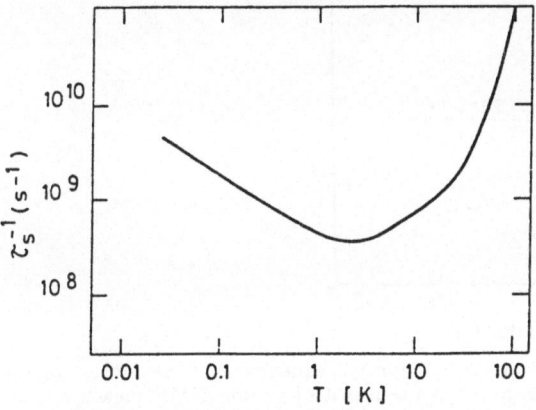

Fig. 10.7. Inverse of the μ^+ mean time of stay, $/\tau_s$, in Al derived from trapping data for various impurities and temperatures. The muons are supposed to self-trap at random; adapted from E. Karlsson in [10.18]

It would be tempting to interpret these data as follows: below $\simeq 2\,K$, the μ^+ diffusion has a band character and above $\simeq 1\,K$ the μ^+ starts diffusing by an incoherent process. Unfortunately, this explanation does not work quantitatively for known diffusion mechanisms [10.6]. Recently it has been noticed that the electrons act as phonons in the sense that they modify the transfer matrix element. In [10.35] a full theory taking into account the *phonons* as in the standard theory *and* the *electron* has been developed. It seems to be able to explain the discrepancies between the experimental results and the standard theory.

Experiments in Cu, Fe, Nb and V have shown that the μ^+ is diffusing by *quantum* mechanisms. Details can be found in a recent conference report [10.36]. In these metals the temperature dependence of τ_s seems to be the same as in Al although in those having the bcc structure (Fe, Nb and V) the formation of a small polaron is not proven definitively.

10.3.4 Classical μ^+ Diffusion in Metals

The method used to study Al cannot work for Ag or Au because the nuclear magnetic moments are too small. However, the diffusion can still be studied by a tagging technique: the metal is doped with paramagnetic rare earth impurities having strong electronic magnetic moments [10.37]. In this way it is found that quantum tunneling is not important in Ag and Au: the μ^+ diffuses by *over-barrier* jumps [10.38]. This is in contrast with the case of Al and Cu where the μ^+ tunnels between sites. Therefore even in metals having the *same crystal structure*, the μ^+ can diffuse by completely *different mechanisms*. In addition, there is some evidence that the μ^+ creates a strong electric field gradient on the neighboring paramagnetic ions which influences the fluctuations of the rare earth magnetic moment[10.39].

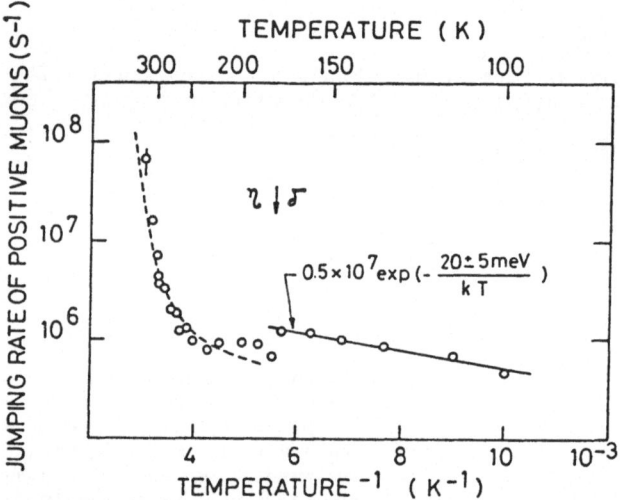

Fig. 10.8. Experimental relation between the μ^+ jump frequency and the inverse of the absolute temperature in V_2H. The broken line is the result of proton NMR measurements [10.8]

10.3.5 μ^+ Diffusion in Hydrides

Following the pioneering zero-field measurements of *Doyama* et al. [10.8] on the hydrides ZrH_x ($1.56 \leq x \leq 1.99$) and V_2H, the number of μ^+SR studies of the properties of hydrogen and deuterium in metals is increasing at a rapid rate. For example, the μ^+ jump rate plotted versus the inverse of the temperature in V_2H is shown in Fig. 10.8. In the same figure, the proton jump frequencies measured by NMR on the proton in the η phase are indicated by the broken line. The jump frequencies of the μ^+ and hydrogen are found to coincide practically above $\simeq 230\,K$. Therefore above this temperature the μ^+ diffusion is controlled by the hydrogen motion in the hydrogen sublattice. Recently, however, it was noticed that the relation between the μ^+ and hydrogen jump rates may not always be straightforward [10.40]. The concentration dependence of the μ^+ correlation time in $\beta - \mathrm{NbH_x}$ indicates a repulsive $\mu^+ - \mathrm{H}$ interaction. This interaction influences the *correlations* between the μ^+ and the *hydrogen atoms*.

10.4 Magnetic Studies by μ^+SR

As predicted by *Garwin* et al. [10.1], the μ^+ has now become a useful probe of magnetism. We shall see in this section that several features are indeed unique to the μ^+SR technique. They bring new insights into the static as well as the dynamic properties of magnetic metals.

10.4.1 Static Properties

Information on the static properties is obtained through both the μ^+ spin precession frequency, ω_μ, directly proportional to the field present at the μ^+ site, B_μ, and the depolarization functions $P_\perp(t)$ and $P_\parallel(t)$. If a quasi-static field distribution exists at the interstitial sites it will necessarily influence $P_\perp(t)$. It will also affect $P_\parallel(t)$ only in zero or very low fields when the magnetic fluctuation rate is lower than ω_μ (Sect. 10.2). The information on the field distribution is of particular interest for disordered systems as shown for instance for the random ferromagnet *Pd*Mn (2% Mn) in [10.41].

Let us now consider only the information obtained from the analysis of the precession frequency. We first emphasize that care has to be taken in the interpretation of the data. If one intends to obtain information on the bulk material, one must be sure that the μ^+ is located at a regular *interstitial* site and is not, for example, trapped by an impurity or a vacancy. Among others, the experiments by *Weidinger* et al. [10.42] show clearly how sensitive the μ^+ precession frequency can be to vacancies or impurities neighboring a μ^+. Let us now describe the different contributions to the magnetic field B_μ. As already seen in (10.3), B_μ is the sum of different terms.

The *dipolar field* B_{dip} is produced by the electronic dipoles surrounding the μ^+. It is strictly zero if the point-like μ^+ is in a site of cubic symmetry as for example in Ni. The effect of the nuclear dipoles is always negligible because they are not magnetically ordered under usual conditions. The calculation of B_{dip} is usually made by dividing the sample into two parts (Fig. 10.9). Within the Lorentz sphere a lattice sum is made over all localized dipoles, providing the term $B'_{dip}(r_\mu)$, which depends on the μ^+ localization site r_μ. The contribution from the dipoles outside the sphere is obtained by performing an integral over the remaining volume giving two terms: the *Lorentz* and the *demagnetizing* field. This last field depends on the shape of the sample. For a spherical sample the Lorentz and the demagnetizing field cancel each other.

In (10.3), the term B_{cond} comes from the *conduction electrons* close to the μ^+ which are spin polarized by the external field and the surrounding magnetic moments. This is often called the Fermi contact field.

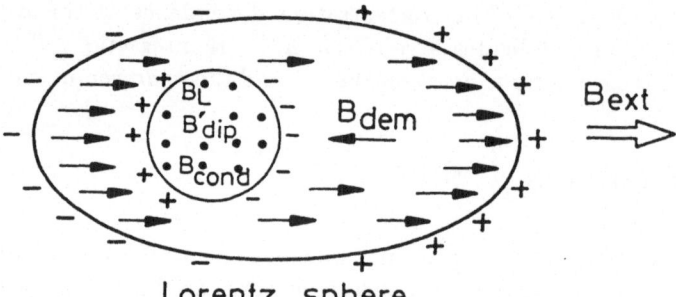

Lorentz sphere

Fig. 10.9. Schematic representation of the different contributions to the magnetic field at the μ^+ site, B_μ, in a metal. B'_{dip} is the lattice sum of the dipoles inside the Lorentz sphere and B_L is the Lorentz field due to magnetic charges at the surface of the sphere

Let us first consider the case of a *diamagnetic metal*. The polarization, proportional to B_{ext}, is weak. It results in a Knight shift due to the Fermi interaction

$$K_\mu = \frac{B_\mu}{B_{ext}} - 1 \tag{10.13}$$

well known in NMR. We suppose the sample to be spherical. K_μ can be regarded as a measure of the local static susceptibility at the μ^+ site which is perturbed by the μ^+ electric charge. The muonic Knight shifts are of the order of 100 ppm [10.43]. They vary strongly from host to host, even changing sign, which reflects the existence of several electronic contributions: direct Knight shift, core polarization, ... This is complicated by the fact that the μ^+ is surrounded by a screening electron cloud. The jellium approximation does not describe the physics because the lattice is not taken into account. Recently a spherical solid model in which the lattice is introduced by a pseudopotential has been shown to explain part of the experimental data [10.44]. Also one must mention the remarkable correlation found by *Schenck* and co-workers between the induced magnetic field in various metals and the molar specific heat which measures the density of states at the Fermi level [10.43].

Two advantages of μ^+SR spectroscopy over NMR in the measurement of Knight shifts must be mentioned. First the skin effect is absent and the μ^+SR Knight shifts are truly representative of the bulk. Second, while proton Knight shifts can only be measured in metallic systems which absorb sufficient quantities of hydrogen, in μ^+SR the muons are easily implanted in a metal and only one μ^+ is present at a time in the sample. Therefore $\mu^+ - \mu^+$ interactions are impossible.

Let us now turn to *magnetic systems*. There the polarization of the conduction electrons can be very strong due to the presence of large magnetic moments of electronic origin. B_{cond} is of the order of a few kG and is written as

$$B_{cond} = -\frac{8\pi}{3} \mu_B \eta(r_\mu)(n_0^\uparrow - n_0^\downarrow) \tag{10.14}$$

where $(n_0^\uparrow - n_0^\downarrow)$ represents the ambient spin density in the absence of the μ^+, as can be obtained from a diffuse neutron scattering experiment. $\eta(r_\mu)$ is a factor which accounts for the perturbation of this density by the μ^+ electric charge. One of the first μ^+SR experiments on a magnetic metal was performed in nickel in 1972 [10.45,46]. It measured $B_{cond} = -0.69\,kG$. Neutron data [10.47] show that the interstitial magnetization is slightly negative and corresponds to a Fermi contact field $B_{cond} = -0.66\,kG$ in excellent agreement with μ^+SR. Thus here $\eta(r_\mu) \simeq 1$. The negative sign indicates that B_{cond} is opposite to the direction of the bulk magnetization, in agreement with other local measurements (Mössbauer, PAC, ...) on diluted non-magnetic impurities in ferromagnetic metals [10.48]. Further μ^+SR experiments on other magnetic metals have, however, shown that nickel is an exception. For example, $\eta(r_\mu)$ is 3.8, 8.4 and 2.5 in Co, Fe and Gd, respectively. In all these cases, the *spin density* is thus *enhanced* by the presence of the μ^+. This clearly shows that, as it might be expected, the strong perturbation due to the positive charge implies a modification of the interstitial spin-density distribution. As for Knight shifts, calculations based on the jellium approximation are not reliable. Recently, good agreement with experiment has been obtained for nickel [10.49,50] by taking into account the hybridization of the s−electrons associated with the μ^+ and the Ni d-electrons. In fact, the physics of these results is contained in the *Daniel-Friedel* model [10.51] which points out that the majority spin states are more confined within the host atom than the minority spins states because of a deeper exchange potential. It has also been noted [10.52] that the μ^+ is a light

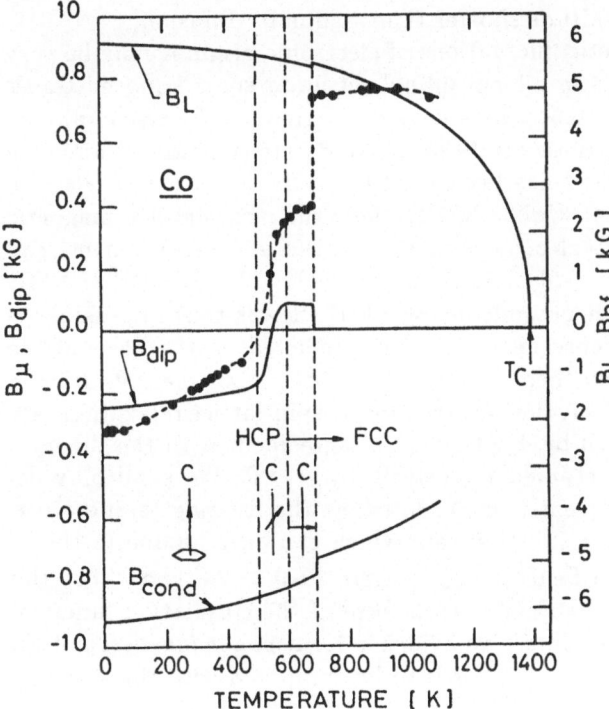

Fig. 10.10. Local field B_μ and various contributions to $B_\mu(T)$ in Co [10.53]. The solid and dash curves for B_{dip} refer to the octahedral and tetrahedral site assignments, respectively

interstitial and therefore has a substantial zero-point motion with an amplitude of about one atomic unit. Thus μ^+ sees a spatially averaged electron charge and spin distribution. The static contact field can be reduced by as much as 30%.

The *dipolar field* B'_{dip} plays an important role in the μ^+SR method essentially in two cases: its value allows the μ^+ *localization* site to be determined; in many cases its temperature dependence provides evidence for *phase transitions* such as crystallographic transformations or spin rotations. In this respect let us take as an example the case of cobalt [10.53]. As shown in Fig. 10.10, the field B_μ has a rather complex temperature dependence due, on one hand, to the crystallographic transition (hcp→fcc at 690 K) and, on the other hand, to the Co spin rotation (spin$\|c$→$\perp c$ between 500 K and 600 K). In both cases B_μ changes drastically. A detailed analysis of the data shows that the μ^+ rests in an octahedral site.

10.4.2 Dynamic Properties

As we have already seen in Sect. 10.2.2 when the fluctuations are fast enough, the depolarization functions are exponential and the damping rate λ can be written $\lambda^2 \simeq \gamma_\mu^2 \langle b^2 \rangle \tau_c$, τ_c being the fluctuation rate and $\sqrt{\langle b^2 \rangle}$ a measure of the width of the field distribution at the muons. As discussed previously, B_μ can fluctuate if the μ^+ diffuses. But in many cases this diffusion does not influence the relaxation rate because either the μ^+ is static or τ_c due to magnetic fluctuations is orders of magnitude shorter than τ_c due to diffusion.

In order to study magnetic fluctuations of electronic origin it may be very useful to make measurements in a longitudinal set up at some hundred Gauss because, as pointed out in [10.54], it is easy to distinguish between random static nuclear fields which, in these conditions, give $P_\|(t) \simeq$ constant $\simeq 1$ and true dynamical effects. This method has been used to study the *itinerant-electron weak helimagnet*, MnSi [10.54]. Unlike NMR, where the large applied magnetic field strongly affects the critical behavior near T_c, the μ^+ measurements can be carried out in a very low field or even in zero field. ^{55}Mn and ^{29}Si NMR experiments in MnSi were not possible below 200 K. It was thus impossible to search for the *critical* divergence near $T_c \simeq 29$ K predicted on the basis of the self-consistent renormalization theory of spin *fluctuations* in itinerant-electron magnets. As shown in Fig. 10.11, the observed muon spin-lattice relaxation rate is indeed represented very well by $T/(T - T_c)$ in agreement with the theory.

Spin glasses have been studied intensively by μ^+SR. We shall only describe the results obtained recently on AuFe below the freezing temperature, T_g [10.55]. The longitudinal relaxation rate which can only be due to the Fe spins varies with the applied field as $B_{ext}^{\nu-1}$ where $\nu \simeq 0.5$. We know that this relaxation rate is related to the Fourier transform of the correlation functions (Sect. 10.2.2). Therefore these data give evidence for a *power law decay* of the Fe *spin correlation*: $\langle S(t)S(0) \rangle \sim t^{-\nu}$. This is to be compared with the standard

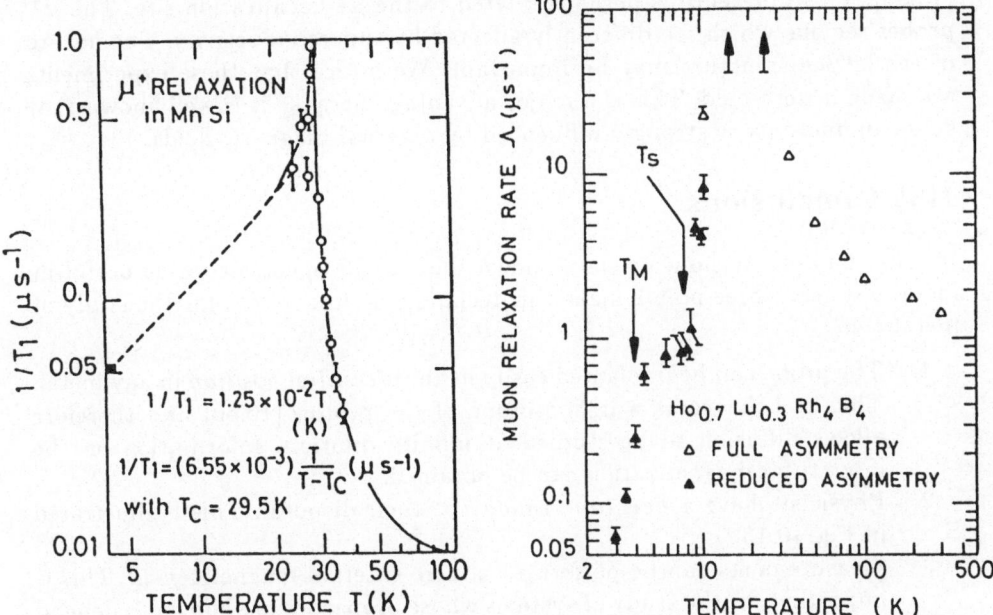

Fig. 10.11. Observed μ^+ lattice relaxation rate plotted as a function of temperature in MnSi. The solid curve for $T \geq T_c$ is a fit as explained in the main text [10.54]

Fig. 10.12. Temperature dependence of the zero-field relaxation rate in $Ho_{0.7}Lu_{0.3}Rh_4B_4$. T_s is the superconducting temperature and T_M the critical magnetic temperature [10.59]

behavior: $\langle S(t)S(0) \rangle \sim \exp(-t/\tau_c)$. In addition, it has been found that these results can explain neutron spin echo experiments [10.56].

Fluctuations in *rare earth* intermetallic *compounds* are currently being examined intensively [10.9,11,57,58]. The study of the magnetic fluctuations in the rare earth aluminium intermetallics ($REAl_2$ [10.11]) clearly shows that two basic relaxation processes of the rare earth (RE) spins are in competition: the *Korringa* scattering of the conduction electrons and the *indirect exchange* (RKKY) coupling between the RE spins. For temperatures high compared to the critical magnetic temperature T_c, the first process dominates. Otherwise, the RKKY mechanism drives the fluctuations which tend to be temperature independent at high temperatures. These measurements show that μ^+SR allows information on the *pair-correlation functions* of the RE spins at very high temperatures compared to T_c to be obtained for most lattice structures. At present these pair correlations which are one of the basic mechanisms leading to critical fluctuations can only be studied in the whole paramagnetic region by μ^+SR.

In Fig. 10.12, we present data obtained in the reentrant ternary alloy $Ho_{0.7}Lu_{0.3}Rh_4B_4$ [10.59]. Among other facts, it is remarkable that the relaxation changes abruptly at the superconducting temperature T_s. On the other hand, no structure is observed at T_s in the magnetic *superconductor* Y_9Co_7

[10.58]. This difference is perhaps related to the μ^+ localization site. The μ^+ probes regions which are differently affected by superconductivity. The nature of the lattice structure may be important. We notice that these experiments are made in zero field. This is a major advantage because it is well known that superconductivity is strongly influenced by external magnetic fields.

10.5 Conclusions

In this chapter, we hope to have shown that μ^+SR spectroscopy is useful in metal physics. Three points make this technique a unique tool for the study of metals:

1. The probe can be implanted easily in an *interstitial position* in any metal. Thus it behaves as a light isotope of the proton [10.60] and therefore allows diffusion to be studied at infinite dilution. Information on the interstitial magnetization can be obtained.
2. Physicists have a *new time window* at their disposal. This is illustrated in Fig. 10.13.
3. Measurements can be performed at *zero* external magnetic *field*. This is important for the study of systems whose properties are known to depend strongly on an external field (spin glasses, superconductors).

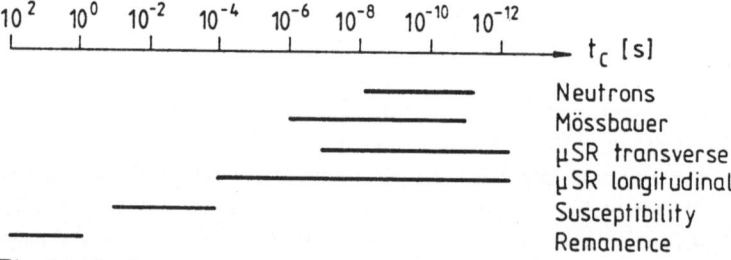

Fig. 10.13. A comparison of time scales accessible by different techniques. NMR on the proton and μ^+SR are comparable but much less data exist for proton-NMR than for μ^+SR

Two problems slow down the development of μ^+SR spectroscopy:

1. It is often necessary to know the *nature of the interstitial site* occupied by the μ^+ for a detailed analysis of the data. A comparison with hydrogen is useful. Unfortunately in some systems, μ^+ seems to be too sensitive to the *impurities* and defects of the metal. This limits the information which can be obtained on the intrinsic properties of the metal.
2. The number of institutes where the measurements can be made is restricted. Therefore the construction of new muon beams is needed. In particular, it is hoped that the very intense *pulsed* muon sources already in operation in Japan and under construction in the United Kingdom will contribute greatly to the development and to the extensive use of μ^+SR spectroscopy by physicists.

Acknowledgements. We thank the μ^+SR group at C.E.R.N. for their introduction to the μ^+SR technique and the numerous discussions along the years. One of us (A.Y.) thanks the μ^+SR group at Los Alamos for useful discussions.

References

10.1   R.L. Garwin, L.M. Lederman, M. Weinrich: Phys. Rev. **105**, 1415 (1957)
10.2   I.I. Gurevich, E.A. Meleshko, I.A. Muratova, B.A. Nikolsky, V.S. Roganov, V.I. Selivanov, B.V. Sokolov: Phys. Lett. **40A**, 143 (1972)
10.3   We take the μ^+ as an isotope of the proton in the sense that for the energy range involved in metal physics, the only difference between these two particles is their mass.
10.4   V.G. Grebinnik, I.I. Gurevich, V.A. Zhukov, A.P. Manych, P.A. Meleshkov, I.A. Muratova, B.A. Nikolsky, V.I. Selivanov, V.A. Suetin: Sov. Phys. JETP **41**, 777 (1976)
10.5   M. Camani, F.N. Gygax, W. Rüegg, A. Schenck, H. Schilling: Phys. Rev. Lett. **39**, 836 (1977)
10.6   K.W. Kehr, D. Richter, J.M. Welter, O. Hartmann, E. Karlsson, L.O. Norlin, T.O. Niinikoski, A. Yaouanc: Phys. Rev. B **26**, 567 (1982)
10.7   E. Yagi, G. Flik, K. Fürderer, N. Haas, D. Herlach, J. Major, A. Seeger, W. Jacobs, M. Krause, M. Krauth, H.S. Mundiger, H. Orth: Phys. Rev. B **30**, 441 (1984)
10.8   M. Doyama, R. Nakai, R. Yamamoto, Y.S. Uemura, T. Yamazaki: J. Less Common. Metals **88**, 405 (1982)
10.9   R.H. Heffner, D.W. Cooke, R.L. Hutson, M. Leon, M.E. Schillaci, J.L. Smith, A. Yaouanc, S.A. Dodds, L.C. Gupta, D.E. MacLaughlin, C. Boekema: J. Appl. Phys. **55**, 2007 (1984)
10.10 D.E. MacLaughlin, L.C. Gupta, D.W. Cooke, R.H. Heffner, M. Leon, M.E. Schillaci: Phys. Rev. Lett. **51**, 927 (1983)
10.11 O. Hartmann, E. Karlsson, R. Wäppling, J. Chappert, A. Yaouanc, L. Asch, G.M. Kalvius; J. Phys. F, Metal Phys., to be published
10.12 H. Wehr, K. Knorr, F.N. Gygax, A. Schenck, W. Studer: Phys. Rev. B **24**, 4041 (1981)
10.13 J.H. Brewer, K.M. Crowe, F.N. Gygax, A. Schenck: In *Muon Physics III*, ed. by V.W. Hughes and C.S. Wu (Academic, New York 1975)
10.14 J.H. Brewer, K.M. Crowe: *Annual Rev. Nuclear Particle Sci.* **28**, 239 (1978)
10.15 A. Seeger: In *Hydrogen in Metals I*: ed. by G. Alefeld, J. Völkl, Topics Appl. Phys., Vol. 28 (Springer, Berlin, Heidelberg 1978)
10.16 Yu. M. Belousov, V.N. Gorelkin, A.L. Mikaelyan, V. Yu. Miloserdin, V.P. Smilga: Sov. Phys. Usp. **22**, 679 (1979)
10.17 E. Karlsson: Phys. Rep. **82**, 272 (1982)
10.18 J. Chappert, R.I. Grynszpan (eds.): *Muons and Pions in Materials Research* (North Holland, Amsterdam 1984)
10.19 The location of these places is listed in [10.18]
10.20 A.B. Denison, H. Graf, W. Kündig, P.F. Meier: Helv. Phys. Acta **52**, 460 (1979)
10.21 C.P. Slichter: *Principles of Magnetic Resonance*, Springer Ser. Solid-State Sci., Vol. 1 (Springer, Berlin, Heidelberg 1980)
10.22 A. Abragam: *The Principles of Nuclear Magnetism* (Oxford U. Press, London 1961)
10.23 O. Hartmann: Phys. Rev. Lett. **39**, 832 (1977)
10.24 R.S. Hayano, Y.J. Uemura, J. Imazato, N. Nishida, T. Yamazaki, R. Kubo: Phys. Rev. B **20**, 850 (1979)
10.25 This is only true if all the muons see the same type of environment; see Y.J. Uemura: Solid State Commun. **36**, 369 (1980)
10.26 Yu. Kagan, L.A. Maksimov: Phys. Lett. **95A**, 242 (1983)
10.27 A.M. Browne, A.M. Stoneham: J. Phys. C**15**, 2709 (1982)
10.28 H. Teichler, A. Seeger: Phys. Lett. **82A**, 91 (1981)
10.29 D. Emin, M.I. Baskes, W.D. Wilson: Phys. Rev. Lett. **42**, 791 (1979)

10.30   M. Borghini, T.O. Niinikoski, J.C. Soulié, O. Hartmann, E. Karlsson, L.O. Norlin, K. Pernestal, K.W. Kehr, D. Richter, E. Walker: Phys. Rev. Lett. **40**, 1723 (1978)

10.31   C. Boekema, R.H. Heffner, R.L. Hutson, M. Leon, M.E. Schillaci, W.J. Kossler, M. Nieman, S.A. Dodds: Phys. Rev. B **26**, 2341 (1982)

10.32   K.P. Arnold, K.P. Döring, M. Gladisch, N. Haas, D. Herlach, W. Jacobs, M. Kruuth, S. Liebke, H. Metz, H. Orth, H.E. Schaefer, A. Seeger: Hyp. Int. **17–19**, 219 (1984)

10.33   J.P. Bugeat, E. Ligeon: Phys. Lett. **71A**, 83 (1979)

10.34   T. Hatano, Y. Suzuki, M. Doyama, Y.J. Uemura, T. Yamazaki, J.H. Brewer: Hyp. Int. **17–19**, 212 (1984)

10.35   J. Kondo: Physica **125B**, 279 (1984)

10.36   Proc. Yamada Conference VII on Muon Spin Rotation and Associated Problems, Hyp. Int. **17–19** (1984)

10.37   J.A. Brown, R.H. Heffner, R.L. Hutson, S. Kohn, M. Leon, C.E. Olsen, M.E. Schillaci, S.A. Dodds, T.L. Estle, D.A. Vanderwater, P.M. Richards, O.D. McMasters: Phys. Rev. Lett. **47**, 261 (1981)

10.38   M.E. Schillaci, C. Boekema, R.H. Heffner, R.L. Hutson, M. Leon, C.E. Olsen, S.A. Dodds, D.E. MacLaughlin, P.M. Richards: In *Electronic Structure and Properties of Hydrogen in Metals*, ed. by P. Jena and C.B. Satterthwaite (Plenum, New York 1983)

10.39   M.E. Schillaci, R.H. Heffner, R.L. Hutson, M. Leon, D.W. Cooke, A. Yaouanc, S.A. Dodds, P.M. Richards, D.W. MacLaughin, C. Boekema: Hyp. Int. **17–19**, 351 (1984)

10.40   D. Richter, R. Hempelmann, O. Hartmann, E. Karlsson, L.O. Norlin, S.F.J. Cox, R. Kutner: J. Chem. Phys. **79**, 4564 (1983)

10.41   S.A. Dodds, G.A. Gist, D.E. MacLaughlin, R.H. Heffner, M. Leon, M.E. Schillaci, G.J. Nieuwenhuys, J.A. Mydosh: Phys. Rev. B **28**, 6209 (1983)

10.42   A. Weidinger: Hyp. Int. **17–19**, 153 (1984)

10.43   F.N. Gygax, A. Hintermann, W. Rüegg, A. Schenck, W. Studer, A.S. Van der Wal: J. Less Common Met. **101**, 97 (1984)

10.44   M. Manninen: Phys. Rev. B **27**, 53 (1983)

10.45   M.L.G. Foy, N. Heiman, W.J. Kossler, C.E. Stronach: Phys. Rev. Lett. **30**, 1064 (1973)

10.46   B.D. Patterson, K.M. Crowe, F.N. Gygax, R.F. Johnson, A.M. Portis, J.H. Brewer: Phys. Lett. **46A**, 453 (1974)

10.47   H.A. Mook: Phys. Rev. **148**, 495 (1966)

10.48   I.A. Campbell: In [Ref. 10.18]

10.49   O. Jepsen, R.M. Nieminen, J. Madsen: Solid State Commun. **34**, 575 (1980)

10.50   J. Kanamori, H.K. Yoshida, K. Terakura: Hyp. Int. **8**, 573 (1981)

10.51   E. Daniel, J. Friedel: J. Phys. Chem. Solids **24**, 1661 (1963)

10.52   J. Rath, M. Manninen, P. Jena, C. Wang: Solid State Commun. **31**, 1003 (1979)

10.53   H. Graf, W. Kündig, B.D. Patterson, W. Reichart, P. Roggwiller, M. Camani, F.N. Gygax, W. Rüegg, A. Schenck, H. Schilling, P.F. Meier: Phys. Rev. Lett. **37**, 1644 (1976)

10.54   R.S. Hayano, Y.J. Uemura, J. Imazato, N. Nishida, K. Nagamine, T. Yamazaki, Y. Ishikawa, H. Yasuoka: J. Phys. Soc. Jap. **49**, 1773 (1980)

10.55   D.E. MacLaughlin, L.C. Gupta, D.W. Cooke, R.H. Heffner, M. Leon, M.E. Schillaci: Phys. Rev. Lett. **51**, 927 (1983)

10.56   R.H. Heffner, D.E. MacLaughlin: Phys. Rev. B **29**, 6048 (1984)

10.57   S.G. Barsov, A.L. Getalov, V.G. Grebinnik, V.A. Gordeev, I.I. Gurevich, V.A. Zhukov, A.I. Klimov, S.P. Kruglov, L.A. Kuz'min, A.B. Lazarev, S.M. Mikirtych'yants, N.I. Moreva, V.I. Selivanov, V.A. Suetin, S.V. Fomichev, G.V. Sheherbakov: Sov. Phys. JETP **57**, 1105 (1983)

10.58   E.J. Ansaldo, D.R. Noakes, J.H. Brewer, C.Y. Huang, R. Keitel, D.R. Harshman, M. Senba, B.V.B. Sarkssian: Solid state Commun. **55**, 193 (1985)

10.59   C. Boekema, R.H. Heffner, R.L. Hutson, M. Leon, M.E. Schillaci, J.L. Smith, S.A. Dodds, D.E. MacLaughlin: J. Appl. Phys. **53**, 2625 (1982)

10.60   D. Richter: In *Neutron Scattering and Muon Spin Rotation*, Springer Tracts Mod. Phys., Vol. 101 (Springer, Berlin, Heidelberg 1983)

11. Perturbed Angular Correlation[1]

Th. Wichert and E. Recknagel
With 31 Figures

Perturbed angular correlation has been established in the past two decades as a powerful tool for the microscopic investigation of solid state properties. It was originally developed and applied to the determination of magnetic dipole and electric quadrupole moments of excited nuclear states, a domain of nuclear physicists. The possibility to get information about internal magnetic fields or electric field gradients, which interact via a hyperfine interaction with the nuclear moments, opened up the field to solid state physics. Today, the precise determination of these electromagnetic fields in the local environment of radioactive probes constitutes the main application of this nuclear method.

11.1 Background

The basic idea of angular correlation stems from the fact that the probability of photon emission from a radioactive nucleus depends, in general, on the angle between the nuclear spin axis and the direction of emission. Normally, the radiation from a radioactive sample is isotropic, because all spins are randomly oriented in space. An anisotropic radiation pattern can only be observed from an ensemble of nuclei whose spins are not randomly oriented. Such a state can be accomplished by applying low temperatures and strong electromagnetic fields, thereby polarizing or aligning the nuclear spins, as it is the case, e.g., in nuclear magnetic resonance (NMR). In the case of the angular correlation technique, an effective spin alignment can be established by picking out only those nuclei whose spins happen to lie in a preferred direction. In a successive emission of two γ-rays the observation direction of the first one determines the preferred spin direction and thereby selects an ensemble of aligned nuclear spins, so that the correlated second γ-quantum of the cascade displays an anisotropic radiation pattern. Interactions of the aligned spins with magnetic fields or electric field gradients effect a precession of the nuclear spins with the result that the anisotropic radiation pattern starts to rotate as well.

In Fig. 11.1 the key elements of a perturbed γ-γ angular correlation (PAC) experiment are collected: An electromagnetic field inside a metal is sampled by a radioactive probe atom, since the field effects a precession of a probe atom's nuclear spin I. During its decay the probe atom emits two phontons γ_1 and γ_2, whose emission probabilities in space are determined by the instantaneous

---

[1] Dedicated to Prof. Karl-Heinz Lindenberger on his sixtieth birthday

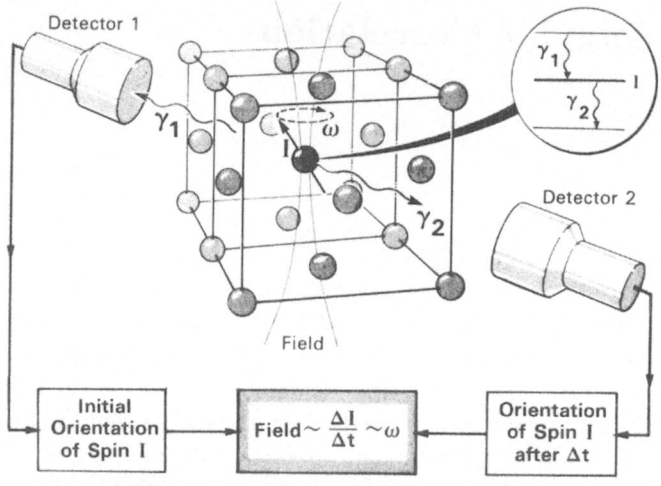

Detector 1

γ_1

ω

I

γ_2

Field

Detector 2

γ_1

γ_2

I

| Initial Orientation of Spin I | Field $\sim \dfrac{\Delta I}{\Delta t} \sim \omega$ | Orientation of Spin I after Δt |

Perturbed $\gamma\gamma$ Angular Correlation (PAC)

Fig. 11.1. PAC in a nutshell – showing a local field which induces a precession of a probe atom's spin **I**. The precession frequency ω modulates the probability to detect γ_1 and γ_2 in coincidence because the directions of the photons are coupled to the instantaneous spin orientation

orientation of I at the respective emission times t_1 and t_2. Thus the initial orientation of I and the new spin orientation after the time $\Delta t = t_2 - t_1$ are sensed by the detectors 1 and 2, respectively. The relative spin reorientation ΔI per time interval Δt reflects the spin precession frequency ω which is proportional to the strength of the electromagnetic field.

In this chapter we intend to introduce the method of perturbed angular correlation to the non-nuclear physicists in a plausible way (Sects. 11.2 and 3). Formulas will be given, which are necessary to understand the experimental results, while for a step-by-step derivation of the theory the reader is referred to the literature [11.1]. Section 11.4 deals with technical problems and the discussion of suitable radioactive probes, while the second part of this chapter is devoted to applications (Sect. 11.5). Since the observation of electric field gradients in metals has played a dominant role in recent years, most of the selected experiments will deal with that interaction.

Due to the limited space we restrict the discussion to typical examples. For further information, some detailed review articles and conference proceedings should be consulted. An extensive article including also applications to chemistry was given by *Rinneberg* [11.2], which, however, is six years old, thus omitting the rapid development during recent years. *Forker* and *Vianden* [11.3] discussed PAC and its application to materials other than metals. The special aspect of defects in metals was extensively reviewed by *Recknagel* et al. [11.4]. In the same volume *Witthuhn* and *Engel* [11.5] summarized the results on electric quadrupole interaction in non-cubic metals. Detailed information can be found in the last three topical conferences on Hyperfine Interactions: Madison (1977) [11.6]; Berlin (1980) [11.7] and Groningen (1983) [11.8].

11.2 Principles

In order to characterize a metal on a microscopic scale, the PAC technique measures its internal fields with the help of radioactive probe atoms. These local fields – magnetic fields or electric field gradients – for their part, effect a precession of the probe atoms' nuclear spins, whereby the precession frequency ω is proportional to the strength of the field at the site of the nucleus. The spin precession, due to the hyperfine interaction between the nuclear moment of the probe atom and an extranuclear field, characterizes the state of the metal. A well-known example of this interaction is the magnetic dipole interaction of the nuclear magnetic moment of a proton with the magnetic field produced by an electronic spin; this interaction results in a precession of the proton spin in the magnetic field generated by the electron. In a similar way, a nuclear electric quadrupole moment can interact with an electric field gradient. The field strength can be measured via the induced spin precession frequency ω by several methods, e.g., nuclear magnetic or quadrupole resonance, and muon-spin rotation. In our case, as outlined already in Fig. 11.1, the change of the spin orientation with time is directly observable by detecting two γ-rays in coincidence, which are emitted during the decay of the unstable (radioactive) probe atom. The conservation of angular momentum connects the orientation of the nuclear spin with the angular distribution of the emitted γ-rays, i.e. the probability to detect γ-rays along a certain space direction. Therefore, the successive detection of a 2-photon cascade allows a direct observation of the change in spin orientation during the time interval between the emission of the γ-rays. The change of the spin orientation within that time interval reflects the spin precession frequency ω. Thus, whenever a radioactive probe atom decays, information about the field strength at its lattice site is delivered to the experimentalist via this frequency ω. However, when discussing the behaviour of a particular nucleus, one should always keep in mind, that in an experiment we are dealing with an ensemble of statistically decaying nuclei, because the decay of a large number of probe atoms has to be detected in order to establish the spin precession frequency.

Using PAC, full information concerning the number of different fields and their strengths, acting on probe atoms, is transmitted by the frequency-modulated radiation field. The information obtained in that way is microscopic and local in its nature, since the field strength decreases rapidly with increasing distance from the probe atom. Therefore, mainly contributions originating from the first shell of surrounding lattice atoms are observed. This holds especially for the case of the electric field gradient.

In the present section we shall explain how the orientation of a nuclear spin, and thereafter its precession, can be inferred from the γ-radiation emitted by a nucleus. In the following section, after having discussed the hyperfine interaction arising from the presence of a magnetic field or an electric field gradient, the theoretical framework needed to describe the measured PAC spectra will be given from which the spin precession frequency is obtained. A more exhaustive coverage of the quantitative description of the PAC technique and the hyperfine interaction can be found elsewhere [11.1,9–11].

11.2.1 Spin Alignment

In order to get an understanding of how the orientation of a nuclear spin I can be monitored via the detection of the emitted γ-radiation, one should recall two phenomena: the conservation of angular momentum, and the angular distribution of electromagnetic radiation with respect to its angular momentum vector L. Considering the latter point first, it is known that electromagnetic radiation is transversely polarized or, using the particle picture, the photon as a massless particle never has its angular momentum vector pointing perpendicular to its flight direction. For a photon with angular momentum L = 1, the two angular distributions are displayed in Fig. 11.2a, showing the probability to find a photon along the direction p_γ enclosing an angle θ with the quantization axis z. Here, M = 0 and M = ± 1 refer to photons having their angular momentum vectors L perpendicular and parallel to the z axis, respectively. The distribution shows that the probability to find a photon in z direction is zero for M = 0, i.e. L is perpendicular to the flight direction. This relationship between the direction of L and the flight direction p_γ of a photon can now supply information about the orientation of the nuclear spin I, because of conservation of angular momentum. In order to illustrate this, for an isotope $_Z^A X$ (A : mass number, Z : atomic number) different states of its nucleus are sketched in Fig. 11.2b, where each state is characterized by its excitation energy E and angular momentum I (the parity will not be considered here). Let us assume that the nucleus is produced via the β-decay of a mother isotope $_{Z+1}^A X$ in its excited state E_i with $I_i = 0$. The nucleus can lose its energy and finally reach its ground state E_f by emitting a γ-γ cascade consisting of two photons with energies $E_{\gamma_1} = E_i - E$ and $E_{\gamma_2} = E - E_f$. (The different energies E_{γ_1} and E_{γ_2} will be used to distinguish between the detection of γ_1 and γ_2). Obviously, in order to conserve angular momentum, the first photon has to carry the spin $L_1 = 1$, since the intermediate state of energy E has I = 1, and for L_1 the possible values are given by $|I_i - I| \leq L_1 \leq I_i + I$. Choosing the momentum direction p_{γ_1} of the photon as quantization axis, only photon states are possible with $M_1 = \pm 1$. For the projection quantum numbers of the three spins involved (Fig. 11.2b) we have to fulfill $m + M_1 = m_i$, the nucleus in its intermediate state has m = ± 1, whereas the state m = 0 cannot be populated by this transition. That means, by detecting γ_1 we produce an *alignment* of the nuclear spins I with regard to p_{γ_1}, whereby in the present case all spins are pointing along p_{γ_1}. (Such a spin alignment exhibits some similarities with a spin polarization where one would have either m = +1 *or* m = −1).

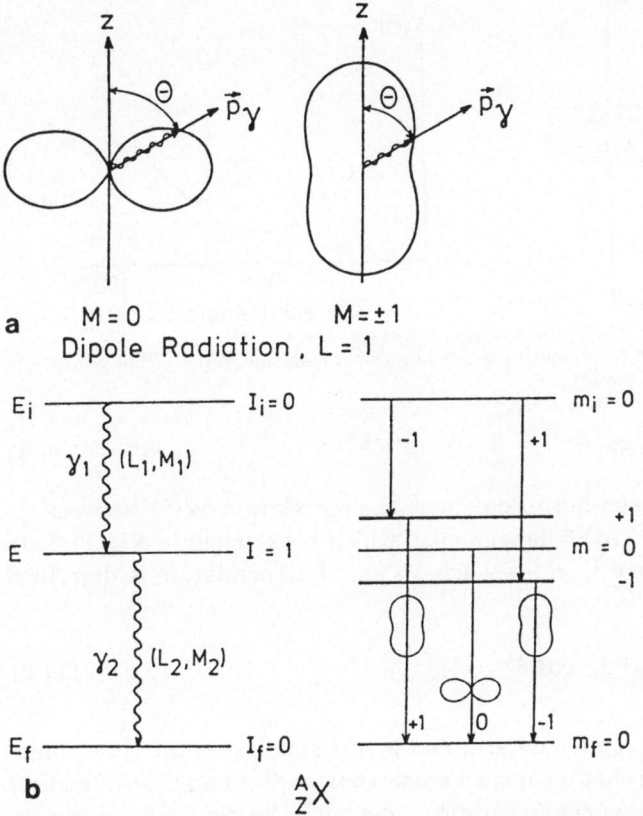

Fig. 11.2. (a) Influence of the spin projection M on the angular distribution of a photon with spin L = 1. The probability to observe the photon along $\boldsymbol{p}_\gamma$ is proportional to $\sin^2\theta$ and to $1/2(1+\cos^2\theta)$ for M = 0 and M = ± 1, respectively. (b) Energies and quantum numbers describing the level scheme of a nucleus that emits a $\gamma_1 - \gamma_2$ cascade. On the right-hand side the emission direction of γ_1 is chosen as quantization axis so that an alignment of the spin $\boldsymbol{I}$ is produced and the angular distribution of γ_2 becomes that one for $M_2 = \pm 1$

The second γ-transition is used to study the consequence of this spin alignment produced by γ_1 : The photon γ_2 connects the nuclear states with spin $I = 1$ and $I_f = 0$, so that its angular momentum has to be $L_2 = 1$. Since a second detector can be placed at any angle θ to the quantization axis, as shown in Fig. 11.3, no restriction regarding M_2 would exist, and without a coincidence with γ_1 a superposition of both radiation patterns, shown in Fig. 11.2a, should be observed which would lead to a constant or isotropic detection probability of γ_2 for all angles θ. However, because of the preceding detection of γ_1, a decay from the state with m = 0 cannot occur, and because of $m_f + M_2 = m$, the photon γ_2 is restricted to $M_2 = \pm 1$. Therefore, the probability W(θ) to detect γ_2 at an angle θ with respect to $\boldsymbol{p}_{\gamma_1}$ and in coincidence with γ_1 becomes anisotropic and depends on the angle θ. In this case, for the second photon, only states with $M_2 = \pm 1$ are allowed and the observed angular correlation W(θ) is given by the corresponding radiation pattern shown in Fig. 11.2a.

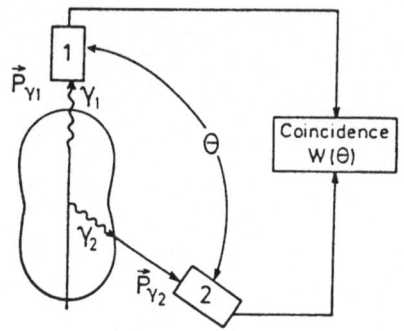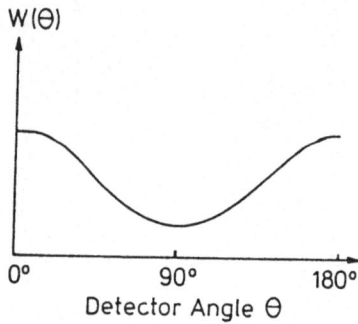

Fig. 11.3. Experimental setup for detecting a γ-γ angular correlation $W(\theta)$. (The conditions outlined in Fig. 11.2b are presumed)

$$W(\theta) \sim 1/2 \left(1 + \cos^2\theta\right). \tag{11.1}$$

Varying the angle θ between both detectors, the correlation $W(\theta)$ between γ_1 and γ_2 shows a $\cos^2\theta$ or $\cos 2\theta$ dependence, which is sketched in Fig. 11.3. In general, for a spin sequence $I_i \rightarrow I \rightarrow I_f$ such an angular correlation is described by

$$W(\theta) = 1 + \sum_{k=2}^{k_{max}} A_{kk} P_k(\cos\theta) \tag{11.2}$$

where $k = 2, 4, \ldots$ and $k_{max}/2$ is determined by the smallest of the three angular momenta I, L_1 and L_2; that means, an anisotropic radiation pattern requires $I \geq 1$. The angular correlation coefficients A_{kk} describe the deviation of the coincidence probability from the isotropic case $W(\theta) = 1$ and their values can be positive, negative or zero, and are governed by the spin sequence and the multipolarity of the γ-radiation. The Legendre polynomials $P_k(\cos\theta)$ reflect the spatial angular distribution of the involved γ-rays. Since the transition probabilities between nuclear states decrease with increasing angular momentum of the emitted ray, we shall usually meet cases with $L = 1$ or 2, so that k_{max} does not become larger than 4. In the case of the above-discussed spin sequence $0 \rightarrow 1 \rightarrow 0$ ($k_{max} = 2$), and using the identity $P_2(\cos\theta) = (3\cos^2\theta - 1)/2$, (11.1) can be rewritten as

$$W(\theta) = 1 + 0.5 \cdot P_2(\cos\theta). \tag{11.1a}$$

We obtain the angular correlation coefficient $A_{22} = 0.5$.

The process discussed above is well-known in nuclear physics and is called an *unperturbed* γ-γ angular correlation. Unperturbed means, that the population of the m states created by γ_1 remains unchanged until the emission of γ_2. An extranuclear magnetic or electric field, however, can change or perturb this population leading to a *perturbed* γ-γ angular correlation, a process labeled by the letters PAC. In this case, the presence of those fields will show up in a change of the angular correlation function $W(\theta)$.

11.2.2 Spin Precession

In classical theory, a field present at the site of the probe atom, will exert a torque on the nuclear spin, resulting in a precession of the nuclear spin about the field direction with a frequency ω, which is proportional to the strength of the field. Such an interaction is well-known for magnetic fields, and ω is the Larmor frequency ω_L. Thus, magnetic fields and also electric field gradients, which behave similarly, manifest themselves in a rotation of the aligned spins. However, for an observation of spin precession through an angle $\Delta\theta$, the emission of γ_2 has to be delayed by a time $\Delta t = \Delta\theta/\omega$ with respect to γ_1. Therefore, the intermediate nuclear state of the probe atom has to exist for some time, which is characterized by a mean lifetime τ. In the case of no precession (no field), the probability to detect γ_2 at a time t after γ_1 becomes

$$I(\theta, t) = I_0 e^{-t/\tau} \cdot W(\theta). \tag{11.3}$$

For two detectors recording γ_1 and γ_2 under a fixed angle $\theta = 180°$ this situation is sketched in the upper part of Fig. 11.4. At t = 0 the coincidence probability is given by $W(\theta)$ times a normalization constant I_0 whose value is determined by the number of radioactive decays per second, the solid angle subtended by both detectors and the detection efficiency. For times t' greater than zero the spin alignment does not change; however, because of the finite lifetime τ of the intermediate state, the coincidence probability decays exponentially with time and its logarithm plotted versus t is a straight line with slope $1/\tau$.

This situation changes, if a magnetic field is present at the nucleus and we expect that $W(\theta)$ now becomes time dependent so that (11.3) reads

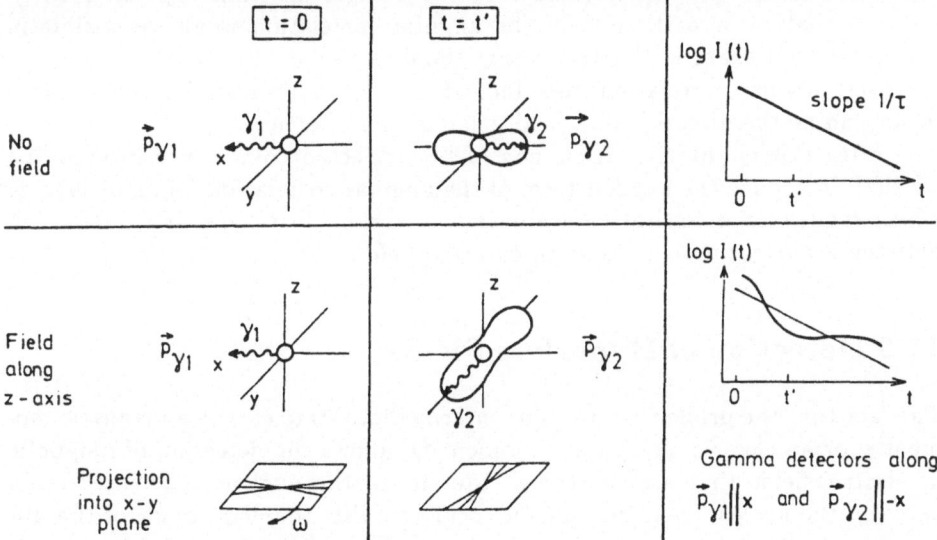

Fig. 11.4. Time dependence $I(\theta, t)$ of the coincidence probability for an unperturbed (*top*) and a perturbed (*bottom*) γ–γ angular correlation, where τ is the mean lifetime of the intermediate nuclear state

323

$$I(\theta, t) = I_0 e^{-t/\tau} W(\theta, t). \tag{11.4}$$

With help of Fig. 11.4 (lower part) we shall discuss this case in more detail. Immediately following the emission of γ_1, the aligned spins I start to precess about the field which is assumed parallel to the z-axis. A projection of the spins into the x-y plane shows the rotation of the spin alignment with angular frequency ω. After at time $t = t'$ the spins are rotated by $\theta' = \omega \cdot t'$. The spatial emission probability of γ_2 at this time can still be described by (11.2), if the x axis is rotated through the same angle θ' about the z axis. Since x along with p_{γ_2} determines the angle θ, this rotation changes θ into $\theta - \omega \cdot t'$; this transformation can be performed for every time t so that the time-dependent angular correlation function $W(\theta, t)$ can be defined by

$$W(\theta, t) = 1 + \sum_k A_{kk} P_k (\cos [\theta - \omega t]). \tag{11.5}$$

Thus, the angular correlation observed by two detectors under a fixed angle θ, but as a function of time, oscillates, as shown in Fig. 11.3, if one simply replaces θ by $\theta - \omega \cdot t'$. The logarithm of the total coincidence probability $I(\theta, t)$ plotted versus t is a straight line, on which an oscillation is superimposed (Fig. 11.4). The amplitude of the oscillation is determined by the angular correlation coefficients A_{kk} and the oscillation frequency by twice the rotation frequency, i.e. 2ω because of the 180° symmetry of the radiation pattern.

From Fig. 11.4, it can be seen that the rotation of the spin alignment is visible, if projected into the x-y plane, whereas the rotation does *not* change the projection of the spins with regard to the z-axis, i.e. the field axis. If, therefore, a field is applied parallel to the x-axis, the spin alignment rotates about this axis, i.e. within itself, and $W(\theta, t)$ does not change with time t. Consequently, no perturbation or oscillation of the angular correlation would be visible in this case. That example illustrates that the direction of the field influences the observed angular correlation, too. Indeed, this spatial sensitivity will be used to determine the direction of a field inside a metal lattice.

After this qualitative discussion of the expected coincidence probability, we shall introduce the explicit form of the angular correlation function $W(\theta, t)$ which requires a quantitative evaluation of the nuclear hyperfine interaction between a nuclear moment and an external field.

11.3 Detection of Hyperfine Fields

The fact that the probe nucleus in its intermediate state carries a magnetic moment μ or an electric quadrupole moment Q, allows the detection of magnetic or electric fields that are present at the site of the nucleus. The interaction between the nuclear moment and the local field lifts the degeneracy of the different m states belonging to the spin I. In order to describe this interaction, the direction of the field instead of p_{γ_1} will be taken as quantization axis so that the m-states are eigenstates with respect to the interaction with energies

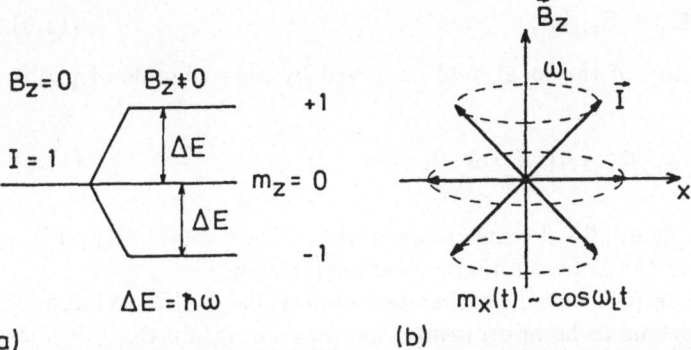

Fig. 11.5a,b. Influence of magnetic field **B** on the substates m_z of a nuclear level with spin I = 1. (a) Energy domain: The level splitting is characterized by the energy ΔE. (b) Frequency domain: The level splitting leads to a spin precession frequency $\omega_L = \Delta E/\hbar$

$$E_m = \langle Im|\mathcal{H}|Im\rangle \tag{11.6}$$

where $\mathcal{H}$ is the Hamiltonian operator describing the respective interaction. Considering again the case of I = 1, as sketched in Fig. 11.2b, and assuming the presence of a magnetic field **B**, Fig. 11.5a shows the familiar splitting into the three m = m_z substates +1,0 and –1. The size of the energy splitting ΔE directly reflects the strength of the local field. Due to the choice of the field direction as new quantization axis, all m-states will be populated equally by the detection of γ_1, as long as p_{γ_1} is not collinear with the field.

However, with regard to p_{γ_1}, we still have a spin alignment and we must discuss the relationship between the energy splitting ΔE and the precession frequency ω of the aligned spins which eventually causes the oscillation of the observable γ-γ angular correlation. Figure 11.5b shows the magnetic interaction in this spin precession picture. By observing γ_1 along the x-axis, we produce a spin alignment with regard to this axis, which rotates about B_z with frequency ω_L. Pursuing the projection m_x of the spins onto the x-axis, we see $m_x(t)$ varying proportional to $\cos \omega_L t$. From NMR experiments we are familiar with the fact that an oscillating magnetic field along the x-axis, having the time dependence $\cos \omega't$, induces a transition between two neighbouring m_z states, separated by ΔE as soon as ω' matches the spin precession frequency ω_L, i.e. $\Delta E = \hbar\omega_L$. Therefore, the relationship between the spin precession frequency ω_L of Fig. 11.5b and the energy splitting ΔE of Fig. 11.5a reads

$$\omega_L = \Delta E/\hbar. \tag{11.7}$$

This relationship is known as the Larmor theorem. The frequency ω_L enters the time-dependent correlation function $W(\theta, t)$ via the perturbation factor $G_{kk}(t)$, whose actual form depends on the type of the particular interaction – magnetic or electric – and on the particular symmetry of the local field. Because of (11.7), this factor contains differences ΔE between energies E_m and $E_{m'}$ belonging to the different substates m and m'

$$G_{kk}(t) \sim \exp\left[\frac{i}{\hbar}(E_m - E_{m'})t\right].\tag{11.8}$$

For a *random orientation* of the local field observed by an ensemble of probe atoms (11.2) becomes

$$W(\theta,t) = 1 + \sum_k A_{kk}G_{kk}(t)P_k(\cos\theta).\tag{11.9}$$

In case of a vanishing field, all substates degenerate and we obtain $G_{kk}(t) = 1$ so that (11.9) describes the known unperturbed correlation.

At this place, a comparison with other techniques like NMR, Mössbauer spectroscopy or μSR seems to be appropriate, for they all obtain the information about a local field via a measurement of the hyperfine splitting energy ΔE. NMR uses an oscillating magnetic field perpendicular to the z or field axis and varies its frequency until it matches the condition $\hbar\omega = \Delta E$ (see above). Mössbauer spectroscopy determines the change of the relative positions of two energy levels with high accuracy by detecting the emitted photon which connects both states (e.g., γ_2 in Fig. 11.2b). Since the hyperfine splitting changes the original energy E_γ by ΔE the hyperfine splitting energy itself is measured. Finally, PAC selects an ensemble of spins via the observation of γ_1 and determines its spin precession frequency ω about the z-axis via γ_2, so that according to (11.7) ΔE will be obtained. The same idea is followed by μSR, which uses a beam of spin-polarized positive muons and detects the angular distribution of positrons emitted during the muon decay.

Since the energy splitting ΔE is given by the product of the probe's momentum and the local field, the momentum has to be determined through a separate experiment, e.g., by using a known external field. In case of the magnetic interaction, this can easily be done because laboratory fields of sufficient strength (ca. 1 kG) can be provided, so that an oscillation can be easily produced, which is observable during the lifetime of the intermediate state. On the contrary, an electric field gradient of sufficient strength cannot be produced by laboratory means in order to determine the electric quadrupole moment in the same way. For most of the suitable PAC probes, however, this problem has been solved by using independent information from nuclear physics.

Whereas magnetic fields are common to us through the variety of magnetic materials, this is not the case for electric field gradients. They are described by a tensor (like the electric quadrupole moment) and occur in materials having lower than cubic symmetry. For example, a probe atom in metals with hexagonal lattice structure, like Mg, Zn, or Cd, experiences an electric field gradient. However, also in metals with cubic lattice structure, like Cu, a probe atom observes a field gradient as soon as the cubic symmetry is broken, e.g., through the presence of an impurity atom. In general, the measured magnetic or electric fields at the site of the nucleus are *effective fields* which can be modified by various sources, last but not least also by the electronic shell of the probe atom itself. All these contributions can fairly well be calculated in the case of magnetic fields, whereas for electric field gradients in metals appropriate theories are still lacking. Thus, a part of the past effort, using PAC in metals, was directed towards a better understanding of this latter problem (see also Sect. 11.5.1).

11.3.1 Magnetic Dipole Interaction

The magnetic field B is sensed by the nuclear magnetic moment $\boldsymbol{\mu} = \gamma\hbar\boldsymbol{I}$, where $\gamma = g\mu_n/\hbar$ ($\mu_N = e\hbar/2cm_p$: nuclear magneton, g : g-factor) is the gyromagnetic ratio and $\boldsymbol{I}$ the spin operator. Taking the magnetic field along the z-axis we obtain for the Hamiltonian

$$\mathcal{H} = -\boldsymbol{\mu}\boldsymbol{B} = -\gamma\hbar B_z I_z \tag{11.10}$$

and the eigenvalues E_m belonging to I_z

$$E_m = -\gamma\hbar B_z m \quad (m = -I, \ldots, +I) \tag{11.11}$$

producing the energetically equidistant Zeeman splitting. The magnetic dipole interaction only mediates transitions between neighbouring states (m = ±1), so that we get

$$\Delta E = E_{m+1} - E_m = -\gamma\hbar B_z. \tag{11.12}$$

As a consequence, only a single transition frequency occurs, the Larmor frequency

$$\omega_L = \Delta E/\hbar = -\gamma B_z. \tag{11.13}$$

For an intermediate state with spin I = 5/2, Fig. 11.6 displays the hyperfine

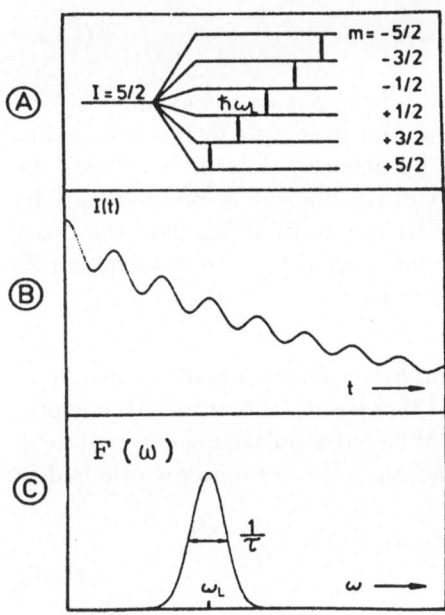

Fig. 11.6. (a) Magnetic level splitting for spin I = 5/2 and (b) the associated time dependent γ-γ coincidence probability; (c) Fourier transform of the time spectrum I(t), where τ is the lifetime of the I = 5/2 state

splitting (part A) that consists of 6 substates according to (11.11). The probability $I(t)$ to detect γ_1 and γ_2 with a time difference t is modulated by the Larmor frequency ω_L (part B). The Fourier transform of the time spectrum $I(t)$ (part C) shows a peak at $\omega = \omega_L$, whose width is mainly determined by the lifetime τ of the intermediate nuclear state. By means of two experiments, special features of the magnetic interaction should be illustrated:

Polarized Sample. In this case, we have as probe atom a ^{100}Rh nucleus inside a copper matrix that is produced through the β-decay of ^{100}Pd (Sect. 11.4). The $I = 2$ spin of its intermediate state ($\tau = 340\,$ns) is polarized by an external magnetic field of 2.22 kG that is perpendicular to the detector plane. The two γ-detectors form an angle $\theta = 225°$ [11.12]. As discussed in Sect. 11.2.2, under this condition one observes a rotation of the static angular correlation by an angle $\omega_L \cdot t$, and we obtain for the angular correlation function (11.5) with $k_{max} = 2$

$$W(\theta, t) = 1 + A_{22} P_2(\cos[\theta - \omega_L t])$$ (11.14)

or setting $\theta = 225°$ and using $P_2(\cos\alpha) = (1/2)(3\cos^2\alpha - 1)$ we get

$$W(225°, t) = 1 + (1/4)A_{22} + (3/4)A_{22}\sin 2\omega_L t.$$ (11.14a)

Figure 11.7 (top) shows log $[I(\theta, t)]$ of two spectra measured for two magnetic fields of the same strength, but applied in opposite directions. According to (11.14a) the visible oscillation reflects $2\omega_L$ and the change of the field direction changes the sign of ω_L or introduces a phase shift of 180°. Designating by $I_\downarrow(\theta, t)$ and $I_\uparrow(\theta, t)$ the two correlation spectra we can form the ratio

$$R(t) = 2\frac{I_\downarrow - I_\uparrow}{I_\downarrow + I_\uparrow} = \frac{3A_{22}}{2 + (1/2)A_{22}}\sin 2\omega_L t$$ (11.15)

which is plotted in the lower part of Fig. 11.7. In this way, the exponential decay and the normalization I_0 are eliminated, and from a fit to the data points according to (11.15) the frequency ω_L can be precisely determined. Since the external magnetic field in copper at the site of the nucleus is modified only by the known diamagnetic correction and the Knight shift, which is of the order of 1% and can be calculated, the magnetic moment of the probe atom can be obtained in this way (if the spin I is also known).

Unpolarized Sample. As a second example the internal field at the site of ^{111}Cd is shown in ferromagnetic nickel [11.13], whereby the probe atom is produced by the β-decay of ^{111}In (Sect. 11.4). Without a polarizing external field, the magnetic fields in a polycrystalline nickel sample are randomly oriented so that the perturbation factor (11.8) becomes

$$G_{kk}(t) = \frac{1}{2k + 1}\sum_{N=-k}^{+k}\cos N\omega_L t.$$ (11.16)

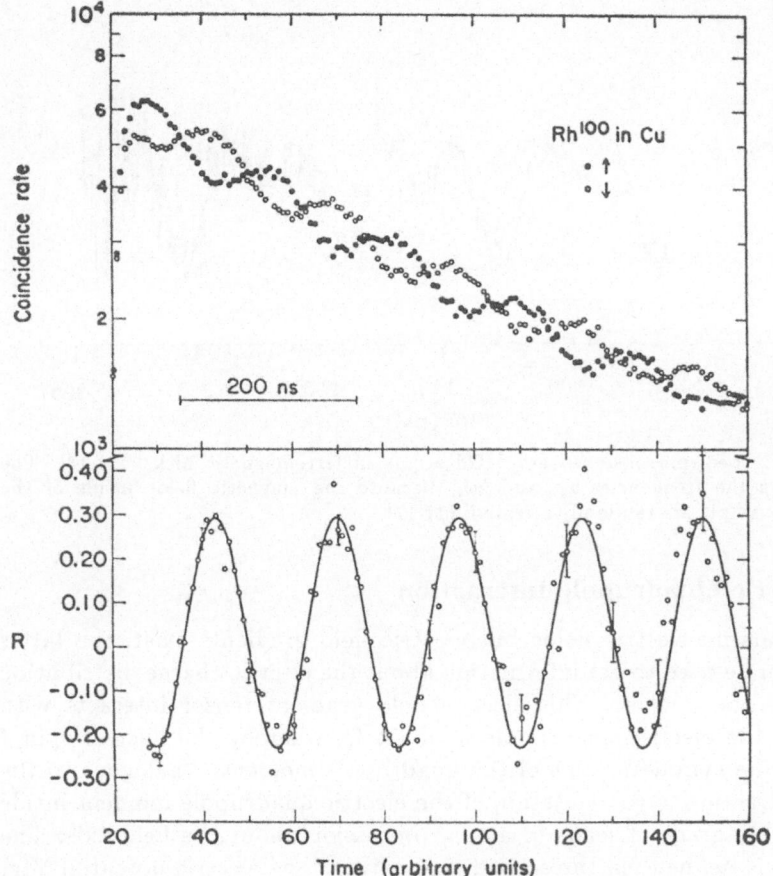

Fig. 11.7. Coincidence spectrum of a γ–γ cascade emitted by ^{100}Rh atoms in copper which are exposed to an external magnetic field applied in opposite directions [11.12] (*top*). Anisotropic part of the above shown spectra extracted according to (11.15) (*bottom*)

With $k_{max} = 2$ we get for the angular correlation (11.9) measured by two detectors under $\theta = 180°$

$$W(180°, t) = 1 + A_{22}G_{22}(t) \tag{11.17}$$

with

$$G_{22}(t) = 1/5(1 + 2 \; \cos \omega_L t + 2 \; \cos 2\omega_L t)$$

where averaging over all field directions leads to the occurrence of ω_L and $2\omega_L$. The presence of both frequencies can easily be seen in Fig. 11.8 showing $R(t) = 3/2 \, [A_{22} \cdot G_{22}(t)]$ measured at room temperature which is far below the Curie temperature T_c. The minimum in the coincidence probability at $t = 0$ is due to the negative value of A_{22} in the case of ^{111}Cd. If an external polarizing field is applied, also the sign of the internal field can be determined as will be discussed in more detail in Sect. 11.5.

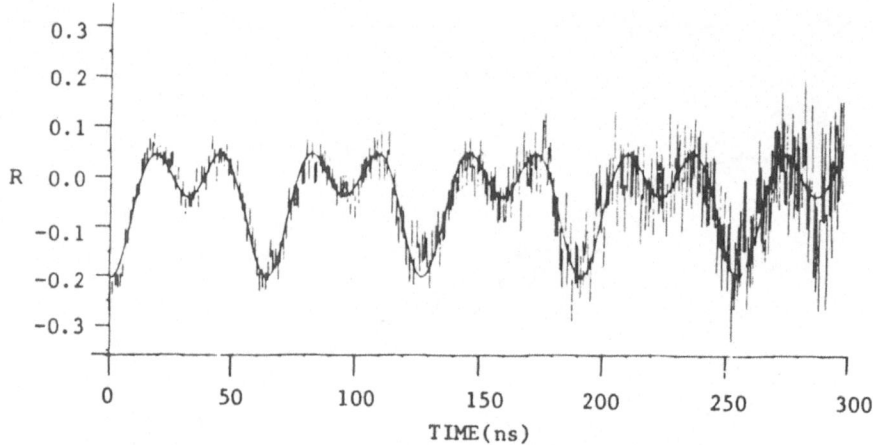

Fig. 11.8. PAC spectrum observed at ^{111}Cd atoms in ferromagnetic nickel [11.13]. The spectrum shows the frequencies ω_L and $2\omega_L$ because the magnetic fields inside of the polycrystalline sample are randomly oriented, (11.17)

11.3.2 Electric Quadrupole Interaction

Since in metals no electric fields but electric field gradients exist, the latter parameter can be used to get information about the electric charge distribution around the probe nucleus. This electric field gradient tensor interacts with the tensor of the electric quadrupole moment Q, whereby the nuclear spin *I* points along the symmetry axis of the quadrupole moment. Analogous to the magnetic interaction, a reorientation of the electric quadrupole moment inside the electric field gradient leads to a spin precession about the field axis. The field gradient is defined via the second derivative of the electric potential $V(r)$ and is described by a tensor whose nine components can be reduced to the three diagonal elements V_{xx}, V_{yy}, and V_{zz} with the convention $|V_{xx}| \leq |V_{yy}| \leq |V_{zz}|$ and $V_{xx} := \partial^2 V / \partial x \partial x$ etc.... With the Poisson equation $V_{xx} + V_{yy} + V_{zz} = 0$ the diagonalized tensor is completely described by two parameters, usually by V_{zz}, the largest of the three principal components, and by η, the asymmetry parameter, defined by

$$\eta = (V_{xx} - V_{yy})/V_{zz} \quad (0 \leq \eta \leq 1) \tag{11.18}$$

which expresses the deviation of the tensor from the axially symmetric case $V_{xx} = V_{yy}$.

The interaction is the product of two tensors and the Hamilton operator becomes

$$\mathcal{H} = \frac{eQV_{zz}}{4I(2I-1)} \left[3I_z^2 - I(I+1) + \frac{\eta}{2}(I_+^2 + I_-^2) \right] \tag{11.19}$$

where I_z, I_+ and I_- designate angular momentum operators.

The orientation of the field and its symmetry are important parameters which determine the final form of the perturbation factor $G_{kk}(t)$. Thus, for

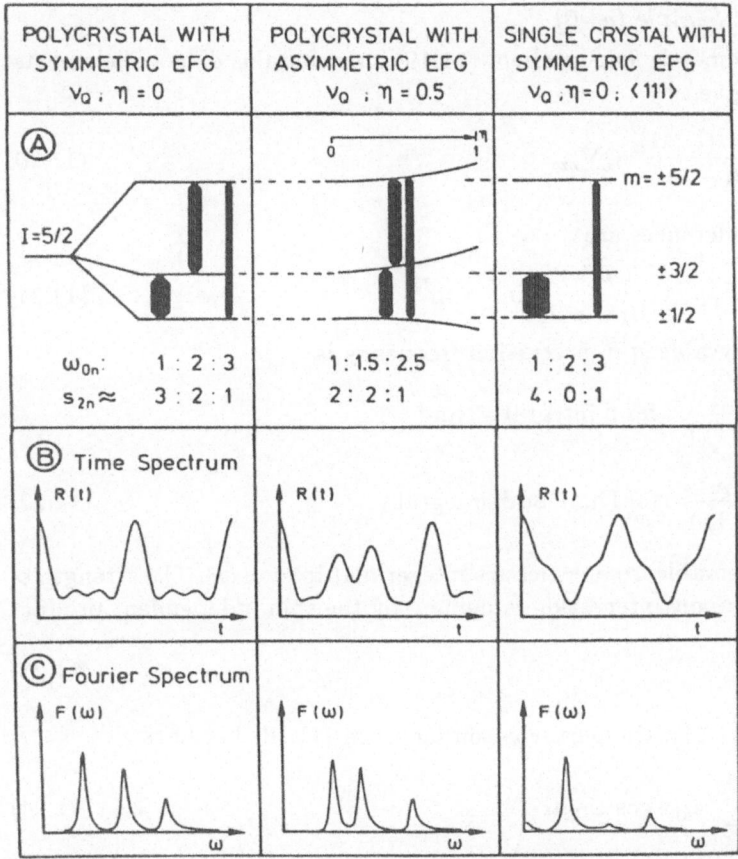

POLYCRYSTAL WITH SYMMETRIC EFG v_Q ; $\eta = 0$	POLYCRYSTAL WITH ASYMMETRIC EFG v_Q ; $\eta = 0.5$	SINGLE CRYSTAL WITH SYMMETRIC EFG v_Q , $\eta = 0$; $\langle 111 \rangle$
ω_{0n}: 1 : 2 : 3	1 : 1.5 : 2.5	1 : 2 : 3
$s_{2n} \approx$ 3 : 2 : 1	2 : 2 : 1	4 : 0 : 1

Ⓐ I=5/2 m=±5/2 ±3/2 ±1/2

Ⓑ Time Spectrum R(t) R(t) R(t) t t t

Ⓒ Fourier Spectrum F(ω) F(ω) F(ω) ω ω ω

Fig. 11.9. (a) Electric quadrupole splitting for spin $I = 5/2$ caused by an electric field gradient with ($\eta = 0$) and without ($\eta < 0$) axial symmetry in a polycrystalline sample. In case of a single crystal, an orientation of the electric field gradient and of the γ-detectors along $\langle 111 \rangle$ lattice directions is assumed. In addition, are shown: **(b)** the spectra R(t) exhibiting the corresponding perturbation factors $G_{kk}(t)$ and **(c)** their Fourier transforms

the following discussion it is useful to distinguish between polycrystalline samples (random field orientation) and single crystals (a particular field direction) because internal fields are studied whose orientations are determined by the symmetry of the metal lattices. Since (11.19) becomes rather simple for an axially symmetric field gradient ($\eta = 0$), cases with $\eta = 0$ and $\eta > 0$ will also be separately dealt with. For these three different conditions the resulting hyperfine splittings in the case of $I = 5/2$ are shown in Fig. 11.9. The m quantum numbers are again the eigenvalues of the I_z operator in (11.19). A comparison with the magnetic interaction reveals two differences: The degeneracy of the m-states is only lifted with regard to $|m|$ and the energy splitting ΔE is not equidistant. That means, the energy splitting and also the number of observable spin precession frequencies ω – being usually more than one – depend on the particular spin value I.

a) Polycrystalline Sample ($\eta=0$)

For an axially symmetric field gradient (11.19) only contains diagonal elements, the energies are given by

$$E_m = \frac{3m^2 - I(I+1)}{4I(2I-1)} eQV_{zz} \qquad (11.20)$$

and the energy differences are

$$\Delta E = E_m - E_{m'} = \frac{3eQV_{zz}}{4I(2I-1)} |m^2 - m'^2|. \qquad (11.21)$$

The smallest observable spin precession frequency is

$$\omega_0 = \frac{3eQV_{zz}}{4I(2I-1)\hbar} \quad \text{for I integral} \quad \text{and}$$

$$\omega_0 = \frac{6eQV_{zz}}{4I(2I-1)\hbar} \quad \text{for I half odd integral} \qquad (11.22)$$

and the other observable frequencies are integer multiples of ω_0. The strength of the electric quadrupole interaction is defined by the spin independent product

$$\nu_Q = \frac{eQV_{zz}}{h}. \qquad (11.23)$$

Using ΔE from (11.21), the perturbation factor in (11.18) becomes

$$G_{kk}(t) = \sum_{n=0}^{n\,max} s_{kn} \cos \omega_{0n} t. \qquad (11.24)$$

Figure 11.9 (left part) presents the hyperfine splitting, the time spectrum $R(t)$ derived from the perturbation factor and its Fourier transform for $I = 5/2$ and $k_{max} = 2$. From (11.21,22) we see that three frequencies occur ($n_{max} = 3$) $\omega_{01} = \omega_0$, $\omega_{02} = 2 \cdot \omega_0$, $\omega_{03} = 3 \cdot \omega_0$, with $\omega_{03} = \omega_{01} + \omega_{02}$. Equation (11.24) becomes explicitly

$$G_{22}(t) = \frac{1}{5} \left(1 + \frac{13}{7} \cos \omega_0 t + \frac{10}{7} \cos 2\omega_0 t + \frac{5}{7} \cos 3\omega_0 t \right). \qquad (11.24a)$$

The amplitudes s_{2n} of the transition frequencies are mainly determined by the spin I of the intermediate state and are normalized to one. Their relative magnitudes are indicated by the width of the arrows in Fig. 11.9. The Fourier transform $F(\omega)$ of the time spectrum $R(t)$ reflects both the ratio of the amplitudes and the 1:2:3 ratio of the three frequencies. The amplitude s_{20} associated with $\omega_{00} = 0$ stems from averaging over all space directions (polycrystalline sample) and takes into account the field directions pointing along either $\boldsymbol{p}_{\gamma_1}$ or $\boldsymbol{p}_{\gamma_2}$, what will be further discussed in the case of a single crystalline sample.

Figure 11.10 shows the electric quadrupole interaction for ^{111}Cd ($I = 5/2$) in polycrystalline cadmium [11.14]. The hexagonal lattice structure produces an electric field gradient with axial symmetry at each lattice site. The experi-

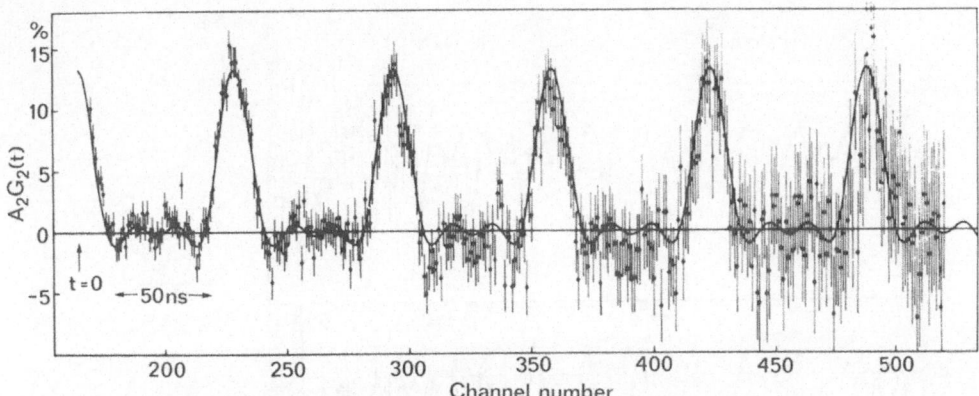

Fig. 11.10. PAC spectrum observed at ^{111}Cd atoms which are exposed to an axially symmetric electric field gradient in hexagonal cadmium [11.14]. (The amplitudes are multiplied by minus one)

mental data show the time spectrum $R(t) = A_{22}G_{22}(t)$, which clearly exhibits the expected periodic structure due to the integer ratio of the three frequencies.

b) Polycrystalline Sample ($\eta > 0$)

In case of an electric field gradient without axial symmetry ($\eta > 0$) the Hamilton operator in (11.19) has to be diagonalized for each η because the angular momentum operators $I_+ = I_x + iI_y$ and $I_- = I_x - iI_y$ mediate transitions between the different m states. As a consequence, the transition frequencies ω_{0n} and their amplitudes s_{kn} change with η as shown in Fig. 11.9 (middle part). Whereas the amplitudes show a weak dependence on η, the absolute values for each ω_{0n} and especially their ratios vary drastically. Thus, for $\eta = 1$ and $I = 5/2$ the frequency ω_{01} increases by a factor of 1.76, ω_{02} becomes equal to ω_{01} and because of $\omega_{01} + \omega_{02} = \omega_{03}$ we get $\omega_{03} = 2\omega_{01}$. With the exception of $\eta = 1$, the modulation in the time spectrum $R(t)$ is now *aperiodic*, as shown for the case of $\eta = 0.5$, because the ratios of the frequencies are no longer integer multiples, as can also be seen in the Fourier transform $F(\omega)$. Therefore, the frequency ratio $\omega_{01} : \omega_{02}$ can be used to determine η from the angular correlation spectrum. The coupling constant $\nu_Q = eQV_{zz}/h$, is deduced from $\omega_{01} = \omega_0$ according to (11.22), where ω_{01}, however, is the corresponding transition frequency for $\eta = 0$.

An electric field gradient without axial symmetry is observed for ^{111}Cd in gallium [11.15], where the orthorhombic lattice structure is the reason for the asymmetry. The experimental data in Fig. 11.11 show a pronounced aperiodic modulation. From a Fourier transform of these data, the ratio $\omega_{01} : \omega_{02} = 0.53$ was obtained, which corresponds to an asymmetry parameter of $\eta = 0.21$.

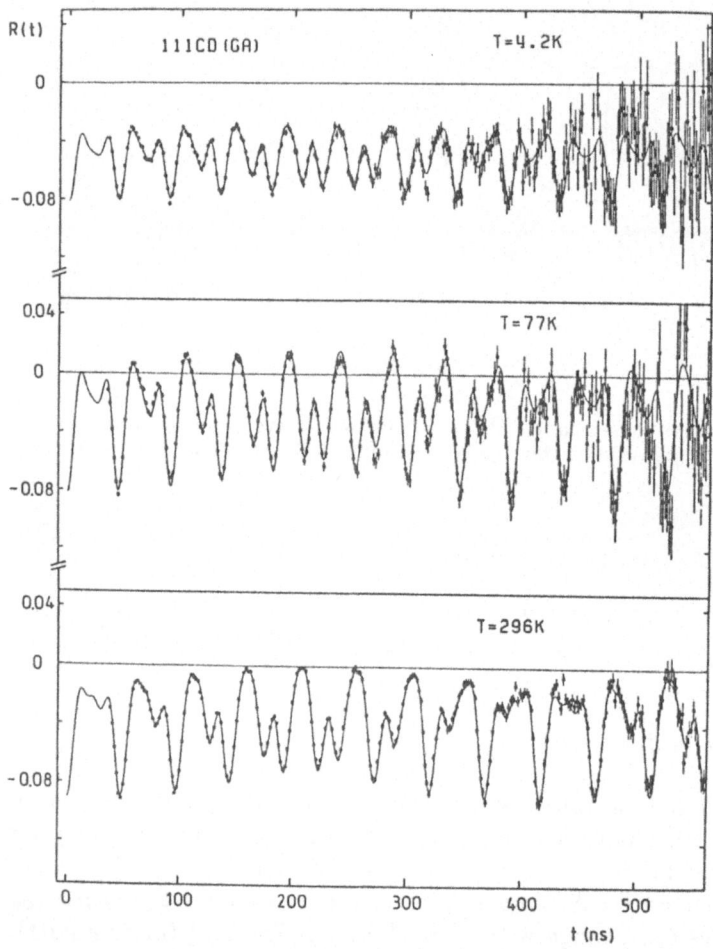

Fig. 11.11. PAC spectra observed at ^{111}Cd atoms which are exposed to an electric field gradient without axial symmetry ($\eta = 0.21$) in orthorhombic gallium[11.15]. Note, because of $\eta > 0$, the modulation is aperiodic, in contrast to the one in Fig. 11.10

c) Single Crystal Sample ($\eta = 0$)

The difference between a polycrystalline sample and a single crystal shows up in the amplitudes s_{2n} but *not* in the frequencies ω_{0n}. The right part in Fig. 11.9 illustrates the hyperfine splitting observed in a single crystal for $\eta = 0$. The amplitudes are drastically changed, so that in this particular case only ω_{01} and ω_{03} are visible in the time spectrum $R(t)$ or the Fourier transform $F(\omega)$. Equation (11.9) is now no longer applicable because the angles, which $\boldsymbol{p}_{\gamma_1}$ and $\boldsymbol{p}_{\gamma_2}$ form with the direction of the V_{zz} component, have to be taken into account, and $W(\theta, t)$ has to be calculated separately for each detector arrangement. On the other hand, using single crystals, the orientation of the field gradient tensor with regard to the crystal lattice can be determined via the measured amplitudes s_{2n}. Taking, e.g., the axially symmetric field gradient observed in cadmium (Fig. 11.10) one can use a single crystal of cadmium and detect $\boldsymbol{p}_{\gamma_1}$ along the c-axis of the hexagonal cadmium lattice in order to prove, that the V_{zz} component lies parallel to this axis. The (axial) field is now parallel to $\boldsymbol{p}_{\gamma_1}$ and therefore, analogous to the magnetic case discussed in Sect. 11.2.2, no perturbation is visible in the angular correlation, i.e., we obtain $s_{21} = s_{22} = s_{23} = 0$.

In addition, Fig. 11.12 illustrates the possibility to determine the orientation of V_{zz} with help of a single crystal [11.16]. In this example, the probe ^{111}Cd observes a field gradient in copper due to a nearest neighbour Rh atom.

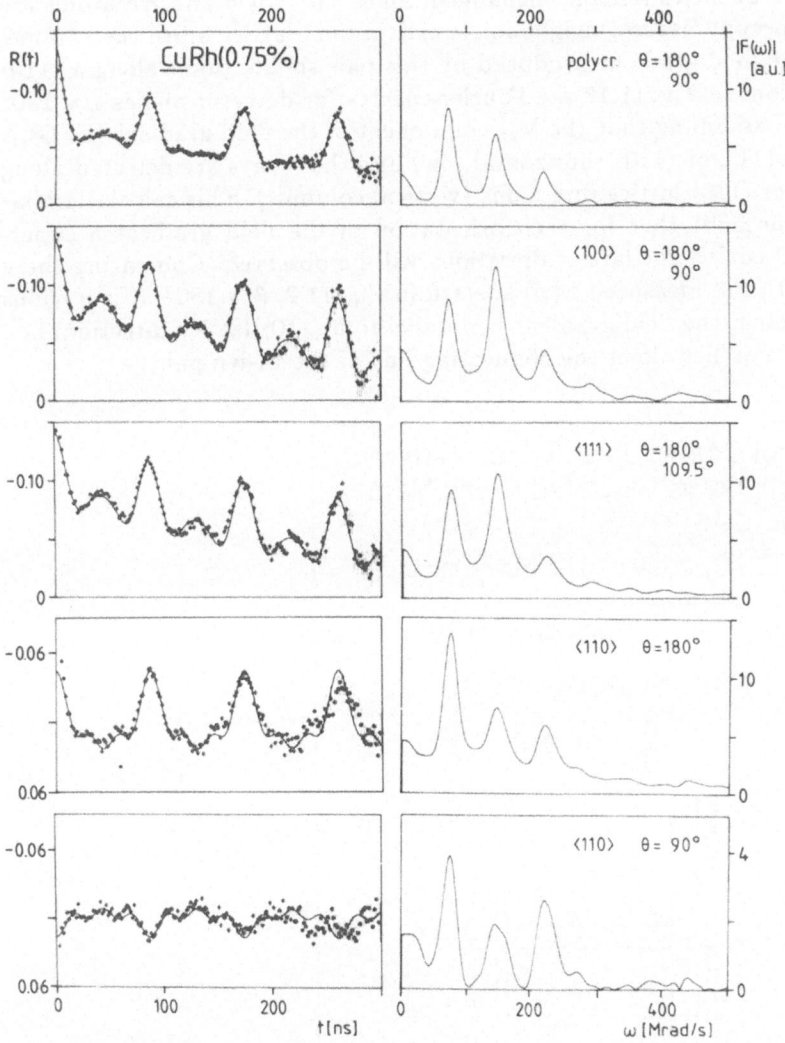

Fig. 11.12. PAC spectra and their Fourier transforms observed at ^{111}Cd in fcc copper doped with Rh atoms, where the axially symmetric electric field gradient arises from a next neighbour Rh atom. The first spectrum was measured using a polycrystalline sample and the following ones using single crystals with γ-detectors oriented along the indicated lattice directions and enclosing the angle θ [11.16]

Whereas for a substitutional Cd atom in Cu the electric field gradient is zero, because of the face-centered-cubic lattice structure, the presence of the impurity atom Rh locally breaks the cubic symmetry and creates a field gradient. The periodic modulation in the first $R(t)$ spectrum and its Fourier transform, measured for a polycrystalline sample, proves that the field gradient is axially symmetric. In the next spectra, the angular correlation was observed with the detectors along different lattice directions ($\langle 100 \rangle$, $\langle 111 \rangle$, and $\langle 110 \rangle$) of a copper single crystal using detector angles of $\theta = 180°$, $109.5°$ and $90°$. It is evident that for the different detector geometries, the amplitudes s_{2n} change, whereas the frequencies remain unchanged. Since the ^{111}Cd and Rh atoms are expected to occupy nearest neighbour substitutional lattice sites, the symmetry axis of the field gradient produced by this pair should point along a $\langle 110 \rangle$ lattice direction. In Fig. 11.13 the Fourier spectra for detector angles $\theta = 180°$ are calculated assuming that the V_{zz} component of the field gradient (EFG) is along $\langle 100 \rangle$, $\langle 111 \rangle$, or $\langle 110 \rangle$ (horizontal row) and the γ-rays are detected along $\langle 100 \rangle$, $\langle 111 \rangle$, or $\langle 110 \rangle$ lattice directions (vertical columns). This calculation has to take into account that for each orientation of the field gradient a superposition of all equivalent lattice directions will be observed. Comparing these calculated with the measured $F(\omega)$ spectra in Fig. 11.2, $\theta = 180°$, it is obvious that, as expected, the field gradient is parallel to a $\langle 110 \rangle$ lattice direction, i.e., its symmetry axis lies along the connecting line of the In-Rh pair.

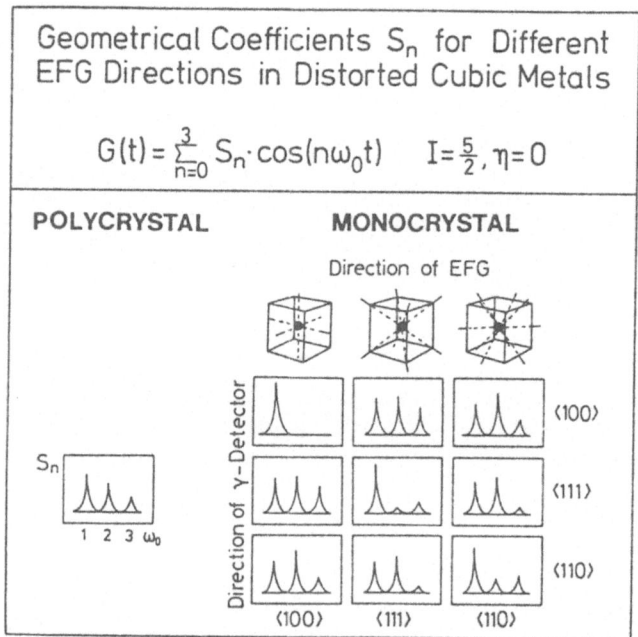

Fig. 11.13. Fourier transforms of the perturbation factor for different orientations of the axially symmetric field gradient and of the γ-detectors in case of a single crystal. For each orientation of the electric field gradient, the average of all equivalent lattice directions $\langle ijk \rangle$ is performed

11.4 Radioactive Probes, Preparation and Techniques

In this section we shall introduce the radioactive probe atoms which are most popular for PAC. Closely connected to the choice and properties of probe atoms is the problem of sample preparation, mainly the doping of metals with radioactive probe atoms, without changing the properties of the metal. Subsequently, the main components of PAC equipment will be discussed where the improvements of analog and digital electronic modules were often inspired by the demands originating from nuclear physics. Meanwhile, many of these components have become quite common in other experimental techniques, e.g., positron annihilation spectroscopy (Chap. 9). We shall close this section with a short comparison of PAC and Mössbauer spectroscopy (Chap. 13) in order to illustrate the complementary character of both techniques. In many cases, it is the more suitable radioactive isotope, which influences the choice of the technique for a given problem.

11.4.1 Probe Atoms and Sample Preparation

a) Probe Atoms

A number of requirements, which have to be met by a PAC probe atom, have already been mentioned in Sect. 11.2. The essential point of a PAC experiment, the detection of a spatially anisotropic γ-γ coincidence, requires the recording of two different photons γ_1 and γ_2, which can be identified via their different energies E_{γ_1} and E_{γ_2}, respectively. It turns out, however, that the properties of the intermediate nuclear state play the central role in answering the question whether a particular isotope is applicable for PAC, because the lifetime τ determines the time window through which the spin precession can be observed. The upper limit for a useful length of the lifetime is imposed by the condition that only one nucleus decays within the time window in order to ensure that both detected γ-rays have been emitted by the same nucleus. Usually, a sample doped with radioactive probe atoms exhibits an activity of about $10\,\mu$Curie (1 Curie stands for 3.7×10^{10} decays/s), so that PAC spectra can be recorded with sufficient accuracy within a reasonable time; in this case, about every $3\,\mu$s a decay of a probe atom occurs, i.e., τ should not exceed $1\,\mu$s. On the other hand, the limited time resolution of the experimental equipment of about 2 ns sets a lower limit for τ. Therefore, frequencies with modulation periods $2\pi/\omega$ between 5 ns and $3\,\mu$s can be detected in this time window. Obviously, a shorter lifetime of a particular probe, which would hinder the observation of low frequencies or small fields, can be compensated by a larger magnetic moment μ or electric quadrupole moment Q, because its magnitude along with the strength of the electromagnetic fields determines the actual spin precession frequency ω (compare (11.13 and 22)). Large angular correlation coefficients A_{kk} are of advantage, because they determine the maximum observable modulation amplitude, and by that, the statistical accuracy of the deduced frequencies. Finally, the initial state E_i of the γ-γ cascade must be populated by a suitable parent isotope. This initial state is usually reached via a β-decay of the parent isotope which very often is called the *probe atom* instead of the daughter atom which provides the γ-γ cascade. Important features of a suited parent isotope are its lifetime τ', which determines upon the time available to perform the PAC experiment, and its availability, i.e., the possibility to produce the respective isotope or to get it commercially.

The most commonly used PAC probes (parent/daughter) are ^{100}Pd/^{100}Rh, ^{111}In/^{111}Cd and ^{181}Hf/^{181}Ta. For these isotopes, the decay schemes and their γ-spectra are shown in Fig. 11.14; the parameters discussed above are listed

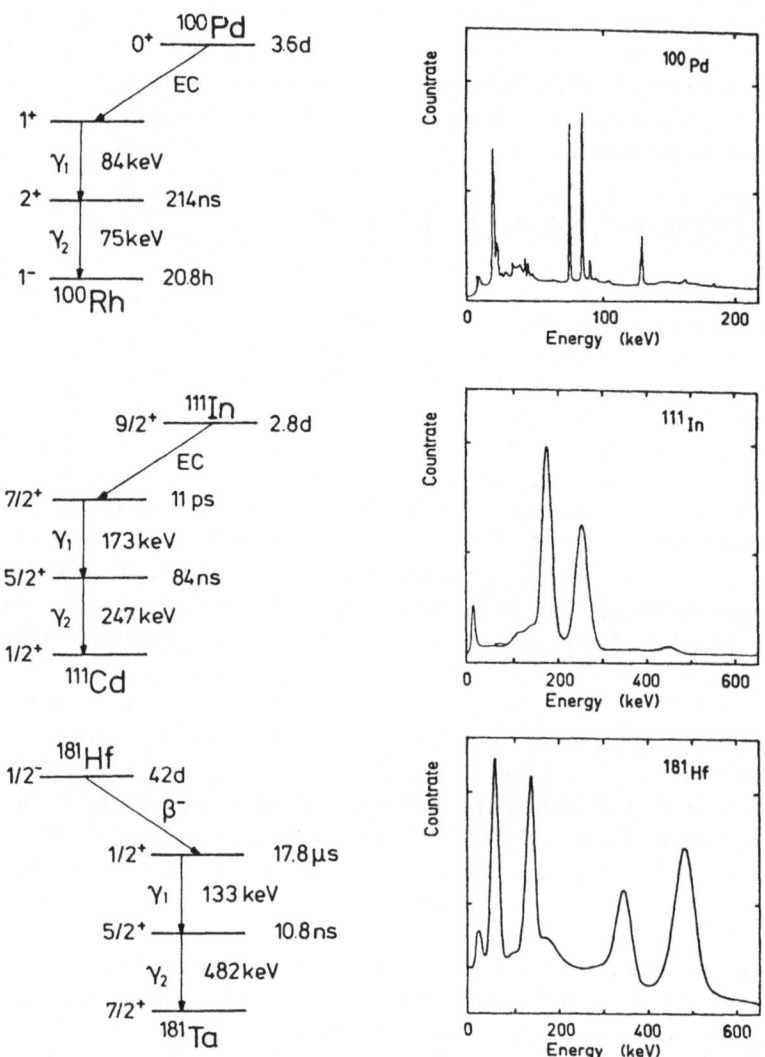

Fig. 11.14. Relevant parts of the decay schemes (*left*) and γ-spectra (*right*) of the PAC probes (parent/daughter): ^{100}Pd/^{100}Rh, ^{111}In/^{111}Cd and ^{181}Hf/^{181}Ta. The ^{100}Pd spectrum is observed with a germanium detector, the ^{111}In and ^{181}Hf spectra with a NaI detector

Table 11.1. Properties of PAC probe atoms

Parent isotope (probe)	$t'_{1/2}$	Daughter isotope	E_{γ_1} [keV]	E_{γ_2} [keV]	A_{22}	A_{44}	I^π	$t_{1/2}$ [ns]	μ [μ_N]	Q [b]	Atoms for 10 μ Curie
^{100}Pd	3.6	^{100}Rh	84	75	0.175	0	2^+	214	4.3	0.076	$1.7 \cdot 10^{11}$
^{111}In	2.81 d	^{111}Cd	173	247	-0.18	0.002	$5/2^+$	84	-0.77	0.83	$1.3 \cdot 10^{11}$
^{181}Hf	42.5 d	^{181}Ta	133	482	-0.288	-0.076	$5/2^+$	11	3.24	2.51	$2.0 \cdot 10^{12}$

$\tau = t_{1/2}/\ln 2$; I^π : spin and parity; $\mu_N :\ = e\hbar/2 cm_p$ (nuclear magneton); $b = 10^{-24} cm^2$ (barn); $10\,\mu$Curie$=3.7 \cdot 10^5$ decays/s

in Table 11.1. Without any doubt, ^{111}In/^{111}Cd is the most convenient probe among these three, comparable to ^{57}Co/^{57}Fe for Mössbauer spectroscopy, because of its favorable parameters: The parent isotope ^{111}In is commercially available in the form of carrier-free ^{111}InCl$_3$-solution and its half life of 2.81 d makes about 2 weeks available for experiments. Both photons, γ_1 and γ_2, are separately detected by NaI crystals without any problems (Fig. 11.14). The large angular correlation coefficient A_{22} along with suitable values of lifetime, magnetic and quadrupole moment of the intermediate state guarantee an accurate determination of the hyperfine interaction parameters.

Compared with ^{111}In the other two isotopes exhibit some inconveniences: radioactive ^{100}Pd is produced by a nuclear reaction in Rh metal, which makes a chemical separation difficult. The two γ-energies cannot be resolved by a NaI detector and in case of a non-axially symmetric electric field gradient ($\eta > 0$) the spin I = 2 effects the occurrence of 10 different frequencies ω_{0n}, compared with only 3 in the case of I = 5/2. ^{181}Hf can easily be produced by thermal neutron capture of ^{180}Hf, its separation, however, is only possible with the help of an isotope separator. The short lifetime of its intermediate state requires a time resolution of less than 1 ns, which lately can be reached by using BaF$_2$ detectors. For this isotope, the initial state of the γ-γ cascade plays a special role because of its long lifetime (Sect. 11.5.3).

b) Sample Preparation

As shown in the last column of Table 11.1, the absolute number of probe atoms required for a PAC experiment lies in the range 10^{11} to 10^{12} atoms, corresponding to an activity of 10 μCurie; thus, all experiments can be performed under extremely dilute conditions. This is not only true for bulk experiments, but also for surface studies, where the probe atoms still represent a small fraction of the 10^{16} surface atoms/cm^2. After an implantation of the probe atoms, where all probes reside within a layer of a few 100 Å, the local concentration is about 10 ppm. Thus, PAC experiments can be performed with concentrations of probe atoms lower than 1 ppm, increasing to any desired higher concentration by adding stable isotopes of the same element.

To dope a metal with the respective probe atom, two procedures can be applied: (a) Implantation of accelerated radioactive atoms into the sample [11.17] or (b) thermal treatment of the sample, so that the probe atoms are introduced via diffusion or melting [11.18]. The advantage of procedure (a) is its applicability to all isotopes even in case of low solubility. With the help of sufficiently high implantation energies, the surface layers of the metal (e.g., Al) can easily be overcome. On the other hand, this procedure requires an expensive facility, and the implantation efficiency lies only around 1%. A further problem represents the radiation damage created during implantation. Procedure (b) leads to a well defined metallurgical state of the impurity in the respective metal. But in the case of ^{100}Pd, a complicated procedure is necessary for extracting it from the Rh metal, and the probe ^{181}Hf can only be introduced along with the inactive component ^{180}Hf under these conditions.

11.4.2 Data Recording and Analysis

a) Data Recording

As sketched in Fig. 11.3, at least two detectors are necessary to measure the angular correlation (11.4) between γ_1 and γ_2 as a function of time elapsed between the arrival of both photons. Thus, the electronic setup has to be able to measure the energy E_γ and the arrival time t_γ for each photon. As γ-detectors usually NaI(Tl) crystals, coupled to a photomultiplier, are used which have a time resolution τ_0 of about 2 ns. They provide an energy resolution of about 10% and accept counting rates up to 10^5 Hz. Figure 11.15 shows the main elements of an angular correlation setup. Let us assume that detector 1 is tuned to record the first photon of the γ-γ cascade, and detector 2 to record the second

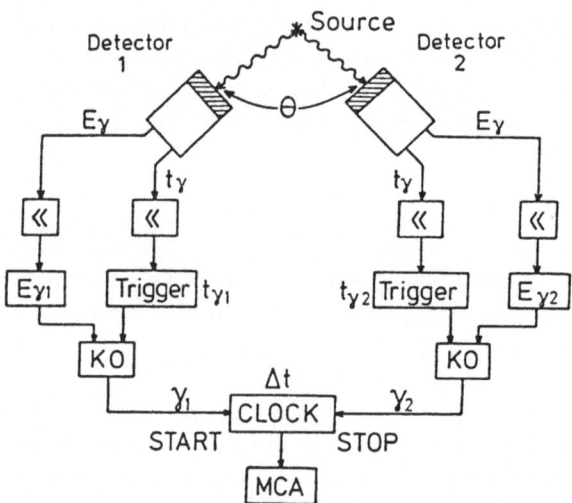

Fig. 11.15. Schematic setup of the electronics used for a PAC experiment with two detectors ($\ll$: analog amplifier, E_γ: energy analyzer, trigger: timing discriminator, KO: digital coincidence, clock: time-to-pulse-height converter, MCA: multichannel- analyzer for recording the time histogram $I(\theta, t)$)

one. After conversion of the high-energy photon into a light flash (NaI(Tl) crystal) which afterwards produces an electron avalanche (photomultiplier), the resulting electronic signal is split and amplified, so that a slow energy signal E_γ with good energy resolution, and a fast time signal t_γ with a fast, jitter-free pulse-rise time, are formed. The trigger module creates a digital pulse which defines the arrival time of the respective photon. The energy signal can only pass through the energy analyzer (single-channel analyzer) if its energy matches the preselected value of E_{γ_1} (detector 1) or E_{γ_2} (detector 2). The digital output signal of the energy analyzer opens the following coincidence unit (KO) for the accompanying time signal t_γ, thereby qualifying it as t_{γ_1} or t_{γ_2}. This procedure is often called a slow (E_γ)-fast (t_γ) coincidence. In order to measure the time difference $\Delta t = t_{\gamma_2} - t_{\gamma_1}$, one uses t_{γ_1} and t_{γ_2} to start and stop a clock (TPC: time-to-pulse-height-converter), which produces an analog pulse whose amplitude is proportional to the elapsed time Δt. Finally, these pulses are stored in the memory of a multichannel-analyzer (MCA) according to their amplitudes, and a histogram "number of events I(t) versus time t" is created, which just reflects (11.4).

It is obvious, that this scheme can easily be improved to higher detection efficiency: If the E_γ signal is analyzed for E_{γ_1} and E_{γ_2} simultaneously, each detector can serve as start and stop detector so that 2 coincidence spectra can simultaneously be recorded. Usually, two more detectors are added, thereby forming an array of 4 detectors, each of them enclosing an angle of $90°$. With each detector, sensitive to E_{γ_1} and E_{γ_2}, 12 coincidence spectra can be measured at the same time. Designating by $I_{ij}(\theta, t)$ the coincidence spectrum recorded by detectors i and j (i, j = 1...4) under an angle θ, the following ratio can be formed using 4 spectra

$$\frac{I_{13}(180, t) I_{24}(180, t)}{I_{14}(90, t) I_{23}(90, t)} = \left(\frac{W(180, t)}{W(90, t)} \right)^2. \tag{11.25}$$

This combination eliminates differences in the detection efficiency of the four detectors so that the factor I_0 in (11.4) along with the exponential decay $\exp(-t/\tau)$ cancel. A similar quotient was already formed in (11.15), where the ratio of two spectra was formed under the condition of identical I_0, because only the direction of the applied field was changed.

From the quotient in (11.25) the perturbation factor $G_{kk}(t)$ multiplied by the angular correlation coefficient A_{kk} is easily obtained: For a random orientation of the local fields, see (11.9), and with $k_{max} = 2$, one obtains

$$R(t) = \frac{2}{3} \left(\sqrt{\frac{I_{13}(180, t) I_{24}(180, t)}{I_{14}(90, t) I_{23}(90, t)}} - 1 \right) \simeq A_{22} G_{22}(t) \tag{11.26}$$

with $P_2(\cos 180°) = 1$ and $P_2(\cos 90°) = 1/2$. This ratio R(t) is usually called the PAC time spectrum which in most cases is shown throughout this chapter. Other possibilities for data reduction using different numbers of detectors have been discussed in [11.19].

b) Data Analysis

In order to extract the hyperfine interaction parameters, a theoretical expression for $A_{kk}G_{kk}(t)$ is used to fit the experimental data $R(t)$ in (11.26). Due to finite solid angles and the shape of the sample, the experimentally observable values of A_{kk} are smaller than those listed in Table 11.1. It is important to know these effective A_{kk} values in order to find out whether all probe atoms contribute to the PAC spectrum.

Probe atoms in the same sample can experience different local fields at different lattice sites. Designating by f_i the different fractions of probe atoms exposed to different electromagnetic fields, a more general ansatz for $G_{kk}(t)$ is used. In case of different internal magnetic fields in an unpolarized sample ($k_{max} = 2$ and $I = 5/2$) one gets

$$G_{22}(t) = \frac{1}{5} \sum_{N=-2}^{+2} \left[\sum_i f_i e^{-N\sigma_i t} \cos(N\omega_L^i t) \right] \tag{11.27}$$

and in the case of a polycrystalline sample and different axially symmetric field gradients $G_{22}(t)$ becomes

$$G_{22}(t) = \sum_{n=0}^{3} s_{2n} \left[\sum_i f_i e^{-n\sigma_i t} \cos(n\omega_0^i t) \right]. \tag{11.28}$$

Here, ω_L^i and ω_0^i are the spin precession frequencies, characterizing the different fractions f_i of probe atoms, and σ_i allows for a (Lorentzian) frequency distribution caused by local fields, which vary slightly from probe to probe. All these parameters can be determined by a least-squares fit of (11.27 or 28) to the experimental spectrum $R(t)$.

In case of an electric quadrupole interaction in a single crystal also the s_{2n} values become unknown parameters. Since they enter the least-squares fit in the form of the product $s_{2n} \cdot f_i$, a separate determination of both parameters – without a second experiment – is not possible.

If a larger number of frequencies are unknown, it is more favorable to start the analysis with a Fourier transform of the experimental data, which shows directly how many frequencies exist and via the frequency ratios the respective η values can be obtained. With the help of this information, a final analysis according to (11.27 or 28) can be performed.

11.4.3 PAC and Mössbauer Spectroscopy

Before closing the discussion of hyperfine parameters and their observation by PAC, a comparison with Mössbauer spectroscopy (MS) may be of interest, since this technique also measures a hyperfine interaction and uses radioactive probe atoms.

In PAC, the time window determines the largest and smallest observable frequency $\omega_0 = 2\pi/T_0$ with

$$2\tau_0 < T_0 \lesssim 3\tau \tag{11.29}$$

where τ_0 is the time resolution of the detection system and τ is the lifetime of the nuclear state. In MS, we have only the condition

$$\hbar\omega_0 > \Gamma \quad \text{or} \quad T_0 < h/\Gamma \approx \tau \tag{11.30}$$

where Γ is the natural line width of the Mössbauer resonance line. Equation (11.30) shows that for MS a limit only for low frequencies exists, which is determined by the width of the experimentally observed resonance line. If the line width increases because of absorber effects or unresolved lines, the low frequency limit will increase. Thus, it can be said that PAC is more sensitive to low ω_0, whereas MS is more suited to detect high ω_0.

The parameter determining the magnitude of the observable effects is – for PAC – the temperature independent angular correlation coefficient A_{kk} and – for MS – the temperature dependent Debye-Waller factor. Because of the temperature dependence of the latter parameter, it becomes more difficult to perform MS at higher temperatures. On the other hand, this temperature dependence can be used to study dynamic effects which is not possible for PAC.

In contrast to PAC, MS is able to detect a third type of hyperfine interaction, the isomer shift, which is caused by the scalar part of the Coulomb interaction; however, because of the variety of possible interactions, Mössbauer spectra can become more difficult to be interpreted in a unique way. The existence of non-resonant probe atoms is difficult to detect in MS, while probes which do not contribute to the PAC spectrum produce a decrease of the observed correlation coefficient A_{kk} and therefore can easily be identified. Thus, in many respects, both techniques can be regarded as complementary hyperfine techniques.

11.5 Applications

In all applications of the PAC technique conclusions with regard to the microscopic state of the metal are drawn from the local magnetic or electric field observed by the probe atom; therefore, we shall first discuss in how far these fields are understood by theory. Theories capable to reproduce the observed hyperfine fields must in most cases deal with the electronic structure of impurities in metals. By combining the efforts of all hyperfine techniques, such as PAC (or PAD), Mössbauer spectroscopy and NMR, experimental data about these fields became available for a large number of impurities in various metals. Thus, in the calculation of magnetic fields considerable progress has been made, whereas an appropriate description of electric field gradients is still lacking.

In the first part of this section, typical examples for the determination of the hyperfine fields are given, while in the following applications these fields are usually taken as tools for research in metals, like the investigation of surfaces, the study of diffusion processes, and the observation of defects and impurities. In most cases, the information will be extracted from the measured electric field gradients, as a consequence of a number of specific properties of this quantity: It is an internal field, which characterizes a metal and it exists in non-cubic lattices, in cubic lattices with local imperfections, and at boundaries; furthermore, it is not restricted to magnetic materials as is the case for the magnetic interaction. Since the electric field gradient is restricted to a very short range, its information is local, which is of specific advantage in many studies. As a tensor it can provide more structural information than the vectorial magnetic field. The examples presented are chosen to illustrate the typical features of the PAC method.

11.5.1 Hyperfine Fields at Impurities

a) Hyperfine Fields in Ferromagnets

For the understanding of ferromagnetism, measurements of hyperfine fields at the site of matrix or impurity atoms represent a sensitive test for theoretical models. With the help of the PAC- and also PAD-technique the magnitude of the magnetic hyperfine field B_{hf}, as well as its sign, and the temperature dependence can be measured. In Sect. 11.3.1 we briefly touched on the magnetic hyperfine interaction, as observed by a [111]Cd atom in ferromagnetic nickel at room temperature. Figure 11.16 shows additional PAC spectra from the same work by *Lindgren* and co-workers [11.13], measured for a series of different externally applied magnetic fields. The sample consists of a polycrystalline nickel sphere, and the external field B_a is applied perpendicular to the detector plane.

The first spectrum (a), measured with a zero external field B_a, corresponds to the case of an unpolarized sample as described by (11.16,17). Proceeding to the next spectra (b–d), the external field increases, but the Larmor frequency ω_L obviously stays constant. However, the amplitude of the modulation belonging to ω_L and $2\omega_L$ changes until – in the third spectrum (c) – ω_L has disappeared and only $2\omega_L$ is visible, which is typical for a polarized sample. By increasing the external field, the magnetization directions of the different domains have been gradually aligned until the whole sample has been completely polarized or magnetically saturated. Beyond that point an increase of B_a also changes the local field B_{loc} at the site of the Cd atoms, which in this experiment is given by $B_{loc} = B_a + B_{hf}$, because the Lorentz field and the demagnetizing field cancel. The spectrum (d) reveals that a further increase in B_a has effected a decrease of $\omega_L \sim B_{loc}$, which means that B_a and B_{hf} are of opposite sign. From an extrapolation of the external field to $B_a = 0$ one obtains, for the hyperfine field at room temperature, $B_{hf} = -68.8\,kG$.

The hyperfine field at the site of the Cd atom is governed by the local spin density that is coupled with the magnetic moments of the surrounding Ni atoms. To study this coupling, the temperature dependence of B_{hf} can be compared with the temperature dependence of the nickel magnetization, which is described by a Brillouin function. *Shirley* and co-workers [11.20] measured B_{hf} between 4.2 K and the Curie point at $T_c = 627.2\,K$; the data points in

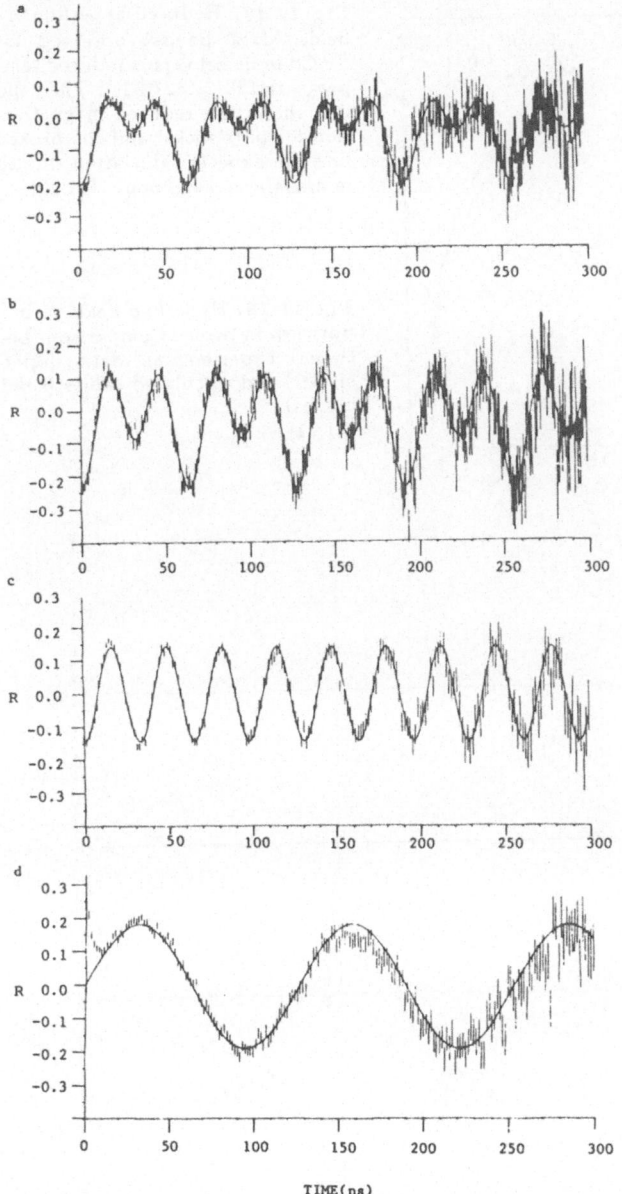

Fig. 11.16a–d. PAC spectra of ^{111}Cd in ferromagnetic nickel for different external fields $\boldsymbol{B}_a$: (a)=0 kG, (b)=1.41 kG, (c)=2.98 kG, (d)=51.7 kG [11.13]

Fig. 11.17 demonstrate that in this case the hyperfine field follows very closely the bulk magnetization.

Using the different available hyperfine techniques (NMR, Mössbauer spectroscopy, PAC and PAD) the magnitude and sign of hyperfine fields at almost all elements of the periodic system can be determined, and a theoretical model,

345

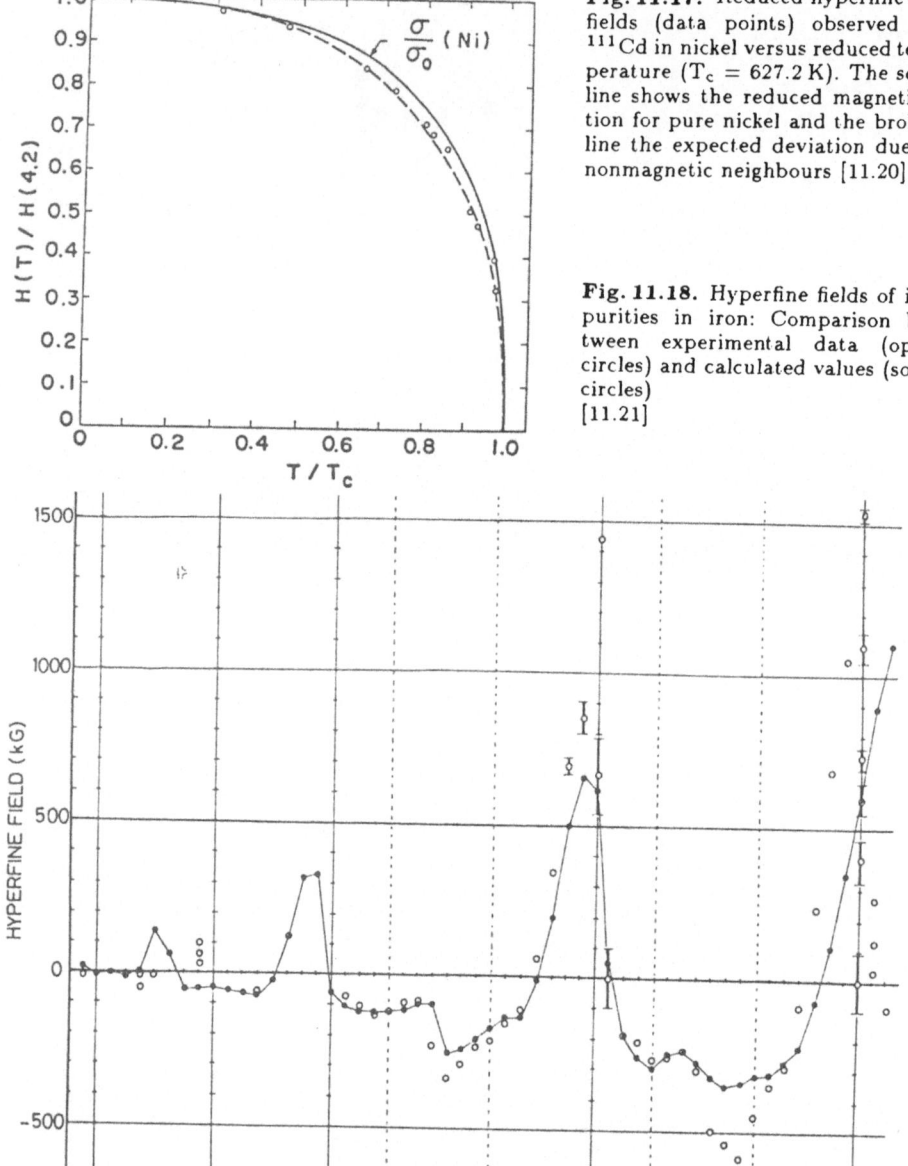

Fig. 11.17. Reduced hyperfine fields (data points) observed for ^{111}Cd in nickel versus reduced temperature ($T_c = 627.2$ K). The solid line shows the reduced magnetization for pure nickel and the broken line the expected deviation due to nonmagnetic neighbours [11.20]

Fig. 11.18. Hyperfine fields of impurities in iron: Comparison between experimental data (open circles) and calculated values (solid circles) [11.21]

capable of describing ferromagnetism in an appropriate way, should reproduce these fields. For ferromagnetic iron Fig. 11.18 shows the hyperfine fields at impurities with atomic numbers Z = 1 to 56, where the open circles represent the experimental data and the solid circles are from recent self-consistent calculations by *Kanamori* and co-workers [11.21]. This comparison covers nonmagnetic as well as magnetic impurities and describes the experimental data in an impressive way, although the influence of a possible relaxation of the lattice surrounding the impurity or a displacement of the impurity has not been taken into account. Similar studies exist for the other ferromagnetic metals cobalt, nickel and gadolinium [11.22]. Since in all cases the lattice site of the respective impurity is decisive for the theoretical description of the hyperfine field, these experiments should be supplemented by techniques that are sensitive to the lattice site of the impurity, like channeling experiments [11.23].

b) Electric Field Gradients in Non-Cubic Metals

Because of their non-spherical charge distribution, the non-cubic metals provide the possibility to investigate electric field gradients [11.24]. From the PAC work by *Christiansen* and co-workers [11.25] Fig. 11.19 shows the effect of the electric quadrupole interaction at the site of ^{111}Cd atoms in cadmium, tin and indium. For all polycrystalline samples the periodic modulation points to an axially symmetric field gradient ($\eta = 0$) at the Cd site. In the R(t) spectra, which are described by (11.24a), the visible modulation is described by $\omega_0 = 3\pi/10 \cdot \nu_Q$, see (11.22). Although the respective electric field gradients V_{zz} (in the order

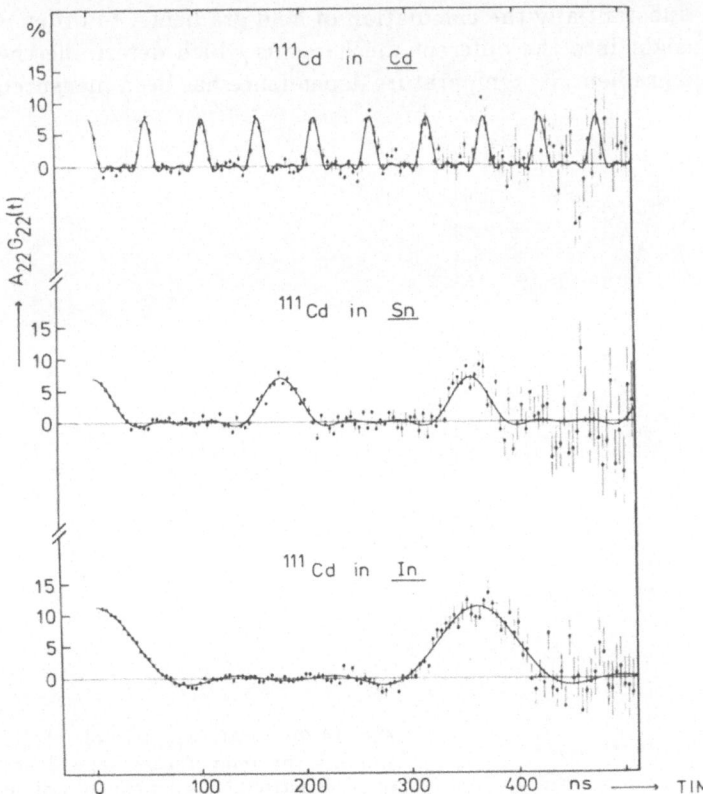

Fig. 11.19. PAC spectra of ^{111}Cd showing the different strengths of the electric field gradient in several non-cubic metals [11.25]. (The amplitudes are multiplied by minus one)

of 1 to $10 \times 10^{17} \mathrm{V/cm^2}$) can be deduced from the coupling constant $\nu_Q = eQV_{zz}/h$ fairly accurately, by using the known quadrupole moment $Q = 0.83\,\mathrm{b}$ of the probe atom, a satisfying theoretical understanding of V_{zz} is presently not available, in contrast to the above-discussed magnetic fields.

There are several sources for the electric field gradient, such as the metal ions (producing V_{zz}^{latt}), the localized electrons at the site of the probe atom, and the conduction electrons (V_{zz}^{el}). Conventionally, the experimentally observed electric field gradient is parametrized in the following way [11.24]

$$V_{zz} = V_{zz}^{latt}(1 - \gamma_\infty) + V_{zz}^{el} \qquad (11.31)$$

where the ionic contribution is amplified by the Sternheimer anti-shielding factor γ_∞ ($=-29.3$ for Cd). The latter represents the effect of the deformed shells of the localized electrons. However, the theoretical problem lies with the calculation of V_{zz}^{el}. It requires a detailed knowledge of the population of the conduction bands and touches also on the calculation of γ_∞. For the sake of simplicity, one usually takes γ_∞ calculated for a free ion rather than calculated for an ion in a metal. The empirical finding by *Raghavan* et al. [11.26] that for many impurity-solvent systems V_{zz}^{el} is linearly related to $V_{zz}^{latt}(1 - \gamma_\infty)$, did not help to improve substantially the calculation of field gradients. In order to contribute further insight into the different mechanisms which determine the magnitude of the field gradient, its temperature dependence has been measured

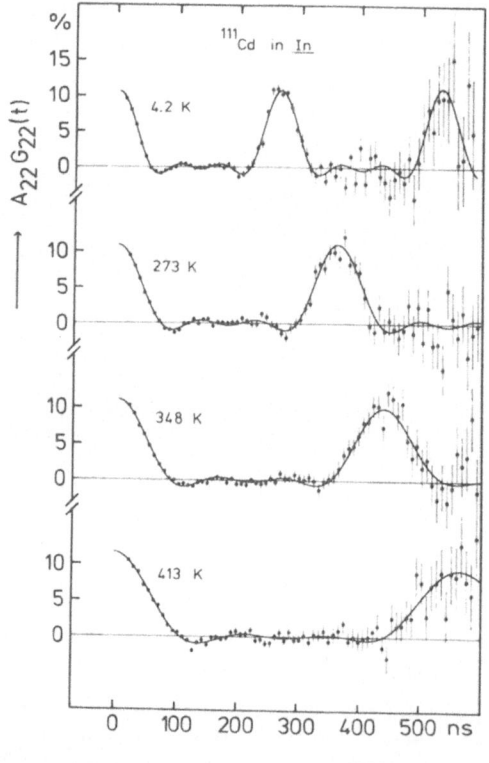

Fig. 11.20. PAC spectra of ^{111}Cd showing the temperature dependence of the electric field gradient in indium [11.25]. (The amplitudes are multiplied by minus one)

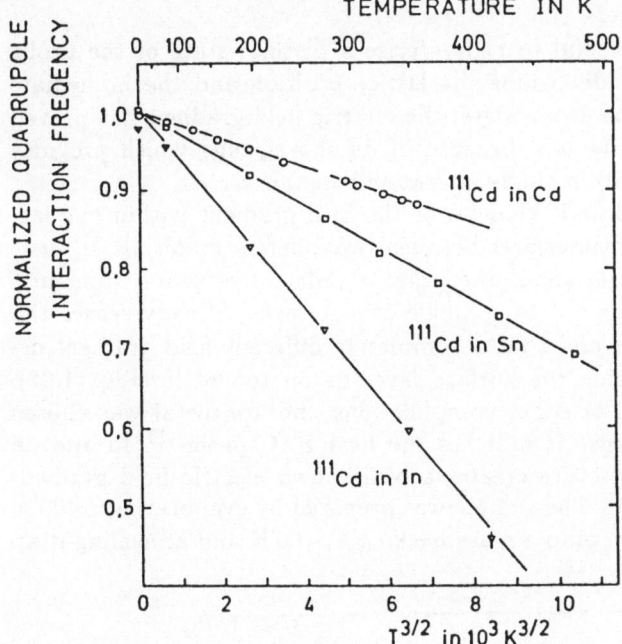

Fig. 11.21. Temperature dependence of normalized electric field gradients plotted against a $T^{3/2}$ temperature scale [11.25]

for several metals. For ^{111}Cd in indium some results are shown in Fig. 11.20. Taking *Raghavan*'s universal correlation [11.26]

$$V_{zz} = V_{zz}^{latt}(1 - \gamma_\infty) \cdot (1 - K) \quad \text{with} \quad K = 3 \tag{11.32}$$

the observed strong temperature dependence should be attributed to V_{zz}^{latt}; the strength of the measured temperature dependence, however, cannot be explained by a variation of the lattice parameters, which produce only a small effect.

Christiansen et al. [11.25] found that the temperature dependence of V_{zz} follows a $T^{3/2}$-law, as shown for ^{111}Cd in cadmium, tin and indium in Fig. 11.21. A model based on the temperature dependent Debye-Waller factor and the influence of lattice expansion is able to reproduce the observed T-dependence, although the underlying mechanism leading to a $T^{3/2}$ power law has not become obvious [11.27]. A different model [11.25] which introduces quadrons to describe the interaction between neighbouring atomic quadrupoles, leads to a Bloch $T^{3/2}$-law at low temperatures in analogy to magnons in ferromagnets. The assumption of quadrupolar elementary excitations, however, must be considered very speculative as long as there is no additional experimental evidence.

Thus, in spite of the universal correlation (11.32) and the intriguing $T^{3/2}$-dependence found for electric field gradients in metals, a successful general description of electric field gradients is still an unsettled problem.

349

11.5.2 Surface Studies

Hyperfine fields are very useful to characterize a distinct state of the probe atom in the metal, i.e. to determine the lattice position and the immediate neighbourhood of the probe atom. Here, the electric field gradient has proved to be an extremely sensitive tool because of its short range which provides microscopic information within single interatomic distances.

Typical examples of drastic changes of the field gradient within one lattice constant are surfaces or interfaces between two different materials. In their experiment *Körner* et al. [11.28] showed that a radioactive probe atom just underneath the surface, covered by a single atomic layer, already senses the bulk electric field gradient, and that a completely different field gradient occurs for a probe atom inside the surface layer or on top of it (Fig. 11.22). In order to avoid chemical or other complications, indium metal was chosen in a first experiment, because it matches the best PAC probe ^{111}In and, in addition, its tetragonal structure creates a well-known electric field gradient inside the metal (Fig. 11.20). The sample was prepared by evaporating a 400 Å thick film of natural indium onto a glass backing at 100 K and annealing it at

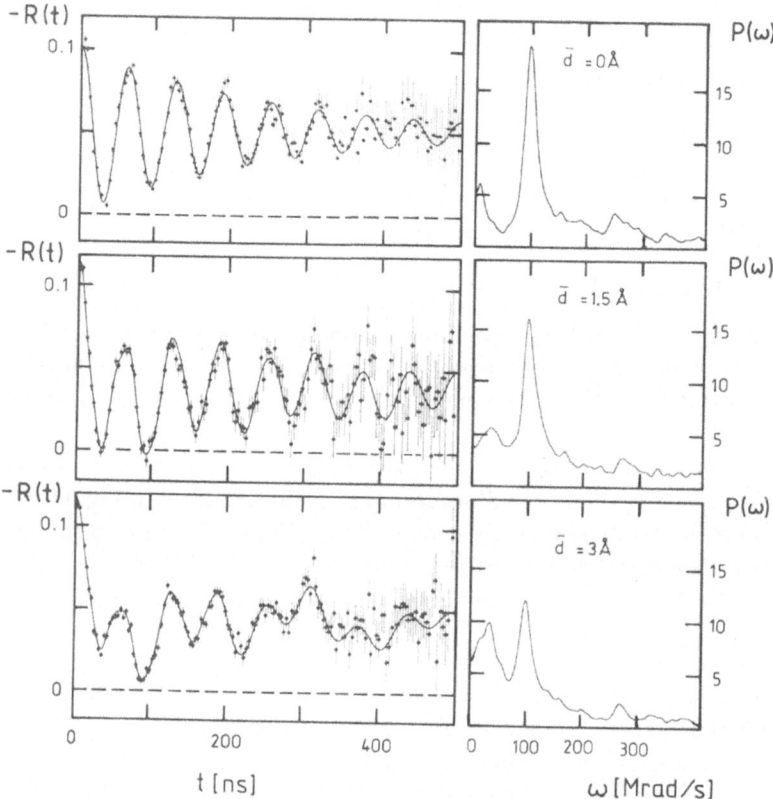

Fig. 11.22. PAC spectra and their Fourier transforms observed at [111]Cd on an Indium (111)-surface. $\bar{d}$ is the average coverage by inactive In atoms [11.28]

200 K. The first spectrum in Fig. 11.22 was measured after evaporating 10^{12} radioactive ^{111}In atoms/cm^2 onto the surface, much less than a monolayer. The time spectrum R(t) and the Fourier spectrum P(ω) clearly reveal that all radioactive ^{111}In atoms occupy a single site, which is characterized by an electric field gradient $V_{zz} = 4.7 \times 10^{17}$V/cm^2 and $\eta = 0.44$ as derived from the spin precession frequency $\omega_{01} = 102$ Mrad/s and $\omega_{02} = 165$ Mrad/s. Because of the single crystalline structure of the indium substrate and the chosen sample-detector geometry in Fig. 11.22, the amplitudes s_{22} and s_{23} in (11.24) (k = 2, $n_{max} = 3$) are close to zero, so that ω_{02} and ω_{03}, the two higher harmonics of ω_{01}, are not visible in the Fourier transform. The second spectrum in Fig. 11.22 was obtained by evaporation of less than a monolayer (about 1.5 Å) in average of natural indium doped with radioactive ^{111}In onto a different substrate. In addition to the surface frequency, an (unresolved) triplet of frequencies with $\omega_0 = 24$ Mrad/s indicating a second site, is now observed. The deduced axially symmetric field gradient $V_{zz} = 1.3 \times 10^{17}$V/cm^2 is well known and character-izes an ^{111}In atom at a substitutional site in indium metal (compare the PAC spectra for ^{111}CdIn in Fig. 11.20). Further evaporation of stable indium re-stores the bulk state and reduces the surface state further, as can be seen from the last spectrum of Fig. 11.22. The results of this experiment are summarized in Fig. 11.23, where the fractions of ^{111}In atoms at the surface and inside the bulk are plotted versus the amount of deposited stable indium. The data show that the coverage of one monolayer is already sufficient to change the surface field gradient into the bulk gradient which is observed throughout the regular indium lattice.

The positive outcome of this pioneering experiment has obviously opened up new prospects for the application of PAC, as, e.g., the study of adsorption of atoms at surfaces or the identification of interface alloy formation down to thicknesses of a few Å. The latter subject was recently pursued by *Keppner* et al., who investigated the compound formation at the interface between Cu and In, proving that a new CuIn$_2$ intermetallic phase is formed by diffusion of Cu into In [11.29].

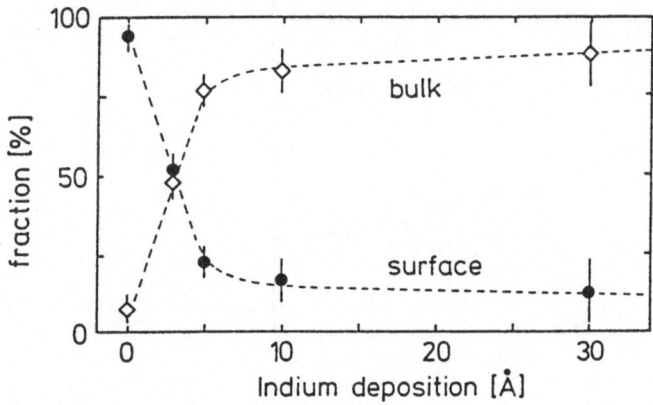

Fig. 11.23. Fractions of ^{111}In atoms exposed to the bulk or surface electric field gradient of indium plotted versus the average coverage by inactive In atoms [11.28]

11.5.3 Diffusion of Light Gases in Tantalum

The fact that radioactive probe atoms usually exist as two different elements in the course of a PAC experiment, can be used to study diffusion processes of gas atoms on a microscopic scale. Taking the radioactive probe ^{181}Hf/^{181}Ta, an experiment can be designed in tantalum metal, doped with ^{181}Hf, where the parent state (Hf) can be used for trapping a gas atom and the daughter state (Ta) allows observation of the future fate of this particular gas atom now residing in a pure Ta lattice. The time scale for this observation is introduced by an isomeric state with a mean lifetime of 25 μs, which is first populated by the β-decay of ^{181}Hf to ^{181}Ta, and which is the initial state of the γ-γ cascade (Sect. 11.4). Therefore, about 25 μs are available for the gas atom to leave the ^{181}Ta probe atom and to escape its detection by the electric field gradient, that characterizes its presence in the originally formed complex. A second feature, important to this type of experiment, utilizes again the short-range nature of the electric field gradient, which has the effect that the trapped gas atom does not have to move off very far, before it becomes invisible via the field gradient. In this way, the time for performing a single diffusion step can be measured in pure tantalum for atoms which have first been trapped or collected by the probe ^{181}Hf. The concentration of Hf atoms can be kept quite low, about 100 ppm, so that the concentration of gas atoms can also be comparably small, as is of importance in the case of the diffusion of H in tantalum.

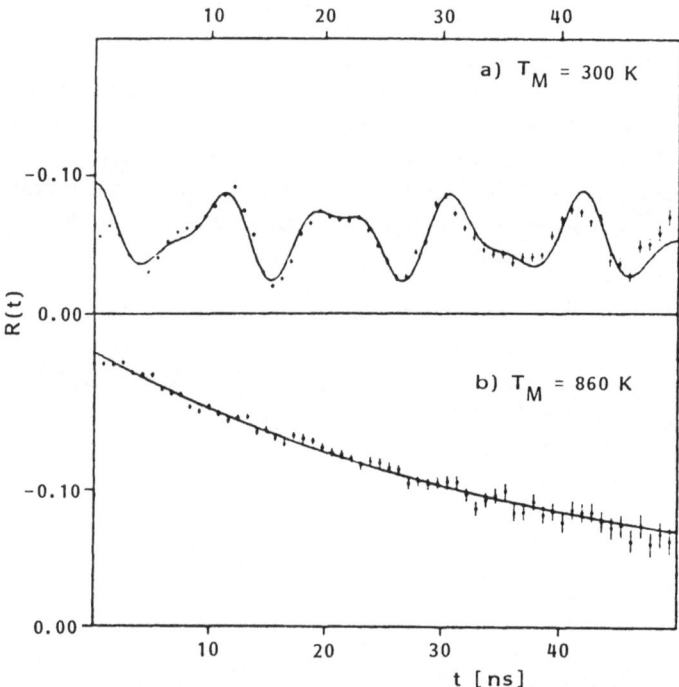

Fig. 11.24. PAC spectra of ^{181}Hf in tantalum showing the presence of oxygen atoms at two temperatures [11.30]

a) Diffusion of Oxygen Atoms

Using this idea *Weidinger* et al. studied the diffusion of O atoms, which – as is well-known – strongly interact with Hf atoms [11.30]. High-purity tantalum was doped with ^{181}Hf by alloying it with neutron-activated hafnium metal to a concentration of about 200 ppm. A detection of the 133-482 keV cascade of ^{181}Ta at 300 K yielded the PAC spectrum at the top of Fig. 11.24. The modulation of the coincidence probability can be fitted, assuming an electric quadrupole interaction, with $\nu_Q = 588$ MHz and $\eta = 0.35$. Since there is no electric field gradient at a substitutional site of a pure bcc lattice, this interaction indicates the presence of an impurity atom near the ^{181}Ta probe atom. The fraction of probe atoms exposed to this field gradient increases if the sample is annealed at 350 K where the macroscopic diffusion of O atoms in tantalum becomes appreciable. Thus, considering the high affinity of oxygen to Hf atoms, the trapped impurity can be identified with an oxygen atom stemming from the residual oxygen contamination of the tantalum sample. Raising the measuring temperature from 300 to 860 K effects a disappearance of the electric field gradient, as is seen in Fig. 11.24 (bottom). Obviously, the temperature is now sufficiently high, so that the O atom can leave its trapped position in a time much less than 25 μs after the ^{181}Hf decay. Figure 11.25a, where the fraction of the probe atoms that still sense O atom neighbors is given as a function of the measuring temperature, shows the reduction of this fraction with the increasing O mobility at higher temperatures. This effect is completely reversible, showing that O atoms can only diffuse away from those radioactive probe atoms which were converted from Hf into Ta.

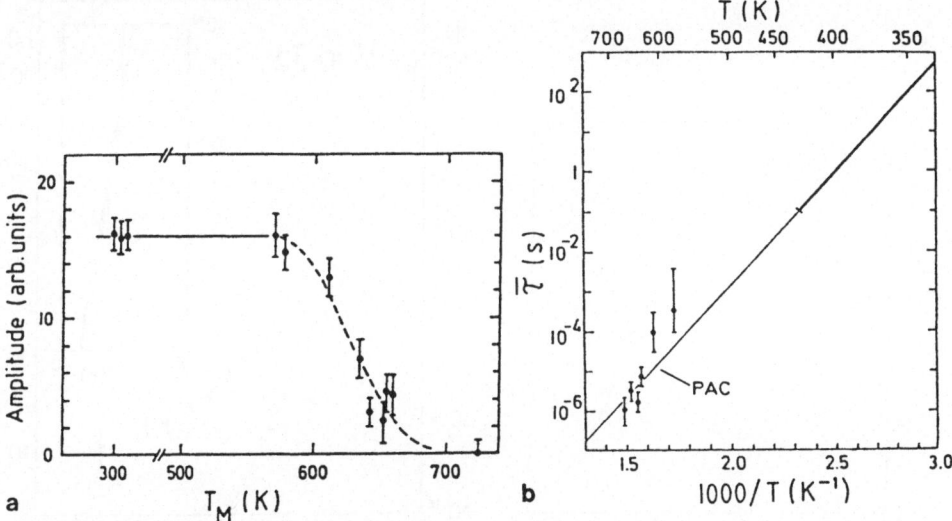

Fig. 11.25. (a) Fraction of ^{181}Ta atoms associated with oxygen atoms in tantalum as a function of measuring temperature. (b) Residence time $\bar{\tau}$ of oxygen atoms deduced from the PAC data in (a). The heavy solid line represents literature data and the thin line is an extrapolation to higher temperatures [11.30]

From the decrease of the observed modulation amplitude f the probability can be inferred that an O atom jumps away from the probe atom within $\tau_N = 25\,\mu s$. (The much shorter lifetime of 15 ns of the isomeric state used to detect the electric field gradient, can be neglected). The relative change of the observed amplitude f/f_0 in Fig. 11.25a can be described by

$$f/f_0 = \tau_c/(\tau_c + \tau_N) \tag{11.33}$$

where f_0 is the static fraction measured around 300 K and τ_c is proportional to the mean residence time $\bar{\tau}$ of the diffusing atom, wich is inversely proportional to the diffusion constant D. Taking into account a correction due to possible jumps back to the probe atom, the temperature dependence of the residence time $\bar{\tau}$ is plotted in Fig. 11.25b. The PAC data indicate an Arrhenius dependence and are in good agreement with an extrapolation of data, measured via the Snoek effect at lower temperatures. The asymmetry of the electric field gradient ($\eta = 0.35$) seems to indicate that the O atom within the O–^{181}Hf complex does not occupy a nearest but a second nearest octahedral interstitial site in the tantalum lattice.

b) Diffusion of Hydrogen Between 15 and 30 K

In the same way, *Weidinger* and *Peichl* [11.31] measured the diffusion of H in Ta below 100 K. Since the solubility of H in the α-phase becomes too small for an observation by classical techniques diffusion data below 100 K were not available. For the PAC experiment, a ^{181}Hf doped tantalum sample was elec-

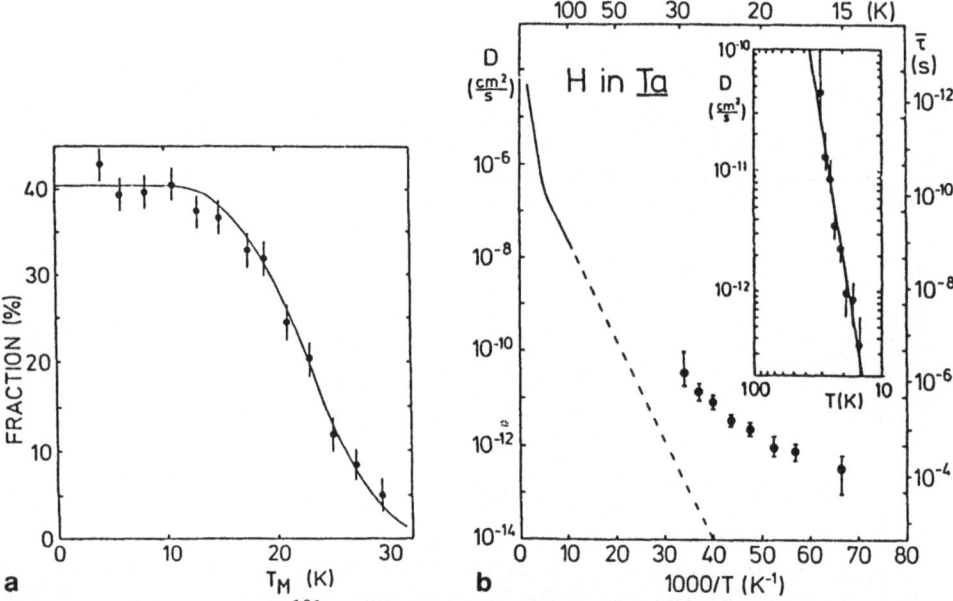

Fig. 11.26. (a) Fraction of ^{181}Ta atoms associated with hydrogen atoms in tantalum as a function of measuring temperature. (b) Diffusion coefficient D and residence time $\bar{\tau}$ deduced from the PAC data in (a). The solid line in the upper left corner represents literature data [11.31]

trolytically charged with hydrogen and subsequently rapidly immersed into liquid He in order to preserve the high-temperature α-phase at low temperatures. The PAC spectra indicate that 40% of the Hf atoms form Hf-H pairs, which are characterized by $\nu_Q = 580\,\text{MHz}$ and $\eta = 0.23$. From Fig. 11.26a it can be seen, that above 10 K the ^{181}Ta-H pairs start to dissociate. As described above, the ratio f/f_0 can be used to deduce the correlation time $\tau_c = 4\bar{\tau}$ where it is assumed that the H-atom occupies a nearest neighbour tetrahedral site and diffuses via tetrahedral sites. Using $D = a_0^2/48\bar{\tau}$ (a_0 is the lattice constant of Ta) the diffusion coefficient D and the residence time $\bar{\tau}$ are plotted in Fig. 11.26b. The solid line represents the existing diffusion data, whereas the dashed line is an extrapolation of these data assuming an Arrhenius dependence. Obviously, the PAC results extend these data to significantly lower temperatures. The inset shows the PAC data on a log-log scale revealing a T^7 temperature dependence.

The extrapolated high-temperature data indicate a first change in the diffusion mechanisms around 100 K. This is ascribed to a change from the classical process of a barrier hopping process to an incoherent tunneling process. Extrapolating these data to lower temperatures, the PAC results indicate a higher diffusivity than predicted by the tunneling process which might indicate a second change of the dominant diffusion mechanism. The observed T^7 dependence would point to a quantum mechanical diffusion process. It can be expected that in the near future these studies will be extended to temperatures below 1 K, where new phenomena might occur. In addition, isotope effects can be studied as well.

11.5.4 Defects and Impurities

Irradiation of metals with energetic particles produces structural defects like self-interstitial atoms, vacancies and their agglomerates, so that several types of defects are simultaneously present which can interact with each other or with impurity atoms. The classical method for monitoring the actual defect concentration consists in measuring the sample's electrical resistivity which is increased by defects. In order to find the mobilities of the different types of defects and the stabilities of their agglomerates, the sample is heated and its residual resistivity is measured at low tempertures. As soon as the annealing temperature reaches a point at which one type of defect becomes mobile, a drop in the resistivity is observed, because the mobile defect can now encounter its anti-defect leading to defect annihilation. Thus, during an annealing sequence, the residual resistivity decreases in recovery stages which indicate the mobilization of either an isolated defect or a defect which is trapped by an agglomerate or an impurity atom. Without assuming a specific defect model, however, it is not possible to infer the defect configuration or to find out whether the defect has been trapped before.

Since a mobile defect can be trapped by an impurity atom, its arrival can be marked by an electric field gradient at the impurity. Thus PAC can be used to identify the type of mobile defect, which causes a particular recovery stage [11.4,32]. The electric field gradient allows recognition of the same defect in different experiments and since the field gradient is short-ranged, only trapped defects are visible rather than defects at greater distance from the probe atoms.

PAC can cope with rather complex systems, consisting of different types of defects and impurities and since it can track the formation of a particular defect under differing conditions, its type can be inferred without assuming a defect model.

a) Identification of Lattice Defects

Once the trapping of a defect is detected by a radioactive probe atom, the next step consists in finding the most favorable conditions for observing the same field gradient or defect by varying the experimental conditions. From

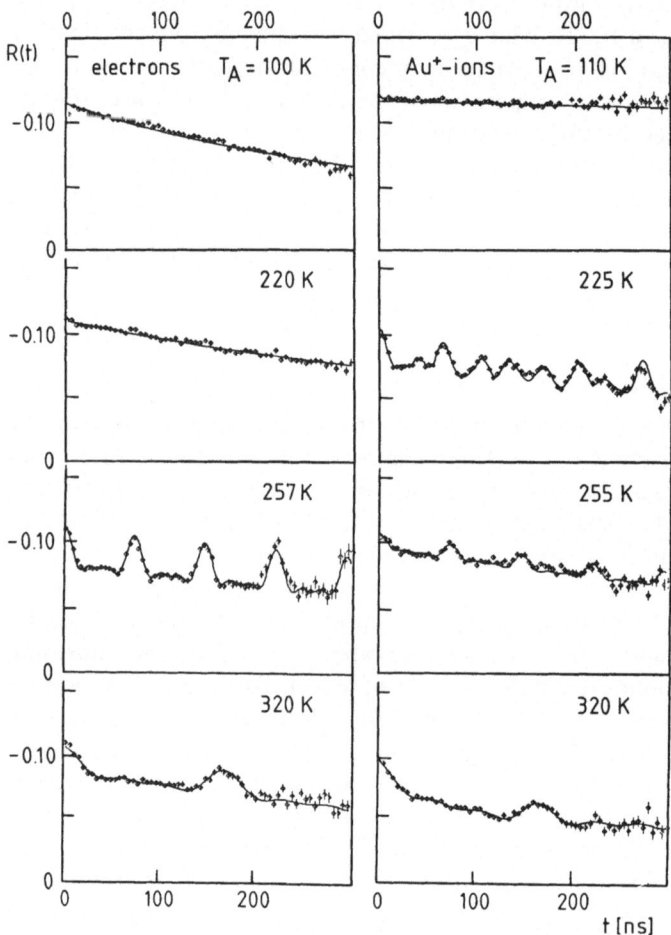

Fig. 11.27. PAC spectra of ^{111}In in gold, which was irradiated with electrons or Au ions and annealed at temperatures T_A. Different frequencies indicate the formation of different defect complexes at ^{111}In atoms [11.33]

these experiments, conclusions can be drawn about the nature of the defect (vacancy or interstitial), its size (how many elementary defects) and, using single crystals, its geometrical arrangement with respect to the probe atom.

For a more detailed illustration, some experiments on ^{111}In-doped gold samples will be discussed. From electrical resistivity measurements a major defect recovery stage is known to occur around 250 K. In order to identify the defects which cause this recovery stage, *Deicher* et al. [11.33] irradiated polycrystalline gold foils with 3 MeV electrons at 4.2 K. Following each annealing step, PAC spectra were recorded at 77 K, a selection of which is displayed in Fig. 11.27 (left side). The first two spectra recorded after annealing at 100 and 220 K, respectively, show no modulation of the coincidence probability, indicating the absence of defect trapping at these temperatures. The defects created through the electron irradiation are still immobile and most of them stay at some distance from the probe atoms, so that the probes only sense a distribution of weak field gradients resulting in an exponential attenuation of the coincidence probability. After annealing at 257 K, however, a pronounced electric field gradient occurs, described by $\nu_Q = 91$ MHz and $\eta = 0$, showing that a particular type of defect has now been mobilized and trapped. This defect complex is replaced by a new one after annealing at 320 K (bottom spectrum), for a different modulation is observed corresponding to $\nu_Q = 40$ MHz and $\eta = 0$. Since we shall meet the formation of several In-defect complexes in the course of this section, for gold and copper Table 11.2 lists their hyperfine parame-

Table 11.2. Defect complexes observed by ^{111}In atoms along with their hyperfine interaction parameter $\nu_Q = eQV_{zz}/h$, $\eta = (V_{xx} - V_{yy})/V_{zz}$ and $\langle ijk \rangle$ (designating the orientation of the V_{zz} component). The given defect assignments are the most probable ones [11.32]

No.	ν_Q	η	$\langle ijk \rangle$	Trapped defect	Visible between [K]
Intrinsic Defects in Au					
1	102 MHz	0.45	–	Divacancy	190–240
2	101 MHz	0	$\langle 100 \rangle$	Divacancy	190–240
3	91 MHz	0	$\langle 110 \rangle$	Monovacancy	210–300
4	40 MHz	0	$\langle 111 \rangle$	Planar vacancy loop	250–600
Intrinsic Defects in Cu					
1	116 MHz	0	$\langle 110 \rangle$	Monovacancy	200–300
2	181 MHz	0	$\langle 100 \rangle$	Divacancy	180–320
3	52 MHz	0	$\langle 111 \rangle$	Planar loop	350–750
4	55 MHz	0.48	–	Defect cluster	350–750
5	47 MHz	0	$\langle 100 \rangle$	Self-interstitial	40– 70
6	19 MHz	1.0	–	Self-interstitial	40– 70
He Defects in Cu					
1	136 MHz	0	–	Substitutional He	10–250
2	225 MHz	0	–	Small He cluster	300–750
3	127 MHz	0.89	–	Small He cluster	300–750
4	297 MHz	0.45	–	Small He cluster	300–750
5	206 MHz	0.18	–	He bubble	750–990
6	57 MHz	0	$\langle 100 \rangle$	Interstitial He	20– 65

ters along with a description of the respective trapped defects. The complexes mentioned above correspond to No. 3 and 4, respectively.

In contrast to electron irradiation, which mainly produces isolated vacancies and interstitials, heavy ions create larger defect agglomerates in their defect cascades. On the right-hand side of Fig. 11.27 PAC spectra are shown for a gold sample irradiated with Au^+ ions. A comparison with the previous spectra reveals both that new defects (1 and 2) have been created and that they are more mobile than defect 3, because these defects create a new frequency and are already visible at 225 K. On the other hand, the trapping of defect 3 at 255 K has become less probable. Since after the first irradiation smaller and after the second one larger defects prevail, it can already be concluded that defect 3 is smaller than either defect 1 or 2.

Figure 11.28 shows the PAC spectrum of a sample which was quenched from 1280 K down to 230 K and subsequently annealed at 257 K. In this way exclusively, vacancies can be studied, which have been formed in thermal equilibrium at 1280 K and are frozen by rapid cooling. The R(t) spectrum exhibits a distinct modulation; comparing its Fourier transform with the Fourier transform of the respective R(t) spectrum measured after electron irradiation (Fig. 11.27), the occurrence of the same frequencies in both cases shows that defect 3 is again observed. Therefore, defect 3 consists of vacancies.

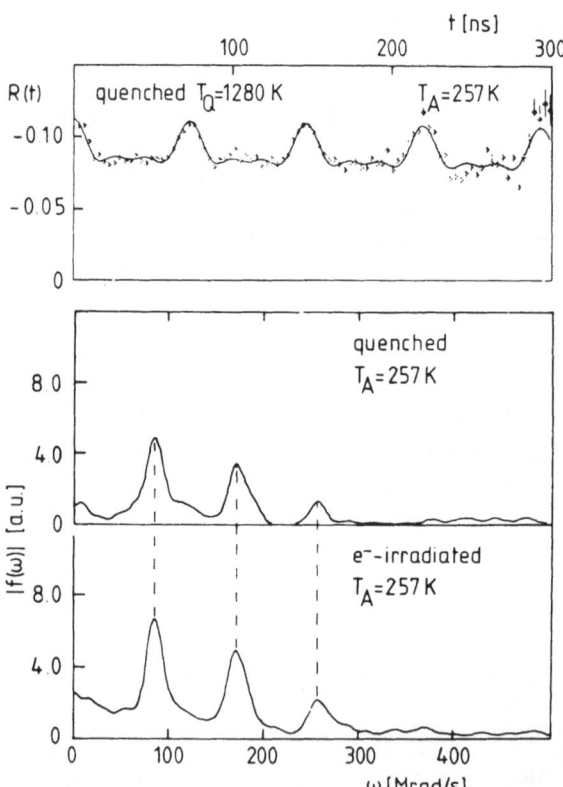

Fig. 11.28. PAC spectrum of ^{111}In in gold, which was quenched from 1280 K and annealed at 257 K. Below, the Fourier transforms belonging to the quenched and electron irradiated sample are compared [11.33]

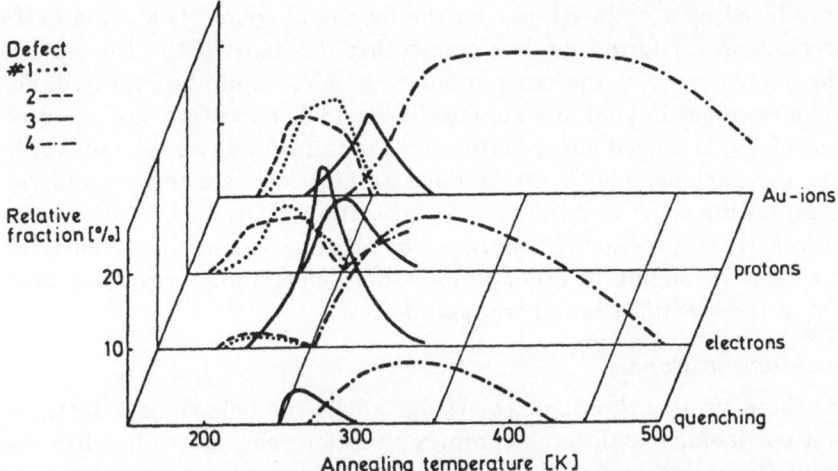

Fig. 11.29. Influence of damaging conditions and annealing temperature on the formation of the four ^{111}In−defect complexes in gold [11.33]. (See Table 11.2 for more details)

The different defect reactions observable by PAC in these recovery stages are compared in Fig. 11.29 for different damaging conditions: Irradiation with Au^+ ions, protons, electrons, or quenching. Eventually, it was shown that defect 3 consists of a monovacancy-In atom pair and defect 4 consists of a larger vacancy cluster. The two other defects, 1 and 2, are due to divacancies trapped by In atoms and forming two geometrically different divacancy-In complexes with different electric field gradients. The latter identification has been mainly supported through measuring the occurrence of defect 1 and 2 as a function of the proton irradiation dose at 205 and 255 K. The absence of divacancy trapping in the quenching experiment seems to be caused by too slow a quenching rate, so that these obviously highly mobile defects could escape detection. Thus, these recovery stages in Au are caused by the migration of two types of defects, the more mobile divacancy and the monovacancy. Their trapping by In atoms leads finally to the formation of thermally very stable vacancy clusters.

At this point, some remarks concerning the limitations of the trapping method should be mentioned: Since the electric field gradient is short-ranged defects, which cannot be trapped by the probe atom, remain invisible, and also different sizes of larger trapped clusters cannot be distinguished; if larger defects exhibit an almost cubic structure, they are invisible in cubic metals, because their field gradients are too small to give rise to a detectable modulation.

Turning back to Fig. 11.29, the growth of the three complexes, formed by trapping mobile mono- or divacancies during annealing (defects 1–3), allows a deduction of the migration energies of the respective defects, whereas the disappearance of the different complexes sets only a lower limit for their binding energies. Another interesting aspect refers to the orientation of the electric field gradient tensor produced, e.g., by a trapped monovacancy: In Sect. 11.3.2 we have discussed that the direction of the V_{zz} component of an electric field gradient produced by a ^{111}In-Rh pair in the fcc metal copper is along a $\langle 110 \rangle$ lattice direction which forms the line connecting this pair. If the Rh atom is replaced by a monovacancy, the same orientation of V_{zz} could be expected. For trapped monovacanies in gold and copper, Table 11.2 shows that the observed orientations of V_{zz} is indeed along $\langle 110 \rangle$, and the same holds for Al and Ag. In bcc metals, the next neighbour site is along a $\langle 111 \rangle$ lattice direction and the experimental results show that here the V_{zz} direction of In-monovacancy pairs is parallel to a $\langle 111 \rangle$ direction. Encouraged by this agreement, the orientation of field gradients belonging to other probe-atom defect complexes, have been used to aid in the identification of trapped defects.

b) Helium Atoms in Copper

As soon as there exists a thorough knowledge about the behaviour of intrinsic defects in a particular metal, more complex situations can be studied like the behaviour of He atoms in Cu. Large fluences of neutrons produce intrinsic defects and can also produce He atoms in a metal. The He atoms are bound by vacancies so that high He concentrations can be accumulated in the form of bubbles which cause severe changes in the properties of a metal. Therefore, the conditions leading to the formation of these bubbles are of great interest.

In order to study this process in copper, *Deicher* et al. [11.34] first investigated the properties of the intrinsic defects along the lines illustrated above. It could be shown that ^{111}In atoms in copper trap self-interstitials (defect 5 and 6) in recovery stage I, mono- and divacancies (defect 1 and 2) in stage III, and that larger defect clusters (defect 3 and 4) are visible in stage V (Table 11.2). Figure 11.30a demonstrates that the six different defect complexes can be easily distinguished by their well defined hyperfine interaction parameters, which are listed in Table 11.2. Underneath each PAC spectrum the fitted fractions belonging to each complex are drawn separately. The growth and decay of the six complexes during annealing is displayed in Fig. 11.31. It reflects a detailed picture of the mobilization and agglomeration of self-interstitials and vacancies in copper.

In a second step 30 keV He$^+$ ions were implanted into ^{111}In-doped copper samples at 300 and 10 K with different doses in order to simulate the effect of a neutron irradiation [11.35]. Figure 11.30b shows the PAC spectra measured after He irradiation at 10 K, and annealing at temperatures similar to the ones in Fig. 11.30a so that the influence of the presence of He atoms can be seen. The He-doped sample exhibits a strong exponential decrease in the R(t) spectrum due to a higher defect concentration but the modulations in the spectra at 260 K are similar in both cases, indicating the trapping of mono- and divacancies. The modulation observed at 52 K is different, indicating additional interactions of In atoms with the implanted He atoms. Instead of self-interstitials, now He atoms occupying trapped vacancies (He-defect 1) as well as He atoms trapped at interstitial sites (He-defect 6) are visible (Table 11.2). More detailed experiments [11.36] show that He-defect 1 is exclusively athermally formed during the He irradiation at 10 K and remains stable up to recovery stage III; He-defect

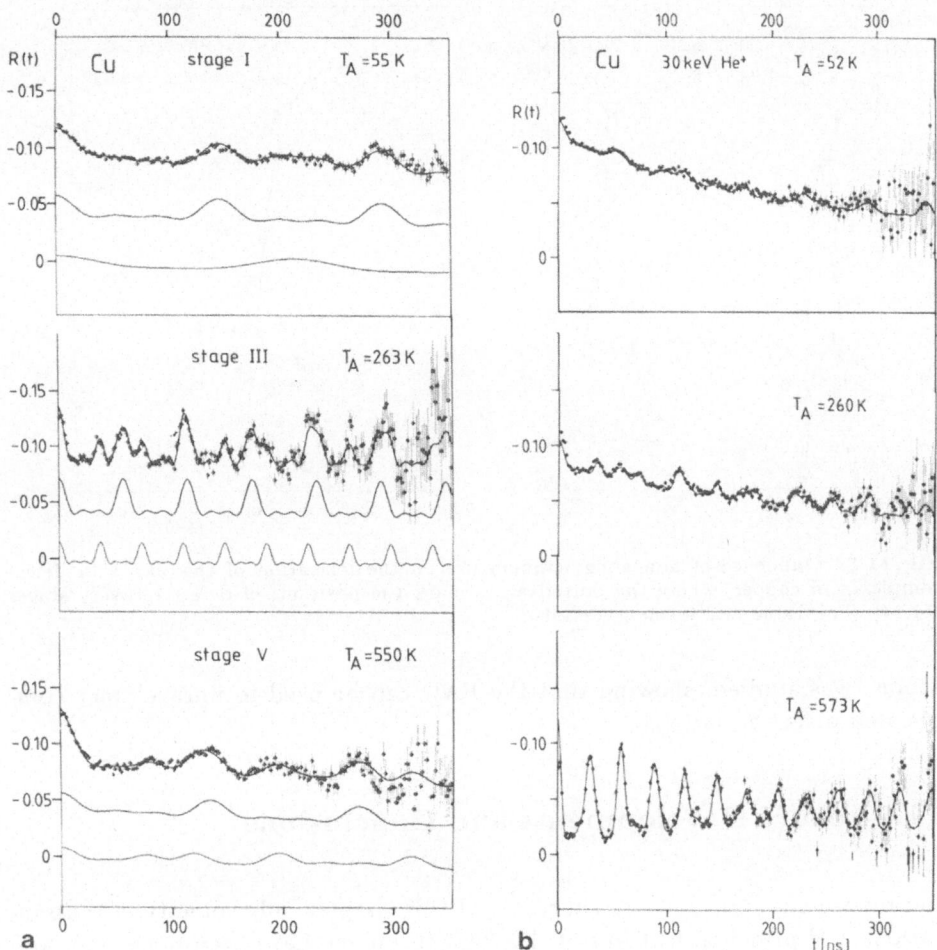

Fig. 11.30. (a) PAC spectra of ^{111}In in copper showing the formation of different defect complexes during annealing in defect recovery stage I, III and V after irradiation. The fitted components belonging to a particular ^{111}In-defect complexes are shown separately [11.34]. **(b)** Additional irradiation with He ions effects the formation of new defect complexes as indicated by the occurrence of new frequencies, especially in stage I and V [11.35]

6 is formed by trapping of interstitially migrating He atoms around 25 K and is visible up to only 65 K which might indicate a low stability of this complex. It is of special interest that no new He complexes are formed before recovery stage III. As soon as the vacancies become mobile in stage III, agglomeration of He atoms occur and larger vacancy clusters are either occupied or their growth is impeded by He atoms; thus, in the PAC spectrum at 573 K the intrinsic defect clusters 3 and 4 are replaced by the He-defect clusters 2, 3 and 4 (Table 11.2). Around 750 K the three different He clusters transform into the same new He cluster (He-defect 5), which is stable up to 990 K.

Thus, in the course of the investigations in copper, the evolution of 12 different defect complexes (Table 11.2), consisting of intrinsic defects and He

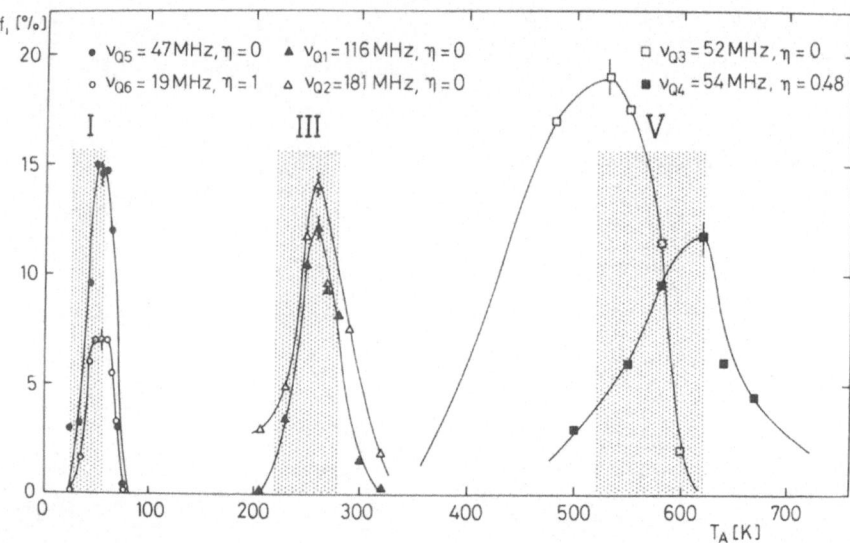

Fig. 11.31. Influence of annealing temperature on the formation of the six ^{111}In defect complexes in copper, where the dotted areas mark the positions of defect recovery stages [11.34]. (See Table 11.2 for more details)

atoms, was studied, showing that the PAC can be used to unravel very complicated defect situations.

11.6 Future Developments and Conclusions

Because of the introductory character of this chapter only some typical applications and principal features of the PAC technique have been presented, and the topics of this volume are restricted to the investigation of metals. Several experiments on semiconductors and insulators, however, have already been carried out in the past and without doubt this field of research will extend rapidly in the future, after some of the technical problems, i.e. doping with radioactive probes and annealing, have been solved.

The PAC technique will also contribute to the understanding of other physical mechanisms as some very promising and pioneering experiments have already shown. For example, the problem of internal oxidation was successfully attacked by several groups [11.37]. The study of clean surfaces, as discussed above, opens up the investigation of adsorption and desorption processes and of catalytic problems. Interfaces and the effect of ion-beam mixing on these structures can be studied by PAC. Long-range diffusion by an internal tracer technique [11.38], as well as other atomic mechanisms are within the scope of application because of the microscopic and local sensitivity of the radioactive probes.

The simultaneous application and combination of other techniques with PAC proves to be necessary in many cases, where a particular method by itself gives only incomplete information. This was successfully demonstrated by combining PAC with residual resistivity measurements in the investigation of defects in metals or by the simultaneous application of PAC and Channeling to study lattice locations of impurities [11.39,40]. In PAC surface studies, the combination with LEED or Auger electron spectroscopy is being used.

The availability of additional suitable radioactive probes as supplied by facilities like ISOLDE at CERN offers the chance to attack specific problems which cannot be solved with the probes presented. In this context the closely related technique of PAD (perturbed angular distribution), where the probing isomeric nuclear states are produced by nuclear reactions, should be mentioned once more. Applications of this technique, e.g., to defect studies, can be found in [11.4].

In conclusion, it can be said that the PAC technique has proved to be a useful and in many cases a unique tool to investigate solid state problems. Although the costs for the required equipment are high and the method seems to be complicated, more and more laboratories are adopting this powerful technique for an increasing number of applications.

Acknowledgements. We would like to thank M. Deicher, G. Froböse, G. Grübel, R. Nussbaum, G. Schatz and M.L. Swanson for valuable help and comments. The support of the "Bundesminister für Forschung und Technologie" and the "Deutsche Forschungsgemeinschaft" is gratefully acknowledged. One of us (Th.W.) is grateful to the Solid State Science Branch of the Chalk River Nuclear Laboratories for their generous hospitality during the writing of the manuscript.

References

11.1   H. Frauenfelder, R.M. Steffen: In *Alpha-, Beta- and Gamma-Ray Spectroscopy*, ed. by K. Siegbahn (North-Holland, Amsterdam 1965) Vol. 2, p.997

11.2   H.H. Rinneberg: Atomic Energy Rev. **17**, 477 (1979)

11.3   M. Forker, R.J. Vianden: Magnetic Resonance Rev. **7**, 275 (1983)

11.4   E. Recknagel, G. Schatz, Th. Wichert: In *Hyperfine Interactions of Radioactive Nuclei*, ed. by J. Christiansen, Topics Current Phys., Vol. 31 (Springer, Berlin, Heidelberg 1983) p. 133

11.5   W. Witthuhn, W. Engel: In [Ref. 1.4, p. 205]

11.6   R.S. Raghavan, D.E. Murnick (eds.): Proc. IV Conf. on Hyperfine Interactions, Madison (1977), Hyp. Int. **4** (1978)

11.7   G. Kaindl, H. Haas (eds.): Proc. V. Intern. Conf. on Hyperfine Interactions, Berlin (1980), Hyp. Int. **9,10** (1981)

11.8   L. Niesen, F. Pleiter, H. de Waard (eds.): Proc. VI. Intern. Conf. on Hyperfine Interactions, Groningen (1983), Hyp. Int. **15/16** (1983)

11.9   S.G. Cohen: In *Hyperfine Interactions*, ed. by A.J. Freeman, R.B. Frankel (Academic, New York 1967)

11.10 R.M. Steffen, K. Alder: In *The Electromagnetic Interaction in Nuclear Spectroscopy*, ed. by W.D. Hamilton (North-Holland, Amsterdam 1975)

11.11 J. Christiansen (ed.): *Hyperfine Interactions of Radioactive Nuclei*, Topics Current Phys., Vol. 31 (Springer, Berlin, Heidelberg 1983)

11.12 E. Matthias, D.A. Shirley, J.S. Evans, R.A. Naumann: Phys. Rev. B **140**, 264 (1965)

11.13 B. Lindgren, E. Karlsson, B. Jonsson: Hyp. Int. **1**, 505 (1976)

11.14 R.S. Raghavan, P. Raghavan: Phys. Lett. A **36**, 313 (1971)

11.15 W. Keppner, W. Körner, P. Heubes, G. Schatz: Hyp. Int. **9**, 293 (1981)

11.16 M. Deicher, R. Minde, E. Rechnagel, Th. Wichert: Hyp. Int. **15/16**, 437 (1983)

11.17 E. Recknagel, Th. Wichert: Nucl. Instr. Meth. **182/183**, 439 (1981)

11.18 H. Haas, D.A. Shirley: J. Chem. Phys. **58**, 3339 (1973)

11.19 A.R. Arends, C. Hohenemser, F. Pleiter, H. de Waard, L. Chow, R.M. Suter: Hyp. Int. **8**, 191 (1980)

11.20 D.A. Shirley, S.S. Rosenblum, E. Matthias: Phys. Rev. **170**, 363 (1968)

11.21 J. Kanamori, H. Akai, M. Akai: Hyp. Int. **17–19**, 287 (1984)

11.22 K.S. Krane: Hyp. Int. **15/16**, 1069 (1983)

11.23 M.L. Swanson: Rep. Prog. Phys. **45**, 47 (1982)

11.24 E.N. Kaufmann, R.J. Vianden: Rev. Mod. Phys. **51**, 161 (1979)

11.25 J. Christiansen, P. Heubes, R. Keitel, W. Klinger, W. Loeffler, W. Sandner, W. Witthun: Z. Physik B **24**, 177 (1976)

11.26 P. Raghavan, E.N. Kaufmann, R.S. Raghavan, E.J. Ansaldo, R. Naumann: Phys. Rev. B **13**, 2835 (1976)

11.27 K. Nishiyama, F. Dimmling, Th. Kornrumpf, D. Riegel: Phys. Rev. Lett. **37**, 357 (1976)

11.28 W. Körner, W. Keppner, B. Lehndorff-Junges, G. Schatz: Phys. Rev. Lett. **49**, 1735 (1982) and private communication by G. Schatz

11.29 W. Keppner, T. Klas, W. Körner, R. Wesche, G. Schatz: Phys. Rev. Lett. **54**, 2371 (1985)

11.30 A. Weidinger, M. Deicher, T. Butz: Hyp. Int. **10**, 717 (1981)

11.31 A. Weidinger, R. Peichl: Phys. Rev. Lett. **54**, 1683 (1985)

11.32 Th. Wichert: Hyp. Int. **15/16**, 335 (1983)

11.33 M. Deicher, E. Recknagel, Th. Wichert: Rad. Eff. **54**, 155 (1981)

11.34 M. Deicher, G. Grübel, R. Minde, E. Recknagel, Th. Wichert: In *Point Defects and Defect Interactions in Metals*, ed. by J. Takamura et al. (University of Tokyo Press, Tokyo 1982) p. 220

11.35 M. Deicher, G. Grübel, W. Reiner, Th. Wichert: Hyp. Int. **15/16**, 467 (1983)

11.36 Th. Wichert, M. Deicher, G. Grübel, E. Recknagel, W. Reiner: Phys. Rev. Lett. **55**, 726 (1985)

11.37 W. Bolse, M. Uhrmacher, K.P. Lieb: Mat. Sci. Eng. **69**, 375 (1985)

11.38 Th. Wichert: Rad. Eff. **78**, 177 (1983)

11.39 H. Hofsäss, G. Lindner, E. Recknagel, Th. Wichert: Nucl. Instr. Meth. Phys. B **2**,13 (1984)

11.40 M.L. Swanson, L.M. Howe, A.F. Quenneville, Th. Wichert, M. Deicher: J. Phys. F **14**,1603 (1984)

12. Nuclear Magnetic Resonance

P. Panissod
With 25 Figures

Nuclear magnetic resonance covers an extremely broad range of applications reaching from pure nuclear physics to the newly introduced medical application, NMR imaging, including atomic physics, condensed-matter physics and chemistry, and biological and chemical analysis.

Like all other spectroscopic methods, it allows the observation of a set of energy levels (here nuclear levels) by some form of spectral absorption or emission (here that of photons). However, among these methods NMR occupies a very special place because it measures directly the paramagnetism of the nuclear ground state. This places the absorbed/emitted photon energies in the low radio frequency (RF) range of $1-10^6$ kHz, which has several important consequences concerning the sensitivity, the resolution and the variety of measurements and techniques that we shall present in this chapter.

12.1 Introductory Comments

Figure 12.1 shows the basic sketch of an NMR experiment. As the sample is placed in a magnetic field H_0 the degeneracy of the nuclear ground state is removed by the Zeeman interaction. NMR refers to the resonant absorption of photons by the nuclear spin system through dipolar transitions between adjacent spin levels. Actually owing to the very low Zeeman energy the photons simply consist of a radio frequency field H_1, perpendicular to H_0, which is produced by an RF generator in the sample coil. Therefore it is more convenient to consider the photons as a coherent classical electromagnetic field which is

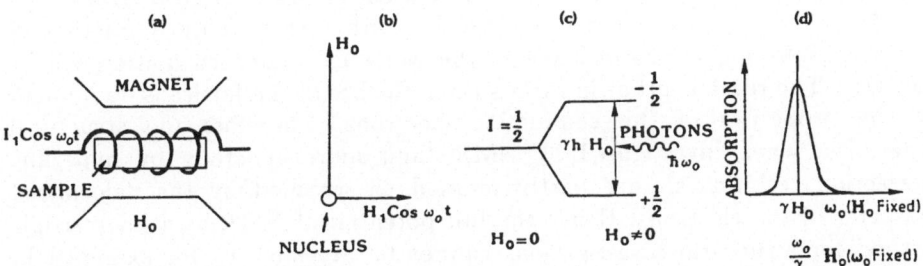

Fig. 12.1a–d. Sketch of an NMR experiment: (a) Laboratory scale; (b) microscopic scale; (c) Nuclear energy levels (I = 1/2); (d) Basic NMR spectrum (frequency or field scan)

part of the experimental setup. The absorption of RF energy can be measured by many means (e.g., directly through the supplementary loss of the coil or indirectly through the subsequently induced emission of photons). Thus an NMR spectrum consists of a plot of some quantity proportional to the energy absorption (e.g., the quality factor of the coil or the electromotive force created in the coil by the re-emitted RF field) as a function of either the RF field frequency at a constant H_0 or as a function of H_0 at a constant frequency (ω_0). Since the energy absorption is proportional to the weak magnetic polarization of the ground state and to the small photon energy, the NMR sensitivity is much lower than that of other related nuclear techniques using triggered detection techniques without the need of polarization (PAC) or excited nuclei involving much higher photon energies (Mössbauer effect). This is the reason why NMR in condensed matter was sought for long before it was discovered in the middle of the 40's.

Actually resonance experiments had been performed in beams by I.I. Rabi and his colleagues in 1939 using a modified beam deflection method (Stern and Gerlach) to measure the magnetic moment of the proton. However, the first successful NMR experiments in bulk matter were achieved in 1945 independently by E.M. Purcell, H.C. Torrey and R.V. Pound and F. Bloch, W.W. Hansen and M.E. Packard each using a different approach. On the one hand E.M. Purcell and coworkers were concerned with the loss of the coil at resonance. On the other hand F. Bloch and his colleagues looked for and observed the electromotive force induced in the coil by the nuclear moments precessing around the magnetic field.

The underlying idea of the former approach is directly related to the conventional description of a spectroscopic method (i.e., photon-energy absorption) while Bloch's idea of "nuclear induction" is related to a complete description of the motion of nuclear moments in steady and alternating fields instead of considering only energy levels. Both schemes have their own merit: according to one's theoretical background or preferences one can understand NMR either in terms of photon-induced transition probabilities between nuclear levels or in terms of the motion of nuclear moments in a radio frequency field. Both descriptions can be found in monographs on NMR [12.1-3]; hereafter we will favour the second approach as it allows a description of pulsed NMR experiments which are presently the most widely used, in quasi-classical terms.

As this book is concerned specifically with metals, only a fraction of the potential applications of NMR in the world of condensed matter will be reviewed. The reason is that in metals the influence of the lattice is dominated by the properties of the conduction electrons. Therefore the structural properties, when investigated by NMR (and more generally by hyperfine interaction techniques), are partly masked or screened by the delocalized character of the electrons. Hence the full potential of NMR as a microscopic method for structural investigations cannot be attained as, for example, in the study of organochemical materials. It should also be noted that most of the recent developments in NMR such as high-resolution Fourier transform

NMR, magic angle spinning or the new sophisticated pulse techniques (spin-flip narrowing, two-dimension NMR) are limited to the case of rather narrow NMR lines, a situation which is not so frequent in solids and even less in metals. They alone will be mentioned here. Some of these techniques are presented in [12.2,4–6] and for readers interested in the applications of NMR to chemistry and related fields it is recommended to consult the recent reviews by *Ben* and *Gunther* [12.7] and by *Clague* [12.8].

In the following we shall first present briefly the principles and raw outputs of an NMR experiment and their origin (Sects. 12.2,3) on a general basis which is usually sufficient for an understanding of the application of the method to structural investigations and phase analysis (examples of which are given in Sect. 12.5). Further details on the microscopic origin of these outputs will be given in Sect. 12.4. These details are necessary to understand NMR studies of the electronic and magnetic properties of metals (examples of which are given in Sect. 12.6).

Because of the limited space allowed the mathematical developments will be restricted to a minimum of formulae. Further details and more complete mathematical and quantum mechanical calculations can be found in the monographs by *Abragam* [12.1] and *Slichter* [12.2] including the case of insulating materials and liquids. Concerning the application of NMR to the investigation of metallic properties a useful introduction can be found in *Winter*'s book [12.3] and pertinent references are given in Sect. 12.4.

Before going into further detail it is worthwhile to define here the various fields and frequencies or angular velocities that are involved in an NMR experiment. These are given in Table 12.1 together with the definitions of the constants we shall use throughout the chapter.

Table 12.1. Various fields, frequencies, and angular velocities that are involved in a NMR experiment

---

Universal constants
k_B: Boltzmann's contant
h: Planck's constant; $\hbar : h/2\pi$
e: electron charge
μ_B: Bohr's magneton

Nuclear constants
I: nuclear spin
γ: gyromagnetic ratio of the nucleus
Q: quadrupole moment of the nucleus

Magnetic fields
H_0: "external" applied field (including classical contributions of matter: demagnetizing field and Lorentz cavity contributions)
H_i: actual field experienced by the nuclei in matter
H_1: radio frequency field (perpendicular to H_0)

Frequencies and angular velocities
ω_0: frequency of the RF field
ω_L: Larmor frequency γH_i (or the angular velocity of nuclear precession around H_i)
ω_1: angular velocity γH_1 of the precession of the nuclear spins around H_1

In the following fat characters will be used for the fields, the angular velocities and the moments each time their orientation is meant; otherwise their magnitude should be understood.

12.2 Physical Background of an NMR Experiment – Hyperfine Interactions

12.2.1 Nuclear Paramagnetism

In condensed matter and for temperatures above the milli-Kelvin range nuclear magnetism is dominated by spin paramagnetism: when a magnetic field H_i is present in nuclei which possess a spin I their ground state is split by the Zeeman interaction into $2I + 1$ levels, indexed by the quantum number $m(-I \leq m \leq +I)$, the energy of which is $E_m = -m\gamma\hbar H_i$. Therefore the energy difference between two adjacent levels is $\Delta E = \gamma\hbar H_i$ and here the NMR spectrum consists of only one absorption line at the Larmor frequency $\omega_L = \gamma H_i$.

According to Boltzmann's law the population of the levels is proportional to $\exp(m\gamma\hbar H_i/k_B T)$ which gives rise to the net nuclear magnetization

$$M_n = N\gamma\hbar \frac{\sum_{m=-I}^{I} m \cdot \exp(m\gamma\hbar H_i/k_B T)}{\sum_{m=-I}^{I} \exp(m\gamma\hbar H_i/k_B T)} \tag{12.1}$$

for a system of N nuclear spins.

Actually the Zeeman energy is almost always much weaker than the thermal energy and M_n can well be approximated by the Curie law

$$M_n = N\gamma^2\hbar^2 H_i \cdot I(I + 1)/3k_B T = \chi_n H_i, \tag{12.2}$$

χ_n being the nuclear spin susceptibility.

Accordingly the NMR absorption signal resulting from transitions between the levels m and $m-1$ is proportional to the difference between their population $(\gamma\hbar H_i/k_B T)$, to the dipole transition probability $[I(I + 1) - m(m - 1)]$ and to the photon energy $(\hbar\omega_L)$, and the whole signal is proportional to $\omega_L M_n$.

While in the early stage of NMR history physicists were interested in χ_n as a means of measuring the nuclear moment they soon realized that the second term intervening in M_n (i.e., the field experienced by the nuclei) was strongly dependent on the material in which the nuclei were imbedded. This observation opened the way for NMR as a method for investigating the properties of condensed matter. Indeed, in most NMR applications χ_n or γ are assumed to be known (as measured by other techniques or by reference to a well known material) and one is interested in the difference between the actual field at the nucleus H_i and the external field H_0 which is called the hyperfine field HF. In non-magnetic materials H_i is proportional to H_0 and it is more convenient to consider the ratio HF/H_0 (called the Knight shift K) as it is independent of H_0. It is obvious from the above equations that HF or K can be measured (once γ is known) not only by measuring the nuclear magnetization (the NMR signal intensity) but much more accurately by measuring the resonance frequency. For the reasons stated in Sect. 12.1 this frequency can be very accurately measured and the ultimate instrumental resolution is limited only by the inhomogeneity of the external field (resolutions down to 10^{10}eV can be achieved).

12.2.2 Thermal Equilibrium and Dynamic Properties of the Spin System

In the foregoing we already implied that the spin system obeyed Boltzmann's law at the temperature T of the sample. At thermal equilibrium the transverse component of the individual spins occupies equally all the directions in the transverse plane because of the energy invariance against rotations around H_i and the net nuclear magnetization lies along H_i. This implies the existence of mechanisms that can thermalize the nuclear spin system after any perturbation such as the onset of H_0 or a H_1 pulse. The mechanisms by which the populations of the nuclear levels equilibrate arise from the coupling between the nuclei and the other degree of freedom of the sample (the other nuclei and the lattice: electrons, phonons). Without yet going into the details of their microscopic origin we can classify these mechanisms in two groups:

i) Mechanisms which allow the longitudinal magnetization $M_{||}$ to reach its expectation value M_n at the temperature T of the lattice (longitudinal or spin-lattice relaxation). This takes place in a time scale T_1 : the spin-lattice relaxation time. The establishment of $M_{||}$ occurs through quantized rotations towards H_i (spin-flips) which require quantized energy transfers $\hbar\omega_L$ to or from an energy reservoir (the lattice). Hence the relaxation rate T_1^{-1} is proportional to the spectral density of energy fluctuations (excitations) in the lattice at the Larmor frequency.

ii) Mechanisms which allow the transverse component $M_\perp$ of the total magnetization to vanish (transverse relaxation). This takes place in a time scale T_2 : the transverse relaxation time. The decay of $M_\perp$ occurs either through quantized rotations towards H_i (obviously) or random rotations around H_i of the transverse components of the individual spins. The former mechanism has the same origin as for T_1, while the latter only needs arbitrarily small energy transfers and is therefore proportional to excitations around zero frequency. One important cause of the transverse relaxation in solids is the dipolar interaction between nuclear spins themselves which do not contribute to T_1 (simultaneous nuclear spin-flips obviously do not change $M_{||}$); this mechanism usually makes T_2 much shorter than T_1 in solids, which explains why T_2 is often called the spin-spin relaxation time.

The relaxation times T_1 and T_2 can be measured directly in a pulsed NMR experiment by the time evolution of the signal intensity and not only through line broadening. This is another advantage of the technique in the sense that it makes it possible to distinguish between dynamic effects (relaxation) and static effects (due to inhomogeneities) in the broadening.

12.2.3 Electric Interaction – Nuclear Quadrupole Moment

So far we have assumed that the nuclear levels were split only by the Zeeman interaction. Actually nuclei with a spin higher than 1/2 possess an electric quadrupole moment Q which interacts with a possible electric field gradient (EFG) in the material (if the point symmetry of the nuclei site is lower than cubic). Although the splitting of the levels due to this interaction alone (i.e., in the absence of any magnetic field) can be measured by the same technique – (renamed nuclear quadrupole resonance (NQR) for this purpose – we shall consider later that the electric quadrupole interaction acts as a perturbation with respect to the Zeeman interaction (high field NMR approximation). The energy difference between two adjacent levels now depends on m and the NMR absorption spectrum consists of 2I lines (Fig. 12.2). From the spacing between these lines the EFG at the nuclei can easily be deduced. The EFG being a second-rank tensor is defined by 6 parameters: 3 Euler angles which define the orientation of its principal axis, and its 3 principal components V_{zz}, V_{yy}, V_{xx}. The tensor being traceless according to the Poisson law ($V_{zz} + V_{yy} + V_{xx} = 0$) the last three can be reduced to two: the stronger one V_{zz} which is measured in NMR by the quadrupole frequency

$$\nu_Q = \frac{3eQV_{zz}}{h \cdot 2I(2I-1)} \tag{12.3}$$

(a definition which differs slightly from that used in PAC) and the asymmetry parameter

$$\eta = |(V_{yy} - V_{xx})/V_{zz}|; \quad 0 \leq \eta \leq 1 \tag{12.4}$$

which measures the deviation from axial symmetry.

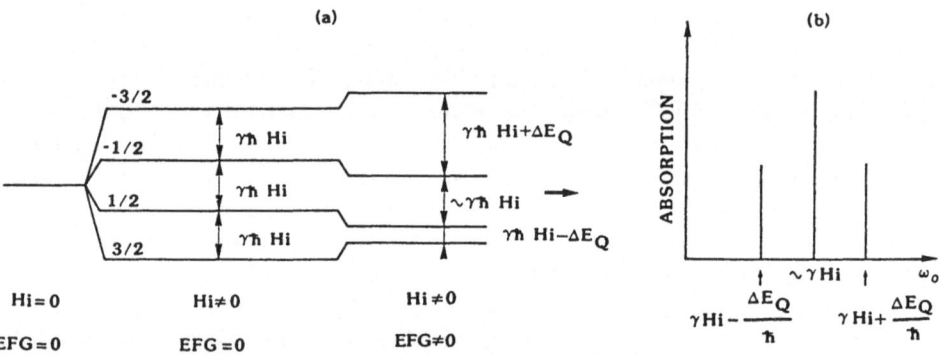

Fig. 12.2. (a) Perturbation of the nuclear energy levels scheme by an electric field gradient (I = 3/2); ΔE_Q is dependent of the relative orientations of the field and the EFG (Sect. 12.4); (b) Basic quadrupole perturbed NMR spectrum

It should be noted that unless very low temperatures are used, so that the population differences between adjacent levels depend on m, the sign of the quadrupole interaction cannot be deduced by NMR contrary to other methods such as the Mössbauer effect.

12.2.4 Summary

We have defined above, on a very general basis, the principal set of raw outputs of an NMR experiment which are related to the interactions between the observed nuclei and their electronic and nuclear environment:

- the hyperfine field HF caused by the magnetic moments of the particles surrounding the nuclei,
- the quadrupole frequency ν_Q and the asymmetry parameter η which measure the EFG at the nuclei caused by the surrounding charges,
- the relaxation times T_1 and T_2 related to the time fluctuations of the HF and the EFG (essentially the former in metals).

All these data originate from interactions which are essentially local, hence they yield information on the site properties on a microscopic scale.

In the application of NMR to phase analysis these quantities as well as the spectrum shapes associated with each interaction (Sect. 12.4) are used directly as "finger prints" of the material in question and the intensity of the NMR signal will be used to measure the abundance of each phase (see examples in Sect. 12.5). As an ultimate possibility from this point of view even structural subunits (submolecules, local atomic configurations) can be identified and counted. This is commonly done in organochemistry.

As far as one is concerned with the electronic and magnetic properties of the materials, the above outputs have to be analyzed in terms of their microscopic origin so that information can be deduced on local densities of electronic states, local magnetic susceptibilities and moments, electronic spin fluctuations...). These topics will be introduced in Sect. 12.4 and examples given in Sect. 12.6.

12.3 Basic NMR Experiment – Principles and Setup

12.3.1 Spin Movement in a Magnetic Field

The peculiarity of NMR already mentioned in Sect. 12.1 results from the fact that the electromagnetic radiation which is involved in a resonant absorption lies in such a frequency range that:

- i) the frequency can be very accurately defined,
- ii) at such a frequency the number of photons per unit time is extremely high even for a small electromagnetic field; hence the amplitude and the phase of H_1 are also accurately defined without violating the uncertainty principle and H_1 can be treated as a classical field,

iii) the number of nuclei involved in the experiment is also very large and the net nuclear magnetization can thus be treated as a classical quantity (for non-interacting spins only – exact quantum mechanical calculations can be found in [12.1,2]).

We are thus introduced to the classical description of the motion of a magnetic moment in a uniform steady magnetic field and a rotating (or alternating) field.

Consider a free spin I with a kinetic moment $\hbar I$ and a magnetic moment $M = \gamma \hbar I$. Placed in a magnetic field H_i it experiences a torque $C = M \times H_i = \hbar \cdot dI/dt$ which from the equation of motion follows:

$$dM/dt = \gamma M \times H_i. \tag{12.5}$$

This equation indicates that M precesses about H with an angular velocity $\omega_L = -\gamma H_i$.

Before introducing the rotating field it is useful to consider a frame rotating around H_i with a velocity ω_0. In this frame M precesses around H_i with a velocity $\omega_r = \omega_L - \omega_0$ as if it had experienced the effective field $H_r = -\omega_r/\gamma = H_i + \omega_0/\gamma$ parallel to H_i.

Let us now apply a field H_1 perpendicular to H_i and rotating around it with a velocity ω_0. In the rotating frame this field is static and the moment now experiences an effective field H_{eff} the magnitude of which is $\sqrt{(H_i + \omega_0/\gamma)^2 + H_1^2}$ and which departs from H_i by an angle Φ defined by

$$\Phi = \arctan[H_1/(H_i + \omega_0/\gamma)] = \arctan[-\omega_1/(\omega_0 - \omega_L] \tag{12.6}$$

(where we use $\omega_1 = -\gamma H_1$).

Therefore, in the rotating frame, M precesses now around H_{eff} as shown in Fig. 12.3. Thus M does not depart significantly from its equilibrium position along H_i unless Φ differs significantly from 0. H_1 being usually much smaller than H_i this condition occurs only at resonance (i.e., when ω_0 is close to ω_L within a scale of the order of ω_1).

It is precisely the non-vanishing misorientation of the moment with respect to H_i at the resonance $\omega_0 = \omega_L$ which is detected in the NMR setup by the observation of a rotating transverse magnetization. Usually H_1 is an alternating

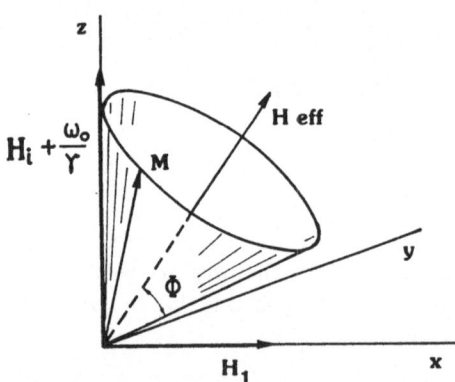

Fig. 12.3. Moment motion about the effective field in the rotating frame $\left[H_{\text{eff}} = (H_i + \omega_0/\gamma) + H_1\right]$

rather than a rotating field. Since any alternating field can be regarded as the sum of two fields rotating with opposite velocities the above discussion is always valid for one component (the other has negligible effects).

This classical description is very helpful in the discussion of transient and dynamic effects in NMR. Particularly at the present time most experiments are performed using the so-called "pulsed NMR" technique in which H_1 is applied only for very short durations and which we shall now basically describe [12.4–8].

12.3.2 Free Induction Decay – Transverse Relaxation Time

The sample is placed in a coil, the axis of which is transverse to the steady field H_0. After a time much longer than T_1 the net nuclear magnetization is along the field. An H_1 pulse at the resonance frequency ω_L is then applied for a duration δ which is assumed to be shorter than T_1 and T_2. Following the above discussion the nuclei experience in the rotating frame the effective field H_1 since $H_i + \omega_L/\gamma = 0$. Hence during the pulse the spins rotate around H_1 with an angular velocity ω_1. After H_1 is switched off M_n departs from H_i at an angle $\Phi = \gamma H_1 \delta$. If δ is such that $\Phi = 90°$ (the pulse is then called a 90° pulse) M_n now lies in the transverse plane, the flux through the coil is maximum and a maximum electromotive force is induced which is proportional to $\omega_L \cdot M_n$ at the Larmor frequency.

If the spins were free, this signal would last forever but actually relaxation mechanisms take place and the transverse magnetization which is observed decays in a time scale T_2. One way to describe this relaxation phenomenon is to consider that, because HF has time fluctuating components, the Larmor frequencies of the individual spins fluctuate randomly. Therefore the individual components of M_n, which were aligned immediately after the pulse, spread progressively in the transverse plane resulting in induction decay. It is also obvious that the fluctuations of the Larmor frequencies correspond to a line broadening. Indeed and more generally it can be proved that free induction decay (FID) is the Fourier transform of the lineshape. Although on a phenomenological basis the decay is often considered as exponential (which gives an operational definition to T_2) this is true only for Lorentzian line shapes which is not always the case (Sect. 12.4).

12.3.3 Spin Echo – Homogeneous and Inhomogeneous Broadenings

The spread of the Larmor frequencies discussed above results from the time-fluctuating fields at the nucleus. In general, the related FID can be considered as irreversible and corresponds to the so-called "homogeneous broadening" of the NMR line. However, there are also time-independent causes for the spread of Larmor frequencies as field inhomogeneities obviously are, (spatial fluctuations). These are responsible for the "inhomogeneous broadening" of the line and also cause a decay of the free induction that can be very short, for example in disordered materials, concentrated alloys etc. This may make the FID unobservable because of the unavoidable dead time of the receiver following the strong RF pulse.

Let us assume that the line broadening is due only to spatial static inhomogeneities. Following a 90° pulse the FID lasts for a time of the order of the inverse line width. After a time τ the static field is instantaneously reversed: this reverses the direction of the various Larmor velocities without changing their magnitude. Therefore the field reversal can be considered as a time reversal for the spin precessions. Consequently, the individual components of the transverse magnetization that had defocused during the first time interval τ will refocus after a second delay τ giving rise to a so-called "spin echo".

If the distribution of Larmor frequencies is due only to static field distributions (no relaxation effect) the amplitude of the spin echo is the same as the initial FID and its shape is that of two FID back to back. In the presence of relaxation effects the amplitude of the echo is reduced by a factor $\exp(-2\tau/T_2)$ with respect to the FID signal. This provides a means of measuring the transverse relaxation (by varying τ) even in the presence of strong inhomogeneous broadening.

Actually the "time reversal" is not achieved by a field reversal but by a second pulse twice as long as the first one (180° pulse) which reverses the whole spin system with respect to H_i. This obviously has the same consequences as reversing the field but it is much more easily achieved in a short time (Fig. 12.4).

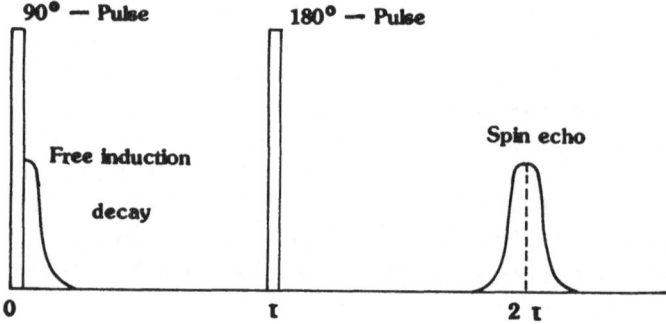

Fig. 12.4. Free induction decay following a 90° pulse and spin echo following a 90°-τ-180° pulse sequence

A great variety of pulse sequences can generate an echo (and even multiple echos) when the spread of resonance frequencies is due to static inhomogeneities (for example, a spatial distribution of EFG) or has slowly varying causes (for example, the slow diffusion of the atoms). Here spin-echo experiments provide a means of measuring in these examples the EFG distribution or the diffusion coefficient, respectively. Lack of space does not allow us to review all multiple-pulse techniques that are nowadays used [12.2,7,8]; however in the case of metals most of them have limited applications.

Both of the NMR signals described above provide two fundamental pieces of information:

i) the magnitude of the longitudinal nuclear magnetization before the application of the pulses (the FID or echo integral intensity),

ii) the spectral shape which is the Fourier transform of the signals shape (this statement has limitations that we shall see later).

At this point we should like to describe briefly the two basic types of NMR measurements: T_1 and spectrum observation.

12.3.4 Spin-Lattice Relaxation Measurement

The nuclear spin system being at thermal equilibrium ($M_\parallel = M_n$; $M_\perp = 0$) at 90° pulse is applied which cancels the longitudinal magnetization ($M_\parallel = 0$; $M_\perp = M_n$). $M_\parallel$ now relaxes towards equilibrium. After a time interval t the longitudinal magnetization $M_\parallel(t)$ is measured by FID or spin-echo production. If we repeat the experiment for various t the full time recovery curve of $M_\parallel$ from 0 to M_n is obtained. This recovery is usually exponential and reads

$$M_\parallel(t) = M_n \cdot [1 - \exp(-t/T_1)] \qquad (12.7)$$

which measures T_1.

12.3.5 Spectrum Measurement

Since the FID (or the echo) signal is the Fourier transform of the line shape, in principle it is easy, by means of back Fourier transformation, to obtain the spectrum. In this method, besides the exernal field inhomogeneity, the resolution is limited only by the procedure used in the time analysis of the signals (high-resolution NMR).

However, we have so far implicitly assumed that all the spins participating in the spectrum have experienced a 90° rotation during the 90° pulse. This is valid only if the spectral power density of H_1 (i.e.,.the Fourier transform of the pulse) is flat in the range of all Larmor frequencies. This condition is fulfilled if $\omega_1 = \gamma H_1$ is large with respect to the spectrum width or in other words if the duration of the pulses is much shorter than the inverse spectrum width. This is not always achievable because of RF power limitations: in solids and particularly in metals, inhomogeneities, quadrupole interaction and other sources of broadening may yield a spectrum width much larger than any technically achievable H_1. In such a case the observed shape of the signals can no longer be simply analyzed by Fourier transformation. However, the signal amplitude is then proportional to, crudely speaking, the nuclear spectral density within a frequency window of width ω_1 centered on the pulse frequency (or any narrower window resulting from band-pass filtering in the signal amplifier). Then the spectrum is obtained by sweeping the frequency of the pulse or, more commonly, by sweeping the external field at a constant pulse frequency and recording the FID or the echo intensity. The energy resolution in such an experiment is of the order of ω_1 or the receiver band pass, whichever is the narrowest.

12.3.6 NMR Techniques and Instruments

To describe all possible NMR detection techniques and instruments would require an entire book. Briefly they can be classified in two groups:

i) Continuous or Quasi-Static Techniques: in this group the fields (both the steady and the RF fields) are continuously applied and have slowly varying amplitudes and/or frequencies. The NMR can be detected either by nuclear induction (continuous wave spectrometers) or by RF power absorption (Q meters, marginal oscillators, superregenerative receivers).

The former are based on the electromotive force created in the sample coil by the small transverse nuclear magnetization which occurs in the presence of a small RF field. They have been widely used during the past 30 years and are still in use for wide line experiments.

The latter are based on the power loss of the coil at resonance and some of them provide convenient ways of sweeping the frequency and exhibit fairly good sensitivities (at the expense of signal purity). They are used for seeking unknown resonances or in situations where it is not convenient to sweep the external field.

ii) Transient Techniques: here again there are two types of setup, some using sudden or rapid changes of H_0 (field cycling, fast passage), others using a pulsed RF field. In addition to the possibility of spectrum observation of the continuous techniques,they offer the possibility of measuring directly time-dependent effects and relaxation times.

The former provide a means of studying the nuclear-spin system in a field which is different from the one in which the NMR is performed (for example, to study the relaxation in zero field).

With the latter a very wide variety of measurements can be made by playing tricks with the spin system (and even several unlike spin systems) the simpler ones have already been seen and therefore we shall now describe only the phase coherent pulsed NMR spectrometer.

12.3.7 Phase Coherent Pulsed NMR Spectrometer

In addition to the electromagnet (conventional or superconducting) which generates the steady field, a pulsed NMR spectrometer consists of five blocks (Fig. 12.5):

- a pulse generator (A), which builds up the pulse sequences, durations and time intervals,
- a multiphase RF generator (B), which provides a continuous highly stable radio frequency wave (CW) with four phases $(0°,90°,180°,270°)$ so that H_1 can be generated along the four directions x, y, −x, −y in the rotating frame,
- a high-power transmitter (C), which energizes the sample coil with the RF field H_1 during the pulses; it is fed by the CW RF generator through gates that are driven by the pulse generator,

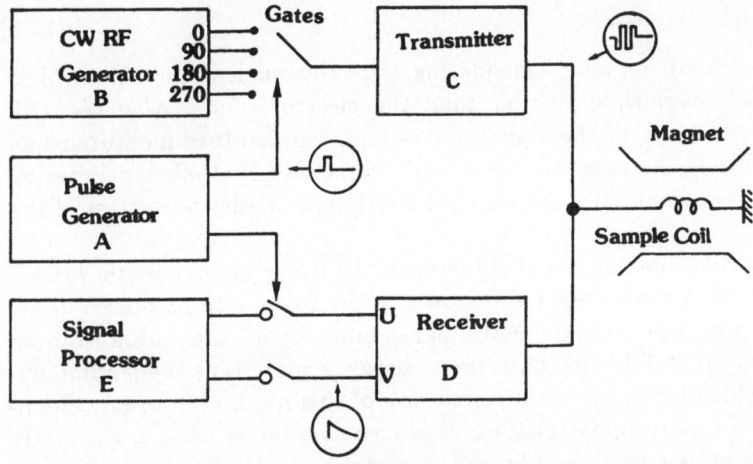

Fig. 12.5. Synoptic diagram of a phase coherent pulsed NMR spectrometer

- a receiver (D) by which the nuclear induction signal is amplified, filtered and detected synchronously in two orthogonal directions u and v of the rotating frame,
- a gated signal processor (E) fed by the detected signals u, v which performs the time analysis and the Fourier transform or a simple sampling for field/frequency/time swept experiments.

The state-of-the-art spectrometers are now operated by minicomputers which perform pulse sequence programming, frequency and/or field setting, data collection and analysis.

Accessories to this standard setup are:

- Field homogeneity and stability regulations (using another narrow NMR signal),
- Sample spinning systems which average out the residual inhomogeneity of the magnet or even the static distributions of Larmor frequencies due to orientation-dependent interactions in polycrystals such as electric quadrupole or magnetic dipole interactions (magic angle spinning [12.8],
- Cryogenic systems: low temperatures are often necessary in NMR to increase the sensitivity.

12.3.8 Feasibility of an NMR Observation

At this stage we have seen that NMR has many interesting features: very high resolution, the possibility of direct measurement of relaxation phenomena, simple instrumentation, no need for nuclear physics facilities, to which it should be added that, with the exception of cerium and argon among the stable isotopes, any element can, in principle, be observed.

In practice, the major drawback of NMR is its low sensitivity: as already seen, the signal is proportional to $\omega_L \cdot M_n$ i.e. to

$$N \cdot \gamma^3 \hbar^2 H_i^2 I(I+1)/3k_B T \qquad (12.8)$$

which is a very small number considering that the nuclear moment $\gamma \hbar I$ is 3 to 4 orders of magnitude smaller than the electronic one. Also the $1/T$ temperature dependence of the signal makes high-temperature measurements difficult. Fortunately the resonance is usually narrow so that efficient filtering and noise reduction techniques can be used which have made the success of the method possible.

Since the achievable external field is up to $10\,\mathrm{T}$ the gyromagnetic ratio γ and to a lesser extent the natural abundance of the isotope to be observed are determining factors. The relevant NMR parameters for all the stable isotopes are listed in Table 12.2. The proton is immediately apparent as the leading nucleus in NMR which explains the universal use of this method in organochemistry, biology and now in body imaging. It is also commonly used in the study of those interesting new materials known as metal hydrides.

Table 12.2. The relevant NMR parameters for all the stable isotopes

A	B	C	D	E	F	G
-1 H	42.57	1/2		99.98	100,00	100.00
-2 H	6.537	1	0.003	0.02	2E-04	
-3 He	32.43	1/2		0.00	E+00	E+00
-6 Li	6.263	1		7.43	6E-02	
-7 Li	16.54	3/2	0.025	92.57	27.18	10.87
-9 Be	5.984	3/2	0.020	100.00	1.39	0.56
-10 B	4.574	3	0.092	18.83	0.37	
-11 B	13.66	3/2	0.036	81.17	13.41	5.37
-13 C	10.70	1/2		1.11	2E-02	2E-02
-14 N	3.074	1	0.045	99.64	0.10	
-15 N	4.314	1/2		0.37	4E-04	4E-04
-17 O	5.771	5/2	0.004	0.04	E1-03	3E-04
-19 F	40.05	1/2		100.00	83.28	83.28
-21 Ne	3.363	3/2		0.26	6E-04	3E-04
-23 Na	11.26	3/2	0.100	100.00	9.26	3.70
-25 Mg	2.606	5/2		10.05	3E-02	7E-03
-27 Al	11.09	5/2	0.150	100.00	20.64	5.31
-29 Si	8.458	1/2		4.70	4E-02	4E-02
-31 P	17.23	1/2		100.00	6.64	6.64
-33 S	3.264	3/2	0.060	0.74	2E-03	7E-04
-35 Cl	4.172	3/2	0.060	75.40	0.35	0.14
-37 Cl	3.472	3/2	0.070	24.60	7E-02	3E-02
-39 K	1.986	3/2	0.100	93.08	5E-02	2E-02
-41 K	1.092	3/2		6.91	6E-04	2E-04
-43 Ca	2.866	7/2		0.13	8E-04	2E-04
-45 Sc	10.34	7/2	0.200	100.00	30.14	5.74
-47 Ti	2.398	5/2		7.75	2E-02	4E-03
-49 Ti	2.403	7/2		5.51	2E-02	42-03
-50 V	4.243	5/2		0.24	3E-03	7E-04
-51 V	11.19	7/2	0.250	99.76	38.07	7.25
-53 Cr	2.407	3/2		9.54	9E-03	3E-03
-55 Mn	10.55	5/2	0.500	100.00	17.77	4.57
-57 Fe	1.381	1/2		2.25	8E-05	8E-05
-59 Co	10.10	7/2	0.500	100.00	28.08	5.35
-61 Ni	3.789	3/2		1.25	4E-03	2E-03
-63 Cu	11.28	3/2	0.155	69.09	6.44	2.58
-65 Cu	12.09	3/2	0.145	30.91	3.54	1.42
-67 Zn	2.663	5/2	0.180	4.12	E1-02	3E-03
-69 Ga	10.21	3/2	0.210	60.20	4.16	1.66
-71 Ga	12.98	3/2	0.130	39.80	5.65	2.26
-73 Ge	1.485	9/2	0.220	7.61	1E-02	2E-03
-75 As	7.289	3/2		100.00	2.51	1.00
-77 Se	8.131	1/2		7.50	5E-02	5E-02
-79 Br	10.66	3/2	0.300	50.57	3.98	1.59
-81 Br	11.50	3/2	0.280	49.43	4.87	1.95
-83 Kr	1.641	9/2	0.155	11.55	E2-02	3E-03
-85 Rb	4.110	5/2	0.280	72.80	0.76	0.20
-87 Rb	13.93	3/2	0.135	27.20	4.77	1.91
-87 Sr	18.44	9/2		7.02	18.85	2.86
-89 Y	2.086	1/2		100.00	1E-02	1E-02
-91 Zr	3.959	5/2		11.23	0.11	3E-02
-93 Nb	10.40	9/2	0.280	100.00	48.21	7.30
-95 Mo	2.772	5/2		15.78	5E-02	1E-02
-97 Mo	2.833	5/2		9.60	3E-02	8E-03
-99 Ru	1.892	5/2		12.81	1E-02	3E-03
101 Ru	2.081	5/2		16.98	2E-02	6E-03
103 Rh	1.338	1/2		100.00	3E-03	3E-03
105 Pd	1.740	5/2		22.23	2E-02	5E-03
107 Ag	1.721	1/2		51.35	3E-03	3E-03
109 Ag	1.982	1/2		48.65	5E-03	5E-03
111 Cd	9.026	1/2		12.86	0.12	0.12
113 Cd	9.442	1/2		12.34	0.13	0.13
113 In	9.309	9/2	0.960	4.16	1.44	0.22
115 In	9.328	9/2	0.980	95.84	33.28	5.04
117 Sn	15.17	1/2		7.65	0.35	0.35
119 Sn	15.87	1/2		8.68	0.45	0.45

A	B	C	D	E	F	G		A	B	C	D	E	F	G
121 Sb	10.18	5/2	0.850	57.25	9.16	2.36		165 Ho	7.237	7/2	2.000	100.00	10.32	1.97
123 Sb	5.515	7/2	1.090	42.75	1.95	0.37		167 Er	1.040	7/2	9.990	22.82	7E-03	1E-03
123 Te	11.15	1/2		0.89	2E-02	2E-02		169 Tm	3.491	1/2		100.00	6E-02	6E-02
125 Te	13.44	1/2		7.03	0.22	0.22		171 Yb	7.512	1/2		14.27	8E-03	8E-02
127 I	8.519	5/2	0.680	100.00	9.35	2.40		173 Yb	2.081	5/2	2.100	16.08	2E-02	6E-03
129 Xe	11.77	1/2		26.24	0.56	0.56		175 Lu	4.872	7/2	6.100	97.40	3.07	0.58
131 Xe	3.491	3/2	0.120	21.24	6E-02	2E-02		176 Lu	5.298	6	8.000	2.60	0.28	
133 Cs	5.586	7/2	0.015	100.00	4.75	0.90		177 Hf	1.277	7/2	3.000	18.39	1E-02	2E-03
135 Ba	4.229	3/2		6.59	3E-02	1E-02		179 Hf	.8042	9/2	3.000	13.78	3E-03	5E-04
137 Ba	4.730	3/2		11.32	8E-02	3E-02		181 Ta	5.109	7/2	5.500	100.00	3.63	0.69
138 La	5.615	5	2.700	0.09	8E-03			183 W	1.750	1/2		14.28	1E-03	1E-03
139 La	6.012	7/2	0.600	99.91	5.91	1.13		185 Re	9.584	5/2	2.850	37.07	4.93	1.27
141 Pr	11.94	5/2		100.00	25.80	6.63		187 Re	9.683	5/2	2.650	62.93	8.64	2.22
143 Nd	2.720	7/2		12.20	7E-02	1E-02		189 Os	3.306	3/2	2.000	16.10	4E-02	2E-02
145 Nd	1.703	7/2		8.30	E1-02	2E-03		191 Ir	.8136	3/2	1.350	38.50	1E-03	5E-04
147 Sm	1.513	7/2	0.720	15.07	E1-02	3E-03		193 Ir	.8609	3/2	1.250	61.50	3E-03	1E-03
149 Sm	1.182	7/2	0.720	13.84	6E-03	1E-03		195 Pt	9.153	1/2		33.70	0.33	0.33
151 Eu	10.48	5/2	1.200	47.77	8.33	2.14		197 Au	.7332	3/2	0.580	100.00	3E-03	1E-03
153 Eu	4.636	5/2	2.500	52.23	0.79	0.20		199 Hg	7.616	1/2		16.86	0.10	0.10
155 Gd	1.182	3/2	1.100	14.86	2E-03	6E-04		201 Hg	2.800	3/2	0.550	13.24	2E-02	8E-03
157 Gd	1.703	3/2	1.000	15.64	5E-03	2E-03		203 Tl	24.32	1/2		29.52	5.51	5.51
159 Tb	7.710	3/2		100.00	2.97	1.19		205 Tl	24.57	1/2		70.48	13.55	13.55
161 Dy	1.182	5/2		18.73	5E-03	1E-03		207 Pb	8.898	1/2		21.11	0.19	0.19
163 Dy	1.608	5/2		24.97	2E-02	4E-03		209 Bi	6.840	9/2	0.400	100.00	13.69	2.07
A	B	C	D	E	F	G		A	B	C	D	E	F	G

A: Isotope; B: Larmor frequency [MHz] in a 1T field; C: Spin; D: Quadrupole moment [barns]; E: Natural abundance [%]; F: NMR sensitivity relative to proton sensitivity (100) in the same field and for one mole of the natural element; G: Same as F but for the central line ($1/2 \rightarrow -1/2$ transition) only (quadrupole split spectrum)

Among the metals the following offer a reasonable sensitivity:

Simple metals: Li, Na, Al, Ga, Rb, Sr, In, Sb, Tl, Bi
(Be, K, Sn, Pb, Cs, Ba)

Transition metals: Sc, V, Mn, Co, Cu, Nb, La, Re
(Y, Zr, Mo, Cd, Ta, Os, Pt, Hg)

Rare earths: Pr, Eu, Tb, Ho
(Nd, Tm, Yb, Lu)

Among the metalloids often present in metallic materials B, P, As and to a lesser extent C, Si, Se, Te are rather easily observed.

Besides purely nuclear parameters, other limitations to NMR observation are caused by the properties of the material. They can be described in terms of accessible time and/or frequency windows:

I.   The length of the relaxation times T_1 and T_2 is critical: if it is too short (10 μs is the lower limit) the nuclear magnetization will have relaxed by the time of observation; if it is too long (1 min is long, 1 hour is exceedingly long) the experiment cannot be performed within a reasonable time.

II. The spectrum width $\Delta\omega$ should not be too broad. Indeed, as we have seen, the apparatus has a limited observation window the width of which is ω_1; hence its sensitivity is reduced by an amount of the order of $\omega_1/\Delta\omega$.

This second category of limitations makes the possibility of NMR observation rather unpredictible since they depend to a great extent on precisely those material properties which are under investigation.

Generally, and taking into account the Curie law of nuclear magnetization, a reasonable estimate of the lowest observable number of nuclei resonating in the instrumental window is 10^{17} at 1 K, in a field of one Tesla and for a frequency of 10 MHz.

Finally there is a category of materials in which the field at the nuclei might be significantly different from the external field. This is the case for magnetically ordered materials in which the spontaneous electronic magnetization creates a hyperfine field even in zero external field, and for non−s state rare earths in which the unquenched orbital moment induces a HF which can be up to many orders of magnitude stronger than the external field. In such cases, owing to the high resolution of NMR and the necessarily wide field or frequency scan in the search for resonance, the signal can easily be missed. It is then more convenient to use other techniques such as the Mössbauer effect, PAC or hyperfine specific heat to have a first estimation of the field.

It should also be noted that because of skin-depth effects in metals the samples must be in powder or thin-film form. This makes any non-destructive investigations and measurements on single crystals very difficult.

12.4 NMR Outputs − Microscopic Origin

So far we have introduced the outputs of an NMR experiment without specifying in detail their microscopic origin and what information on the material properties can be obtained. It is the aim of this section to specify these points as an introduction to the examples given in Sect. 12.6 on the study of the electronic and magnetic properties of metals.

12.4.1 Hyperfine Fields

As we have seen before, the nuclei of the solid experience a field which differs from the external field by the hyperfine field HF created by its electronic − and possibly its nuclear − surroundings [12.9,10]. The characteristic time between electronic transitions usually being much shorter than the nuclear Larmor period, HF can reasonably be separated into its thermal average $\langle HF \rangle$ and fluctuating components hf. In this approximation $\langle HF \rangle$ is responsible for resonance-frequency shifts and hf which, crudely speaking, induces random rotations or spin-flips of the nuclei moments, is responsible for the relaxation phenomena.

Considering the variety of possible interaction mechanisms between the nucleus and the other magnetic moments in the material HF is often written as a sum

$$HF = \sum_i HF_i = \sum_i A_i M_i \tag{12.9}$$

where the A_i's are hyperfine coupling constants and the M_i's the relevant magnetic moments associated with the different interaction mechanisms i. Quite generally, the A_i's are tensors which can be separated into traceless symmetrical tensors and scalars corresponding, respectively, to the anisotropic and the isotropic parts of the interaction. Like HF, M_i can often be separated into its thermal average $\langle M_i \rangle$ and fluctuating m_i parts.

The first interaction to consider between a nucleus and another distant magnetic moment $\gamma_m \hbar J$ is the direct dipole interaction:

$$\boldsymbol{HF_m} = \gamma_m \hbar [3(\boldsymbol{J \cdot r})\boldsymbol{r}/r^5 - \boldsymbol{J}/r^3] \tag{12.10}$$

which is purely anisotropic. Although they are formally different, other anisotropic interactions have basically the same effects on the NMR (they are often called pseudo-dipole).

The most important electron-nucleus interactions in metals are those involving conduction electrons, the isotropic – scalar – part of which we shall now review.

For s electrons the dominant hyperfine mechanism is associated with the difference between the spin up and down populations in the nucleus volume, the contact interaction:

$$A_s = 8\pi/3 \langle |\psi(0)|^2 \rangle \tag{12.11}$$

where $|\psi(0)|^2$ is the electronic density at the nucleus. It can be compared with the isomer shift of the Mössbauer effect although it does not involve the core electrons.

Non−s electrons have a vanishing electronic density at the nucleus and therefore do not contribute to the HF via the contact interaction but rather through the field created by their angular momentum, the orbital interaction:

$$A_{orb} = 2 \langle r^{-3} \rangle. \tag{12.12}$$

Also non−s electron spins can interact with the nucleus in an indirect way through the spin polarization of the closed s shells (due to the overlap between polarized non−s wave functions and those of core electrons), the core polarization interaction

$$A_{cp} = -8\pi/3 \langle |\psi_{cp}(0)|^2 \rangle \tag{12.13}$$

where $|\psi_{cp}(0)|^2$ involves the Coulomb exchange integrals between non−s and all s electrons.

The spin and orbital interactions can only be considered separately in the case of negligible spin-orbit coupling which is not true for f electrons (rare earth elements) but these electrons are usually localized.

The order of magnitude of the hyperfine coupling constants A_s and A_{orb} is about $100\,T$ per Bohr magneton, A_{cp} being much smaller ($-10\,T$ for $3\,d$ elements).

In addition to these interactions, localized nuclear or electronic moments also create a field at the nucleus site.

As far as other nuclei are concerned the dipole interaction is the only important one. The case of localized electronic moments is more complicated. Actually several interactions with the nuclei can take place together with the direct dipole interaction. These are indirect interactions mediated by the conduction electrons and can be scalar or pseudo-dipole. The relative importance of the direct versus the indirect interactions depends strongly on the observed element: for light elements it is expected and commonly observed that the dipole interaction plays the dominant role while for heavy elements indirect interactions dominate. The reason for this is the high electronic density at the nucleus for the latter elements.

For all these interactions – if they are not local – a summation over the lattice sites or a wave number average should be used (lattice of localized moments, conduction electrons).

12.4.2 Frequency Shifts

In non-magnetic substances the thermal average $\langle HF \rangle$ is usually small as compared with the external field H_0. Here the only efficient component of $\langle HF \rangle$ is that along $\boldsymbol{H}_0$ (say the z direction). It leads to a resonance frequency shift with respect to that of the nucleus in vacuum $\langle HF \rangle_z / H_0$ [12.11].

This shift is very weak in non-magnetic insulators (a few tens of ppm); it is caused by the very small magnetic response of the closed electronic shells (chemical shifts). In metallic systems, on the contrary, the presence of open shells leads typically to shifts in the range of a few per cent (Knight shifts). In accordance with the above notations these shifts (K_i) can be written as

$$K_s = A_s \langle M_s \rangle_z / H_0 = A_s \chi_s \tag{12.14}$$

for s conduction electrons, χ_s being their spin contribution to the static susceptibility, and

$$K_{cp} = A_{cp}\chi_p \quad \text{and} \quad K_{orb} = A_{orb}\chi_{vv} \tag{12.15}$$

for non−s conduction electrons with χ_p being their spin contribution and χ_{vv} the Van Vleck orbital contribution to the static susceptibility.

In this section χ_i's refer to a magnetic susceptibility per atom.

It should be noted that considering the huge magnitude of the A_i's the weakness of K is due only to that of the electronic susceptibilities. Spin susceptibilities are related to the partial (s, p, d) electronic densities of states at the Fermi level and thus K (and T_1 as we shall see) measurements provide a means of tracing them in metals (electronic structure studies). The orbital susceptibility has a more complex origin: in condensed matter the orbit of any electron precesses in the electric field of the other charges which averages out its orbital moment; then the orbital magnetism arises only from secondary effects of the external magnetic field.

As far as other electronic moments are concerned the associated shifts can similarly be written

$$K_i = A_i \chi_i^{zz}(0) \tag{12.16}$$

where $\chi_i^{zz}(0)$ is the real longitudinal static susceptibility corresponding to the moment [(0) stands for zero frequency]. Hence K_i measures selectively the susceptibility of the observed atoms.

Until now we have treated only the shifts arising from the scalar part of the interactions (isotropic shift K_{iso}). In cubic crystals the shifts do not depend on the field orientation (for symmetry reasons). For lower symmetries the anisotropic interactions give rise to an anisotropic Knight shift K_{an}, the principal components of which are K_z, K_y, K_x so that

$$K_z + K_y + K_x = 0; \quad K_{an} = K_z; \quad \varepsilon = (K_y - K_x)/K_z. \tag{12.17}$$

The resonance frequency then reads

$$\omega_L = \gamma H_0 \cdot [1 + K_{iso} + \frac{1}{2} K_{an}(3 \cos^2\theta - 1) + \frac{1}{2}\varepsilon K_{an} \sin^2\theta \cdot \cos 2\varphi] \tag{12.18}$$

where θ and φ are the polar and azimuthal angles, respectively, of the external field $\boldsymbol{H_0}$ in the principal axis frame of the anisotropic interaction. Since most measurements are performed on polycrystals the orientation of the field in the crystals is randomly distributed and the so-called "powder patterns" shown in Fig. 12.6 are observed from which the values for K_{iso}, K_{an} and ε are deduced.

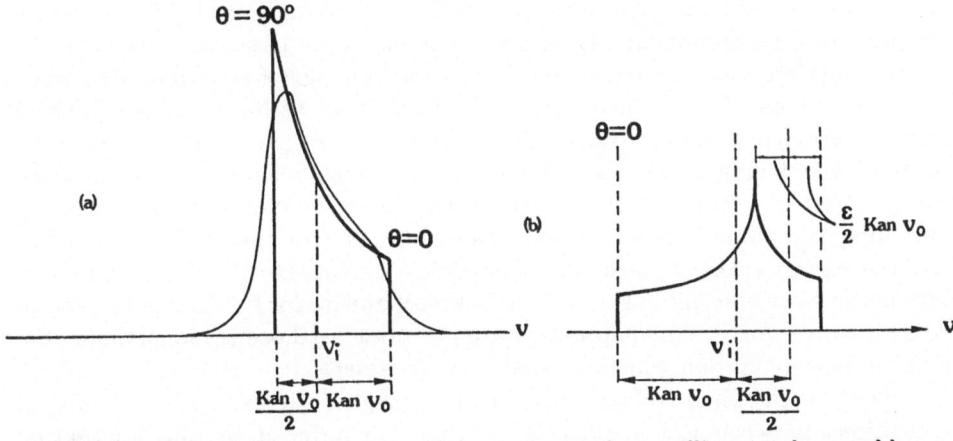

Fig. 12.6a,b. NMR spectra – "powder patterns" – in a polycrystalline sample caused by an anisotropic magnetic interaction. $[\nu_0 = \gamma H_0/2\pi; \nu_i = \nu_0(1 + K_{iso})]$. (a) Axial site symmetry (shown for $K_{an} > 0$); *thin line:* broadened spectrum; (b) non-axial site symmetry (shown for $K_{an} < 0$)

In magnetically ordered materials the existence of a spontaneous electronic magnetization means that at the nuclei scale there exists a polarization of the core and conduction electron spins and orbital momenta even without an external field. This polarization creates a hyperfine field which, as for the Knight shifts, is caused by the contact interaction with s electrons (HF_s), the core polarization (HF_{cp}) and the orbital interaction (HF_{orb}). This field is usually very strong and NMR can be observed in it (zero-field NMR). The resonance frequency then yields directly the hyperfine field. If there is a distribution of HF (several crystallographic and/or magnetic sites, several phases) the NMR spectrum represents this distribution directly. This situation is simpler than in the case of the Mössbauer effect which requires deconvolution procedures to obtain the HF distribution from the spectra. However, in ferromagnetic materials H_1 and the NMR signals are enhanced by the bulk permeability of the sample which leads to difficulties in the quantitative determination of phase or site abundances.

12.4.3 Relaxation Times

As we have seen in Sect. 12.2 longitudinal relaxation concerns the return of longitudinal nuclear magnetization towards its thermal equilibrium value M_n at a lattice temperature T; it involves quantized rotations (spin-flips) towards the static field H_i (say, the z direction) and energy transfers $\hbar\omega_L$. The transverse relaxation relates to the decay of the transverse magnetization through spin-flips or arbitrary rotations around H_i [12.12].

Here we are concerned essentially with relaxation mechanisms arising from the fluctuations hf of the hyperfine field (fluctuations of the EFG are dominant only in non-metallic systems).

A simple sketch of the transverse relaxation that results from random rotations around H_i has already been described in Sect. 12.3: it results from the random modulation of the individual Larmor frequencies of the nuclei by the fluctuating component hf_z of HF. It is easily understood that only the fluctuations around zero frequency (i.e., which change the Larmor frequency for long periods of time) have a significant effect as we have already inferred from the absence of energy transfer in such mechanisms (Sect. 12.2). Among the particles which create a field at a nucleus are the neighboring nuclei of the same isotope; these nuclei play a special role in transverse relaxation because mutual spin-flips (i.e., simultaneous dipole transitions of two spins) can take place without energy transfer to the other particles. This makes the dipole spin-spin interaction very efficient in solids. The last mechanisms for transverse relaxation which involve uncorrelated spin-flips are the same as those responsible for the longitudinal relaxation which we shall now consider.

The longitudinal relaxation can be roughly described in terms of transitions between the nuclear levels that are induced by the fluctuating transverse components hf_x and hf_y of the HF (which act in the same way as H_1 except that they are not coherent). Of course, the only efficient components are those which (roughly) fluctuate at the Larmor frequency. If hf_z and hf_y arise from the fluctuating components m_z and m_y of the moment M_i it can be shown that the relaxation rate (T_{1i}^{-1}) can be written in terms of a generalized frequency dependent susceptibility as

$$T_{1i}^{-1} = k_B T \gamma^2 A_i^2 \text{Im}\{\chi_i^{\perp}(\omega_L)\}/\omega_L \qquad (12.19)$$

where $\text{Im}\{\chi_i^{\perp}(\omega_L)\}$ is the imaginary part of the transverse susceptibility of the moment M_i at the nuclear frequency. A wave-number average should be added for non-local interactions. Thus, while K measures the local static susceptibility, T_1 measures the dynamic susceptibility (thermal or quantum fluctuations of the moments). It should be noted that according to this formula T_1 gives basically the same information as quasi-elastic neutron scattering.

The above expression for T_1 is quite general; however, it is often possible to find a simple relation between the transverse susceptibility and the static longitudinal susceptibility $\chi_i^{zz}(0)$ which leads to expressions of T_1 that are more easily handled:

-   For localized paramagnetic impurities (if their fluctuation spectrum is Lorentzian in shape with a correlation time) one finds

$$T_{1imp}^{-1} = k_B T \gamma^2 A_{imp}^2 \chi_{imp}^{zz}(0) \cdot 2\tau_e/(1 + \omega_L^2 \tau_e^2). \qquad (12.20)$$

-   This relation provides a means to estimate the electronic correlation time from the measurements of the nuclear relaxation time and the static susceptibility.
-   For conduction electrons the s contact interaction and the core polarization interaction lead to the following expressions (if electronic correlations are neglected)

$$T_{1s}^{-1} = k_B T \gamma^2 A_s^2 \chi_s^2 h/2\mu_B^2, \quad T_{1cp}^{-1} = k_B T \gamma^2 F_1 A_{cp}^2 \chi_p^2 h/2\mu_B^2. \qquad (12.21)$$

As far as the orbital interaction with conduction electrons is concerned, there is no simple relation between T_{1orb} and the static orbital susceptibility χ_{vv} but rather between T_{1orb} and the static spin susceptibility χ_p

$$T_{1orb}^{-1} = k_B T \gamma^2 F_2 A_{orb}^2 \chi_p^2 h/2\mu_B^2, \qquad (12.22)$$

(F_1 and F_2 are factors depending on the orbital degeneracy which it is not feasible to specify here). In non-magnetic metals the various contributions to the susceptibility are temperature-independent and T_1 follows the so-called Korringa law $T_1 T = $ constant.

Thus both K and T_1 provide means of tracing independently the various contributions to the susceptibility. Since the hyperfine coupling constants are only roughly estimated, it is often useful to eliminate them by the expressions for T_1 and K. For a given mechanism this yields

$$K_i^2 T_{1i} T = S_i \tag{12.23}$$

which is known as the Korringa product. For s electrons S_s is the Korringa constant $S = 2\mu_B^2/h\gamma^2 k_B$. For non-s electrons $S_{cp} = S \cdot F_1$ which gives information on the orbital occupancy. The Korringa product is also used to measure the importance of electronic correlations which are responsible for the occurrence of magnetism [12.10]. Hence the value of the Korringa product yields information about the dominant contribution to the susceptibility.

In the above general expression of T_1 only the scalar interactions have been considered. The anisotropy of the dipole interaction makes nuclear spin-flips possible without the necessity of electronic spin-flips; such relaxation mechanisms involve longitudinal electronic susceptibility (or the longitudinal electronic correlation time $\tau_{e\parallel}$). As we have seen above, this direct interaction may be important for light elements.

The question of longitudinal relaxation due to the interactions with other nuclear species will not be dealt with here although it has important applications in chemical analysis (cross relaxation, cross polarization).

12.4.4 Electric Field Gradient

The problem of EFG's in metals is rather complicated. Even the field gradient V_{zz} at the nucleus of an isolated atom is modified by the response of the closed electronic shells with respect to that of the bare nucleus V_{zz}^0 according to [12.13]

$$V_{zz} = V_{zz}^0 [1 - \gamma_S(r)]. \tag{12.24}$$

$\gamma_S(r)$ is the Sternheimer antishielding factor which is close to 0 when the charges responsible for the gradient are close to the nucleus but can be as large as 100 for distant charges. Thus the computation of the EFG in a metal requires not only knowledge of the spatial charge distribution but also a precise knowledge of $\gamma_S(r)$ both of which are difficult tasks.

The results of EFG measurements are often compared to a point-charge model which assumes the charges to be located at the lattice sites. This, of course, has limited validity in metallic materials.

In an NMR experiment the EFG is given by two parameters: the quadrupole freqency ν_Q and the asymmetry parameter η.

In the presence of an EFG the NMR frequency for the transition between the levels m and $m - 1$ reads

$$\nu_{m \leftrightarrow m-1} = \gamma H_i + \frac{1}{2}\nu_Q\left(m - \frac{1}{2}\right)(3\cos^2\theta - 1)$$

$$- \frac{1}{2}\eta\nu_Q\left(m - \frac{1}{2}\right)\sin^2\theta\,\cos 2\varphi, \tag{12.25}$$

where θ and φ are the polar and azimuthal angles of the field H_i in the principal axis frame of the EFG.

Since the experiments are performed on powdered samples the orientation of the field is random with respect to the EFG principal axis and powder

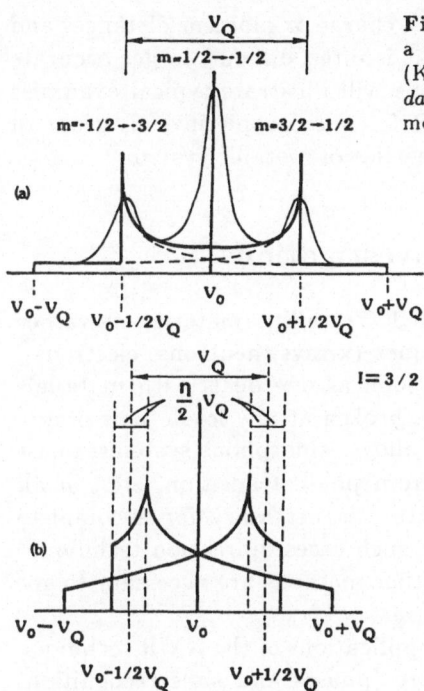

Fig. 12.7a,b. NMR spectra – "powder patterns" – in a polycrystalline sample caused by quadrupole effect (K = 0). ($\nu_0 = \gamma H_0/2\pi$). **(a)** Axial site symmetry; *dashed line:* broadened spectrum **(b)** non-axial site symmetry

patterns are observed, as shown in Fig. 12.7. These spectrum shapes and the deduced values for ν_Q and η are rather unambiguous signatures of particular phases which can be used in phase analysis.

Also the spectral shapes indicate clearly the point symmetry at the nucleus: $\nu_q = 0$ for cubic symmetry, $\eta = 0$ for an axial symmetry and $\nu_Q \& \eta \neq 0$ for symmetries lower than axial. This can be used for structural investigations when the standard diffraction techniques fail to give information about spatial correlation functions higher than pair correlations (amorphous solids, liquids).

12.4.5 Summary

The major advantage of NMR (and other hyperfine interaction techniques) is its local character which allows the study of selected contributions to the bulk properties while the macroscopic techniques yield only average properties. This is particularly useful for the study of inhomogeneous systems (impurities, concentrated alloys, chemically and topologically disordered systems).

For example: K and T_1 measurements can reveal the susceptibility and the spin dynamics of a diluted impurity; in transition metals, they trace the various contributions to the bulk magnetism. T_1 and T_2 measurements in non-magnetic systems can yield information about distances and atomic diffusion. EFG measurements reveal symmetries. Magnetic structures can be dedcued from the orientation dependence of the HF.

The drawback of NMR as of any nuclear method is that the hyperfine parameters which couple the nuclei with such interesting data as magnetic susceptibilities, electronic moments, charge or moment distances and orientations are only vaguely known. Thus it is often difficult to get accurate quantitative data. However, the next sections will illustrate typical examples to show the interesting possibilities of NMR for microscopic investigations of the electronic, magnetic and structural properties of metallic systems.

12.5 Applications – Structural Investigations

The question of the structure of well defined crystalline materials is rather easily solved by standard diffraction techniques (x-rays, neutrons, electrons). Problems concerning this question arise as soon as one deals with materials in a form where the translation invariance is broken at any level. This occurs around impurities, in chemically disordered alloys, amorphous systems, phase admixtures resulting from preparation or from phase transition, etc., in all cases where a knowledge of the microstructure is necessary for a complete understanding of the material properties. In such cases diffraction techniques can provide us only with pair correlations: other methods are necessary to get a more complete view of the detailed structure.

Thus we shall present here examples of applications of the NMR technique to the investigation of local atomic structure (phases and sites recognition, chemical and topological short-range order). Some examples of investigations of atomic diffusion or collective movements will also be given.

12.5.1 Phase Analysis

The two examples given here are typical of the use of NMR in the identification of phases and sites. They are relevant to the heterogeneous catalysis problem on small particles of metals and metallic compounds. An understanding of the catalysis properties requires the identification of the active phase and site. This is gained by comparing the abundance of each identified phase or site in the catalyst with the sample activity. Additionally, the information given by NMR on the electronic structure of the various phases or sites might be of help in the understanding of the catalytic process.

The first example concerns small platinum particles where the differences in Knight shifts made it possible to distinguish between bulk atoms, surface atoms and also atoms decorated with adsorbed molecules [12.14]. The NMR spectra of small size particles exhibit new lines distinct from the bulk peak which are identified as being surface Pt atoms on clean samples and to decorated Pt atoms on samples exposed to various atmospheres. Pt atoms coated by molecules are also distinguished by their long spin-lattice relaxation time due to their non-metallic character.

The second example concerns cobalt sulfide samples dispersed on silica and carbon substrates [12.15]. For heavy cobalt loadings a stable Co_9S_8 phase is formed in which Co occupies two crystallographic sites: one is octahedrally coordinated with sulfur (OC) with no Co nearest neighbor and eight are tetrahedrally coordinated with sulfur (TC) and have 3 Co nearest neighbors. OC's are characterized by the single line of their NMR (cubic site symmetry) and their long T_1 (360 ms) due to their poor metallic environment whereas TC's are characterized by a split quadrupole spectrum (axial site symmetry) and a short T_1 (10 ms). When the dispersion of Co on the substrates increases a new "phase" is observed which corresponds to a strongly broadened spectrum reminiscent of the TC spectrum and is consequently identified to a highly distorted – quasi-amorphous – assembly of tetrahedrally coordinated Co atoms (Fig. 12.8). This new phase is the only one observed on carbon substrates and the OC sites are completely absent. The abundance of this phase is shown

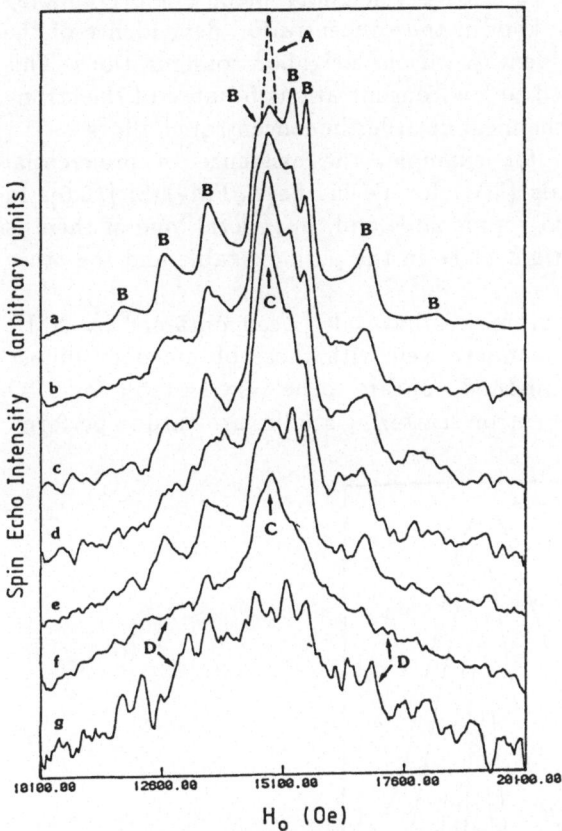

Fig. 12.8. ^{59}Co NMR spectra in sulfided cobalt based and cobalt-molybdenum based catalysts on SiO_2 and active C substrates [12.15]. *(a)* Heavy loading: Co38%wgt/SiO_2, *(b)* Co8%/SiO_2, *(c)* Co2.5%/SiO_2, *(d)* Co4.6%Mo7.4%/SiO_2, *(e)* Co8.4%/C, *(f)* Co2.7%Mo8.3%/ SiO_2, *(g)* Co2.8%/C. *A:* regular octahedral Co sites in Co_9S_8: the dashed line shows about 1/3 of the full intensity; the full line spectra are obtained in such experimental conditions that the NMR of these slowly relaxing nuclei vanishes; the delay between successive pulse sequences is too short for these nuclei to relax towards thermal equilibrium with the lattice, *B:* regular tetrahedral Co sites in Co_9S_8: quadrupole split spectrum, *C:* rapidly relaxing octahedral Co sites: CoS phase or Co-Mo bondings, *D:* distorted – quasi-amorphous – tetrahedral Co sites: active catalyst phase

to correlate with the catalytic activity of the samples. On silica substrates or in mixed Co-Mo sulfides the regular OC sites of Co_9S_8 are also removed but they are replaced by other octahedrally coordinated Co atoms (single NMR line) with a much shorter T_1, which implies metal-metal bondings. These new OC sites are probably similar to CoS sites (an unstable phase where Co is octahedrally coordinated with sulfur and has two Co nearest neighbors); they may also result from Co-Mo bondings in mixed CoMoS catalysts.

12.5.2 Chemical Short-Range Order

NMR of the matrix and the impurity in diluted binary alloys exhibit satellite lines which result from different Knight shifts of hyperfine fields of nuclei having 1, 2, 3 ... impurities in their 1st, 2nd, 3rd ... neighbor shell. The overall alloy crystallographic structure being known, the concentration dependence of the satellite intensity allows us to identify various neighbor configurations. This kind of analysis is usually limited to low concentrations because of the strong broadening which results from chemical disorder in concentrated alloys.

Such studies have shown, for example, the existence of preferential substitutions of transition metals (TM) for Fe in $(Fe_{1-x}TM_x)_3Si$ [12.6]. In Fe_3Si iron occupies two crystallographic sites and, in general, one of them is substituted by elements to the right of Fe in the periodic table and the other by elements to the left.

Short-range order (SRO) parameters have also been obtained by NMR in FeV alloys (Fig. 12.9) which compare well with those obtained by diffuse neutron scattering [12.17]. The method appears to be very suitable for SRO studies in the dilute limit where neutron scattering studies are hard to perform.

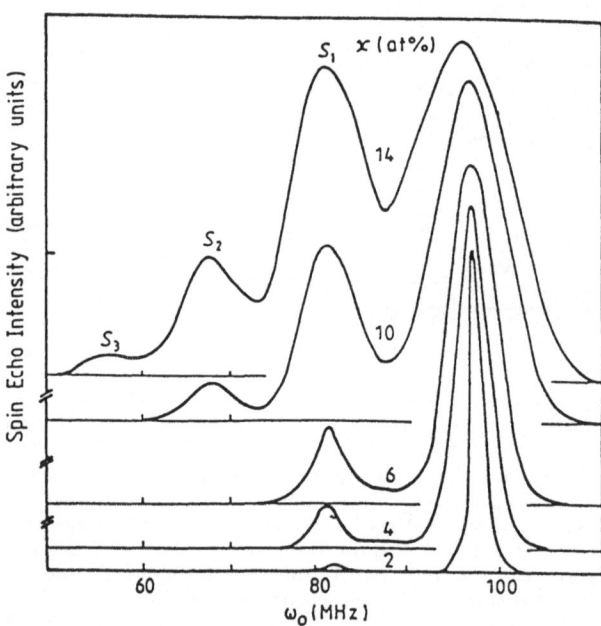

Fig. 12.9. Zero-field NMR spectra of ^{51}V in ferromagnetic $Fe_{1-x}V_x$ alloys [12.17] showing satellites due to V pairs (S_1) and other V configurations

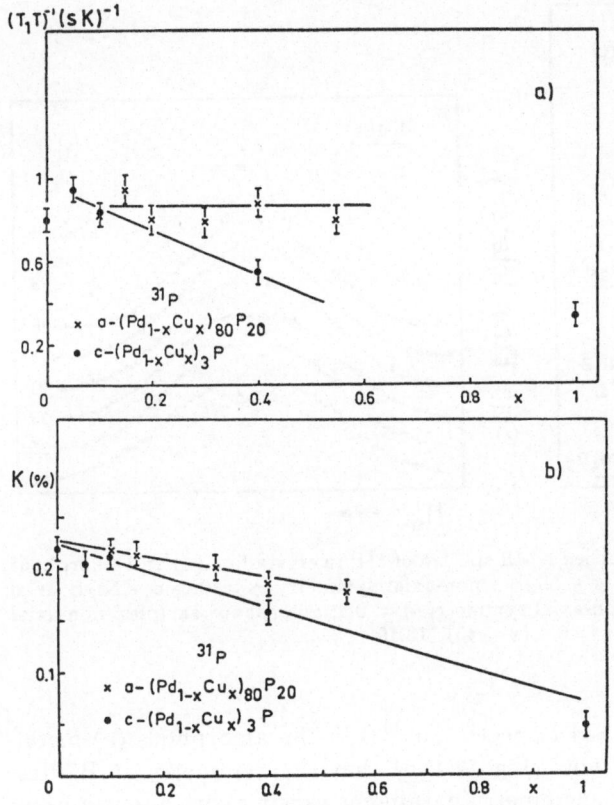

Fig. 12.10a,b. Relaxation rate (a) and Knight shift (b) of ^{31}P in amorphous $(Pd_{1-x}Cu_x)_{80}P_{20}$ and crystalline $(Pd_{1-x}Cu_x)_3P$ showing Cu-P avoidance in the amorphous compound: T_1 and K are independent of x [12.18]

Finally, let us cite the NMR study on ^{31}P in the diluted amorphous alloy $(Pd_{1-x}Cu_x)_{80}P_{20}$ which shows that T_1 and K (Fig. 12.10) are nearly concentration independent and maintain the value measured in crystalline Pd_3P. This shows that in the amorphous system P atoms tend to avoid Cu neighbors whereas in the crystalline alloy $(Pd_{1-x}Cu_x)_3P$ the substitution of Cu for Pd implies that P atoms have an increasing amount of Cu neighbors and a related concentration dependence of K and T_1 is observed [12.18].

12.5.3 Structure of Amorphous Metals

The lack of translational invariance in amorphous and liquid materials makes the determination of their structure a difficult task. Standard diffraction techniques can at best give information on distances, number and nature of the atoms in successive coordination shells and give no information on bonding angles and symmetries.

On the other hand, NMR and Mössbauer effect can, via the EFG at the nuclei, yield some information on local symmetries and may possibly identify local atomic configurations by comparison with well-known materials.

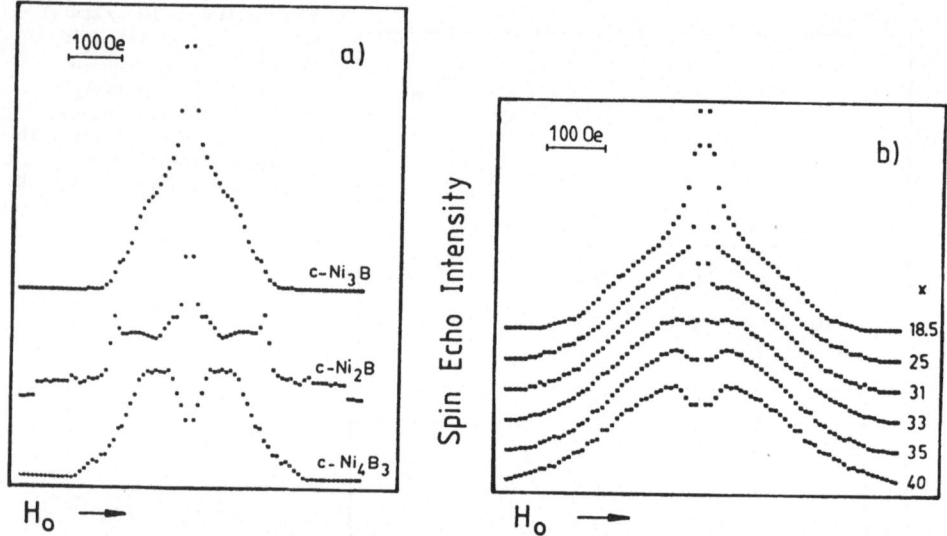

Fig. 12.11a,b. Quadrupole perturbed NMR spectra of ^{11}B in crystalline (a) and amorphous (b) nickel borides ($Ni_{100-x}B_x$); $c - Ni_3B$: non-axial symmetry $\eta = 0.6$; $c - Ni_2B$ axial symmetry: $\eta = 0$; $c - Ni_4B_3$: non-axial symmetry $\eta = 0.4$; amorphous samples: non-axial symmetry $\eta = 0.6$, ($x = 18.5$) to $\eta = 0.4$ ($x = 40$) [12.19]

Quadrupole perturbed NMR spectra on ^{11}B in the amorphous (a) metal $a - Ni_{100-x}B_x$ ($18 \leq x \leq 40$) show (Fig. 12.11b) that the symmetry at B sites is non-axial with an average asymmetry parameter varying continuously from $\eta = 0.6$ to 0.4 with increasing B content. These values at the low and high ends of the concentration range correspond exactly to those measured in the crystalline (c) borides $c - Ni_3B$ and $c - Ni_4B_3$, respectively (Fig. 12.11a). Furthermore the distribution of the quadrupole frequency which measures the fluctuation of the EFG at the B sites is quite narrow at both ends. It is then concluded that the local atomic arrangement around B for $x = 18$ and $x = 40$ fluctuates little from site to site and is very similar to the arrangement which prevails in $c - Ni_3B$ and $c - Ni_4B_3$, respectively [12.19]. However, the axial symmetry of the B site in $c - Ni_2B$ is not observed in the amorphous modifications. Thus the amorphous nickel borides would consist of a disordered assembly of Ni trigonal prisms centered on B which are the building units of $c - Ni_3B$ and $c - Ni_4B_3$. The absence of sites with an axial symmetry suggests that the Ni antiprisms (twisted cubes) surround the B atoms in $c - Ni_2B$ are absent in amorphous Ni borides whereas, in $a - Mo_2B$, B sites do exhibit an axial symmetry ($c - Mo_2B$ and $c - Ni_2B$ have the same structure).

T_2 measurements have also proved useful in the investigation of the amorphous structure. In the above example T_2 is dominated by the spin-spin interaction between the B nuclei which carry the only significant moment. The $1/r^3$ range of the dipole interaction makes T_2 roughly proportional to the cube of the mean B–B distance in the compounds. Comparisons of T_2's show that this mean B–B distance is the same in the amorphous samples as in the crystals of the same composition. From this measurement one can conclude that in the high boron content range, B–B bondings exist as in $c - Ni_4B_3$ contrary to the well known metalloid-metalloid avoidance found at low concentrations. This conclusion is also supported by neutron and x-ray diffraction studies.

It is clear from this study that no unique structural model of amorphous metallic systems can be conceived although these systems were initially described by a dense random packing of spheres expected from the isotropic character of the metallic bonding. Instead, a variety of chemical and topological short-range orders are found as in crystalline materials. However, there is not necessarily a direct correspondance between the crystalline compounds and their compositionally equivalent amorphous modification, as suggested by quasi-crystalline or microcrystalline models.

12.5.4 Atomic Motion in Metals

As soon as atoms are able to move in a material, the hyperfine interactions to which their nuclei are submitted vary randomly with a correlation time τ_d related to the jump frequency and hence to the diffusion parameters. The most spectacular effect of the modulation of the hyperfine interactions by atomic motion is the so-called motional narrowing of the NMR line at high temperatures which results from the averaging of the traceless dipole interaction by the motion of the nuclei (Fig. 12.12). In general, the motion of the atoms gives rise to a fluctuation spectrum of the hyperfine interactions (usually spin-spin dipole but also quadrupole) which is often considered as Lorentzian in shape (the Bloembergen, Purcell and Pound model "BPP" [12.20]). Then the relaxation times follow the temperature dependence of τ_d according to

$$T_1^{-1} = \alpha \cdot 2\tau_d / (1 + \omega_L^2 \tau_d^2), \qquad T_2^{-1} = \alpha \cdot 2\tau_d \qquad (12.26)$$

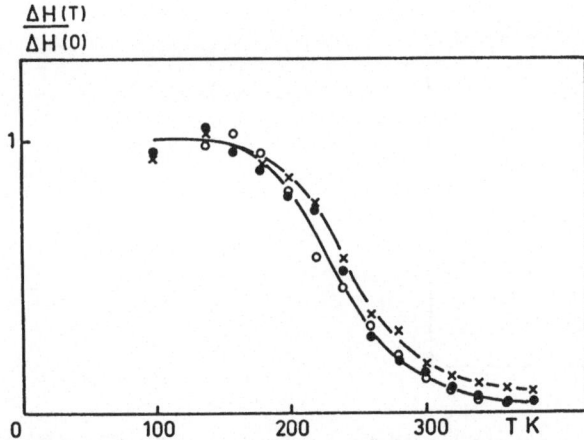

Fig. 12.12. Temperature dependence of the proton dipole line width (motional narrowing) in amorphous $a - Pd_{0.35}Zr_{0.65}H_{0.7}$ (○), $a - Pd_{0.35}Zr_{0.65}H_{1.7}$ (●), and crystalline $c - PdZr_2H_{4.4}$ (×) [12.22]

(α is a constant that we do not specify here). It should be noted that contrary to the cases discussed in Sect. 12.4 it is the fluctuation spectrum of the hyperfine coupling which is involved here and not that of the moments. A third relaxation time $T_{1\varrho}$ is often considered in these studies which characterizes the relaxation in the rotating frame when a weak H_1 is maintained after a $90°$ pulse: in this case the relaxation involves quantized energy transfers $\hbar\omega_1 = \gamma\hbar H_1$ and

$$T_{1\varrho}^{-1} = \alpha{\cdot}2\tau_d/(1 + \omega_1^2\tau_d^2). \tag{12.27}$$

Thus these three relaxation times provide means of measuring the correlation time of the atom motion in quite a reasonable time range around $1/\tau_d = 0, \omega_1, \omega_L$ for $T_2, T_{1\varrho}, T_1$, respectively, where the atomic motion usually dominates over other relaxation mechanisms: typically an overall range from 10^{-3} to 10^{-10}s for τ_d can be explored [12.21]. Although the absolute values for τ_d from the BPP model can be in error by as much as 50% their temperature dependence yields reliable diffusion activation energies that agree within 10% with other techniques (Fig. 12.13) [12.22].

As far as the collective motion of atoms is concerned let us mention the peak of relaxation observed at $55\,\mathrm{K}$ on P and B in $a - $ NiPB metallic glasses [12.23]. This peak could be related to internal friction peaks observed at lower temperatures for lower frequencies in similar glasses (Fig. 12.14).

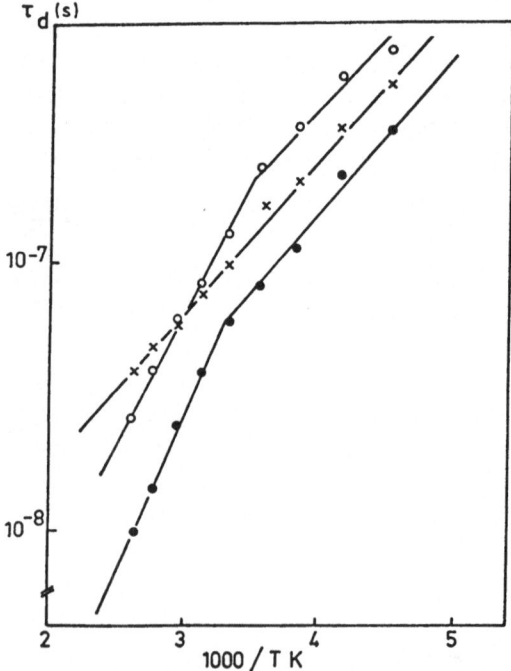

Fig. 12.13. Temperature dependence of the proton self-diffusion time in $a - \mathrm{Pd}_{0.35}\mathrm{Zr}_{0.65}\mathrm{H}_{0.7}$ ($\circ$), $a - \mathrm{Pd}_{0.35}\mathrm{Zr}_{0.65}\mathrm{H}_{1.7}$ ($\bullet$) and $c - \mathrm{PdZr}_2\mathrm{H}_{4.4}$ (x) (Arrhenius plot obtained from $^1\mathrm{H}T_2$: the slope of the curves is the activation energy of the diffusion process) [12.22]

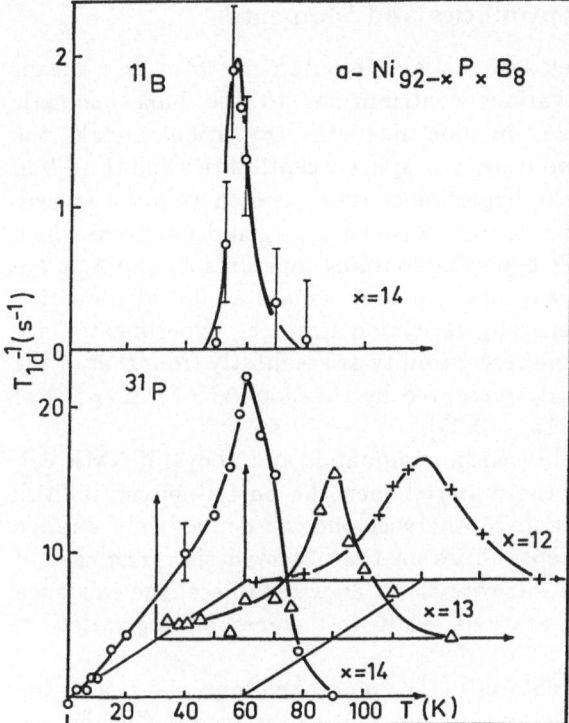

Fig. 12.14. Spin-lattice relaxation of ^{11}B and ^{31}P in amorphous NiPB alloys due to collective motions of atoms [12.23]

12.6 Applications – Electronic and Magnetic Properties

The raw outputs of the NMR (Knight shift, hyperfine fields and more seldomly relaxation times and quadrupole frequency) are sometimes computed in theoretical works: this provides a direct means of checking theoretical band structure calculations against the experiment. However these calculations are almost always limited to the band structure itself and some basic electronic properties such as magnetic susceptibility or moment, spin or charge dynamic properties, magnetic or crystallographic phase stability. Then it is necessary for the experimentalist to fill the gap between the NMR output and the theoretical result by estimating the magnitude of the hyperfine interactions from experiments, from calculations or from phenomenological models.

In the following we shall thus give examples of NMR studies which yield information about the electronic and magnetic properties of metallic materials. We shall however keep the discussion on a phenomenological basis witout going into quantum mechanical details.

Again the full interest of a local magnetic probe will be found in the ability of NMR to measure selectively the partial contributions to the bulk properties.

12.6.1 Local Magnetic Susceptibilities and Moments

Knight shifts and spin-lattice relaxation time measurements provide a means of tracing independently the various contributions to the bulk magnetic susceptibility χ_B. For example, in non-magnetic transition metals the susceptibility arises from the s and d electron spin susceptibilities and the d Van Vleck susceptibility. Assuming the hyperfine coupling constants to be known, χ_B, K and T_1 yield independent relations between χ_s, χ_p and χ_{vv} from which their values can be computed. The hyperfine coupling constants A_s and A_{orb} are usually deduced by calculation. A_{cp} has also been computed for 3d transition metals [12.24] but it is more generally estimated from the experiment: since the temperature dependence of the susceptibility arises mostly from that of the d spin contribution, A_{cp} is directly measured by the slope $\Delta K(T)/\Delta \chi_B(T)$ of the K(T) vs $\chi_B(T)$ plot (Fig. 12.15) [12.25].

The situation is more complicated in compounds and alloys: if NMR can be observed all the elements of the material then the partial susceptibilities on each site can be roughly estimated, otherwise one can deduce only a range of consistent values for the susceptibilities on the observed site. The case of concentrated alloys is even more intricate since, as we shall see, the existence of a variety of local atomic configurations results in a corresponding variety of local properties.

In magnetically ordered materials the more relevant data are the spontaneous hyperfine fields which exist at the nuclei sites even in the absence of an external field. Relaxation times are less readily related to the electronic structure because the nuclear relaxation is driven mostly by magnetic domains

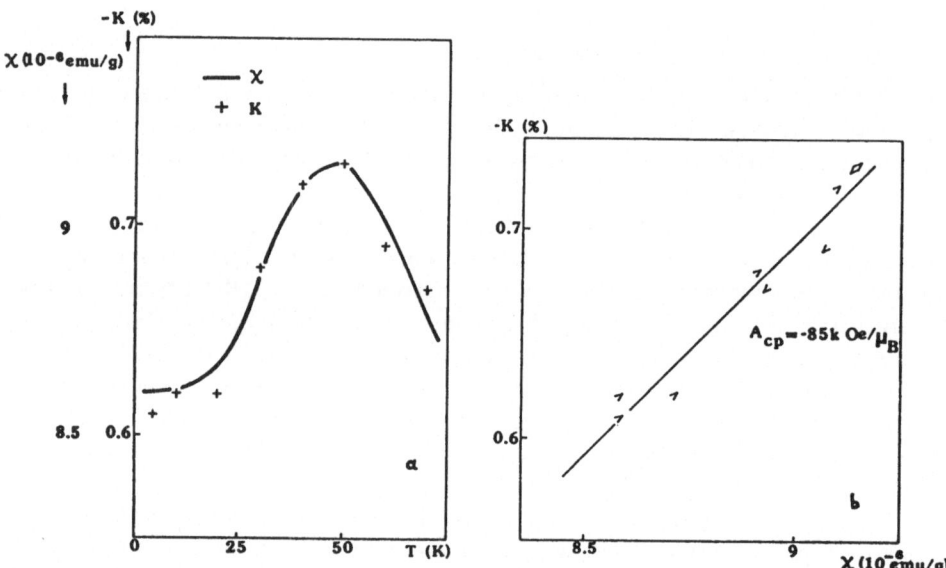

Fig. 12.15. (a) Co Knight shift (+) and bulk susceptibility χ (*full line*) in $CoSe_2$; (b) K vs χ plot in the same compound: the slope of the curve gives the hyperfine coupling constant A_{cp} [12.25]

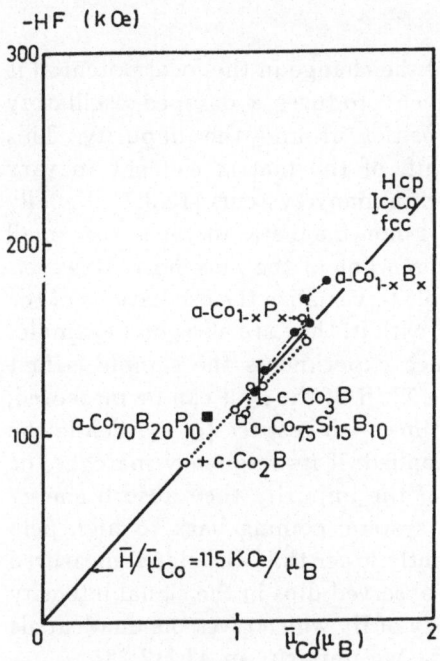

Fig. 12.16. Co hyperfine field vs Co moment in ferromagnetic cobalt based compounds showing the proportionality between the Co moment and the Co HF; *(c)* crystalline; *(a)* amorphous [12.26]

and domain wall dynamics (spin waves, walls thermal fluctuations). Like the Knight shift the HF is directly related to the electronic magnetization, that is to the magnetic moments. Although, as we have seen, the HF can be theoretically described as a sum of contact, core polarization and orbital contributions, the difficulty of computing and handling them had led the experimentalists to a more phenomenological description in terms of a local contribution due to the on-site moment (i.e., the moment on the site of the proble nucleus) and a non-local one due to the polarization of the conduction electrons by the surrounding moments. If the local contribution is dominant for the magnetic elements, which is generally the case, except perhaps for nickel, the measurement of the HF can reasonably give the magnitude of the moment. As shown in Fig. 12.16, for example, there is a very good proportionality between the moment and the HF in many cobalt based crystalline and amorphous compounds [12.26]. On the basis of such proportionality one can then estimate the moments of the various elements in an alloy at less expense (but with less accuracy) than by neutron diffraction. This is particularly useful for amorphous metals studies since the latter technique cannot be used because of the lack of translational invariance. In the case of a strong polarization of conduction electrons their contribution to the hyperfine field (called transferred HF) can be estimated by the HF measurement on a substituted non-magnetic element but of course such determinations have to be handled with great care.

12.6.2 Impurities in Metals

When an impurity is introduced in a metal the change in the local potential is screened by the conduction electrons, which produces a damped oscillatory distribution of the charge and spin densities around the impurity. This modulation causes the average Knight shift of the matrix element to vary and this effect has been thoroughly studied in many systems [12.3,7]. Usually the effect of a non-magnetic impurity in a non-magnetic metal is too small to extract satellite NMR lines out of the main line of the pure host. However, interesting experiments have made it possible to visualize the oscillations of the charge density through the EFG associated with it; they are also good examples of a field cycling technique [12.27]. In such experiments the sample is first placed in a high magnetic field in which the NMR of the host can be measured; then the field is switched off for a short time with respect to T_1. While the field is off, a radio frequency field H_2 is applied: if its frequency matches the quadrupole frequency of some neighbors of the impurity they absorb energy which raises the temperature of the spin system: coming back to high field the observed NMR signal (S_{on}) is consequently lower than the signal measured with $H_2 = 0$ (S_{off}). Figure 12.17 shows the observed dips in the signal intensity ratio (S_{on}/S_{off}) as a function of the frequency of H_2 which gives the quadrupole frequency of the various neighbor shells of a Mg impurity in Al [12.28].

The effects of a non-magnetic impurity in a magnetic matrix are much stronger and satellites due to the different HF's on the impurity neighbors are resolved (see the FeV case in Sect. 12.5). The same situation occurs when a magnetic impurity is diluted in a non-magnetic host.

12.6.3 Magnetic Impurities – Occurrence of Magnetism

The case of magnetic impurities is relevant to the general problem of the occurrence of magnetism, moment formation and interaction of localized moments

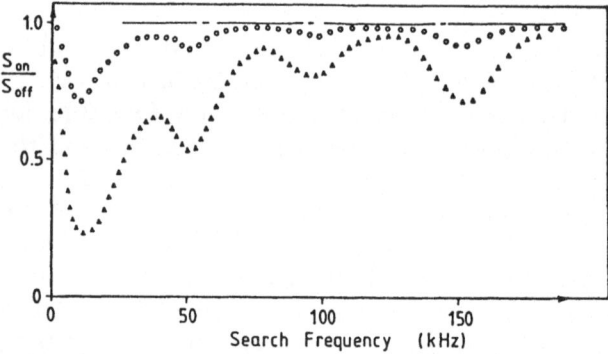

Fig. 12.17. Field cycling experiment in A̲lMg (S_{on}/S_{off} : see text), the first dip corresponds to the average spin-spin interaction, the other dips correspond to the quadrupole frequency of the various Al shells around a Mg impurity [12.28]; the two curves are for different magnitudes of H_2

with the conduction electrons in metals. To this problem many studies have been devoted and a detailed discussion does not fall within the scope of this book although NMR experiments on the impurities, on their neighbors and on the host nuclei have proved very useful for the characterization of impurity magnetism. An excellent although now dated review of a number of these studies has been given by *Narath* [12.10] including measurements by other hyperfine techniques (Mössbauer effect, nuclear orientation, PAC) which offer much higher sensitivities than NMR for the study of very diluted impurities.

Particular emphasis should be given to relaxation time measurements which provide a potentially powerful tool to probe impurity spin dynamics, on the basis given in Sect. 12.4, although several problems arise from uncertainties on the functional form of the dynamic susceptibility (often assumed Lorentzian) or on the origin of the relaxation mechanism (direct, indirect, isotropic or anisotropic hyperfine coupling). Though it is not directly relevant to the impurity problem an example of an NMR study of spin fluctuations in rare earth based compounds is given later which illustrates the potentiality of the method.

These studies emphasize the obvious advantage of local hyperfine techniques over macroscopic ones since they not only permit a separate probe of the behavior of the impurity and the host but also they provide a means of studying spatially inhomogeneous properties resulting from various impurity interactions or from alloying.

Indeed, the inhomogeneous occurrence of a moment on impurities in a diluted alloy was first pointed out by *Jaccarino* and *Walker* [12.29] after their NMR measurements on Co diluted in $Rh_{1-x}Pd_x$. To explain their results they proposed a model in which the impurity is assumed to be non-magnetic unless it has a critical neighbor configuration: in RhPd alloys Co does not carry a moment unless two Pd atoms at least are in its nearest neighbor shell.

12.6.4 Concentrated Alloys – Local Environment Effects

Since the pioneer work of *Jaccarino* and *Walker* a number of experiments have been performed not only on diluted impurities but also in concentrated systems which have emphasized the role of local environment effects on moment formation. These effects can lead to the formation of giant moments (statistical clusters of ferromagnetically coupled impurities) or to the coexistence of magnetic and non-magnetic sites for the same element. Such magnetic inhomogeneities are rather easily observed and studied by NMR and the Mössbauer effect. Figure 12.18 shows an example of the coexistence of magnetic and non-magnetic cobalt atoms in $Co(S_{1-x}Se_x)_2$ alloys due to different chalcogen surroundings as evidenced by their very different HF's [12.30].

These effects of the local environment distribution in alloys are generally neglected in the studies of macroscopic properties. Since alloying is the most commonly employed means of modifying the material properties it is obvious that the information given by local techniques is extremely useful for a complete

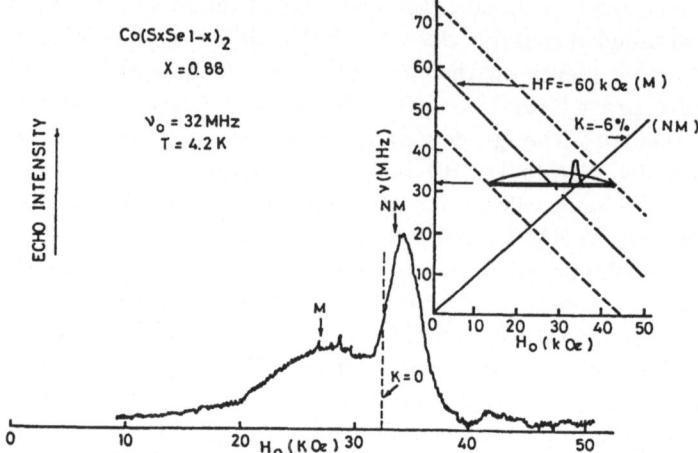

Fig. 12.18. ^{59}Co NMR spectrum observed in $Co(S_xSe_{1-x})_2$ showing the coexistence of magnetic (M) and non-magnetic (NM) Co atoms: magnetic Co atoms experience an internal field $H_0 - 60kOe$ (*dashed-dotted line*) non-magnetic Co atoms experience an internal field $0.94H_0$ ($K = -0.06$) (*full line*) [12.3]

understanding of these properties. It should be noted that in the same spirit the effects of lattice defects such as vacancies or interstitial atoms can be studied in the same way.

Increasing the concentration of magnetic impurities can also lead to the occurrence of a spin glass state, that is a state in which the local moments freeze at random in contrast to the magnetically ordered ferro, antiferro or ferrimagnetic states. On a local scale however there exists a spontaneous HF in which the NMR can be performed as in the ordered states [12.31].

12.6.5 Magnetic Structure and Phase Transition

Although the study of magnetic structures and magnetic phase transitions is mostly performed by neutron diffraction, NMR allows at least preliminary studies at much less expense. Indeed, qualitative conclusions in this field can easily be inferred from simple NMR measurements, as shown in the following examples.

First, in anisotropic materials the easy magnetization direction can be deduced from the spectrum analysis: as shown in Fig. 12.19, the Co spectrum in Co_2B is a singlet at low temperatures and becomes a doublet above 77 K [12.32]. Given the fact that in the tetragonal structure of Co_2B there is only one crystallographic Co site, the observed doublet implies the existence of two magnetically unequivalent Co sites and hence that the magnetization lies in the basal plane at high temperatures while it is along the c (four-fold) axis at low temperatures.

Also the existence of ferrimagnetic behavior can sometimes be inferred from the study of the hyperfine fields. For example, the Mn NMR spectra in

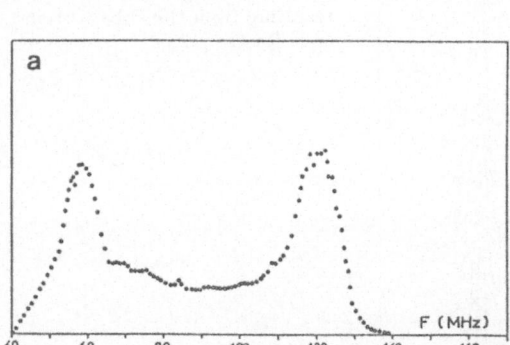
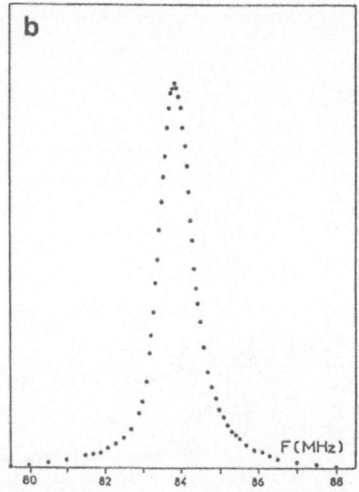

Fig. 12.19a,b. Co NMR spectra in zero field in Co_2B : **(a)** T>77 K : the doublet shows the existence of two magnetic sites (easy magnetization axis in the basal plane of the tetragonal structure); **(b)** T = 4.2 K a singlet spectrum implies the existence of only one magnetic site (the easy magnetization axis is the four fold c axis) [12.32]

$Ni_{1-x}Mn_x$ alloys exhibit several high frequency peaks for low Mn content and a low frequency peak that grows with increasing Mn content (Fig. 12.20). From the field dependence of the position of the peaks the sign of the HF was found to be positive for the low frequency peak (the external field adds to the HF) whereas it is negative for the others. Therefore it was concluded [12.33] that those Mn atoms responsible for the low frequency NMR peak carry a moment which is antiferromagnetically coupled to the others.

As far as magnetic transitions are concerned, it is obvious that they can be easily observed and traced by the measurement of the hyperfine field and its dependence on external parameters (temperature, composition, etc.).

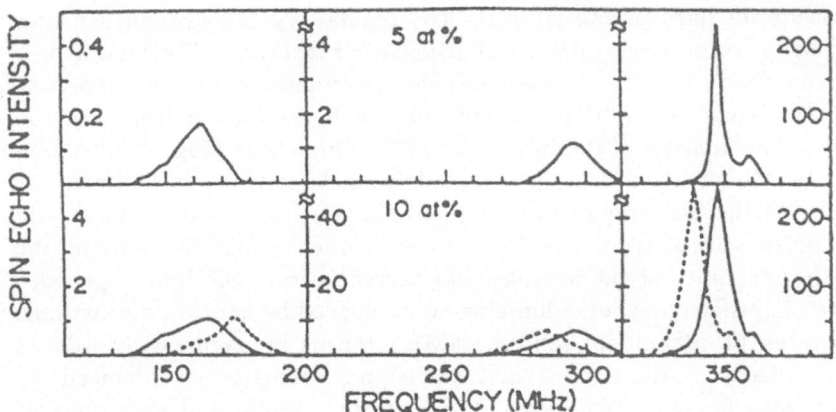

Fig. 12.20. ^{55}Mn NMR spectra in $Ni_{1-x}Mn_x$: *full line:* $H_0 = 0$, *dotted line:* $H_0 = 15kOe$: the opposite field dependence of the line positions suggests the existence of Mn atoms with opposite moments [12.33]

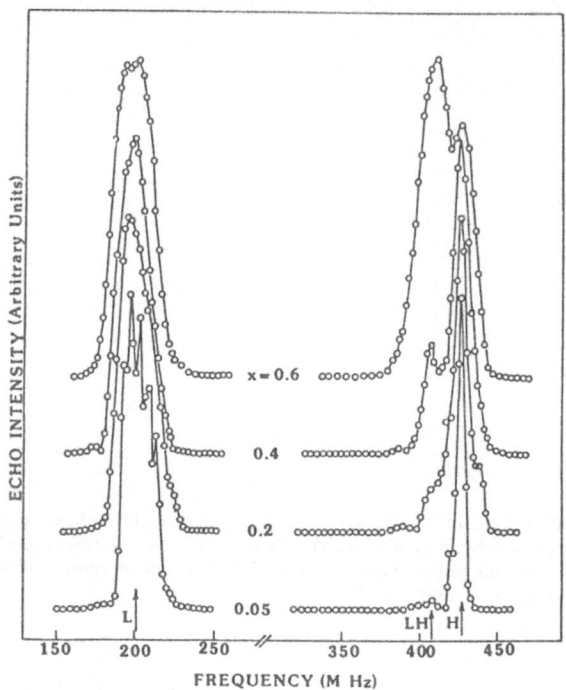

Fig. 12.21. ^{55}Mn NMR spectra in $Mn_5(Ge_{1-x}Si_x)$: L→Mn_I sites; H→Mn_{II} sites (regular Mn sites in Mn_5Ge_3); LH→ new magnetic sites resulting from the substitution of Si for Ge [12.34]

The case of $Mn_5(Ge_{1-x}Si_x)_3$ alloys provides a good example of an NMR study of a transition from a ferromagnetic phase (Mn_5Ge_3) to an antiferromagnetic phase (Mn_5Si_3) [12.34]. This study shows that the transition occurs inhomogeneously in three steps.

At the early stage of the substitution of Si for Ge ($x<0.5$) the moment of some Mn atoms vanishes when they have two or more Si atoms in their neighborhood. These atoms are identified by a new line (LH) in the NMR spectra beside the lines (L and H) of the two regular crystallographic Mn sites (Mn_I and Mn_{II}, respectively) present in Mn_5Ge_3 (Fig. 12.21). Their moment is deduced from the value of their hyperfine field by comparison with the hyperfine fields at the regular sites (the moments of which are known from neutron diffraction). The analysis of the intensities of the three lines (Fig. 12.22) shows that the new Mn sites are Mn_I sites which have at least two Si neighbors.

For $x>0.4$ the behavior of the samples changes dramatically. The strong variation of the intensity of the three lines is due to modifications of the magnetic permeability of the samples which results from the inhomogeneous occurrence of antiferromagnetic domains (as evidenced by neutron diffraction). This illustrates the difficulty of handling NMR intensity in magnetically ordered materials: indeed, in these materials, the signal intensity is enhanced by the initial (zero field) susceptibility of the sample [12.35] and variations of the macroscopic magnetic properties of the material induce corresponding variations of the NMR signal intensity. This feature of NMR could be

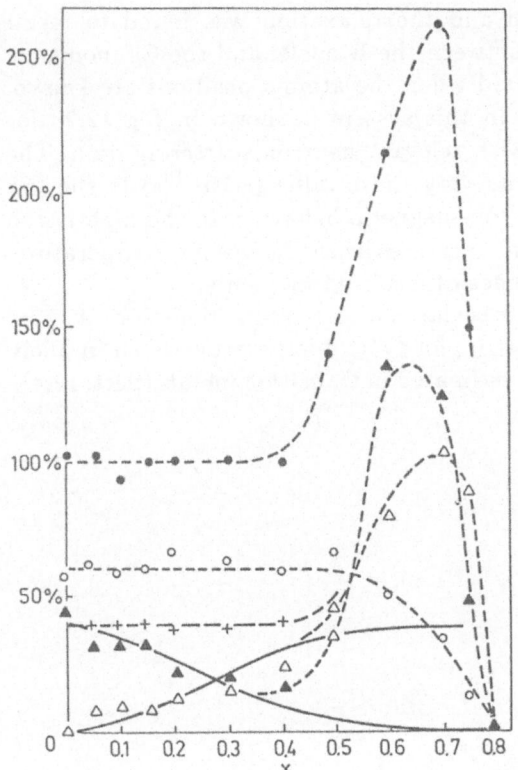

Fig. 12.22. NMR line intensities in $Mn_5(Ge_{1-x}Si_x)_x$: (●) total intensity (L + H + LH); (▲) L line; (△) LH lines; (○) H line; + (L + LH) intensity; *full lines:* calculated statistical number of Mn_1 atoms with less than 2 Si neighbors (decreasing) and with more than 2 Si neighbors (increasing) which identify the origin of LH [12.34].
– above x = 0.4 the strong variation of the intensities result from modifications of the magnetic structure
– above x = 0.8 the samples are fully antiferromagnetic

used to investigate the "technical" magnetic properties of ferromagnets on a microscopic scale but there is still a lack of detailed understanding of the phenomenon which precludes quantitative analysis.

Finally for x>0.8 the transition to the antiferromagnetic state has been completed, the permeability of the samples vanishes and the intensity of the NMR signals drops below the minimum observable level.

12.6.6 Spin Fluctuations in Rare Earth Based Compounds

The f electrons of rare earth elements are normally localized and their behavior can be described in terms of ionic states; however in some cases, particularly at both ends of the series (Ce, Yb), they show a tendency to delocalize, which has drastic consequences on their magnetic behavior. In many aspects this situation is similar to that encountered in the case of magnetic impurities. The example of $CePd_3B_x$ compounds shows how the electronic spin dynamics of Ce is traced by susceptibility and relaxation time measurements [12.36]. As indicated in Sect. 12.4, the electronic correlation time τ_e is measured by T_1 once the static susceptibility is known and providing the relevant hyperfine coupling constant can be estimated:

$$\tau_e^{-1} = 2\gamma^2 k_B A_i^2 T_1 T \chi_i^{zz}(0).$$ (12.28)

In $CePd_3B_x$ boron NMR was used and the relaxation was found to result mostly from the dipole interaction between the B nuclei and the Ce moments. In such a case A_{dip} is easily computed when the atomic positions are known. The temperature dependence of τ_e in this system is shown in Fig. 12.23 for various B concentrations together with relevant neutron scattering data. The temperature dependence of τ_e and its very short value ($\sim 10^{-13}$s) in the low B content range is characteristic of non-magnetic behavior. In the high boron concentration τ_e is much longer and increases with decreasing temperature, which is an indication of the occurrence of localized moments.

If the dominant interaction is a scalar indirect interaction then A_i can be estimated by the slope of the K vs χ plot (with temperature as an implicit parameter) in a similar way as A_{cp} is estimated in transition metals (Sect. 12.4).

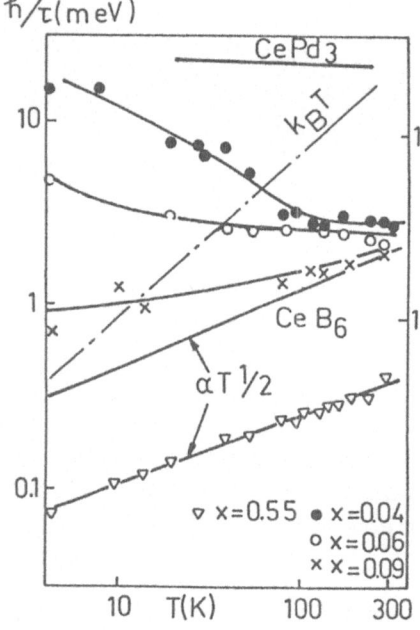

Fig. 12.23. Thermal and concentration dependence of the Ce electronic relaxation rate in $CePd_3B_x$ (in units of energy $\hbar/\tau$) [12.36] as deduced from ^{11}B T_1 measurement; *solid line:* neutron quasi-elastic line width ($CePd_3$ after *Holland-Moritz* et al. [12.38]; CeB_6 after *Horn* et al. [12.39]

12.6.7 Electronic Phase Transitions

Under this heading we are concerned with those phase transitions which are neither crystallographic nor magnetic in origin: superconductivity, metal-insulator transition, etc. The NMR studies of such phenomena remain marginal and we shall limit ourselves to a brief description of two cases: superconductivity and charge density waves.

Superconductivity in metals results from the onset of new electronic states which correspond to the formation of coupled pairs of conduction electrons of opposite spins (the so-called Cooper pairs). The pairing of the electrons consequently reduces their spin susceptibility, which is observed in NMR by

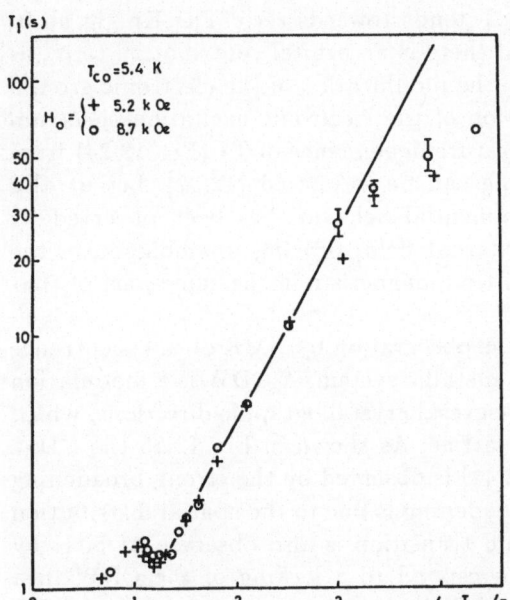

Fig. 12.24. Thermal behavior of T_1 in a superconductor $a - (MoRu)_{80}P_{20}$. T_{co} is the transition temperature in zero field [12.39]. In the normal state $(T > T_{co})$ T_1 obeys the Korringa law: $T_1 T = $ constant

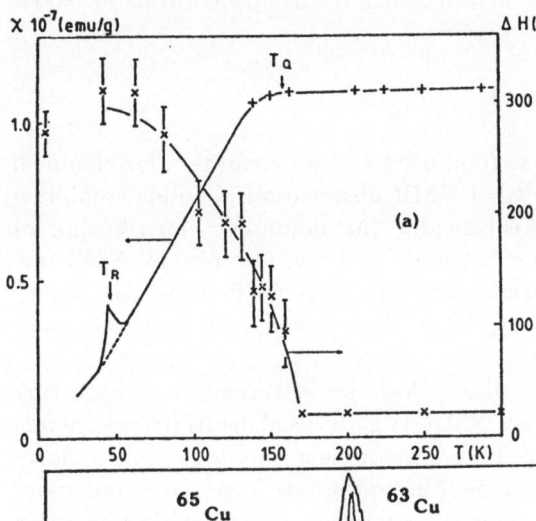

Fig. 12.25a,b. ^{63}Cu and ^{65}Cu NMR in CuS_2 : (a) temperature dependence of the bulk magnetic susceptibility and of the Cu NMR line width (in field units); (b) NMR spectra above and below the occurrence of a charge density wave (at 150 K) [12.41]

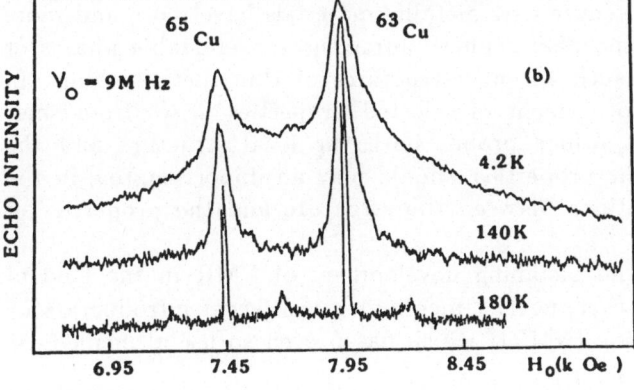

a decrease of the Knight shift when T tends towards zero. The Knight shift, however, does not vanish for $T = 0$ if there is an orbital contribution to it (as is the case in transition metals). Also the modification of the electronic ground state is accompanied by a modification of the electronic excitation spectrum which leads to an exponential temperature dependence of T_1 (Fig. 12.24) from which the importance of the coupling can be estimated [12.37]. Let us also mention that this characteristic exponential behavior has been observed by NMR in $(Ce_{1-x}Gd_x)Ru_2$ in zero external field, proving unambiguously the coexistence of superconductivity and ferromagnetism in the same part of that compound [12.40].

The second example illustrates the observation by NMR of the occurrence of a charge density wave (CDW) in a metallic system. A CDW is a modulation of the electronic density, along one or several crystallographic directions, which does not have the periodicity of the lattice. As shown in Fig. 12.25 the CDW which settles in at 150 K in CuS_2 [12.41] is observed by the strong broadening of the NMR spectrum of Cu. This broadening is due to the spatial distribution of the EFG at the Cu sites. A second transition is also observed at 50 K by a susceptibility peak which could correspond to a locking of the CDW in a commensurate state (the period of the CDW is then a multiple of the period of the lattice) as suggested by the constant width of the spectrum below 50 K.

12.7 Conclusion and Outlook

While in organic materials NMR was soon used as a powerful tool for chemical and structural analysis, the difficulty of NMR observation in solids combined with the difficulty of handling theoretically the dominant contribution of conduction electrons in metals has strongly limited the use of NMR for structural analysis in metallic materials. Actually most NMR studies of metals, from the beginning of NMR history, have been devoted to electronic and magnetic properties.

The aim of this chapter is to show that, besides the very interesting studies of local magnetic properties, NMR can be used fruitfully in metals to investigate structural properties. This development should be particularly important in the future because new metallic materials have more and more complex structures, i.e. amorphous, phase admixtures, metastable phases or even artificial structures such as superstructures of thin metallic films. In all these systems the improvement of selected properties arises from their microstructure and NMR, which probes both the local structure and the local magnetic or electronic properties, should play an important role in the understanding of the relations between the structure and the properties of interest.

Without going into the blooming development of NMR in the field of chemistry, we cannot, however, neglect mentioning the recent introduction of the NMR imaging technique (NMRI) which has proved so useful in medical applications that, in non-scientific circles, NMR is automatically associated with the field of medicine. In contrast to standard NMR experiments, where the external field must be homogeneous, in NMRI use is made of linear magnetic field gradients to select planes or spots in the three-dimensional object to be imaged. The selection is made by the position dependence of the Larmor frequency of the nuclei in the inhomogeneous field. The image contrast can be obtained by the signal intensity, the frequency shift or the relaxation times of the nuclei which differ from spot to spot. As has been clearly demonstrated in the context of medical applications, NMRI is an ideal technique for mapping the distribution of fluids in an object. Therefore, NMRI has potential applications in a number of fields (absorption and diffusion of fluids in composites and in porous materials, imaging of solids), which makes its future very promising.

References

12.1 A. Abragam: *Principles of Nuclear Magnetism* (Oxford Univ. Press, London 1961)
12.2 C.P. Slichter: *Principles of Magnetic Resonance*, Springer Ser. Solid-State Sci. Vol. 1 (Springer, Berlin, Heidelberg 1980)
12.3 J. Winter: *Magnetic Resonance in Metals* (Oxford Univ. Press, London 1971)
12.4 T.C. Farrar, E.D. Becker: *Pulse and Fourier Transform NMR Introduction to Theory and Methods* (Academic, New York 1971)
12.5 U. Haeberlin: *High Resolution NMR in Solids Selective Averaging* (Academic, New York 1976)
12.6 M. Mehring: *High Resolution NMR Spectroscopy* (Springer, Berlin, Heidelberg 1976)
12.7 R. Benn, H. Günther: Modern pulse methods in high resolution NMR spectroscopy. Angew. Chem. Int. Ed. Engl. **22**, 350 (1983)
12.8 A.D.H. Clague: High resolution solid state NMR: Theory and Applications. In *Catalysis*, (Royal Society of Chemistry 1985) p. 75
12.9 V. Jaccarino: Hyperfine interactions in transition metals. In Proc. International school of physics Enrico Fermi (Course XXXVII) (Academic, London 1967)
12.10 A. Narath: *Magnetic Hyperfine Interaction Studies of Magnetic Impurities in Metals*. CRC Crit. Rev. in solid state sci. **3**, 1 (1972)
12.11 G.C. Carter, L.H. Bennet, D.J. Kahan: *Metallic Shifts in NMR*, Prog. in Material Sci. Vol. 20 (Pergamon, London 1977)
12.12 T. Moriya: General theory of relaxation, Prog. Theoretical Phys. **16**, 23 and 64 (1956); and **28**, 371 (1962)
12.13 M.H. Cohen, F. Reiff: Nuclear quadrupole effects in solids. In *Solid state physics* Vol. 5 (Academic, New York 1957) p. 321
12.14 H.E. Rhodes, P.K. Wang, H.T. Stokes, C.P. Slichter: Phys. Rev. B**26**, 3559 (1982)
12.15 M.J. Ledoux, O. Michaux, G. Agostini, P. Panissod: J. Catalysis **96**, 189 (1985)
12.16 V. Niculescu, J.I. Budnick, W.A. Hines, K. Raj, S. Pickart, S. Skalski: Phys. Rev. B**19**, 452 (1979) and references therein
12.17 I. Mirebeau, M.C. Cadeville, G. Parette, I.A. Campbell: J. Physics F**12**, 25 (1982)
12.18 P. Panissod: Helvetica Physica Acta **58**, 60 (1985)
12.19 P. Panissod, I. Bakonyi, R. Hasegawa: Phys. Rev. B**28**, 2374 (1983)
12.20 N. Bloembergen, E.M. Purcell, R.V. Pound: Phys. Rev. **73**, 679 (1948)
12.21 R.M. Cotts: *Hydrogen in Metals*, ed. by G. Alefeld, J. Völkl, Topic Appl. Phys., Vol. 28 (Springer, Berlin, Heidelberg 1978) Chap. 9
12.22 P. Panissod, T. Mizoguchi: Proc 4th Intern. Conf. on Rapidly Quenched Metals, ed. by T. Masumoto, Y. Susuki (Sendai 1982) p. 1621
12.23 D. Alliaga-Guerra, P. Panissod, J. Durand: J. Physique C**8**, 674 (1980)
12.24 A.J. Freeman, R.E. Watson: In *Magnetism* (Academic, New York 1965) p. 217
12.25 P. Panissod, M. Larhichi, M.F. Lapierre-Ravet: Sol. St. Commun. **31**, 273 (1979)
12.26 P. Panissod, J. Durand, J.I. Budnick: Nuc. Inst. Method. **199**, 99 (1982)

12.27   A.G. Redfield: Phys. Rev. **130**, 589 (1963)
12.28   M. Minier: Phys. Rev. **182**, 437 (1969)
12.29   V. Jaccarino, L.R. Walker: Phys. Rev. Lett. **15**, 258 (1965)
12.30   N. Inoue, H. Yasuoka, M. Matsui, K. Adachi: J. Phys. Soc. Jap. **50**, 1180 (1981)
12.31   H. Alloul, F. Hippert: J. Mag. Mag. Mat. **31-34**, 1321 (1983)
12.32   B. Lemius: Thesis Strasbourg (1978)
12.33   Y. Kitaoka, K. Ueno, K. Asayama: J. Phys. Soc. Jap. **44**, 142 (1978)
12.34   P. Panissod, A. Qachaou, G. Kappel: J. Phys. C**17**, 5799 (1984)
12.35   A.M. Portis, A.C. Gossard: J. Appl. Phys. **31**, 2058 (1960)
12.36   E. Beaurepaire, P. Panissod, J.P. Kappler: J. Mag. Mag. Mat. **47-48**, 108 (1985)
12.37   D. Alliaga-Guerra, J. Durand, W.L. Johnson, P. Panissod: Sol. St. Commun. **31**, 487 (1979)
12.38   E. Holland-Moritz, D. Wohlleben, M. Loewenhaupt: Phys. Rev. B**25**, 7482 (1982)
12.39   S. Horn, F. Steglich, M. Loewenhaupt, E. Holland-Moritz: Physica **107**B, 103 (1981)
12.40   Y. Kitaoka, N.S. Chang, T. Ebisu, M. Matsumura, K. Asayama, K. Kumagai: J. Phys. Soc. Jap. **54**, 1543 (1985)
12.41   F. Gautier, G. Krill, P. Panissod, C. Robert: J. Phys. C**7**, L170 (1974)

13. Mössbauer Spectroscopy

U. Gonser

With 19 Figures

The method to be discussed here can be described in two ways: – either as Recoil-Free Gamma Resonance Absorption (RGRA) or as Mössbauer Spectroscopy. The latter term is the one now generally accepted. The reason for the use of the name of a particular person is the fact that this method has a unique history. It is to be hoped that the wonderful and even romantic story of its discovery will one day be told in full detail in a book which could, of course, only be written by that person himself, Rudolf Mössbauer.

13.1 History

Chance, skill and ingenuity are ingredients of great discoveries. For Mössbauer, the chance was provided by the isotope ^{191}Ir. In his search for nuclear resonance fluorescence in this isotope he stumbled over an unexpected phenomenon, actually only a peculiar irregularity in his γ-ray counting device. He demonstrated his experimental skill by not simply ignoring this minor effect, but by proceeding to find its origin consistently and systematically. The right solution was an ingenious idea which developed into one of the great methods of modern science.

The development of the underlying concepts of the discovery can be divided into three stages, as shown schematically in Fig. 13.1.

In the pre-Mössbauer period nuclear reactions – for instance, the emission and absorption of γ-radiation – were treated in a classical fashion by the equations of motion which govern the movement of bodies. Thus, in a nuclear transition with energy E_0, a recoil of the nucleus with energy E_R occurs and the energy of the emitted γ-ray E_γ is reduced by the recoil according to the conservation of energy

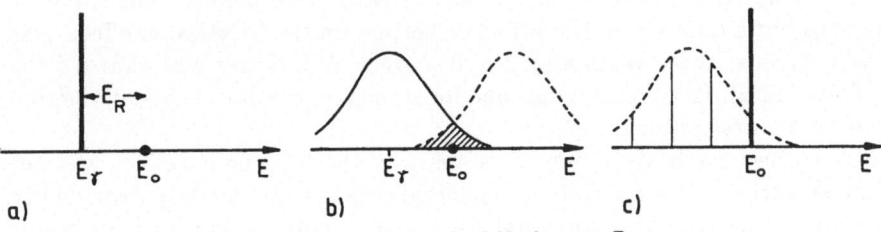

Fig. 13.1. Energy relationship relevant to the Mössbauer effect

$$E_0 = E_R + E_\gamma. \tag{13.1}$$

This situation is shown in Fig. 13.1a. However, in the pre-Mössbauer period it was possible to perform γ-resonance fluorescence experiments. This was accomplished by using prior nuclear reactions, fast rotating devices or high temperatures in order to provide *large* Doppler effects to compensate for the recoil loss. In such experiments the absorber line and/or the emission line can be broadened and in the overlap region resonances are possible (shaded area in Fig. 13.1b). Thus, it was expected that absorption would increase with higher temperatures. Mössbauer's thesis work proceeded along this line to find nuclear resonance fluorescence. Unexpectedly he observed an increase in absorption at lower temperatures, and in a careful analysis he was able to deduce the physical principle of this phenomenon [13.1,2]. He showed that an earlier paper by *Lamb* [13.3] on neutrons was applicable to his problem. He realized that the emission and absorption of γ-radiation can occur in a recoil-free fashion. Without recoil the γ-ray carries the full energy of the transition ($E_\gamma = E_0$) and in the schematic representation (Fig. 13.1c) a line appears at E_0 : the Mössbauer line. The personal accomplishment of discovering recoil-free γ-ray resonance absorption justifies the name Mössbauer Effect. If the method had been found by theoretical predictions, experimental observations and engineering skill – a combination sometimes leading to new methods – we would probably speak of Recoil-Free Gamma Resonance Absorption (RGRA). Instead, we take the name Mössbauer and use it in connection with the method, isotopes, lines, spectrometer, etc.

The early history of the Mössbauer effect has been well described by *Lipkin* [13.4]: in the prehistoric period (before 1958) the effect might have been discovered, but was not. In the early iridium age (1958) the effect was discovered in ^{191}Ir, but taken no notice of. During the middle of the iridium age the effect was noticed, but not believed. In the latter part of the iridium age the effect was believed, but aroused no interest. And then nature helped by providing the iron isotope ^{57}Fe. Actually Mössbauer himself had already realized the favorable properties of this isotope, but at that time an appropriate source was not available to him. With the iron age (after 1959) the effect experienced a breakthrough of dramatic proportions. All previous doubts were dismissed and the wonderful applications such as hyperfine interactions and relativistic effects were demonstrated. Particularly in the United States a fierce fight began in an effort to obtain some priority in the remaining glamour of the new effect. In the editorial of Phys. Rev. Lett. (April 1960) we read: "We believe that the time has come to put a damper on the influx of Letters on the Mössbauer effect". As early as 1961, just three years after the discovery, *Mössbauer* was awarded the Nobel Prize [13.5]. It is virtually unique for someone to achieve this distinction de facto by his first publication.

In retrospect we have to ask ourselves why the Mössbauer effect was not found much earlier. Theoretically, the general concept was already provided in the twenties. It seems surprising that we are genuinely so wrapped up in our macroscopic environment, where we experience recoil everywhere, that it took the analysis of an accidental observation to open our eyes to this microscopic recoil-free quantum effect.

The Mössbauer effect entered with ease all disciplines of natural science: physics, chemistry, biology, metallurgy, engineering, geology, and even such fields as archaeology, art, medicine and others. The Mössbauer Effect Data Index (MEDI) [13.6] compiles all the work done with this method. Several books on the Mössbauer effect have been published, the emphasis being mainly on certain aspects and applications [13.7–19]. Unfortunately, the effect is still considered exotic by many and it is only with reluctance that text books and field conferences take note of the effect. The people using the Mössbauer effect form a kind of close community with regular conferences and news letters. It is very rare for a community to exist on the basis of a scientific method alone. Perhaps the special terminology of the method has some psychological effect – could it be that the constant references to resonances and hyperfine interactions help to create a friendly atmosphere? It is also encouraging that the man who started it all is still actively engaged in the method which has been given his name.

13.2 Principles

Mössbauer's discovery was substantially the realization that a certain probability for recoil-free events exists in the γ-ray emission and absorption processes. When there is no energy loss in the transition from the excited nuclear state to the ground state the γ-ray carries the total energy of the nuclear transition. When this γ-ray strikes, on its path, an isotope with the same nuclear transition there is a certain probability that excitation will take place. The phenomenon of resonance involves two bodies. Resonance between the two nuclei of source and absorber is achieved by the γ-rays. as indicated schematically by the bold arrows in Fig. 13.2. In this figure the situation for the isotope ^{57}Fe is shown

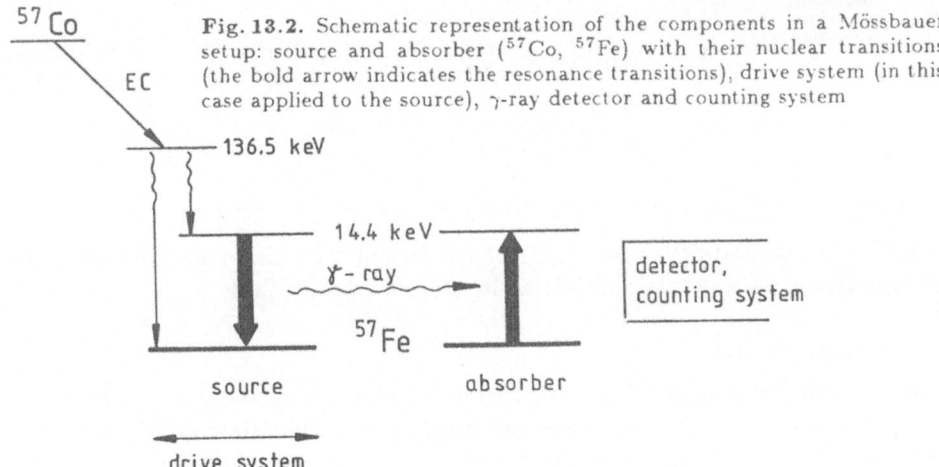

Fig. 13.2. Schematic representation of the components in a Mössbauer setup: source and absorber (^{57}Co, ^{57}Fe) with their nuclear transitions (the bold arrow indicates the resonance transitions), drive system (in this case applied to the source), γ-ray detector and counting system

Table 13.1. Mössbauer parameters and effects

Physical parameter	Mössbauer line		Relativistic effects	Isomer shift
^{57}Fe level diagram of nuclear transition energy E_0 including shift and splittings in source (S) and absorber (A)	source	absorber	source absorber	source absorber
Schematic representation of observation (resonance absorption vs. velocity)	2Γ f		δ_R	δ
Cause of the effects	Relative intensity of resonance line (f) — Probability of recoil-free events Number of recoil-free γ-ray events (Debye-Waller factor)	Full width at half maximum (Γ) — Mean lifetime of the excited state, Line width: spread in wave length, Uncertainty: spread in energy	Change in temperature, change in pressure, acceleration and gravitational fields	Interaction of the nuclear charge distribution with the electron density at the nucleus in source and absorber (electric monopole interaction)
Formulation	$f = \exp - k^2 \langle x^2 \rangle$	$\Gamma = \dfrac{\hbar}{\tau}$	$\delta_R = \dfrac{v^2}{2c^2} E_\gamma$	$\delta = C \dfrac{\delta R}{R} [|\psi_A(0)|^2 - |\psi_S(0)|^2]$
Angular dependence of the hyperfine interactions; Change in the relative line intensities, polarization of the γ-radiation				

including the transitions occurring in the source of ^{57}Co. Information regarding all Mössbauer parameters and effects is summarized in Table 13.1.

13.2.1 Line Width

If no exchange of energy with the surroundings occurs in these nuclear transition processes, the resulting line is extremely sharp. The narrow width of the Mössbauer line or the spread in energy is basically governed by the Heisenberg

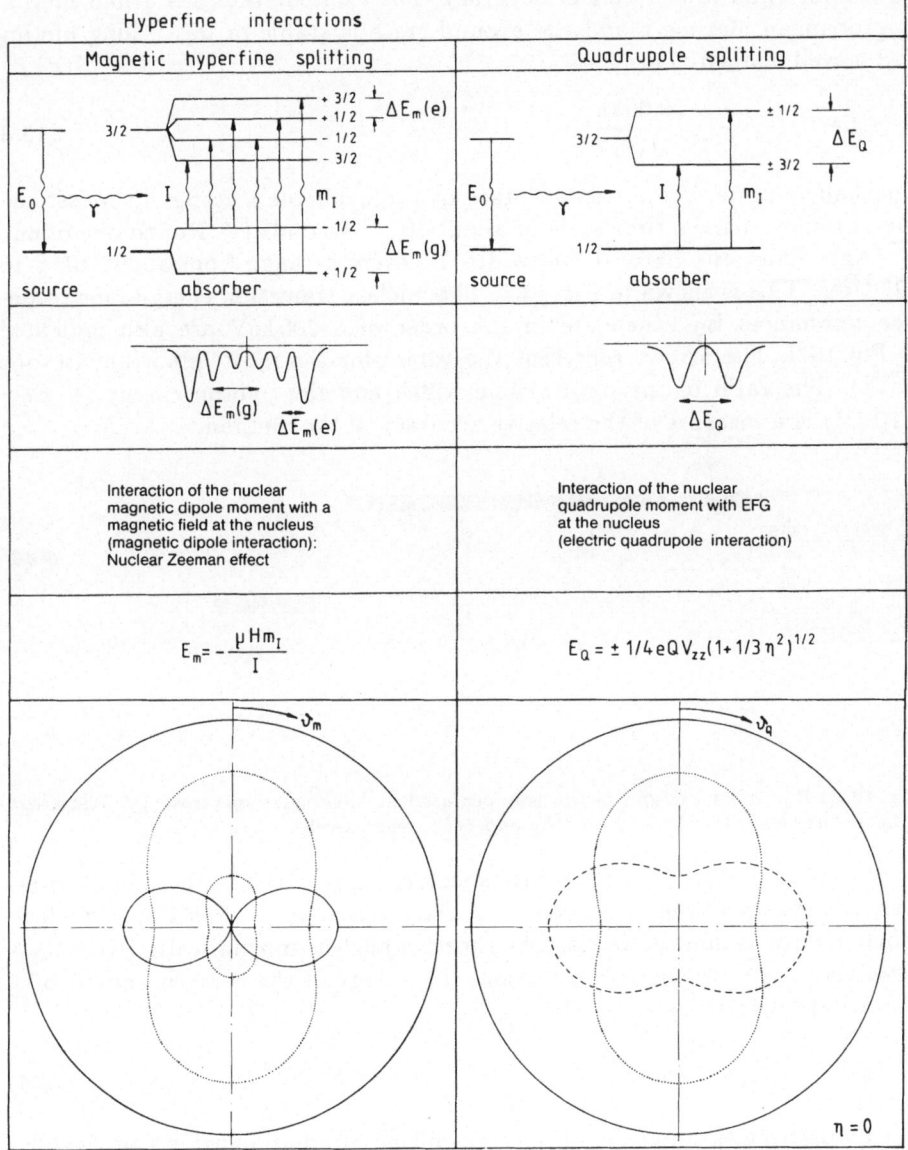

uncertainty principle

$$(\Delta E)\cdot(\Delta t)\geq\hbar. \tag{13.2}$$

The product of the conjugate variables of energy and time are related to Planck's constant. The consequence of this principle is the fact that no measurement of observation of the two conjugate variables can be made simultaneously with higher accuracy than stated here. Physics has to live with this inherent uncertainty and lower limit of accuracy. The excited state has a half lifetime $t_{1/2}$ or mean lifetime τ and the ground state is stable or has a long lifetime and is well defined. Thus

$$\Delta E = \Gamma = \frac{\hbar}{\tau} = \frac{0.693\hbar}{t_{1/2}}. \tag{13.3}$$

The half lifetime of the excited state of the available isotopes in Mössbauer spectroscopy spans a time scale of about 10^{-11} seconds (^{187}Re) to one minute (^{109}Ag). Thus, the natural line width Γ covers a range from about 10^{-5} to 10^{-17}eV. This is shown in Fig. 13.3. The nuclear transition energies for recoil-free resonances E_0 which are in the order of 5–200 keV are also indicated in Fig. 13.3. The arrows represent the values for the most important isotope (^{57}Fe). The ratio of the natural line width and the photon energy ($\Gamma/E_0 \simeq 3 \cdot 10^{-13}$) is a measure of the relative accuracy of the method.

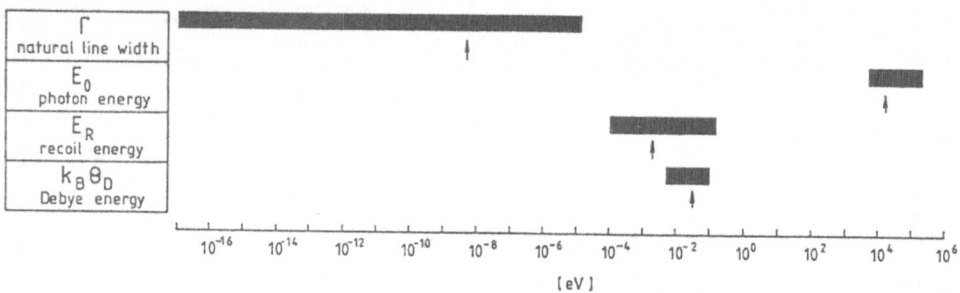

Fig. 13.3. Ranges of energies of the isotopes used in Mössbauer spectroscopy. The arrows indicate the values for the isotope ^{57}Fe and α-Fe, respectively

Obviously, even the best γ-ray detectors cannot resolve these lines. However, the energy of the γ-ray can be slightly varied by a *small* Doppler effect. When a γ-ray is emitted or absorbed from a nucleus moving with a velocity v along the γ-ray propagation direction, the energy of the γ-ray is shifted by a first-order linear Doppler effect

$$E_D = \frac{v}{c} E_\gamma. \tag{13.4}$$

By plotting the intensity of the transmitted γ-radiation behind an absorber as a function of the relative velocity of source and absorber, a resonance line can be traced. In this way the Mössbauer effect becomes spectroscopy.

13.2.2 Line Shape

The energy distribution of the recoil-free nuclear transition processes is governed by the Breit-Wigner formula which can be derived classically by considering the exponential decay. The resulting line shape is Lorentzian and is represented by the total cross section

$$\sigma = \sigma_0 \frac{\Gamma^2}{\Gamma^2 + 4(E - E_0)^2}. \tag{13.5}$$

The maximum cross section

$$\sigma_0 = \frac{\lambda^2}{2\pi} \frac{2I_e + 1}{2I_g + 1} \frac{1}{\alpha_t + 1} \tag{13.6}$$

contains the wavelength of the γ-ray λ, the nuclear spins of the excited and the ground state I_e and I_g, respectively, and the total internal conversion coefficient α_t, taking into account the competing modes of the transitions. If a resonance emission line of natural line width Γ is moved over an absorption line also of natural line width, the sum of both resonance line widths (2Γ) will be observed.

13.2.3 Line Intensity (Recoil-Free Fraction)

The probability of phonon creation or annihilation by the emitting or absorbing atom determines the line intensity. The recoil-free fraction (ratio of recoil-free events to the total number of events) where no exchange of energy with the surroundings occurs is given by

$$f = \exp\left(-k^2 \langle x^2 \rangle\right). \tag{13.7}$$

Thus, f decreases with increasing magnitude of the wave vector of the γ-ray, k, and also decreases with the projection of the mean square of vibrational amplitude, $\langle x^2 \rangle$, averaged over all amplitudes in the half lifetime in the direction of the γ-ray. For a certain γ-ray transition $\langle x^2 \rangle$ effectively controls the intensity of the recoil-free line. No further restrictions on the condensed system are necessary, that is, the crystalline state is not required and a Mössbauer effect can be observed in liquids, amorphous and glassy systems.

Assuming a Debye model for the solid, one has to consider the large number of oscillator levels and their distribution. Each of these levels has a certain probability of being excited by the recoil. Quantitatively from the Debye model the following expression for the recoil-free fraction can be deduced assuming that the mass of the resonance atoms corresponds to the mass of the host atoms:

$$f = \exp\left\{-\frac{3E_R}{2k_B\theta_D}\left[1 + 4(T/\theta_D)^2 \int_0^{\theta_D/T} \frac{x}{e^x - 1} dx\right]\right\} \tag{13.8}$$

where k_B is the Boltzmann constant, T the absolute temperature, and θ_D the Debye temperature. The recoil-free fraction f decreases as the recoil energy of the free atom E_R becomes larger, i.e. with higher energy of the γ-rays or k. On the other hand, higher Debye temperatures θ_D or Debye energies $k_B\theta_D$ involve smaller effective $\langle x^2 \rangle$ and cause f to increase. The ranges of E_R of the available Mössbauer isotopes along with the Debye energies $k_B\theta_D$ of the typical matrices in which the isotopes are embedded, are shown in Fig. 13.3.

The arrows indicate the values for the 14.4 keV excited state of ^{57}Fe and α-Fe, respectively. From the magnitude of the experimentally determined absorption one obtains information on the phonon spectrum, on vibrational modes of the resonating atom as a function of direction, temperature, pressure, in different lattices and phases, near critical temperatures, at the surface or close to other defects and, in special cases, the lattice vibrational anisotropy is revealed.

By comparison two important features can be read off from Fig. 13.3. The large difference of the natural line widths Γ and the nuclear transition energies of the γ-radiation E_0 marks the high relative accuracy and resolution of this method while the recoil energies of the free atoms E_R and the Debye energies of typical solids are close together and overlap to a great extent.

The recoil-free fraction f is known as the Debye-Waller factor in the evaluation of x-rays scattered from a crystal. In the analysis of Mössbauer data in certain cases the Debye-Waller factor has to be replaced by a Lamb-Mössbauer factor to take into account the temporal aspects in the interaction of the γ-rays with the solid and allowing to differentiate the static and dynamic contributions [13.20].

It had also been noted by quantum mechanical treatment that the mean energy transferred to the lattice in the emission (or the absorption) process of γ-quanta is just equal to the recoil energy of the free atom E_R (Lipkin's sum rule [13.21]).

13.3 Mössbauer Isotopes

In Mössbauer spectroscopy two nuclei are in resonance by virtue of recoil-free γ-radiation. So far the Mössbauer effect has been found in a total of about 100 isotopes. Because some isotopes have more than one transition available for resonances, altogether 120 transitions have been successfully tested and used. Fortunately, the Mössbauer Effect Data Index (MEDI) regularly published by John and Virginia Stevens does statistics and book-keeping on the ever-increasing number of new isotopes and resonances available to the Mössbauer community.

Relevant properties of the 25 most popular Mössbauer isotopes are listed in Table 13.2. The last column represents the percentage of publications using the specific isotope. With 66.2%, ^{57}Fe has the lion's share. The prevalence of ^{57}Fe in this type of spectroscopy is, on the one hand, due to the favorable nuclear parameters and, on the other, to the importance of the element iron in technology, science and life in general. The natural abundance of only 2.2% is sometimes a handicap, but this can often be overcome by enrichments. As we can see from the table, in the case of the second and third places (^{119}Sn and ^{151}Eu) we are down in the usage to 14.6 and 2.0%. The 25 isotopes listed represent 96.8% of the total work. In other words, all the papers dealing with the remaining isotopes represent only a fraction of 3.2%.

Table 13.2. Relevant properties of the 25 most popular resonance transitions [a: isotopic abundance, E_0: nuclear transition energy, $t_{1/2}$: half lifetime, I_e and I_g: nuclear spin quantum number of excited (e) or ground (g) state, respectively, σ_0: maximum resonance cross section in barn, Γ: natural line width, theoretical full width at half maximum, E_R: recoil energy of the free atom, U: usage in percentage]. The isotopes listed represent 96.8% of the experimental work (courtesy John and Virginia Stevens)

	Isotope	a [%]	E_0 [keV]	$t_{1/2}$ [ns]	I_e	I_g	σ_0 [10^{-20}cm^2]	2Γ [mm/s]	E_R [10^{-3}eV]	U [%]
1.	^{57}Fe	2.19	14.4125	97.81	3/2	1/2	256.6	0.1940	1.957	} 66.2
		2.19	136.46	8.7	5/2	1/2	4.300	0.2304	175.4	
2.	^{119}Sn	8.58	23.871	17.75	3/2	1/2	140.3	0.6456	2.571	14.6
3.	^{151}Eu	47.82	21.64	9.7	7/2	5/2	11.42	1.303	1.665	2.0
4.	^{125}Te	6.99	35.46	1.48	3/2	1/2	26.56	5.212	5.401	1.7
5.	^{121}Sb	57.25	37.15	3.5	7/2	5/2	19.70	2.104	6.124	1.6
6.	^{129}I	(radioactive)	27.77	16.8	5/2	7/2	40.32	0.5863	3.210	1.5
7.	^{197}Au	100	77.35	1.90	1/2	3/2	3.857	1.861	16.31	1.2
8.	^{161}Dy	18.88	25.65	28.1	5/2	5/2	95.34	0.3795	2.194	} 1.1
		18.88	43.84	920	7/2	5/2	28.29	0.006782	6.410	
		18.88	74.57	3.35	3/2	5/2	6.755	1.095	18.55	
9.	^{237}Np	(radioactive)	59.537	68.3	5/2	5/2	32.55	0.06727	8.031	0.8
10.	^{170}Yb	3.03	84.262	1.60	2	0	23.93	2.029	22.43	0.7
11.	^{166}Er	33.41	80.56	1.82	2	0	23.56	1.866	20.99	0.6
12.	^{99}Ru	12.72	89.36	20.5	3/2	5/2	14.28	0.1493	43.31	0.5
13.	^{155}Gd	14.73	60.012	0.155	52	3/2	9.989	29.41	12.48	} 0.5
		14.73	86.54	6.32	5/2	3/2	34.40	0.5002	25.94	
		14.73	105.308	1.16	5/2	3/2	24.88	2.239	38.42	
14.	^{193}Ir	62.7	73.028	6.3	1/2	3/2	3.058	0.5946	14.84	} 0.5
		62.7	138.92	0.080	5/2	3/2	5.833	24.61	53.69	
15.	^{169}Tm	100	8.42	3.9	3/2	1/2	21.17	8.330	0.2253	0.4
16.	^{61}Ni	1.25	67.40	5.06	5/2	3/2	72.12	0.8021	39.99	0.4
17.	^{181}Ta	99.99	6.23	6800	9/2	7/2	167.6	0.006457	0.1151	} 0.4
		99.99	136.25	0.0406	9/2	9/2	5.968	49.45	55.07	
18.	^{182}W	26.41	100.102	1.37	2	0	25.17	1.995	29.56	0.3
19.	^{83}Kr	11.55	9.40	147	7/2	9/2	107.5	0.1980	0.5716	0.3
20.	^{133}Cs	100	89.997	6.30	5/2	7/2	10.21	0.5361	26.49	0.3
21.	^{153}Eu	52.18	83.3652	0.82	7/2	5/2	6.705	4.002	24.39	} 0.3
		52.18	97.4283	0.21	5/2	5/2	17.97	13.37	33.31	
		52.18	103.1774	3.9	3/2	5/2	5.417	0.6798	37.36	
22.	^{127}I	100	57.60	1.9	7/2	5/2	21.37	2.500	14.03	0.3
23.	^{67}Zn	4.11	93.31	9150	3/2	5/2	10.12	0.000320	69.78	0.2
24.	^{195}Pt	33.8	98.857	0.170	3/2	1/2	6.106	16.28	26.91	} 0.2
		33.8	129.735	0.620	5/2	1/2	7.425	3.401	46.35	
25.	^{73}Ge	7.76	13.263	4000	5/2	9/2	361.2	0.005156	1.294	} 0.2
		7.76	68.752	1.86	7/2	9/2	22.28	2.139	34.77	

The Mössbauer spectra reflect the properties of source and absorber. In most cases the absorber represents the specimen under investigation, but in certain instances the spectral parameter of the source gives interesting information.

13.3.1 Sources

In the sources the excited state is populated and the transition to the ground state must occur in a recoil-free fashion.

The sources must conform to stringent conditions in order to be considered good candidates for the method. The nuclear reaction in reactors or accelerators needed to produce the parent isotopes –and also the chemical separation, if necessary – should be relatively simple. Long lifetime of the parent isotopes avoids frequent replacement of the source. However, the lower rate of decay necessitates that more isotopes be present to have a certain activity available. A simple mode of decay and simultaneous high population of the excited state from which the recoil-free γ-rays are emitted is preferred. For sources it is most usual to choose metallic matrices in which the parent or excited isotopes are diffused, implanted, Coulomb-excited or produced by nuclear reaction. Some of the unique advantages of metal sources are as follows: high coordination symmetry and in non-magnetic cubic metals (fcc and bcc) single-line emission spectra are produced. The effective Debye-Waller factor is relatively high for most metals. Electronic relaxation processes are extremely fast contrary to insulators where localized charge states are found as a result of the foregoing decay existing over long time periods, as compared to the lifetime of the excited states. In metals the nature of defects created by the recoil of the processes leading to the parent isotopes or to the excited nuclear resonance level is relatively better understood than that of defects created in insulators and semiconductors. Uniform distribution of the isotope in the metal matrix can be achieved by an appropriate heat treatment and, finally polarized γ-rays can be produced by magnetizing ferromagnetic sources, for instance, α-Fe.

^{57}Co with a mean half lifetime of about 270 days populates the 14.4 keV excited states of ^{57}Fe, as indicated in Fig. 13.2. The matrices for ^{57}Co commonly used have an interesting historical development: In the early days of the method paramagnetic fcc stainless steel was used as a single-line source material. The resulting line is broadened considerably by a distribution of small shifts and quadrupole splittings due to a variety of different local environments. ^{57}Co in Cu had the drawback of possible precipitation of the supersaturated solid solution. ^{57}Co in Pd and Pt showed relatively high electronic absorption and interferences of the K x-rays or L x-rays, respectively. ^{57}Co in Cr sources are difficult to produce in foil form due to the brittleness of this element. In recent years most researchers have used ^{57}Co in Rh sources. Usually sources contain some millicuries of ^{57}Co; however for very strong sources in the order of Curies ^{57}CoO and Sb$_3$ ^{57}Co have been suggested. In these cases the accurate stoichiometry is important and frequent renewal of the source is required when broadening occurs due to the self-absorption of the ^{57}Fe produced [13.22].

13.3.2 Absorbers

There are favorable isotopes and others where resonances are difficult to achieve. Considering Table 13.2 the preferable isotopes are relatively easy to distinguish: The isotope in question should have a reasonable isotopic abundance a. Radioactive absorbers are difficult to handle. The nuclear transition energy E_0 should not be less than 5 keV (to get the γ-rays through or out of the sample) and not more than 200 keV in order to have reasonable Debye-Waller factors. The half lifetime of the excited state $t_{1/2}$ should be in the order of about 10^{-5}–10^{-9}s. Half lifetimes beyond this range cause problems: the natural line width Γ of the isotopes with long lifetimes is very small and becomes broadened by their sensitivity to various contributions and, on the other side, the isotopes with very short lifetimes lack in resolution. The relative change in the nuclear radii of the excited and ground state $\delta R/R$, the nuclear magnetic dipole moment μ and the nuclear electric quadrupole moment eQ should not be too small in order to produce large shifts and splittings, respectively. The maximum resonance cross section σ_0 should be large.

The purist working only in the γ-ray transmission mode might find the total internal conversion coefficient of the ^{57}Fe 14.4 keV transition ($\alpha_t \simeq 8$) to be high. However, the competing conversion electrons are extremely useful when used in scanning surfaces in the scattering mode.

The absorber thickness is important when a quantitative analysis of the spectra is desired. The effective absorber thickness t_A (for single line spectra) is usually defined by

$$t_A = \sigma_0 f_A d_A na \qquad (13.9)$$

where σ_0 is the maximum absorption cross section, see (13.6), f_A the recoil-free fraction and d_A the physical thickness in cm. The subscript A designates the absorber, n is the number of atoms/cm^3 of the particular element and a the isotopic abundance. Most experiments are carried out with a thickness of $t_A \simeq 1$. If t_A is very large the resonance lines become rather broad and if t_A is reduced the intensity of the resonance line becomes small.

13.4 Methodology

While in the nuclear fluorescence experiments of the pre-Mössbauer time very large Doppler velocities were required for compensating the recoil, in contrast, only very small relative Doppler velocities between source and absorber are used to trace the narrow recoil-free Mössbauer line.

The Doppler variation technique had already been recognized by Mössbauer in his first experiments which were accomplished by using a toy. In the early years mechanical and electromechanical systems at various levels of sophistication were used. Technical details and theoretical background have been treated by a number of authors [13.23]. We shall, therefore, give only a brief discussion on some of the important features.

13.4.1 Classical Setup

A basic Mössbauer spectrometer consists of the following elements: source and absorber, drive system for moving the source relative to an absorber and detector with a counting system. The arrangement in transmission geometry is shown schematically in Fig. 13.2.

In most cases the source is moved by an electromechanical drive, in principle similar to a loudspeaker. By electronics an appropriate velocity range and mode can be controlled and adjusted. Quite common are motions of constant velocities or constant accelerations. The radiation is detected and counted by γ-ray detectors. Various detectors are available: scintillation counters, proportional counters, channeltrons, solid-state detectors (lithium drifted germanium and silicon). These detectors are considerably different in performance, lifetime, reliability and price. Usually the detected γ-ray pulses are stored in a multi-channel analyzer where each channel corresponds to a specific relative velocity between the source and the absorber. An accurate calibration of the relative velocity between source and absorber is required. For this purpose the spectrum of $\alpha - Fe$ is very useful. The convention has been adopted that positive velocity is defined as the motion of source and adsorber towards each other and negative velocity as their motion away from each other. In plotting a Mössbauer spectrum, for the sake of convenience, the γ-radiation is usually normalized to the off-resonance count rate, thus one obtains the relative transmission.

Mössbauer spectra of an $\alpha - Fe$ foil are exhibited in Fig. 13.4. The data points show a certain scatter due to the statistical nature of the count rate of nuclear events. The statistical fluctuation becomes smaller as the number of counts increases. The appearance becomes smooth by using least-squares fittings and by doing so the Mössbauer parameters can be determined with high accuracy.

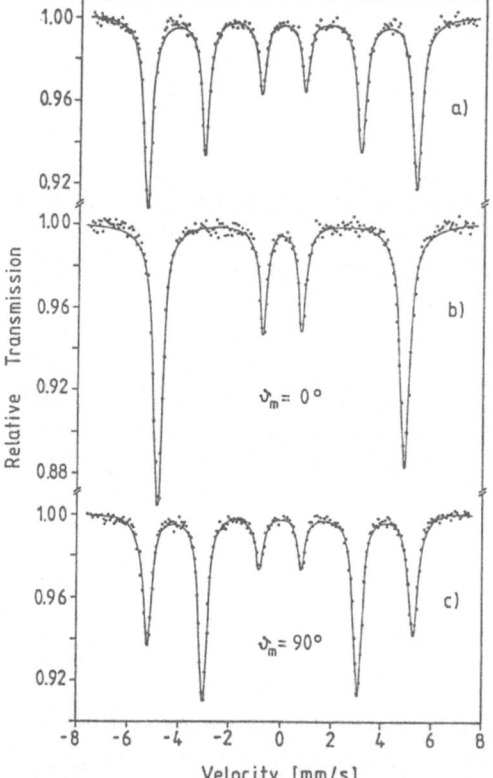

Fig. 13.4. Mössbauer spectra of α-Fe at room temperature: *(a)* $H_{ext} = 0$, *(b)* $H_{ext} = 50\,kOe$, $\vartheta_m = 0°$, *(c)* $H_{ext} = 3.5\,kOe$, $\vartheta_m = 90°$. H_{ext} indicates the external magnetic field

13.4.2 Scattering Geometry

The conventional transmission geometry requires a certain thickness of the absorber as regards both nuclear resonance absorption and electronic absorption. This requirement might be circumvented by scattering geometry. The scattering has the advantage that the competing radiation accompanying the nuclear transition can be used: re-emitted γ-rays, x-rays associated with internal conversion and conversion electrons. When the resonance condition exists (bold arrow in Figs. 13.2 and 5) the deexcitation from the excited state in the absorber causes an *increase* for the various radiations, which results in a peak instead of a dip in the Mössbauer spectrum. The penetration depths of the various radiations (γ-rays, x-rays and electrons) are quite different. Thus, the spectra differ depending on which type of radiation is measured. The choice of one type of radiation selects a particular range of depth in the sample. The scattering geometry is particularly useful in probing with 7 keV conversion electrons resulting from the first excited state of ^{57}Fe. Since only the conversion electrons from nuclei to a depth of about 1000 Å emerge, this method is applicable for studying the atoms near the surface. This technique is often abbreviated as CEMS (conversion electron Mössbauer spectroscopy).

Recently the method has been extended by analyzing the energy of the emerging electrons, thus gaining depth selective information, DCEMS (depth-selective conversion electron Mössbauer spectroscopy [13.24]).

The combination and simultaneous measurements of γ-ray absorption in transmission geometry and conversion electron emission in scattering geometry have proven to be of great value [13.25,26]. In such cases the absorber is placed within an electron counter, as shown in Fig. 13.5. Whenever nuclear resonance

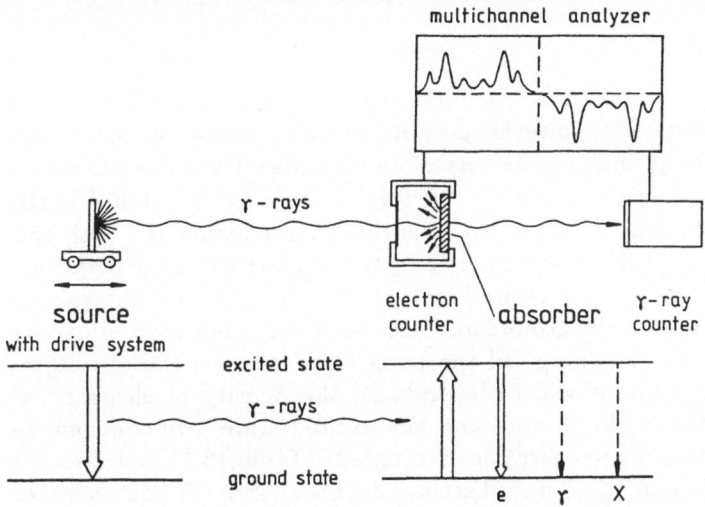

Fig. 13.5. Schematic representation of a combined experimental setup of γ-ray absorption and conversion electron emission geometry. In the lower part the relevant transitions between the nuclear excited and the ground state are shown

absorption has taken place, it is witnessed by conversion electron emission from the excited state. The count rate of electrons might be stored in the first half of a multichannel analyzer. The emission spectrum reflects the properties of the surface where the electrons originate. The γ-rays transmitting absorber and electron counter might be stored in the second half of a multichannel analyzer. This absorption spectrum represents and reflects the bulk properties.

13.5 Hyperfine Interactions

The phenomena of recoil-free γ-resonance was in itself a most interesting discovery. However, the true greatness and wide applicability of the Mössbauer effect came with the realization and resolution of the hyperfine interaction. The sharpness of the lines made it possible to measure small shifts and splittings. Usually the electromagnetic environment of a nucleus is at least partly determined by the electrons of the same atom so that the effects of electromagnetic environment are considered to be hyperfine effects.

The hyperfine coupling mechanisms consisting of interaction between a nuclear (moment) property and an electronic property are of great significance. There are three main hyperfine interactions corresponding to the nuclear moments determining the nuclear levels:

i)   electric monopole interaction   isomer shift,
ii) magnetic dipole interaction   nuclear Zeeman effect,
iii) electric quadrupole interaction   quadrupole splitting.

The nuclear level scheme, the transitions, the spectral observations and their angular dependence are schematically represented in Table 13.1.

13.5.1 Isomer Shift

Most atomic electrons have zero probability of being within the nucleus. The wave functions of s electrons, however, do not have zero amplitude at the nucleus and therefore the probability density of an s electron has a definite small, but non-zero value within the nucleus. The origin of the isomer shift δ is the result of the Coulomb interaction of the negative charge associated with this small electron density within the nucleus and the nuclear charge distribution over finite nuclear radius R in the excited and ground state. Because the volume of the nucleus is different in the ground and excited states, the energy shifts are also different. Therefore the energy of the transitions between the two states is also shifted by an amount which depends on the density of electrons at the nucleus. The levels of the ground and the excited states are changed, as schematically indicated in the energy level diagram (Table 13.1) and also the transition energies E_S and E_A. The subscripts denote source (S) and absorber (A). E_0 can be considered as a hypothetical case of a point nucleus with the same charge. In the resonance experiment the difference between E_S and E_A is shown by a shift which is called the isomer shift [13.27,28]. It is also called the chemical shift because a change in the chemical state of the atoms has a direct or indirect effect on the density of electrons at the nucleus and thus, finally, on the resulting shift in the transition energy. This shift can be expressed (in the non-relativistic approximation) as

$$\delta = C\frac{\delta R}{R}[|\psi_A(0)|^2 - |\psi_S(0)|^2] \qquad (13.10)$$

where C is a constant for a given isotope containing nuclear parameters, $\delta R/R$ is the relative change of nuclear radius between excited state and ground state, and the term in parentheses represents the difference in the total electron density evaluated at the nucleus $|\psi(0)|^2$ between the absorber and source isotopes. From the direction of the resonance line shift one can deduce that δR has a negative sign for ^{57}Fe, i.e. the nuclear radius of the excited state is smaller than the radius of the ground state. With the isotope ^{119}Sn the opposite result is found.

The experimental isomer shifts are relative. It is desirable to establish unique reference standard materials for each resonance isotope so that various measurements can be easily compared. Metallic $\alpha -$ Fe at room temperature has customarily been chosen for the resonance transition in ^{57}Fe.

13.5.2 Magnetic Hyperfine Interaction

The interaction of the nuclear magnetic dipole moment μ with a magnetic field H at the site of the nucleus splits the nuclear state with spin I(I>0) into (2I+1) sublevels with eigenvalues

$$E_m = -\frac{\mu H m_I}{I} \qquad (13.11)$$

where m_I is the magnetic quantum number with values $m_I = I, I-1, \ldots - I$. The isotope ^{57}Fe in its nuclear ground state has I = 1/2 and in the 14.4 keV excited state has I = 3/2. A magnetic field at the site of the nucleus causes a splitting of the nuclear states, as shown in the energy level diagram (Table 13.1). Of the eight possible transitions from the four upper sublevels to the two lower sublevels only six are allowed in the absence of other perturbations leading to the typical six-line hyperfine pattern shown in Fig. 13.4 and in Table 13.1. The ordering of the sublevels m_I indicates the opposite of the nuclear magnetic moments in the ground state and in the excited state [13.29,30].

The splitting of the magnetic sublevels is directly proportional to the magnetic field present at the nucleus which is often called the effective field or internal field H_i. Thus the splitting of the lines of the spectrum is a direct measure of the internal field. For $\alpha -$ Fe at room temperature $H_i = 330$ kOe has been deduced.

A hyperfine-splitting spectrum contains information on the orientations of the internal fields of the source and/or absorber. This is because both the intensity and polarization of the radiation associated with each of the six transitions depend on the angle ϑ_m between the propagation direction of the radiation and the direction of the magnetic field, as seen in Fig. 13.4. The relative intensities of the first three transitions (the other three are equivalent) are given in Table 13.3 and graphically shown in Table 13.1. When the direction of the internal fields at different atoms is isotropically distributed the Mössbauer spectrum produced exhibits the 3:2:1:1:2:3 intensity ratios. In all cases the sum of the intensities of the six lines is isotropic, as indicated by the circle (or sphere) in Table 13.1. By making measurements of the angular dependence and polarization it is possible to determine the direction and degree of alignment of the hyperfine fields in a sample.

13.5.3 Electric Quadrupole Interaction

The interaction of the nuclear electric quadrupole moment eQ with the principal component of the diagonalized electric field gradient (EFG) tensor $V_{zz} = \delta^2 V/\delta z^2$ at the site of the nucleus splits the nuclear state into sublevels with the eigenvalues

$$E_Q = \frac{eQV_{zz}}{4I(2I-1)} [3m_I^2 - I(I+1)] \left(1 + \frac{\eta^2}{3}\right)^{1/2}. \tag{13.12}$$

The asymmetry parameter η is given by

$$\eta = \frac{V_{xx} - V_{yy}}{V_{zz}} \tag{13.13}$$

with $|V_{zz}| > |V_{yy}| \geq |V_{xx}|$; $V_{zz} + V_{yy} + V_{xx} = 0$, thus $0 \leq \eta \leq 1$.

The electric quadrupole interaction splits the first nuclear excited state of ^{57}Fe and ^{119}Sn (I = 3/2) into sublevels, as indicated in the energy-level diagram (Table 13.1) with the eigenvalues

$$E_Q = \pm \frac{1}{4} eQ V_{zz} \left(1 + \frac{1}{3}\eta^2\right)^{1/2}. \tag{13.14}$$

Typical quadrupole split spectra of the intermetallic phases Zr_2Fe and Zr_3Fe are shown in Fig. 13.6a and b [13.31].

A uniform and sufficiently strong electric field gradient cannot be applied by external means, and observed quadrupole patterns therefore reflect only the influence, on each nucleus, of all the charged particles in the sample.

The angular dependence of the radiation pattern produced at the ^{57}Fe nucleus by an EFG with axial symmetry ($\eta = 0$) is given in Tables 13.1 and 3. In the this case ϑ_q represents the angle between the principal axis of the EFG and the propagation direction of the γ-ray. In the extreme case the relative line intensities of the two quadrupole split lines (in the thin absorber approximation) are for $\vartheta_q = 0°$ 6:2 and for $\vartheta_q = 90°$ 3:5. For a randomly oriented polycrystalline material the relative line intensities are 1:1. Again in all cases the sum of the line intensities is isotropic, as indicated by the circle (or sphere) in Table 13.1.

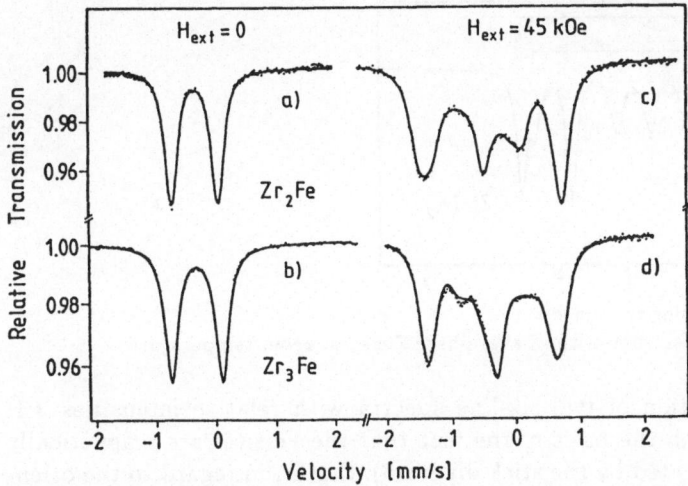

Fig. 13.6. Mössbauer spectra of Zr_2Fe (*a* and *c*) and Zr_3Fe (*b* and *d*) at room temperature. The spectra on the right (*c* and *d*) were obtained in an external magnetic field of $H_{ext} = 45\,kOe$

Table 13.3. Angular dependence of the magnetic hyperfine six-line pattern and of the electric quadrupole interaction ($\eta = 0$) of ^{57}Fe. ϑ_m or ϑ_q represents the angle between the propagation direction of the γ-radiation and the direction of the magnetic field or the direction of the principal axis of the electric field gradient at the site of the ^{57}Fe nucleus, respectively. These relationships are also shown in Table 13.1

Transition	Δm	Angular dependence	Transition	Angular dependence
$\pm 3/2 \rightarrow \pm 1/2$	± 1	$3/4(1 + \cos^2 \vartheta_m)$	$\pm 3/2 \rightarrow \pm 1/2$	$1 + \cos^2 \vartheta_q$
$\pm 1/2 \rightarrow \pm 1/2$	0	$\sin^2 \vartheta_m$	$\pm 1/2 \rightarrow \pm 1/2$	$2/3 + \sin^2 \vartheta_q$
$\mp 1/2 \rightarrow \pm 1/2$	∓ 1	$1/4(1 + \cos^2 \vartheta_m)$		

13.5.4 Mixed Interactions

Additional information is obtained when magnetic-dipole and electric-quadrupole effects act simultaneously on the same nucleus, although splittings and intensities of the pattern may be difficult to predict or interpret [13.30,32,33]. If one of these hyperfine effects is only a small perturbation on the other, both the prediction and the interpretation are – at least qualitatively – simpler. The Zr-Fe system containing the intermetallic compounds Zr_3Fe, Zr_2Fe and $ZrFe_2$ is useful to demonstrate this point.

The similarity in appearance of the two spectra in Fig. 13.6a and b might also indicate a similarity in structure. However, if an external magnetic field of 45 kOe is applied to these alloys the similarity is removed (Fig. 13.6c and d). The analysis reveals opposite signs of the main components of the EFG's and an asymmetry parameter of $\eta \simeq 0$ for Zr_2Fe and $\eta \simeq 0.6$ for Zr_3Fe. Another example is given by the ferromagnetic Laves phase $ZrFe_2$ shown in Fig. 13.7. Although all Fe sites in this structure are crystallographically equivalent, the spectrum

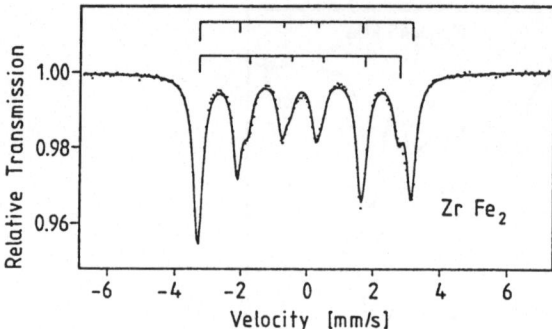

Fig. 13.7. Mössbauer spectrum of the Laves phase $ZrFe_2$ at room temperature

exhibits a superposition of two six-line spectra with relative intensities 3:1. This was explained on the basis of the fact that the Fe sites are magnetically inequivalent (as indicated by the stick diagram), that is, in regard to the orientations of the spins and the principal axis system of the electric field gradient [13.34].

13.5.5 Polarimetry

The magnitude of the hyperfine splittings reveals the values of the internal magnetic field (H_i) or the electric field gradient (V_{zz}), respectively. The relative line intensity and orientation dependence of the emitted (or absorbed) radiation are determined by the conservation of angular momentum in the system of nucleus plus γ-quantum [13.35–37]. The multipolarity of the 14.4 keV γ-rays is almost exclusively M1 with the selection rule $\Delta m = 0, \pm 1$. Intensities of typical hyperfine patterns are shown in Table 13.1.

Three methods are used to produce polarized γ-radiation [13.38,39], again exemplified by the resonance transition in ^{57}Fe.

1. The emitted radiation pattern of a magnetized ferromagnetic source of ^{57}Co in α – Fe consists of six lines, all linearly polarized if observed perpendicular to the direction of H_i ($\vartheta_m = 90°$). The plane of polarization of the γ-rays depends on the transitions corresponding either to $\Delta m = \pm 1$ or $\Delta m = 0$. If observed longitudinally parallel to the direction of H_i ($\vartheta_m = 0°$), however, the radiation consists of only four lines; the transitions corresponding to $\Delta m = 0$ are missing. The γ-radiation of these four lines is circularly polarized. The helicity depends on the sign of the transition; thus the reversal of the magnetic field reverses the helicity.

2. In certain cases the quadrupole interaction can be useful in producing linearly polarized γ-radiation.

3. Using a filter technique polarized γ-rays are also produced by the phenomenon of dichroism in which certain polarization components are selectively absorbed out of a beam of electromagnetic radiation. The transmitted γ-rays will be partially linearly or circularly polarized depending on the absorption in the filter.

In such experiments the polarization may be complete or partial depending on the kind of hyperfine interaction and also on the direction of observation relative to the hyperfine fields or the axes of the electric field gradient. As an example we might consider a ferromagnetic source and a ferromagnetic absorber, both exhibiting a six-line hyperfine pattern. In general, a 36-line spectrum is produced by the absorption of each of the six absorption lines at various velocities. However, if the internal fields of source and absorber are equal the number of lines is reduced by degeneracy to 15.

Furthermore, if source and absorber (for instance, ^{57}Co in α − Fe versus α − Fe) are magnetized perpendicular to the γ-ray propagation directions all lines are linearly polarized. When the magnetic fields of source and absorber are parallel, nine resonance lines appear, and when these fields are perpendicular to each other, six lines become evident.

If source and absorber are magnetized longitudinally in the direction of the γ-ray propagation all lines are circularly polarized. In this case, we are moving a four-line source pattern over a four-line absorber pattern. Provided the magnetic fields are the same in magnitude and anti-parallel six lines appear, but only three appear in the case of parallel magnetic fields. The latter situation is demonstrated in Fig. 13.8. The spectrum was taken with a source of ^{57}Co in α − Fe and an α − Fe absorber both in an external magnetic field of 52 kOe applied perpendicular to the plane of the foils but longitudinally with regard to the γ-rays. The external applied field in conjunction with the demagnetizing

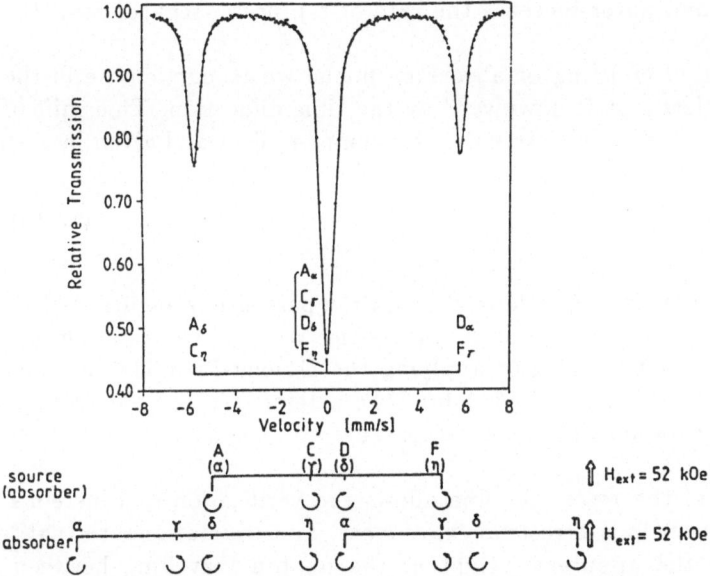

Fig. 13.8. Mössbauer spectrum obtained with a source of ^{57}Co in α-Fe and an α-Fe absorber. Both source and absorber are in a longitudinal magnetic field ($\vartheta_S = \vartheta_A = 0°$), $H_{ext} = 52$ kOe. The stick diagrams indicate the positions, relative line intensities and helicity of the radiation. A, C, D, F are source lines while α, γ, δ, η represent the absorber lines in ascending order of energy

field and the negative hyperfine interaction reduces the internal field to about 300 kOe. If both source and absorber are stationary strong resonance occurs at zero velocity because both resonance line *and* polarization are perfectly matched. By moving the absorber at a velocity of ±5.6 mm/s the condition of resonance is fulfilled in terms of resonance energy as well as helicity. This is indicated by the stick diagram, the letters and the symbols for the polarizations.

By analogy with optics a fully automatic birefringence rotation polarimeter with polarizer (source) and analyzer (absorber) has been built [13.40]. The dispersion associated with Mössbauer resonance absorption in transmission can be measured, in particular the Mössbauer-Faraday rotation [13.38,41,42].

13.6 Relativistic Effects

In general, during the lifetime of the nuclear excited state ($\simeq 10 - 10^{-11}$s) the vibrating atoms in a solid pass through many oscillations with typical frequencies of $\simeq 10^{13}$Hz. The average velocity vanishes and a first-order Doppler or classical Doppler effect is not to be expected. However, the thermal vibrations cause a change in the frequency or shift of the resonance line as a result of time dilatation. Time dilatation is the well-known slowing down of all processes within a moving object and it stems from the factor $\sqrt{1 - v^2/c^2}$ in the Lorentz transformation, v being the velocity of the object and c the velocity of light. Because this factor contains only the square of v, the result of motion is always to decrease and never increase the rate of a process, regardless of the sign or direction of v.

The γ-radiation of emitting or absorbing nuclei which participate in the thermal motions suffers a shift downward by the time dilatation. This shift of the resonance line is obviously a second-order Doppler effect and given by

$$\delta_R = \frac{\overline{v^2}}{2c^2} E_\gamma. \tag{13.15}$$

The average square velocity $\overline{v^2}$ is dependent on temperature, pressure, Debye temperature of the surrounding matrix and lattice defects. Considering the kinetic part of the specific heat and applying the Dulong-Petit law one obtains at elevated temperatures ($T \geq \theta_D$) for ^{57}Fe a temperature shift of $\delta_T \simeq 7 \cdot 10^{-4}$ mm(s·K)$^{-1}$ or the shift becomes comparable with the natural line width at $\Delta T \simeq 150$ K.

The sharpness of the resonance line allows the verification of Einstein's mass-energy equivalence in a terrestrial experiment. The gravitational red shift measures effectively the apparent weight of the photon travelling between source and absorber at different gravitational potentials. Such experiments have been carried out [13.45,46] with ^{57}Fe and also with the much sharper resonance line of ^{67}Zn.

13.7 Time-Dependent Effects

Every method has its characteristic time, effectively the time in which the information on a certain property is gathered and measured with an appropriate instrument. In dynamic processes timing is essential and a clock is needed for the measurements. In Mössbauer spectroscopy the clock is located right in the microscopic resonating nucleus. The most basic characteristic time is the mean lifetime for spontaneous transitions from the excited state to the ground state. This characteristic time starts at t_0 which is the instant when the excited state is reached from above in source experiments or, in absorber experiments, when the leading edge of the photon arrives at the nucleus.

Considering the isotopes available one has a wide range to choose from: $\sim 1 \min$ (^{109}Ag) to 10^{-11}s (^{187}Re). In general, the resonance line has a Lorentzian shape and its width in energy units is determined in the ideal case by the Heisenberg uncertainty principle or natural line width Γ. The natural line width has been verified for ^{57}Fe, but for the isotopes with lifetimes of about two orders of magnitude larger (^{67}Zn, ^{73}Ge and ^{181}Ta) natural line widths have not yet been obtained. Broadening of the natural line width often occurs, and various reasons can be given. Without violating the uncertainty principle the observed resonance line might also become narrower than the natural line width. This becomes clear by recalling that all the nuclear transitions contributing to an observed resonance line have different individual lifetimes of which only the mean τ is known. By observing all the decays one obtains the natural line width Γ. However, the decay to the nuclear excited state in the source can be used as a signal in a delay coincidence experiment [13.47]. Thus, in the following transition to the ground state only those γ-rays emerging from long-lived excited states can be selected; thus the effective mean lifetime τ_{eff} of the events producing the resonance has been artificially increased and therefore Γ_{eff} is decreased. Of course, the coincidence technique can also be used to observe selectively only those transitions with shorter lifetimes, which results in resonance lines being broader than the natural width.

Whenever a fluctuating hyperfine interaction is present (magnetic dipole, electric monopole or quadrupole) the characteristic time of the associated nuclear moment becomes important. Considering the hyperfine splitting ΔE and the Larmor (or quadrupole) precession time τ_L the uncertainty principle can be written as

$$\Delta E = \hbar/\tau_L. \tag{13.16}$$

The question as to whether a spectrum is resolved depends on the time relationship of the nuclear lifetime, the Larmor (or quadrupole) precession time and the dynamic or electronic relaxation time of the corresponding atomistic process. Typical examples are given in the section on relaxation phenomena.

13.8 Applications

In this section some typical examples are cited to demonstrate the success as well as the limitations of the method. The selection should be considered as somewhat arbitrary, reflecting the author's interest. In particular, it will be shown that Mössbauer spectroscopy makes it possible to identify and quantify phases; here the term phases is used in a rather general way, thus including crystallographic, magnetic, order-disorder phases. In addition, properties of atoms and their surroundings can be studied, giving information on the characteristic behavior of defects, sublattices, surfaces, dynamics, magnetic order and others.

Iron plays a leading rôle in physical metallurgy and is the star performer in Mössbauer spectroscopy; with the advent of the new method intimate love and the embrace of method and field occurred instantly.

13.8.1 Iron

Elementary iron has two structures: body-centered cubic (bcc) which exists at low temperatures (α-phase) and at high temperatures (δ-phase). In the temperature range of 1183–1663 K, the face-centered cubic (fcc) structure (γ-phase) is stable. The α-phase exhibits ferromagnetism with a Curie temperature of 1046 K. The paramagnetic state of bcc Fe is called the β-phase.

The spectrum of α-Fe is the most common one. It serves as a standard for the isomer shift and also for the calibration of the relative velocity between source and absorber.

"Why is iron magnetic?" [13.48]. This is an intricate question and also, "why is the ordering ferromagnetic in α-Fe and antiferromagnetic at low temperatures in γ-Fe?" Moreover, "why is the internal magnetic field in α-Fe at room temperature 330 kOe and in γ-Fe at low temperatures only 18 kOe?" and "what are the various contributions to the internal magnetic fields?". In a general and simplified way we can write

$$H_i = H_{intra} + H_{inter} + H_{ext}. \tag{13.17}$$

The first term contains the contribution of the resonance atom itself, in particular, the core polarization. It is a spin-dependent Coulomb interaction between the d electrons and the s electrons of the various shells. This leads to the so-called Fermi-contact interaction, an interaction between the nuclear moment with the polarized spins of the s electrons which have a certain probability at the site of the nucleus. With regard to the term H_{inter} we might consider the dipole contribution of the neighboring atoms and, finally, external fields and demagnetizing fields contribute to H_{ext}. It should be noted that an external field reduces the splitting, indicating that the hyperfine interaction has a negative sign: atomic moment and internal magnetic field are oppositely oriented.

There are three ways to stabilize fcc γ-Fe in order to make investigations at lower temperatures possible:

1. coherent precipitation of supersaturated Cu-Fe alloys,
2. alloying with elements extending the γ-phase region, and
3. epitaxial growth of Fe-vapour condensed on appropriate substrates such as Cu.

Antiferromagnetic ordering of γ-Fe at low temperatures becomes evident by broadening of the resonance line. The splitting corresponds to only $H_i =$ 18 kOe which is insufficient to resolve the six-line pattern. By means of a thermal scan technique and measurement of the count rate at one velocity only the Néel temperature was first measured of γ-Fe coherent precipitates in Cu and found to be 67 K [13.49]. The size of the precipitates or the thickness of the films and, in the fcc stainless steel, the composition, have a strong influence on the Néel temperature.

For a laboratory course the following simple and instructive experiments are recommended: Cu containing a small amount of ^{57}Fe (0.2–3.0 at.%) is solution-annealed in a reducing atmosphere at ~1200 K and then held at ~900 K for about one day. By this procedure coherent (fcc) γ-Fe precipitates are formed. The spectrum at room temperature exhibits a paramagnetic single line (Fig. 13.9a). Below 67 K broadening of the line is observed due to antiferromagnetic ordering; however, the small internal field will not lead to a resolved six-line pattern. Upon plastic deformation with a hammer or a rolling device the following phase transformation occurs:

$$\gamma - \text{Fe (fcc)} \rightarrow \alpha - \text{Fe (bcc)}.$$

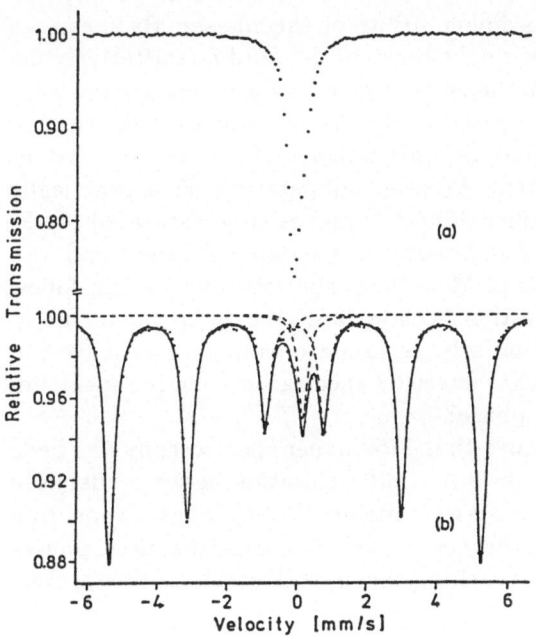

Fig. 13.9a,b. Mössbauer spectra at room temperature (a) before and (b) after cold rolling of a Cu sample originally containing coherent γ-Fe precipitates

431

In the spectrum a mixture of a six-line pattern (α-Fe) and a center line can be distinguished (Fig. 13.9b). The latter is a combination of a retained fcc phase and iron still in solution. If one assumes the same Lamb-Mössbauer factor for all the iron, an evaluation of the areas allows one to determine the fraction which has transformed to α-Fe.

13.8.2 Phase Analysis

Stoichiometry in chemistry is governed mainly by ionic (charge compensating) bondings and by covalent (directional) bondings. In contrast and in general, metal bonds are unsaturated and non-directional; instead we find crystallographic high symmetry, high coordination, many defects (such as dislocations) and a relatively wide composition range of so-called intermetallic compounds. The unlimited variety is best exemplified by the steel alloys which after certain treatments (temperature, cooling rate, annealing, deformation, etc.) inherit a wide range of mechanical, electrical and magnetic properties.

In metallurgical investigations the constitution diagrams showing the phases existing at various temperatures normally serve as a starting point and guideline. The phase diagrams give information on the range of existence of the different atomic structures, on ordered structures and on magnetic ordering. In addition to the thermodynamically stable phases appearing in the phase diagrams, a number of metastable and intermediate phases are often formed under certain conditions. In some cases the final stable phases can be produced only with great difficulty, such as various ordered phases which actually form only at relatively low temperatures where diffusion practically ceases.

Phases containing Mössbauer isotopes exhibit characteristic hyperfine patterns which might be described as finger prints of the phases. By means of these typical spectra certain phases might be identified and quantitatively determined. However, microscopic methods have given us a detailed knowledge of the local atomic structure, and consequently the word phase has acquired a new meaning, or the definition has become blurred. As an example let us consider the copper-rich Cu-Fe system. At lower temperatures Fe is practically insoluble in Cu. However, by fast quenching techniques supersaturated Cu-Fe alloys can be produced in which Fe is present as a monomer, dimer and also in small clusters of γ-Fe. By means of Mössbauer spectroscopy we can follow the growth of the cluster. The question is: how large must the cluster be in order to constitute a new phase? Similarly, certain atoms might participate in Guiner-Preston zones and exhibit characteristic spectra. Is a unique spectrum sufficient to call the agglomerate a phase?

Numerous examples exist to prove that Mössbauer spectroscopy has been most useful in phase analysis. To name a few: differentiation between austenite and ferrite and determination of carbides and nitrides in steel, phases appearing in internal and external oxidation, hydrogen in various storage matrices, phases formed in amorphous metals by approaching the crystallization temperature.

432

The Al-Fe system might serve as a master example because extensive work has been done on this system and, in addition, it has a number of features typical of metal systems and appropriate for Mössbauer spectroscopy:

stable phases: $Al(Fe)$, $Al_{13}Fe_4$ (Al_3Fe), Al_5Fe_2, Al_2Fe, $AlFe$, $AlFe_3$, $Fe(Al)$,

metastable phase: Al_6Fe,

order-disordered phase: $AlFe$, $AlFe_3$,

solubility: Fe in Al,

defects: vacancies, interstitials, dimer, antiphase domain boundary.

Typical spectra of Al-rich Al-Fe samples are shown in Fig. 13.10. The treatments were chosen in such a way that typical features dominate the spectra [13.50]. The spectra (a) can be decomposed into a fraction corresponding to Fe monomers and Fe dimers. The quadrupole splitting of the latter indicates that the local cubic symmetry has been removed. In (b) the spectrum of the metastable phase Al_6Fe is exhibited and in (c) the spectrum $Al_{13}Fe_4$ indicates that Fe occupies crystallographically different lattice sites.

In certain cases a precision-phase analysis is possible [13.51]. The accuracy of the method will be demonstrated by the determination of the solubility of Fe

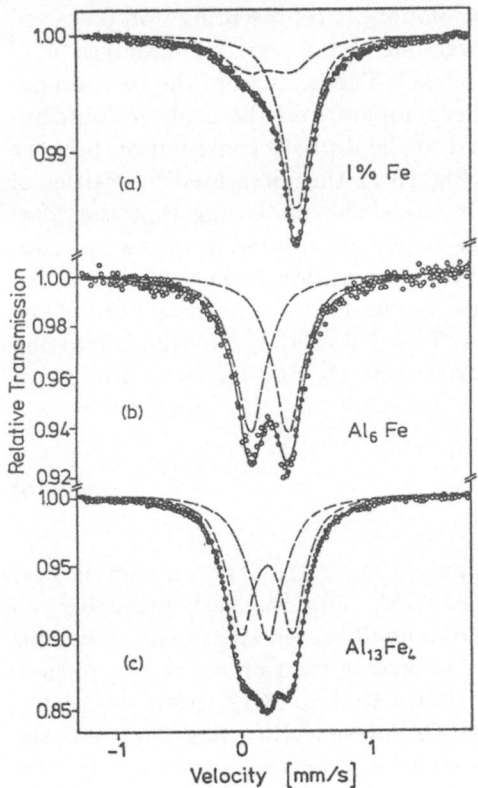

Fig. 13.10. Mössbauer spectra at room temperature of: *(a)* splat-quenched Al-1 at.% Fe specimen. The single line represents the monomers and the doublet the dimers. *(b)* Metastable Al_6Fe precipitates and *(c)* stable $Al_{13}Fe_4$ precipitates

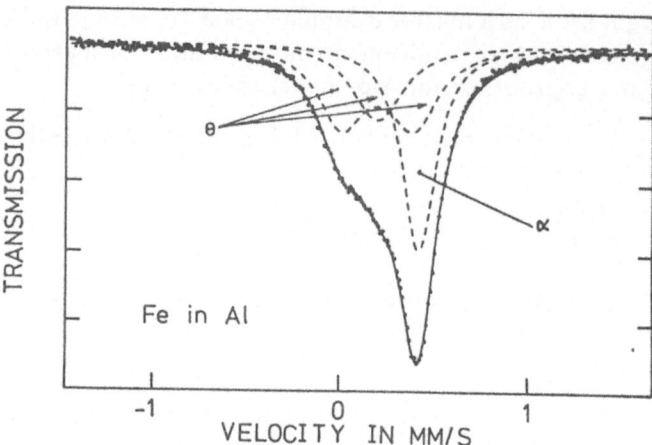

Fig. 13.11. Mössbauer spectrum from a dilute 0.019 at.% ^{57}Fe in Al alloy at room temperature. Dotted curves are least-squares fits of Lorentzian lines representing Fe in solution (α) and $Al_{13}Fe_4$ precipitates (θ)

in Al. Specimens with a starting composition C_0 of about 0.003–0.05 at.% Fe in Al – within the two phase regions – were annealed in the relevant temperature range (450–650°C) and then quenched. A typical spectrum of a specimen containing 0.019 at.% Fe is shown in Fig. 13.11. As expected, the spectrum shows two characteristic patterns: a single-line component representing iron dissolved in the Al matrix (α-phase) while the three-line component is characteristic of the intermetallic compound $Al_{13}Fe_4$ (θ-phase). The fraction of the two components is temperature dependent. At higher temperatures the α-phase contribution becomes more pronounced compared to the θ-phase contribution because more Fe dissolves in the Al matrix. In Fig. 13.11 the integrated intensities of both components (α and θ) are roughly the same, indicating that the total number of resonance atoms ^{57}Fe is rather evenly distributed in the two phases. On the other hand, because the actual Fe concentration is much higher in the θ-phase, as compared to the α-phase, this implies that the near equality of the resonance intensities could only be achieved by deliberately choosing a starting composition C_0 very close to the equilibrium line C_α and far away from the θ composition: $|C_0 - C_\alpha| \ll |C_\theta - C_0|$. Now, applying the lever relationship

$$\frac{C_0 - C_\alpha}{C_\theta - C_0} = \frac{M_\theta}{M_\alpha} \tag{13.18}$$

it follows for the amount of the two phases: $M_\alpha \gg M_\theta$. The amount is best expressed by the number of primitive cells M_α and M_θ each primitive cell containing one atom. Spectra obtained from equilibrated specimens at various temperatures allow the evaluation of the resonance fraction of the two phases. The solubility curve can be drawn, as shown in Fig. 13.12 (note the scale). It is the inequality of the amount of phases present (M_α and M_θ) and the closeness to the solubility line C_α which make high precision possible. In fact,

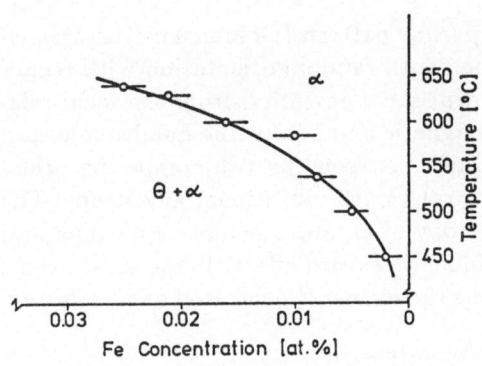

Fig. 13.12. Phase boundary at the Al-rich side of the Al-Fe phase diagram

the solubility limit of Fe in Al can be determined to an accuracy of more than 0.001 at.%.

13.8.3 Texture

Texture is the preferred orientation of an assembly of molecules, crystals, spins etc., a common phenomenon in nature. In physical metallurgy the individual grains become oriented by means of various processes in a certain direction where the alignment is usually intimately connected with the main direction of the flow of the material and the temperature (gradient). One might distinguish among: deformation texture, hot and cold rolling texture, recrystallization texture, annealing texture, tension and compression texture, surface texture from machining and polishing. Indeed it is difficult to produce a piece of metal without a texture. From a technical point of view texture in a material may be desirable. Similarly, in magnetic materials the spin texture depends on the magneto-crystalline anisotropy and can be influenced by magneto-elastic effects.

Texture is reflected in the angular dependence of the hyperfine interaction and observed by the relative line intensities of the split spectrum [13.52–54]. However, texture is not the only effect to influence intensity. In general, the relative line intensities in the hyperfine interactins are governed by

1. intrinsic effects (characteristic of the material): a) angular dependence of the hyperfine interaction, b) lattice vibrational anisotropy (Goldanskii-Karyagin effect).
2. extrinsic effects (accidental): a) texture (preferred orientations), b) thickness effects.

Three cases are considered where texture is a competing and interesting phenomenon:

a) Angular Dependence of the Hyperfine Interactions Versus Lattice Vibrational Anisotropy

If the mean-square displacement of the resonating atom is orientation dependent the corresponding anisotropy of the Debye-Waller factor is reflected in the relative line intensities of the hyperfine pattern (Goldanskii-Karyagin effect) [13.55,56]. For instance, a specimen with random orientation with regard to spins or principal axes of the EFG will show deviation from the ideal relative line intensities 3:2:1:1:2:3 for the magnetic and 1:1 for the quadrupole split case. However, it is difficult to distinguish between the two competing orientation effects: lack of randomness (texture) versus vibrational anisotropy. The intrinsic effect of the vibrational anisotropy of an atom is more appealing and significant for a scientist than the accidental texture effect. It therefore seems that in many evaluations and discussions the former is overrated and preferred.

b) Evaluating Texture from Relative Line Intensities

With isotropic mean-square displacements the preferred orientations are reflected in the relative line intensity. However, the situation is rather complex when the magnetic hyperfine pattern is perturbed by quadrupole interactions or vice versa, and the hyperfine interactions have independent textures. In general, the magnetic hyperfine interaction in its orientation is governed by the long-range domain structures while the quadrupole interaction is ruled by the short-range arrangement of the immediate surroundings of the resonance atom. Furthermore, in the case of quadrupole interactions, the determination of the principal-axes texture requires information on the sign and on the asymmetry parameter. In favorable cases, for instance, spectra with pure magnetic hyperfine interaction or quadrupole interaction of axially symmetric electric field gradients ($\eta = 0$) the texture can be evaluated by changing the angle between the γ-ray propagation direction and the plane of the sample. One proceeds in a similar fashion as with an x-ray texture goniometer, but instead of the preferred orientations of planes one obtains the preferred orientation of spins or principal axes of the EFG. A texture pole figure in polar coordinate representation can be constructed; however, it should be considered only as a minimum texture which consists of those low-order moments of the texture that can be determined by the method. The real texture may contain additional moments. This means that the distribution of extrema in the true texture may be sharper than shown by the minimum texture. It is also of interest to compare the texture of the bulk obtained by transmission γ-ray absorption spectra with the surface texture resulting from conversion electron emission spectra. Of course, these methods cannot compete with the common x-ray texture determinations; however, there are some cases where they are useful because x-rays cannot be applied, as for instance, in amorphous metals or certain biological tissues.

c) Evaluating Orientations (Spins or Principal Axes of EFG) from Textures Obtained by X-rays

This method is the reverse of the former method and particularly useful if single crystals are not available. It consists of correlating the texture obtained by x-rays with the angular dependence of the hyperfine parameters. This technique has been applied in determining the easy direction of the spins in ferromagnetically ordered cementite (Fe_3C) [13.57].

13.8.4 Defects

In general, metals contain a large number of defects and in important metallurgical processes such as plastic deformation, transformation, diffusion, recrystallization, etc. defects play an important rôle. Also, impurities in source or absorber are involved in most Mössbauer experiments and the hyperfine interactions reflect the impurity state (as parent and/or daughter isotope) in a foreign matrix.

Depending on the dimension, the following distinctions can be made: point defects, line defects, interfaces and disorder. In addition, one might distinguish between stationary defects and transient or dynamic defects. The dividing line between stationary defects and transient or dynamic defects is coupled with the characteristic time of the method, as discussed in Sects. 13.7 and 13.8.6.

Defects are associated with changes in one or several of the following physical properties in the vicinity of the defect as compared with perfect crystals: symmetry, effective mass, force constant, elastic strain, charge and spin density, texture, disorder .... If resonant probes are placed near defects the spectra might provide information on these properties [13.58,59]. In certain cases not only defects in crystals but also defects in molecules can be identified.

In crystals, point defects form rather unique associations with impurity probes, that is, particular spectra reflect the particular associations. This has been observed for ^{57}Fe associations with interstitials [13.60] and vacancies [13.61] and also for ^{57}Fe pairs (dimer). Defects can be introduced by irradiation, cold rolling or fast quenching. In subsequent annealing experiments at elevated temperatures they become mobile and can be trapped at the impurity sites (I), as schematically shown in Fig. 13.13a. In a metal with cubic symmetry

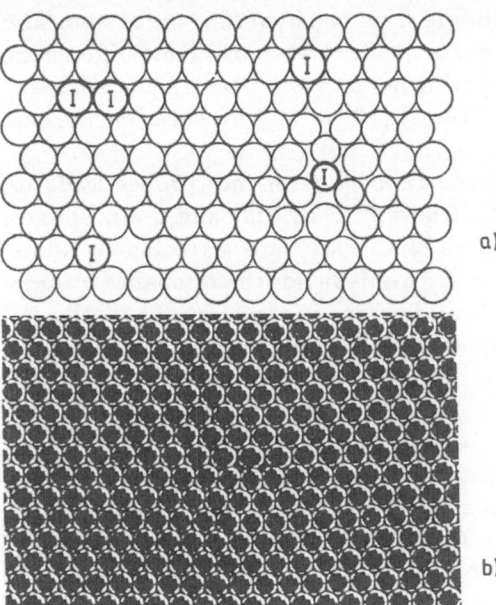

a)

b)

Fig. 13.13. (a) Point defects: schematic representation of impurity (I) as monomer and dimer, and associated with vacancy and interstitial. (b) Edge dislocation simulated by bubbles

437

this symmetry is removed locally, as indicated by the appearance of quadrupole splitting, and the local electronic structures show up in characteristic isomer shifts. The dynamics of the association are reflected in the intensities. At elevated temperatures a break-up may occur and the defects dislodge themselves and diffuse to deeper sinks (edge dislocations or boundaries).

The most successful experiments were carried out with aluminum [13.60]. One advantage is that, of all metals, aluminum can be produced with the highest purity and another is that electronic absorption of the ^{57}Fe 14.4 keV γ-rays is rather small. Specimens with extremely small concentrations of impurities can be investigated, in particular experiments where the number of ^{57}Fe impurity atoms become of the same order of magnitude as such introduced defects as interstitials and vacancies. Another possibility is to implant resonance atoms into a lattice. In conjunction with the implantation process defects are introduced which can be observed in their geometry and annealing behavior [13.62].

Also, the reverse is of interest, that is, the lattice consists of resonant atoms and contains, in addition, nonresonant impurities in interstitial or substitutional lattice sites. For instance, C, N, B, H, etc. dissolve as interstitial atoms in α-Fe and γ-Fe. The distortion of the lattice caused by the interstitials will influence the hyperfine parameters of the neighboring atoms. Particularly, in the important Fe-C system it was possible to decompose the spectra into subspectra which reflect different near-neighbor shells. In addition, the early stages of the precipitation of the carbide phases could be followed. Different phases simultaneously present and different environments within these phases could be distinguished [13,63,64].

Similarly, numerous experiments have been carried out in order to measure the influence of substitutional impurities on the hyperfine parameters of the neighboring atoms. At surfaces and interfaces the hyperfine parameters are changed: particularly the internal magnetic fields show remarkable differences from the bulk value [13.65]. Such experiments are important for the understanding of superstructures (layer structures) [13.66] and for such phenomena as catalysis and corrosion [13.67,68].

In contrast to cases where the association of resonance probes leads to unique spectra, one expects that dislocations (Fig. 13.13b) and grain boundaries produce rather complicated spectra. Impurities are attracted by these defects and around them numerous sites are available for the resonance probes, exhibiting their own spectra with slightly different parameters. Thus, the resulting spectra represent a rather complex distribution. The situation becomes worse in the case of amorphous metals.

13.8.5 Amorphous

Whatever the starting material, physicists are always tempted to look for single crystals. The motivation for this desire has best been formulated by *Kittel* [13.69] in his famous book *Introduction to Solid State Physics*: "Measurements on single crystals are always much more significant and informative than are measurements on polycrystalline specimens". However, in recent years we have

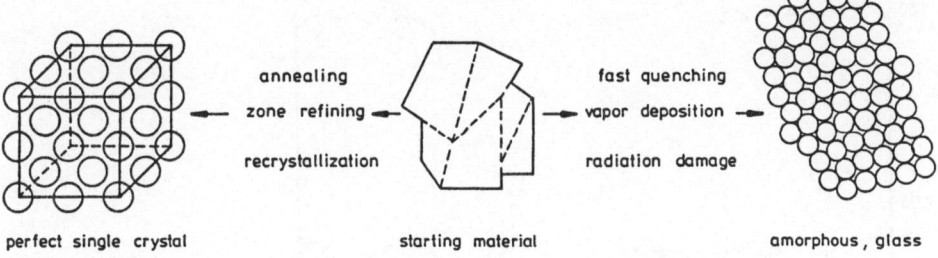

Fig. 13.14. Two alternatives

also learned to look in the opposite direction, as schematically indicated in Fig. 13.14, and the above statement can be reversed: "A fool can handle the crystalline state, but it takes a genius for the amorphous state".

In the cohesion of condensed matter we differentiate between four types of binding forces: ionic, covalent, van der Waals and metallic. We are well aware that the first three types occur in the amorphous state, as a matter of fact the biological material of our own bodies is the best evidence of this fact. In metals it was realized rather late that they can also be produced in the amorphous state, so their history is short and falls in the second half of this century. Most prominent in this development were *Buckel* and *Hilsch* [13.70], who condensed metal vapor on a cold substrate, and *Duwez* et al. [13.71], who developed various techniques for the rapid quenching of molten alloys. In their work they found ferromagnetism, and also brought some systematics into the field. With these new amorphous alloys some principles of physical metallurgy and such well-known terms as lattices, defects, Fermi surface, Brillouin zones, intermetallic phases and other properties acquire new meanings or have to be redefined.

In physical metallurgy we normally start with the appropriate phase diagram showing the thermodynamically stable phases and their ranges of existence. The multitude of these phases and their properties are the basis for their wide technological applicability. While the number of stable phases is limited, the variety of amorphous alloys has no bounds. It seems then that any alloy can be produced in the amorphous state, although some already crystallize at rather low temperatures.

The most important class of amorphous alloys is the so-called $T_{80}M_{20}$ type (T : transition elements such as Fe, Co, Ni, Pd ... and M : metalloid atoms such as B, C, N, P ...). Their composition ranges of high relative stability coincide with their pronounced eutectics. Most of these materials are ferromagnetically ordered [13.72]. These alloys have won our fascination mainly because of their lack of magneto-crystalline anisotropy and their lack of ordinary dislocations, which enables them to combine magnetic softness with mechanical strength and hardness.

It was thought that by proceeding in two steps the microscopic resonance probe might reveal the atomic scale structure of these alloys: in the first step

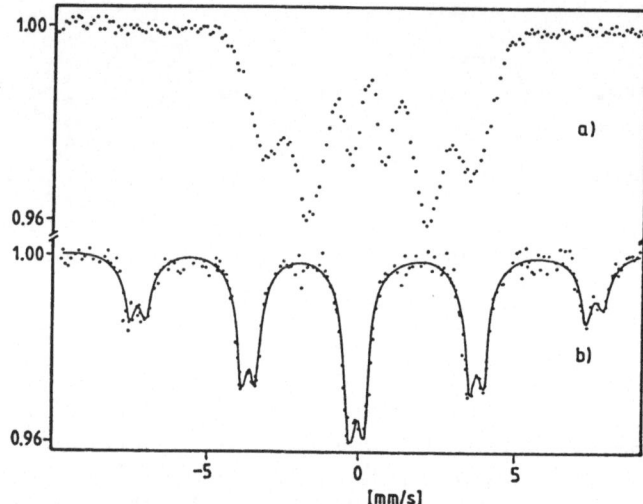

Fig. 13.15. Mössbauer spectra of amorphous $Fe_{40}Ni_{40}P_{14}B_6$ at room temperature *(a)* in zero magnetic field, *(b)* in radio frequency magnetic field of 44.5 MHz and 13 Oe

one attempts to derive the Mössbauer parameter and in the second step one tries to correlate these parameters with appropriate models. Unfortunately the common approach so far has been the reverse, that is, one assumes a model and tries to fit the data.

A typical spectrum of an amorphous ribbon is shown in Fig. 13.15a. The relative strength of lines 2 and 5 ($\Delta m = 0$) indicates that the spins are preferentially oriented in the plane of the ribbon. More important is the fact that the lines are very broad. The difficulty in fitting becomes evident when one looks for the exact cause of the broadening. Contrary to the usual findings for crystalline materials with their unique values of shift, splitting and field, there is a wide variety of distributions in amorphous metals which reflect the multiplicity of the surroundings of the resonating atoms and must be taken into account: distribution in isomer shift, distribution in the magnitude of magnetic hyperfine fields, distribution of the magnitude, sign, asymmetry parameters of the quadrupole interactions, distribution of orientation (texture) of the hyperfine fields and of the electric field gradient principal axes. In addition, thickness effects, polarizations, dichroism, lattice anisotropy and other factors might have some influence.

Recently a method has been developed which effectively eliminates the magnetic hyperfine interaction and allows the direct determination of the hidden quadrupole splitting in ferromagnetic amorphous metals [13.73]. By applying radio frequency (rf) magnetic fields the magnetic hyperfine structure can be brought to collapse. If the frequency of an external rf field is larger than the nuclear Larmor precession frequency and the magnetic reversal follows the oscillation of the switching field, the magnetic hyperfine fields at the nuclei are reduced and finally averaged to zero, as seen in Fig. 13.15b. The remaining spectral lines contain only the distribution of isomer shifts and quadrupole splittings. Due to magneto-acoustic coupling magneto-strictive materials exhibit another effect: the collapse is accompanied by side bands which appear symmetrically on both sides. Their separation depends on the rf frequency and they disappear at the Curie point. An evaluation of the collapse spectra enables one to obtain information regarding the distribution of isomer shifts and quadrupole splittings.

In the discussion concerning the structure of amorphous metals there is general agreement that a certain type of short-range order is present; however, controversies exist as to what the elementary units or building blocks are and how the assembly as a whole is constructed. As many scientists have obtained reasonable fittings regardless of the model chosen, this clearly indicates that this approach does not lead us to the information sought regarding the atomic structure. Despite great endeavours in the field it must be objectively stated that no unique and generally accepted atomic-scale model of any amorphous alloy has as yet evolved. Too much intuition is used in the game. It has become evident that we have reached a certain limitation of Mössbauer spectroscopy, but other microscopic methods also share this fate.

While the output of research on the atomic scale structure of amorphous metals has so far been rather poor, Mössbauer spectroscopy has its great merit – and is even unmatched – when it comes to crystallization. Especially problems such as the question where crystallization commences and which metastable and stable phases are formed have been tackled.

In particular, the technique of simultaneously measuring electron emission – scanning the surface – and γ-ray transmission – measuring the bulk – as seen in Fig. 13.5, have been successfully applied [13.25,26]. Figure 13.16 shows spectra of the melt-spun amorphous ribbon $Fe_{83}P_5C_{12}$ after annealing in air at 630 K for 15 min. The electron spectra show clearly that the material has fully crystallized and oxidized on both sides, and by means of characteristic lines corresponding to octahedral and tetrahedral sites the spinel Fe_3O_4 could be identified. The bulk was measured in transmission by γ-resonance absorption and no oxidation was detectable. However, the $\Delta m = 0$ lines (second and fifth) exhibit significant changes. Originally they were strong showing that the spins were aligned in the plane of the ribbon (Fig. 13.17a). After annealing the lines became suppressed. The weakening of these lines indicates that by magneto-elastic phenomena the spins became preferentially oriented normal to the ribbon, probably forming a closure domain pattern, as shown in Fig. 13.17b. Considering the fact that the magneto-striction of this material is positive ($\lambda > 0$), the sample must have been under compression stress as a result of the formation of oxides at the surface. Removal of the oxides from the surface or applied tensile stress brings the spins back to the plane of the ribbon, as observed by the strong $\Delta m = 0$ lines.

The technique of simultaneously measuring γ-ray transmission and electron emission has also been applied to amorphous metal-metal alloys [13.74], for instance to $Zr_{91}Fe_9$. After fast quenching of the alloy, paramagnetic spectra

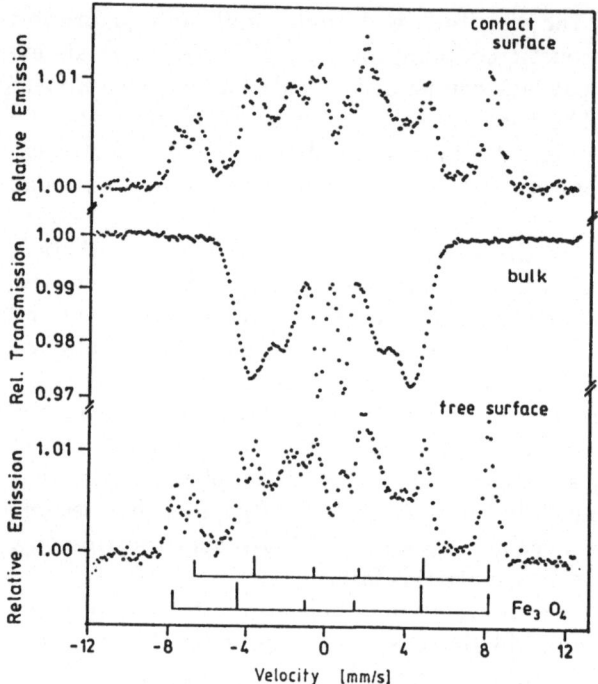

Fig. 13.16. Mössbauer electron emission (from contact and free surfaces) and γ-ray absorption spectra (bulk) of amorphous $Fe_{83}P_5C_{12}$ at room temperature after annealing in air at 630 K. Stick diagram indicates line positions of octahedral and tetrahedral sites in Fe_3O_4

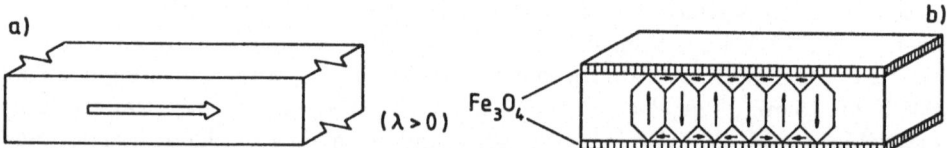

Fig. 13.17a,b. Domain structures of amorphous $Fe_{83}P_5C_{12}$ (a) as freshly prepared, (b) annealed sample

with an indication of quadrupole splitting are obtained. Slow quenching leads to precipitation of α-Fe as indicated by the characteristic six-line pattern in the γ-ray transmission spectrum as well as in the electron emission spectrum. By quenching at an intermediate speed one can produce situations where the γ-ray transmission spectrum and the electron emission spectrum from the contact surface (the contact with the wheel produces relative fast quenching) exhibit paramagnetic hyperfine spectra, but the electron emission spectrum of the free surface shows a six-line pattern (Fig. 13.18). This indicates that precipitation of α-Fe has taken place on the free surface where the quenching rate is slower. Annealing of the specimen for one hour at 680 K leads to precipitation of α-Fe on the contact surface, too. In this sequence α-Fe precipitation occurs last in the bulk, as detected by the γ-transmission spectrum.

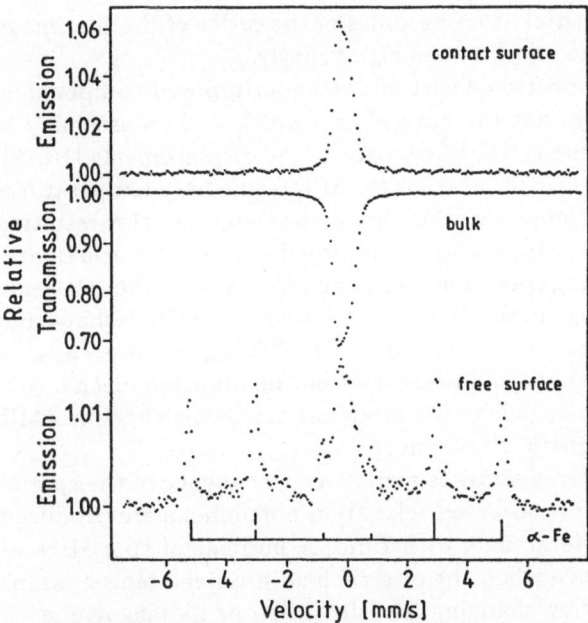

Fig. 13.18. Mössbauer electron emission (from contact and free surfaces) and γ-ray absorption spectra (bulk) of amorphous $Zr_{91}Fe_9$ at room temperature. The quenching rate was chosen in such a way that precipitates of α-Fe occurred only on the free surface. The positions of the α-Fe spectrum are indicated by the stick diagram

13.8.6 Relaxation Phenomena

In Sect. 13.7 we described the characteristic times (or the clocks) of the method making themselves known by the line width and by time-dependent relaxation effects. These phenomena are related to the lifetime of the excited state, the nuclear precession time and magnetic, electronic or atomic relaxations. An example of the latter is given below.

Soon after the discovery of the Mössbauer effect it was already predicted that the diffusive motion of atoms would cause random phase shifts of Mössbauer radiation so that the line width would increase [13.75,76]. This line broadening occurs when the jump frequency of the atoms becomes of the order of the reciprocal of the lifetime of the excited state. In the succeeding years the diffusion of resonating atoms (^{57}Fe) into neighboring vacant lattice sites in Cu and Au matrices was observed [13.77]. Broadening can also occur in interstitial solid solutions where the Mössbauer atoms do not leave their lattice sites. Instead the interstitial atom (hydrogen, for instance) passes, that is, approaches and leaves the resonating atom from time to time, thereby causing sudden small distortions of the surrounding lattice or displacements of the equilibrium positions. The diffusion of the interstitial atoms will govern the dynamics of the displacements. If the changes in the displacements occur during the emission or absorption of a γ-quantum the wave train suffers small phase shifts. When the mean dwell time of the interstitial atom becomes of the order of the lifetime of the excited state the spectrum can broaden significantly.

It was shown that such spectra consist of two superimposed components: a Lorentzian line which always has the natural line width and an assembly of Lorentzians whose width depends on the frequency of the displacements [13.78]. At low temperatures all lines have natural width. At intermediate temperatures the temperature-dependent component has broadened so that the resulting spectrum is broad. At higher temperatures the broad component apparently disappears so that the remaining spectrum is once again narrow, although weak, and exhibits an apparent change in the Mössbauer-Lamb factor. Line shape and broadening depend on the interstitial sites which the diffusing atoms occupy. In principle, one should be able to obtain information on the diffusion mechanism. The narrowing of the line is analogous to the motional narrowing effect in NMR (see Chap. 12 on Nuclear Magnetic Resonance).

Whereas the effect considered above is merely a consequence of the spatial motion of resonance atoms, most observed relaxation phenomena are produced by a variation of hyperfine interactions with time. A number of special cases have been analyzed, for instance when the electric field gradient tensor jumps between discrete values either by changing its orientation or its magnitude.

The introduction of hydrogen into the lattice can destroy the local cubic symmetry, resulting in an electric field gradient (EFG) at the site of the resonance nucleus and leading to a quadrupole splitting ΔE_Q. One might assume that the various hydrogen arrangements produce the same quadrupole splittings but with different orientations of the principal axes. At elevated temperatures and when hydrogen diffuses there will be a succession of different arrangements of hydrogen atoms around each Mössbauer isotope. This causes jump-like fluctuation in the orientations of the EFG tensor. When, with increasing temperature, the rate of reorientation of the EFG axes system reaches the order of the ΔE_Q, relaxation effects are observed. At a higher reorientation rate the resonance atoms experience a higher time-average symmetry so that the quadrupole splitting begins to decrease. Finally, the splitting may collapse completely when within the quadrupole precession time the orientation fluctuation average corresponds effectively to cubic symmetry. This effect has been observed in the β-phase of TiFe-H [13.79].

Another case consists of two EFG's, the magnitude of which jumps with an average rate W between two states of different populations [13.80]. The ratio W/Γ is a measure of the mean number of jumps during the lifetime of the excited state. In the static case (W = 0) the spectrum exhibits two superimposed incoherent well-resolved doublets with different splittings and intensities, as shown in Fig. 13.19. With increasing jump frequency the lines broaden and furthermore the positions of the lines on the left-hand side and on the right-hand side move towards the centre of gravity. With very high jump frequencies the time-dependent nuclear states are determined by the temporal average of the hyperfine fields. Accordingly, the spectrum collapses to one doublet of natural line width. The extreme sensitivity of the sharp resonance line of ^{181}Ta has

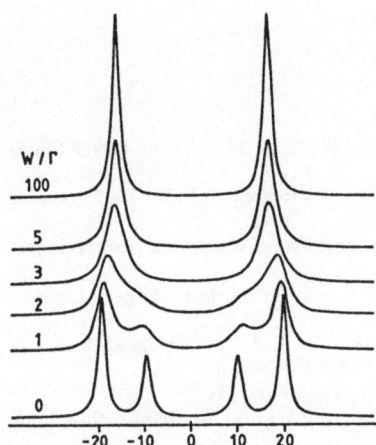

Fig. 13.19. Relaxation spectra of two ^{57}Fe EFG's the magnitude of which changes at an average rate W. The unit of the energy scale is Γ

been used to measure, via the isomer shift, the fluctuation of the electronic density [13.81].

Numerous relaxation processes have been studied and their mechanisms elucidated. Electronic relaxations play an important rôle in ionic and covalent bonded materials. In metals, where electronic relaxation is usually fast compared to the characteristic time of the method, effects of this kind are not so common. However, superparamagnetism and spin glass behavior have been found in metals and have attracted greater interest in recent years.

13.9 Outlook

Once the resonance in ^{57}Fe was discovered it was remarkable how quickly the scientific aspects were realized and appropriate experiments were carried out in all disciplines of natural science. Nowadays, Mössbauer spectroscopy is considered to be a standard and routine method and numerous Mössbauer scientists are waiting for their chance to enter upon new endeavours. On the other hand, the practical and technological applications were, as usual, tackled relatively slowly. More adaptations, greater patience and careful analyses are required. The first conference on technological applications of the Mössbauer effect was held in 1984 (in Honolulu) [13.82]. Moreover, there are other encouraging recent developments which may open up new horizons:

1. Mössbauer spectra have been produced with synchrotron radiation [13.83].
2. The gamma-ray laser (gaser) still has a chance to be realized one day [13.84].
3. One would like to transfer the progress which this method has made in solids particularly to biological problems in order to deepen the understanding of structure, dynamics and functions [13.85]. In certain cases spectral components indicate stereochemical and pathological behavior, and attempts have been made to use this spectroscopy as a diagnostic tool in connection with molecular diseases [13.86].

References

13.1   R.L. Mössbauer: Z.Physik **151**, 124 (1958)
13.2   R.L. Mössbauer: Naturwissensch. **45**, 538 (1958)
13.3   W.E. Lamb, Jr: Phys. Rev. **55**, 190 (1939)
13.4   H.J. Lipkin, private communication in H. Frauenfelder: *The Mössbauer Effect* (Benjamin, New York 1962)
13.5   R.L. Mössbauer: Les Prix Nobel en 1961 (Nobel Foundation, Stockholm 1962) p. 136; and Science **137**, 731 (1962)
13.6   A.H. Muir, Jr., K.J. Ando, H.M. Coogan: *Mössbauer Effect Data Index* (Interscience, New York 1958–1965);
        J.G. Stevens, V.E. Stevens: *Mössbauer Effect Data Index* (Adam Hilger, London 1966–1976)
        J.G. Stevens, V.E. Stevens: *Mössbauer Effect Reference and Data Journal*, Vol. 1–9 (Mössbauer Effect Data Center, Asheville, NC, USA 1977–1986)
13.7   H. Frauenfelder: *The Mössbauer Effect* (Benjamin, New York 1962)
13.8   G.K. Wertheim: *Mössbauer Effect, Principles, and Applications* (Academic, New York 1964)
13.9   H. Wegener: *Der Mössbauer-Effekt und seine Anwendungen in Physik und Chemie* (Bibliographisches Institut, Mannheim 1965)
13.10   I.J. Gruverman (ed.): *Mössbauer Effect Methodology*, Vol. 1–10 (Plenum, New York 1965–1976)
13.11   V.I. Goldanskii, R.H. Herber (eds.): *Chemical Applications of Mössbauer Spectroscopy* (Academic, New York 1968)
13.12   L. May (ed.): *An Introduction to Mössbauer Spectroscopy* (Plenum, New York 1971)
13.13   N.N. Greenwood, T.C. Gibb: *Mössbauer Spectroscopy* (Chapman & Hall, London 1971)
13.14   C. Janot: *L'Effet Mössbauer et ses Applications a la Physique du Solid et a la Métallurgie Physique* (Masson, Paris 1972)
13.15   U. Gonser (ed.): *Mössbauer Spectroscopy* I and II, Topics Appl. Phys., Vol. 5 and Topics Current Phys., Vol. 25 (Springer, Berlin, Heidelberg 1975 and 1981)
13.16   R.L. Cohen (ed.): *Application of Mössbauer Spectroscopy* I and II (Academic, New York 1976 and 1980)
13.17   P. Gütlich, R. Link, A. Trautwein: *Mössbauer Spectroscopy and Transition Metal Chemistry* (Springer, Berlin, Heidelberg 1978)
13.18   A. Vértes, L. Korecz, K. Burger: *Mössbauer Spectroscopy* (Elsevier, Amsterdam 1979)
13.19   B.V. Thosar, P.K. Iyengar, J.K. Srivastava, S.C. Bhargava (eds.): *Advances in Mössbauer Spectroscopy* (Elsevier, Amsterdam 1979)
13.20   R.L. Mössbauer: J. Physique **C-6**, 11 (1976)
13.21   H.J. Lipkin: *Quantum Mechanics* (North-Holland, Amsterdam 1973)
13.22   E. Huenges, J. Loock, H. Morinaga, F. Parak: Nucl. Instr. Methods **203**, 527 (1982)
13.23   G.M. Kalvius, E. Kankeleit: *Mössbauer Spectroscopy and its Applications*, Panel Proc. Series (IAEC, Vienna 1972)
13.24   T. Shigematsu, H.-D. Pfannes, W. Keune: Phys. Rev. Lett. **45**, 1206 (1980)
13.25   S. Nasu, U. Gonser: J. Physique **C-8**, 690 (1980)
13.26   U. Gonser, M. Ackermann, H.-G. Wagner: J. Magn. Magn. Mater. **31–34**, 1605 (1983)
13.27   D.C. Kistner, A.W. Sunyar: Phys. Rev. Lett. **4**, 412 (1960)
13.28   L.R. Walker, G.K. Wertheim, V. Jaccarino: Phys. Rev. Lett. **6**, 98 (1961)
13.29   S.S. Hanna, J. Heberle, G.J. Perlow, R.S. Preston, D.J. Vincent: Phys. Rev. Lett. **4**, 513 (1960)
13.30   K. Ono, A. Ito: J. Phys. Soc. Japan **19**, 899 (1964)
13.31   F. Aubertin, U. Gonser, S.J. Campbell, H.-G. Wagner: Z. Metallkde. **76**, 237 (1985)
13.32   W. Kündig: Nucl. Instr. Methods **48**, 219 (1967)
13.33   N. Blaes, H. Fischer, U. Gonser: Nucl. Instr. Methods B **9**, 201 (1985)
13.34   G.K. Wertheim, V. Jaccarino, J.H. Wernick: Phys. Rev. **135**, A151 (1964)
13.35   S.S. Hanna, J. Heberle, C. Littlejohn, G.J. Perlow, R.S. Preston, D.H. Vincent: Phys. Rev. Lett. **4**, 177 (1960)
13.36   H. Wegener, F.E. Obenshain: Z. Physik **163**, 17 (1961)
13.37   H. Frauenfelder, D.E. Nagle, R.D. Taylor, D.R.F. Chochran, W.M. Visscher: Phys. Rev. **126**, 1065 (1962)

13.38 P. Imbert: J. Physique **27**, 429 (1966)
13.39 U. Gonser, H. Fischer: In *Mössbauer Spectroscopy*, ed. by U. Gonser, Topics Current Phys., Vol. 25 (Springer, Berlin, Heidelberg 1981) p. 99
13.40 H.-D. Pfannes, U. Gonser: Nucl. Instr. Methods **114**, 297 (1974)
13.41 M. Blume, O.C. Kistner: Phys. Rev. **171**, 417 (1968)
13.42 R.M. Housely, U. Gonser: Phys. Rev. **171**, 480 (1968)
13.43 R.V. Pound, G.A. Rebka, Jr.: Phys. Rev. Lett. **4**, 274 (1960)
13.44 B.D. Josephson: Phys. Rev. Lett. **4**, 341 (1960)
13.45 R.V. Pound, J.L. Snider: Phys. Rev. **140**, B788 (1965)
13.46 T. Katila, K. Riski: Hyp. Int. **13**, 119 (1983)
13.47 F.J. Lynch, R.E. Holland, M. Hamermesh: Phys. Rev. **120**, 513 (1960)
13.48 M.B. Stearns: Phys. Today **34** (1978)
13.49 U. Gonser, C.J. Meechan, A.H. Muir, H. Wiedersich: J. Appl. Phys. **34**, 2373 (1963)
13.50 S. Nasu, U. Gonser: Proc. Int. Conf. Mössbauer Spectroscopy, Bratislava, Czechoslovak AEC, Prague (1973) p.311
13.51 S. Nasu, U. Gonser, A. Bläsius, F.E. Fujita: J. Physique **C-2**, 619 (1979)
13.52 U. Gonser, H.-D. Pfannes: J. Physique **C-6**, 113 (1974)
13.53 H.-D. Pfannes, H. Fischer: Appl. Phys. **13**, 317 (1977)
13.54 J.M. Greneche, F. Varret: J. Phys. C 15, 5333 (1982)
13.55 V.I. Goldanskii, E.F. Makarov, V.V. Khrapov: Phys. Lett. **3**, 334 (1963)
13.56 S.V. Karyagin: Dokl. Akad. Nauk. SSSR **148**, 1102 (1963)
13.57 U. Gonser, M. Ron, H. Ruppersberg, W. Keune, A. Trautwein: phys. stat. sol. (a) **10**, 493 (1972)
13.58 U. Gonser, H. Wiedersich: J. Phys. Soc. Japan **18**, Suppl. 11, 47 (1963)
13.59 G.K. Wertheim: In *The Electronic Structure of Point Defects*, ed. by S. Amelinckx, R. Gevers, J. Ninoul (North-Holland, Amsterdam 1971)
13.60 G. Vogl, W. Mansel, P.H. Dederichs: Phys. Rev. Lett. **36**, 1497 (1976)
13.61 R.S. Preston, S. Nasu, U. Gonser: J. Physique **C- 2**, 564 (1979)
13.62 H. De Waard: Phys. Scr. **11**, 157 (1975)
13.63 F.E. Fujita: Metall. Trans. **8A**, 1727 (1977)
13.64 M. Ron: In *Applications of Mössbauer Spectroscopy II*, ed. by R.L. Cohen (Academic, New York 1980 p. 329
13.65 I. Tyson, A.W. Owens, J.C. Walker: J. Appl. Phys. **52**, 2487 (1981)
13.66 T. Shinjo: *Proc. Intern. Conf. Application of the Mössbauer Effect*, Alma Ata, USSR (Gordon Breach, New York 1983)
13.67 S. Morup, J.A. Dumesic, H. Topsoe: In *Applications of Mössbauer Spectroscopy*, ed. by R.L. Cohen (Academic, New York 1980) p. 1 and 56
13.68 W. Meisel: Proc. 12th Seeheim Workshop, ed. by P. Gütlich, G.M. Kalvius (1983) p. 15
13.69 C. Kittel: *Introduction to Solid State Physics*, 3rd ed. (Wiley, New York 1967)
13.70 W. Buckel, R. Hilsch: Z. Phys. **132**, 420 (1952); and **138**, 109 (1954)
13.71 P. Duwez, R.H. Willens, W. Klement: J. Appl. Phys. **31**, 1136 (1960)
13.72 C.C. Tsuei, P. Duwez: J. Appl. Phys. **37**, 435 (1966)
13.73 M. Kopcewicz, H.-G. Wagner, U. Gonser: Sol. Stat. Commun. **48**, 531 (1983)
13.74 M. Ackermann: Dissertation, Universität des Saarlandes (1985)
13.75 K.S. Singwi, A. Sjölander: Phys. Rev. **120**, 1093 (1960)
13.76 M.A. Krivoglaz, S.P. Repetski: Sov. Phys. Solid State **8**, 2325 (1967)
13.77 R.C. Knauer, J.G. Mullen: Phys. Rev. **174**, 711 (1968)
13.78 A. Bläsius, R.S. Preston, U. Gonser: Z. Phys. Chem. **115**, 187 (1979)
13.79 N. Blaes, R.S. Preston, U. Gonser: *Proc. Intern. Conf. Application of the Mössbauer Effect*, Alma Ata, USSR (Gorden Breach, New York 1983)
13.80 H. Fischer: Dissertation, Universität des Saarlandes (1984)
13.81 A. Heidemann, G. Kaindl, D. Solomon, H. Wipf, G. Wortmann: Phys. Rev. Lett. **36**, 213 (1976)
13.82 G.J. Long, J.G. Stevens (eds.): The Industrial Applications of the Mössbauer Effect (Plenum, New York 1986)
13.83 E. Gerdau, R. Rüffer, H. Winkler, W. Tolksdorf, C.-P. Klages, J.P. Hannon: Phys. Rev. Lett. **54**, 835 (1985)
13.84 V.I. Goldanskii, R.N. Kuzmin, V.A. Namiot: In *Mössbauer Spectroscopy II*, ed. by U. Gonser, Topics Current Phys., Vol. 25 (Springer, Berlin, Heidelberg 1981) p. 49

13.85 H. Hartmann, F. Parak, W. Steigemann, G.A. Petsko, D. Ringe Ponzi, H. Frauen-
       felder: Proc. Natl. Acad. Sci. **79**, 4967 (1982)
13.86 E.R. Bauminger, S. Ofer: *Proc. Intern. Conf. on the Applications of the Mössbauer
       Effect* (Jaipur, India 1981) p. 61

Additional References with Titles

Chapter 3

Barry, J.C., Bursill, L.A., Sanders, J.V.: Electron microscope images of icosahedral and cuboctahedral (fcc packing) clusters of atoms. Aust. J. Phys. **38**, 437 (1985)

Bovin, J.O., Wallenberg, R., Smith, D.J.: Imaging of atomic clouds outside the surfaces of gold crystals by electron microscopy. Nature **317**, 47 (1985)

Bursill, L.A., Lin, P.J.: Penrose tiling observed in a quasicrystal. Nature **316**, 50 (1985)

Guyot, P., Audier, M.: A quasicrystal structure model for Al-Mn. Philos. Mag. B**52**, L15 (1985)

Hiraga, K., Hirabayashi, M., Inoue, A., Masumoto, T.: Structure of Al-Mn quasicrystal studied by high-resolution electron microscopy. J. Phys. Soc. Jpn. **54**, 4077 (1985)

Hiraga, K., Hirabayashi, M., Sagawa, M., Matsuura, Y.: A study of microstructures of grain boundaries in sintered $Fe_{77}Nd_{15}B_8$ permanent magnet by high resolution electron microscopy. Jpn. J. Appl. Phys. **24**, 699 (1985)

Ichinose, Y. Ishida: High resolution electron microscopy of grain boundaries in fcc and bcc metals. J. Phys. (Paris) C4, **46**, 109 (1985)

Knowles, K.M., Greer, A.L., Saxton, W.U., Stobbs, W.M.: High-resolution electron microscopy of an Al-Mn alloy exhibiting icosahedral symmetry. Philos. Mag. B**52**, L31 (1985)

Loiseau, A., Lasalmonie, A., van Tendeloo, G., van Landuyt, J., Amelinckx, S.: A model structure for the $Al_{5-x}Ti_{3+x}$ phase using high-resolution electron microscopy. Acta Crystallogr., Sect. B **41**, 411 (1985)

Mayer, J., Urban, K.: Observation of Ni_8Mo ordered phase in Ni-Mo alloy. Phys. Status Solidi A**90**, 469 (1985)

Portier, R., Shechtman, D., Gratias, D., Cahn, J.W.: High resolution electron microscopy of the icosahedral quasiperiodic structure in Al-Mn system. J. Microsc. Spectrosc. Electron. **10**, 107 (1985)

Schryvers, D., van Tendeloo, G., Amelinckx, S.: On the ordering mechanism of Ni_3Mo. Phys. Status Solidi A**87**, 401 (1985)

Smith, D.J.: Atomic-resolution studies of surface dynamics by electron microscopy. J. Vac. Sci. Technol. B**3**, 1563 (1985)

Smith, D.J., Camps, R.A., Freeman, L.A., O'Keefe, M.A., Saxton, W.O., Wood, G.J.: Approaching atomic-resolution electron microscopy. Ultramicroscopy **18**, 63 (1985)

van Tendeloo, G., Amelinckx, S.: On the nature of the short-range order in 1 1/2 0 alloys. Acta Crystallogr., Sect. B**41**, 281 (1985)

Ye, H.Q., Wang, D.N., Kuo, K.H.: Fivefold symmetry in real and reciprocal space. Ultramicroscopy **16**, 273 (1985)

Chapter 10

Abragam, A.: Spectrométrie par croisements de niveaux en physique de muon. C.R. Acad. Sci. Ser. 2, **299**, 95 (1984). This proposes a new experimental method

Barsov, S.G., Getalov, A.L., Grebinnik, V.G., Gordeev, V.A., Gurewich, I.I., Zhukov, V.A., Ivanter, I.G., Kirillov, B.F., Klimov, A.I., Kruglov, S.P., Kuz'min, L.A., Lazarev, A.B., Mikirtych'yants, S.M., Moreva, N.I., Morozova, V.A., Nikol'skii, B.A., Pirogov, A.V., Ponomarev, A.N., Selivanov, V.I., Suetin, V.A., Sherbakov, G.V.: Muon diffusion in dysprosium. JETP Lett. **40**, 889 (1984). This and the following reference propose a new experimental method for the study of μ^+ diffusion in ordered magnetic metals

Barsov, S.G., Getalov, A.L., Grebinnik, V.G., Gordeev, V.A., Gurevich, I.I., Zhukov, V.A., Kirillov, B.F., Klimov, A.I., Kruglov, S.P., Kuz'min, L.A., Lazarev, A.B., Mikirtych'yants, S.M., Mikhailov, B.P., Nikol'skii, B.A., Pirogov, A.V., Ponomarev, A.N., Selivanov, V.I., Suetin, V.A., Shilov, S.N., Scherbakov, G.V.: Muon spin relaxation in crystalline and amorphous states of $Cu_{10}Zr_7$. JETP Lett. **42**, 242 (1985)

Kreitzman, S.R., Brewer, J.H., Harshman, D.R., Keitel, R., Williams, D.L., Crowe, K.M., Ansaldo, E.J.: Longitudinal-field μ^+ spin relaxation via quadrupolar level-crossing resonance in Cu at 20 K. Phys. Rev. Lett. **56**, 181 (1986). This demonstrates the Abragam method

Schenck, A.: *Muon Spin Rotation Spectroscopy* (Adam Hilger, Bristol 1985)

Subject Index

Quadrupole frequency 370, 386, 392
Quadrupole interaction 370
Quadrupole moment 367
Quadrupole splitting 422
Quantitative analysis 205, 433
Quantitative analysis by AES 227
Quantization axis 320
Quantum diffusion 306
Quantum μ^+ diffusion 306
Quantum tunneling 304
Quasicrystal 54

Radial distribution function 125, 158, 162
Radiation induced defects 276
Radio frequency field 367, 440
Radio frequency pulses 13
Radioactive probes 317, 337
Radiotracer 242
Random substitutional alloys 268
Range of positrons 263
Rapid quenching 55
Rare earth intermetallic compounds 313
Reactor-cladding materials 280
Reciprocal lattice vector 254
Recoil-free fraction 415
Recoil-free gamma resonance absorption 409
Recovery stages 355
Re-emitted positrons 263
Relative correlation parameter 175
Relative free energy 98
Relativistic effects 428
Relaxation energy 207
Relaxation in the rotating frame 394
Relaxation phenomena 297, 443
Relaxation times 371, 384
Relaxation times T_1 and T_2 369
RE-metal amorphous metals 189
Residence time 354
Residual stresses 118, 144
Resonance condition 421
Resonance transitions 417
Resonance tunneling 100
Rest mass 252
Retarder work function 104
Retarding potential 101
Rigid-band model 268
RKKY 300
Rotating anode 163
Rotating frame 372

Scalar or pseudo-dipole interactions 382
Scanning acoustic microscopy 7
Scattered intensity per atom 122
Scattering amplitudes 168
Scattering experiments 118, 419
Scattering geometry 421
Scattering phase shifts 168
Scherzer defocus condition 36
Scherzer focus 41
Scherzer resolution limit 42

Schottky-type barrier 109
Secondary ion mass spectroscopy 220, 234
Self-diffusion 92
Self-interstitial atoms 355, 360
Self-trapping 304
Semiconductor 236
Shake-up transitions 161
Shape of Fermi surface 266
Shape-memory alloys 66
Short-range order 388, 441
Single-photon annihilation 252
Singularity index 212
Site mapping 98
Skin effect 310, 380
Slow-positron moderation 289
Small atom approximation 157, 160
Small polaron 304
Small-angle scattering 139
Spallation neutrons 130
Spatial resolution in Auger microscopy 231
Spatial resolution in XPS 235, 245
Spherical aberration 37
Spherical aberration coefficient 35
Spherical solid model 310
Spin alignment 317, 320
Spin density 311
Spin dynamics 399, 403
Spin echo 373
Spin fluctuations 403
Spin glasses 312, 314, 400
Spin movement 371
Spin precession frequency 318, 323
Spin susceptibilities 382
Spin-flip narrowing 367
Spin-lattice relaxation 369, 375
Spin-orbit splittings 201
Spin-spin interaction 392
Spin-spin relaxation 369
Sputter-depth profile 222
Sputtering yields 230
Stacking disorder 51
Stacking fault 34, 144, 220
Stacking modifications 51
Stacking sequence 66
Stacking-fault fringes 34
Stacking-fault tetrahedron 46
Stair rod dislocations 49
Static concentration wave-packet model 63
Stationary defects 437
Sternheimer antishielding factor 348, 386
Structural parameters 168
Structure of amorphous metals 441
Sublimation energy of rhodium 111
Substitutional impurity 51
Substrate-layer interfaces 26
Superconductor 313, 404
Superstructure 60
Superstructure image 62
Surface 287, 343, 350
Surface acoustic waves 16

Surface atoms 67, 213
Surface defects 19
Surface diffusion of adatoms 90
Surface energy 220
Surface plasmons 198
Surface segregation 222, 240
Surface-atom core-level shifts 214
Surface-field gradient 351
Surface-terrace image 67
Susceptibility 314
Symmetrical Laue case 36
Symmetry axis 330
Synchrotron radiation 129, 163, 445

Temper brittleness 241
Tetrahedrally close-packed structure 64
Texture 435
Thermal energy of positron 270
Thermal equilibrium 251
Thermal expansion 269
Thermalization time 251
Thermionic cycle 101
Three-photon annihilation 252
Tilted illumination 36
Time resolution 262
Time spectrum 341
Time window 314, 337, 379
Time-dependent effects 429
Time-of-flight neutron diffractometer 131
Time-of-flight technique 132
T-M metallic glasse 181
Topological short-range order (TSRO) 118, 124, 393
Total annihilation rate 254
Total diffracted power 129
Total scattered amplitude 33, 58
Total structure factor 122
Transfer Hamiltonian 80
Transferability 168
Transmission EXAFS 164
Transmission function 39
Transmission geometry 421
Transverse relaxation 369, 373, 384
Trapping model 271, 352, 356
Trapping probability 272, 281
Trapping rate 278
Traps 306
T-T amorphous alloys 188
Tunneling 80, 104
Tunneling of valence electron 77
Tweed pattern 66
Two-beam dynamical theory 37
Two-beam lattice fringe 37

Two-body correlation function 55
Two-dimension NMR 367
Two-photon annihilation 252

Ultrafine particle 67
Umklapp processes 254, 259
Uncertainty principle 412
Under-focus 40
Universal correlation 349
Universal curve 199
Unperturbed γ-γ angular correlation 322
Untilted illumination 36

Vacancies 250, 306, 355, 437
Vacancies in thermal equilibrium 269
Vacancy cluster 279, 359
Vacancy migration 278
Vacancy-impurity binding energy 273
Vacancy-like defects 250, 280
Valence electrons 77, 258
Valence-band spectra 215
Van Vleck static linewidth 301
Variable λ-method 130
Variable 2θ-method 130
Vibration frequency 110
Vibrational anisotropy 435
Virtual bound-state 208
Void nucleation 280
Voids 250

Warren-Averbach method 141
Warren-Cowley short-range order parameter 127
Wave aberration 40
Weak-beam method 34
Weak-phase approximation 40
Weak-phase object 43
WKB approximation 81
Work function 85, 200

X-ray absorption edge anomaly 212
X-ray absorption edge spectroscopy 203
X-ray diffraction 117
X-ray emission spectroscopy 204
X-ray fluorescence analysis 234
X-ray microanalysis 234
X-ray neutron diffraction 117
X-ray photoelectron spectroscopy 193, 220, 234
X-ray satellites 201
X-ray synchrotron 117

Zeeman interaction 368
Zero-point motion 312

Topics in Current Physics

Founded by Helmut K. V. Lotsch